FIRE OFFICER'S GUIDE TO DISASTER CONTROL

SECOND EDITION

FIRE OFFICER'S GUIDE TO DISASTER CONTROL

WILLIAM M. KRAMER, Ph.D.
District Fire Chief, Cincinnati Fire Division
Director of Fire Science, University of Cincinnati

CHARLES W. BAHME, J.D.
Deputy Chief Los Angeles Fire Department (Retired)
Captain, United States Naval Reserve (Retired)
Attorney at Law

Fire Engineering Books & Videos
A PENNWELL PUBLISHING COMPANY

Park 80 West, Plaza Two, 7th floor, Saddle Brook, NJ 07662

LIBRARY OF CONGRESS CATALOGING-IN-PUBLICATION DATA

Kramer, William M. (William Michael), 1944–
Fire officer's guide to disaster control / William M. Kramer and Charles W. Bahme. — 2nd. ed.
p. cm.
ISBN 0-912212-26-8
1. Emergency management—United States—Handbooks, manuals, etc.
2. Fire departments—United States—Handbooks, manuals, etc.
3. Fire extinction—United States—Handbooks, manuals, etc.
I. Bahme, Charles William, 1914– .
HV551.3.K73 1992
363.37'8'0973—dc20 91-43911

Printed in the United States of America

Book design by Bernard Schleifer

10 9 8 7 6 5 4 3 2 1

To my Father, John William Bahme, whose fine example as a concerned fire officer inspired me to become a fire fighter and make a career of the fire service.

—C.B.

To my two loving daughters, Cari and Jennifer, who have often had to sacrifice time away from their father at the sound of a bell, be it a fire alarm or the signal of a college class, as I have pursued a dual career in the fire service and in the academic community. I would also like to dedicate this book to my co-author, Chuck Bahme, who has been a mentor, a source of inspiration, and a great friend during our association, which spans two decades.

—W.K.

CONTENTS

PREFACE TO SECOND EDITION

MANY YEARS HAVE GONE BY since disaster control was my responsibility at such catastrophes as the Bel Air fire, the Watts Riots, and the Baldwin Hills flood, and more than a decade has elapsed since my suggestions for coping with such major emergencies were offered in this guide book. Though I have attempted to provide some later insights on these problems in the disaster course syllabus of the Open Learning Fire Service Program, the publication of an updated edition of the original work is long overdue.

In order to more fully pursue other interests and yet continue my instruction in this national program, I am finding co-authors to assist in the revision of my various text books. Bill Kramer, an old shipmate and good friend, is helping me revise this disaster study. In view of my maturing years (almost four score) Bill will one day have sole responsibility for future editions, and I am confident that this task is in good hands. Bill is not only an internationally recognized education administrator but also an active chief officer in one of our nation's outstanding fire departments, and has a down-to-earth outlook on fire service problems.

Since the causes of disasters have not changed, nor their frequency and magnitude, why do we need a new edition? Mass casualties in earthquakes, floods, hurricanes, and plane crashes still occur almost daily. (As I write these words, hundreds of victims of a Boston subway train wreck are being hospitalized, and in a nearby suburb of that same city a conflagration in multistory residences is raging). Disasters are no less a problem today then they were in the time of that distinguished fire fighter Benjamin Franklin, who warned us to "either write things worth reading or do things worth writing."

Rewriting this text is probably worth doing because of the technological, legislative, and organizational changes that have occurred in recent years, making it possible to more easily identify the risks, evaluate the scope and mitigate the damage of impending or actual disasters. Systems for command and communications are greatly improved. Efficiently designed apparatus and equipment for response and control are more readily available. Specially trained and equipped fire fighters can more effectively handle plane crashes, chemical spills, wildland fires and other major emergencies in both urban and rural communities than ever before. More colleges and fire service instructors throughout the nation now augment the National Fire Academy staff in providing the information needed by fire fighters wherever they are located. And finally, there are more fire officers today who have obtained the education and experience to provide needed guidance in fire service publications. While it is true that some of their nomenclature may be barely recognizable to old "smoke-eaters," we are gradually acknowledging that a fire chief can be an "on-scene manager," and that "combustion degradation with damage mitigation" are just fancy terms for old-fashioned "fire suppression and overhauling" operations. However, I am confident that neither the "old school" nor new generation fire fighters will have any difficulty understanding Dr. Kramer's terminology. We "old timers" are certainly impressed with all the high-tech changes taking place today in both systems and semantics, but we are sometimes reminded of the old fire chief's comment, when the efficiency expert came back to express his gratification for all the extra personnel that had been hired since his new system had been installed. "Yes," he replied, "I hired them to take care of the system."

A wise old philosopher, in his *American Bible*, said long before I was born that the best service any book can offer is to make you think for yourself. His name was Elbert Hubbard, and I agree with him. In so doing, you are bound to come up with even better ideas for disaster control than we have offered here. We hope that you will share them for the next edition.

Chuck Bahme

FOREWORD

SINCE THE FIRST EDITION of the *Fire Officer's Guide to Disaster Control* was published in 1978, the face of the Fire Service in North America has changed significantly. We provide services that were once considered outside of our scope. One thing that has not changed is that the Fire Department will be the first agency called when disaster strikes—no matter what type of disaster.

No single individual can be well versed in the skills needed to manage every type of disaster—there are not enough of them, thank goodness, for each of us to practice on. This book provides the reader, and potential participant in disaster mitigation, with a basis from which to work when confronted with emergencies that are well outside of the norm.

Charles Bahme's original edition of this publication became a vanguard publication, and now it has been updated and enhanced.

Bill Kramer is a well known leader and educator in the fire service. As coordinator of the University of Cincinnati's Open Learning Fire Service Program and regular feature for the *American Heat* video series, his name and face are familiar to many members of the service. Chief Kramer's eighteen years of experience with the Cincinnati Fire Department are backed up by impressive educational credentials. He brings current practicing common sense to a complicated subject.

Enjoy the book—while you are reading try to apply what you have learned to your own community. Try as we may, we still have not been able to schedule disasters, or fires, in our communities.

Alan V. Brunacini

CHUCK BAHME

BILL KRAMER

ABOUT THE AUTHORS

DR. WILLIAM M. KRAMER has baccalaureate degrees in Industrial Management and Business Administration, a master's degree in Industrial Relations, and a Ph.D. in Administrative Management from the University of Cincinnati; and a master's degree in Business Administration from Xavier University. He has edited numerous course guides for the Open Learning Fire Service Program and published numerous fire service journal articles. He is chairman of the seven-University National Open Learning Educational Consortium, Associate Professor and Director of the Open Learning Program at the University of Cincinnati, a District Fire Chief for the City of Cincinnati and the educational commentator for *American Heat* video productions.

CHARLES (CHUCK) BAHME is the author of many fire service articles as well as books, including the *Handbook of Disaster Control*, predecessor to the *Fire Officer's Guide to Disaster Control, Fire Service and the Law, Fire Officer's Guide to Dangerous Chemicals, Fire Protection for Chemicals, Fire Officers Guide to Emergency Action, Firemen's Law Book*, and *Fire Officer's Guide to Extinguishing Systems*. After responding to a devastating chemical explosion in the City of Los Angeles he wrote its *Dangerous Chemicals Code*, the first comprehensive set of regulations ever adopted by any municipality pertaining to hazardous materials. Following his naval service in World War II he was called upon to write *U. S. Navy: Structural Fire Fighting* for the Chief of Naval Operations.

In thirty years of service with the City of Los Angeles Fire Department he advanced through the ranks from recruit to deputy fire chief, the rank he held at the time of his retirement. During that period he augmented his A.B. degree from UCLA with a Juris Doctor degree from Southwestern University. Through the same years he served with the U. S. Navy: six years of active duty during World War II and the Korean conflict, and twenty-eight years with the Naval Reserve, in which he retired with the rank of Captain.

Chief Bahme is an attorney at law with admission to practice in California, before federal district courts, the Supreme Court of the United States, and the highest court of military appeals. He has taught courses in fire protection engineering a UCLA, and fire administration at USC and various state, national, and international conferences. He has served in Europe and the Far East for the Department of Defense and the U. S. State Department. In recent years he has been teaching a course in Political and Legal Foundations of Fire Protection as a faculty member of Cogswell College's Open Learning Fire Service Program. He has also served as the NFPA's Fire Extinguishing Specialist, its Western Representative, and as chairman of its Committee on Hazardous Chemicals Fire Fighting. Upon his moving to Lake County, California, he served as the chairman of his local Kelseyville fire district's board of directors and as a disaster control advisor to his local Red Cross chapter.

ACKNOWLEDGMENTS

I WISH TO ACKNOWLEDGE the contributions of Joseph Weber, Steve Collier, Robert Obermayer, and James Craig. Each of these authors wrote major portions of the National Fire Academy's Open Learning Course entitled "Disaster and Fire Defense Planning." Some of their current research and literary endeavors appear with credit in the chapters of this new text. I acknowledge their input as follows: Joseph Weber, for contributions to chapters 2, 4, and 9; Steve Collier, for contributions to chapters 6, and 15; Robert Obermayer, for contributions to chapters 7 and 10; and James Craig, for contributions to chapter 13.

I am indebted to Ed Kaplan at the National Fire Academy for providing current information and to the Open Learning Fire Service Program for producing some of the most current and relevant research which rightfully belongs in any new text on disaster planning.

Sincere thanks also to Ron Buckingham, Gary Auffart, and Dennis McCloskey for providing the majority of the photos in this book; and to Barbara Barkley, Minnie Easley, and Jennifer Cobb for their never-ending ability to type and retype.

Finally, a special tribute to the many courageous fire service professionals whose exploits are documented herein, and with whom I feel a close kinship, whether or not we have met.

WILLIAM M. KRAMER

FIRE OFFICER'S GUIDE TO DISASTER CONTROL

CHAPTER 1

DISASTER PLANNING

DISASTERS OCCUR DAILY

HARDLY A DAY passes that a disaster endangering many lives and much property does not occur somewhere in our nation. Train wrecks, plane crashes, hazardous cargo spills involving dangerous chemicals, tank truck collisions, overturned buses, and similar transportation accidents are common, even in the heart of densely populated areas. Every year recurring conflagrations, floods, hurricanes, tornadoes, and other natural disasters cause great destruction, and occasionally one of the thousands of earthquakes that jar our countryside each year grows to devastating magnitude in a populated region, producing tragic results. As we entered the decade of the 1990s the United States suffered a serious calamity when hurricane *Hugo* devastated the east coast and shortly thereafter, an earthquake measuring 7.1 on the Richter scale caused widespread destruction on the West Coast.(1) Also, despite a lessening of tensions between superpowers on the global political scene and much improvement in East–West relations, world tensions continue, especially in the Mideast, as evidenced by the 1991 war against Iraq by the United States and our allies. There remains a need for a workable plan to cope with worldwide terrorist attacks and possible future global war.(2)

In the aftermath of the six-week war that ended the occupation of Kuwait by Iraq, one of the most serious challenges facing the government of Kuwait and its American allies was the extinguishment of hundreds of oil well fires. These environmental disasters were not only costing millions of dollars a day in petroleum lost to flames but were polluting the atmosphere at a catastrophic rate. As the American government hastened to help Kuwait rebuild its country and restore order, the containment of all oil well fires would take little more than a year.(3) The last one was extinguished by November 1, 1991.

The American Insurance Association was indeed correct when it pointed out, over three decades ago in its Special Interest Bulletin No. 310, that "Disasters can and do occur anywhere. It is the duty of large and small fire departments, well equipped or poorly equipped, to be thoroughly prepared to assist in operations at the scene of a disaster."(4) Add to this statement the often noted concept that "fire fighters rush in where others are rushing out," and you will begin to have an appreciation of why the fire department is the logical agency to respond to the many types of disasters confronting humankind, even though most of these have nothing to do with flames and fire. Disaster planning is one of the most important tasks that a fire chief performs, for the fire service is basically a disaster control organization. Although fire fighters are usually the first to arrive at the scene of a calamity, their ability to cope with problems will largely depend upon the extent that advance plans have been made, resource availability identified, and key personnel trained for execution of the plans.

In the broadest sense, a "disaster" is a sudden and extremely unfortunate event that affects many people. It could also be defined as a misfortune that reaches large proportions. The words *catastrophe* and *calamity* are often used as synonyms for disaster. Table 1-1 identifies some of the world's greatest disasters. In the narrowest sense, "disaster" is a relative term, depending upon the number of casualties and the extent of property damage, the length of time involved, and the capacity, size, and strength of forces available to cope with it.

Consider the number killed or injured as a factor in classifying a mishap; to constitute a disaster, the victims need not necessarily be human. The newspaper *USA Today* reported that the damage from the Valdez Alaskan

This vintage pumper witnessed the transformation of the fire service from a fire suppression force to one capable of handling complex disasters of all sorts. (Photo courtesy of Jim Lanagan.)

oil spill killed 580,000 birds, including 144 bald eagles, and polluted the sea to the depth of 300 feet. Along the Alaskan coastline 1,200 miles were coated in thick petroleum goo.(5)

Although the Persian Gulf oil spill created during the Iraqi war was considered ten times worse, it was contained quickly and covered a less sensitive area so that it did not cause as much damage.

Disasters that we will consider can be expected to involve fatalities. An explosion of railroad tank cars that wipes out half the population of a small community or a bus accident that kills all the graduating class of the local high school would certainly be considered a disaster in a regional sense. However, these incidents might not receive the nationwide publicity that would result following a nuclear reactor accident causing thousands of radiation casualties.

If all the fish in Lake Michigan were to be suddenly exterminated by a gigantic poison spill, most people would recognize this as a disaster. On the other hand, lesser pollution occurring over a long period, but achieving the same result, would not receive such recognition, nor would a lesser but immediate spill that killed only half the fish. Note that although technically a disaster can occur without the loss of life, the disasters listed in Table 1-1 all involved significant loss of human life.(6)

Certainly in the eyes of the victims and the bereaved it is immaterial whether the mishap is large enough to be publicized nationally or not. An unvented gas heater could cause a disaster in a single household if carbon monoxide were to kill all members of a family as they lay sleeping. Whether the incident would be listed in the annual tabulation of the *World Book Encyclopedia* would probably depend upon the size of the family; where the death toll was ten, it has been so recorded.(7)

Now let us consider the capacity of forces available to handle the problems. No matter how great the threatened danger, if there is warning and available forces are sufficient to act quickly, it might not result in a disaster. For example, if a tank truck of anhydrous ammonia were to rupture in a densely populated area as a result of falling off of a freeway exit ramp (as occurred in Houston, Texas, in 1976), and if this toxic gas were to spread out through the neighborhood to engulf a nursing home or hospital of nonambulatory people, emergency forces could do little to avert a large number of casualties. The result would be a disaster.

On the other hand, in a large industrial plant, storage containers of chlorine gas (commonly kept in ton containers in water chlorination buildings) might suddenly blow out their soft plugs because of excessive temperature. If a warning alarm is sounded immediately and all employees quickly don gas masks and leave the zone of danger no disaster will be caused by this poison gas. Nor would a disaster occur under the above circumstances if equipment is in place to immediately discharge the leaking container contents into a neutralizing chemical solution to prevent its release to the atmosphere.

Finally, whether a specific mishap will be classed as a disaster may depend upon the period of time that the classifier uses in compiling the list. For example, when an airplane crash wiped out a row of houses, killing all twenty passengers and some residents, this was classed as a major disaster

Table 1-1. Major Disasters

Year	Location	Dead	Disaster
64	Rome	Unknown	City fire
79	Pompeii & Herculaneum	2,000	Vesuvius eruption
1556	Shensi, China	830,000	Earthquake
1666	London, England	Unknown	City fire
1669	Sicily	20,000	Mount Etna eruption
1737	Calcutta, India	300,000	Earthquake; tornado
1755	Lisbon, Portugal	60,000	Earthquake
1865	Mississippi River	1,653	Steamboat explosion
1871	Chicago, Illinois	300	City fire
1871	Peshtigo, Wisconsin	800	Forest fire
1883	Indonesia	36,000	Krakatoa eruption; tidal wave
1887	Hanan, China	900,000	Hwang Ho flood
1889	Johnstown, Pennsylvania	2,200	Burst dam; flood
1900	Galveston, Texas	6,000	Hurricane; storm tide
1902	Martinique	38,000	Mont Pélee eruption
1903	Chicago, Illinois	575	Iroquois Theater fire
1904	New York City, New York	1,030	Excursion boat fire
1906	San Francisco, California	700	Earthquake; fire
1907	Monongah, West Virginia	361	Coal mine explosion
1911	Eastern China	100,000	Flood
1912	North Atlantic	1,500	Liner *Titanic* collision with iceberg
1913	Indiana; Ohio	467	Ohio River basin flood
1914	Saint Lawrence River	1,029	Ship collision; explosion
1917	Modane, France	550	Train derailment
1917	Halifax, Nova Scotia	1,635	Ship explosion
1918	Nashville, Tennessee	101	Train collision
1920	Kansu, China	180,000	Earthquake; landslide
1923	Tokyo; Yokohama	143,000	Earthquake; fire
1925	Illinois; Indiana; Missouri	689	Tornadoes
1928	Florida	1,836	Hurricane
1934	Hakodate, Japan	2,015	Widespread fire
1937	New London, Texas	296	Gas explosion in school
1938	New England	600	Hurricane
1942	Bengal, India	11,000	Cyclone; storm wave
1942	Manchuria, China	1,572	Coal mine explosion
1942	Boston, Massachusetts	499	Nightclub fire
1944	Hartford, Connecticut	160	Circus fire
1947	Texas City, Texas	512	Ship explosion
1951	Manchuria	1,810	Flood
1951	Papua	3,000	Volcanic eruption
1953	North Sea	2,000	Storm and floods

Year	Location	Dead	Disaster
1975	China	2,000	Earthquake
1975	New York City, New York	113	Aircraft crash
1975	Turkey	2,000	Earthquake
1975	India	372	Mine flood
1976	China	650,000	Earthquakes
1976	Italy	1,000	Earthquake
1976	Guatemala	23,000	Earthquake
1976	Indonesia	9,000	Earthquake
1976	Yugoslavia	176	Aircraft collision
1976	Philippines	4,000	Earthquake
1976	Turkey	4,000	Earthquake
1977	Romania	1,000	Earthquake
1977	Tenerife, Canary Islands	577	Aircraft crash
1977	Southgate, Kentucky	162	Nightclub fire
1978	Bombay, India	213	Aircraft explosion
1978	San Diego, California	135	Aircraft collision
1978	Colombo, Sri Lanka	202	Aircraft crash
1978	Palermo, Italy	108	Plane fell into sea
1978	Tabas, Iran	25,000	Earthquake
1978	Southern Spain	250	LP gas truck explosion
1978	Afghanistan-Pakistan border	122	Flood
1978	Acajutla, El Salvador	100	Tsunami wave
1978	Southeast Asia	111	Flood
1978	Midwest U.S.	100	Blizzard
1978	Northeastern U.S.	100	Winter storm
1978	West Bengal, India	100	Tornado
1978	Sri Lanka	500	Cyclone & floods
1978	Bangladesh	100	Boat capsized
1978	Malaysia	330	Boat sank
1978	Delhi, India	200	Heat wave
1979	Chicago, Illinois	272	Aircraft crash
1979	Jidda, Saudi Arabia	156	Aircraft crash
1979	Antarctica	257	Aircraft crash
1979	Eastern Iran	200	Earthquake
1979	Yugoslavia-Albania border	129	Earthquake
1979	Northeastern Iran	500	Earthquake
1979	Colombia-Ecuador border	400	Earthquake
1979	Southeast Brazil	700	Flood
1979	Midwestern U.S.	100	Blizzard
1979	Southeast India	600	Hurricane & flood
1979	Caribbean & Eastern U.S.	2,000	Hurricane *David*
1979	Indonesia and Java	175	Volcanic eruption
1979	Northern India	230	Snowstorm & avalanche
1979	Lomblen Island, Indonesia	539	Landslide

1954	Japan	[illegible]	[illegible]
1955	India; Pakistan	1,700	Flood
1956	Cali, Colombia	1,290	Explosion; fire
1956	China	1,960	Typhoon
1956	Off Nantucket Island, Massachusetts	51	Liner *Andrea Doria* collision
1959	Japan	4,253	Typhoon
1960	New York City, New York	134	Airliner collision
1960	Agadir, Morocco	12,000	Earthquake; tsunamis
1960	Chile	5,700	Earthquake; tsunamis
1961	Pakistan	2,000	Typhoon. floods
1962	Peru	3,500	Huascaran avalanche
1963	Bali	1,500	Mount Agung eruption
1963	Cuba; Haiti	6,000	Hurricane *Flora*
1963	Skopje, Yugoslavia	1,100	Earthquake
1964	Alaska	131	Earthquake
1965	East Pakistan	12,762	Cyclone
1966	Aberfan, Wales	144	Coal mine avalanche
1966	Eastern Turkey	2,400	Earthquake
1967	Venezuela	240	Earthquake
1969	Alabama; Louisiana; Mississippi	170	Hurricane *Camille*
1970	East Pakistan (now Bangladesh)	200,000	Cyclone; tidal wave
1970	Peru	70,000	Earthquake
1970	East Pakistan	500,000	Cyclone; storm wave
1970	India	1,200	Two floods (July & Sept.)
1970	Buenos Aires, Argentina	139	Train wreck
1971	Los Angeles, California	64	Earthquake
1971	Marioka, Japan	162	Aircraft collision
1971	South Korea	159	Hotel fire
1971	Afghanistan	1,000	Landslide; flood
1971	Orissa State, India	10,000	Cyclone; storm wave
1972	Managua, Nicaragua	10,000	Earthquake
1972	Moscow, USSR	176	Airliner crash
1972	Osaka, Japan	116	Nightclub fire
1972	Rapid City, South Dakota	239	Burst dam, flood
1972	Southern Iran	5,000	Earthquake
1972	Man (Buffalo Creek), West Virginia	118	Dam break
1972	Philippines	427	Flood; landslide
1972	Saltillo, Mexico	208	Train wreck
1973	Kano, Nigeria	180	Airliner crash
1973	Mexico	539	Earthquake
1973	Paris, France	122	Aircraft crash
1974	Bangladesh	2,000	Flood
1974	Central & Southern U.S.	336	Tornadoes
1974	Bogatá, Colombia	200	Landslides
1974	Peru	250	Landslides
1974	Near Paris, France	345	Aircraft crash
1974	Honduras	8,000	Hurricane *Fifi*
1974	Pakistan	5,200	Earthquake
1979	Indonesia	100	Tidal wave
1980	Northern Iran	128	Aircraft crash
1980	Tenerife, Canary Islands	146	Aircraft crash
1980	Alma-Ata, USSR	163	Aircraft crash
1980	Riyadh, Saudi Arabia	301	Aircraft fire
1980	Central India	100	Bus plunged into culvert
1980	El Asnam, Algeria	3,000	Two earthquakes
1980	Southern Italy	3,000	Earthquake
1980	Kingston, Jamaica	157	Fire in home for aged
1980	Northern India & Bangladesh	1,500	Flooding & mudslides
1980	North Sea, off Norway	123	Oil platform capsized
1980	China	151	Severe winter storm
1980	Central America & Southeastern U.S.	273	Hurricane *Allen*
1980	Bangladesh	230	Ferryboat capsized
1980	Tablus Strait, Philippines	116	Two ships collided
1980	Andra Pradesh, India	100	Boat sank
1980	Sincelejo, Colombia	222	Bleachers collapsed
1980	Southeastern & Midwestern U.S.	1,265	Prolonged heat wave
1981	Kansas City, Missouri	113	Hotel walkway collapsed
1981	Kerman, Iran	2,500	Two earthquakes
1981	Taipei, Taiwan	110	Aircraft crash
1981	Ajaccio, Corsica	178	Aircraft crash
1981	Irian Jaya, Indonesia	250	Earthquake
1981	Northern Pakistan	212	Earthquake & landslide
1981	Sichuan Province, China	753	Floods
1981	Bay of Bengal, India	120	Tornado
1981	Philippines	120	Flood & mudslides
1981	Arabian Sea off India	470	Hurricane sank 11 boats
1981	Near Macapa, Brazil	311	Overloaded ferry sank
1981	Java Sea	500	Liner sank in storm
1981	Obidos, Brazil	300	Riverboat sank in port
1981	Mansi, India	400	Train fell into river
1981	Northern India	270	Cold wave & snowstorms
1982	Near Kabul, Afgahnistan	1,200	Trapped in tunnel
1982	Thirty-eight States in U.S.	300	Violent storms
1982	Kenner, Louisiana	145	Aircraft crash
1982	Krasnoyarsk, USSR	150	Aircraft crash
1982	Fortaleza, Brazil	137	Aircraft crash
1982	Guilin, China	112	Aircraft crash
1982	Dhamar Province, Yemen	2,800	Earthquake
1982	Caracas, Venezuela	145	Oil truck fire
1982	Lima, Peru	600	Flood
1982	Guangdong Province, China	430	Flood
1982	India	700	Flood
1982	South Sumatra, India	225	Flood

Table 1-1. Major Disasters (continued)

Year	Location	Dead	Disaster
1982	Nagasaki, Japan	307	Flood & mudslides
1982	El Salvador & Guatemala	1,160	Flood & mudslides
1982	Orissa State, India	200	Hurricane
1982	Gujarat State, India	300	Hurricane & tidal waves
1982	Rangoon, Burma	130	Ferryboat capsized
1982	Rangoon, Burma	160	Ferry hit a sandbar
1982	Near Sigulu, Uganda	200	Train derailed
1982	Near Tepic, Mexico	120	Train fell into gorge
1983	Midwest U.S.	200	Heatwave
1983	Madrid, Spain	270	Two plane crashes in ten days
1983	Popayan, Colombia	300	Earthquake
1983	Honshu, Japan	106	Tidal wave
1983	Turkey	2,000	Earthquake
1983	Cuenca, Ecuador	119	Aircraft explosion
1983	Abu Dhabi	112	Aircraft crash
1983	Lubango, Angola	150	Aircraft crash
1983	Near Guailor, India	100	Bus swept into river
1983	Eastern Turkey	2,000	Earthquake
1983	Northwestern Guinea	300	Earthquake
1983	Bolivia & Northern Peru	260	Floods & mudslides
1983	Gujarat State, India	900	Flash floods
1983	Anhwei Province, China	100	Floods
1983	Northern & Eastern India	700	Floods
1983	India	365	Floods
1983	Guangdong Province, China	147	Ferry capsized
1983	Lake Nassar, Egypt	300	Ship explosion
1983	Near Ulyanovsk, USSR	240	Cruise ship hit bridge
1983	Kansu Province, China	270	Landslide
1983	Chimborazo Province, Ecuador	100	Landslide onto autos
1983	Gachala, Colombia	160	Landslide
1984	Kabul, Afghanistan	240	Aircraft crash
1984	Severormsk, USSR	300	Naval ammunition explosion
1984	Tlainepantla, near Mexico City	490	LP gas explosion & fire storm
1984	Tbilisi, USSR	100	Apartment house gas explosion
1984	Mozambique, South Africa	100	Floods
1984	Bangladesh and India	200	Floods
1984	South Korea	190	Floods
1984	Nepal	150	Floods & landslides
1984	Andra Pradesh & Tamil Nadu states,	393	Floods & cyclones

Year	Location	Dead	Disaster
1986	Cameroon, Africa	1,700	Poison gas from volcanic lake
1986	San Miguel el Alto, Mexico	167	Aircraft hit mountain
1986	Zehedan, Iran	103	Aircraft hit mountain
1986	San Salvador, El Salvador	1,200	Earthquake
1986	Kantalai, Sri Lanka	100	Dam burst
1986	Northern & Eastern India	200	Floods
1986	Solomon Islands	1,000	Typhoon
1986	Sichuan Province, China	100	Hailstorm
1986	Philippines & China	409	Typhoon
1986	Bethal, South Africa	177	Mine fire
1986	Shansi Province, China	129	Ferry sank
1986	Munshiganj, Bangladesh	500	Storm hit ferry
1986	Leyte, Philippines	170	Ship sank
1986	Bhola, Bangladesh	600	Ferry sank
1986	Narail, Bangladesh	200	Ferry capsized
1986	Novorossiysk, USSR	398	Freighter rammed cruise liner in Black Sea
1986	Port Harcourt, Nigeria	100	Two ships collided
1986	Off Isle de la Gonave, Haiti	200	Ferry sank
1986	Northern India	220	Coldwave & winter storm
1986	Southern Colombia	200	Mudslides
1987	Zeebrugge, Belgium	134	Ferry capsized
1987	Northeastern Ecuador	4,000	Earthquake
1987	Warsaw, Poland	183	Aircraft crash
1987	Eastern Siberia, Chinese Manchuria	250	Forest fire
1987	Detroit, Michigan	156	Airliner crash
1987	Manila, Philippines	1,500	Ferry & tanker collide
1987	Southeastern Zaire	128	Trailer & truck collide
1987	Luapula River, Zaire	390	Passenger-carrying barge hits sandbar
1988	Soviet Armenia	25,000	Earthquake
1988	Jamaica, Caymans, Mexico	400	Hurricane *Gilbert*
1988	North Sea near Scotland	167	Oil platform explodes
1988	Yellowstone Park, Wyoming		Forest fires destroy 1.2 million acres
1988	Carrollton, Kentucky	27	Church bus crash
1988	Lockerbie, Scotland	270	Terrorist bomb exploded airliner
1989	Overtown, Miami, Florida	2	Three-day race riot, motorist killed by policeman

1984	Western USSR near Ivanovo	400	Tornadoes
1984	Philippines	1,360	Typhoon
1984	Philippines (one month later)	515	Typhoon
1984	Taipei, Taiwan	103	Coal mine fire
1984	Dhaka, Bangladesh	150	Two boats collided
1984	Malaysian coast	200	Ferry capsized
1984	Western Nigeria	100	River ferry sank
1984	Marindaque Island, Philippines	100	Ferry sank
1984	Barasai, Bangladesh	100	Boat capsized
1984	Dongchuan, China	100	Rain & landslide
1984	Mumar, India	125	Rope bridge collapsed
1984	Bhopal, India	2,500	Pesticide gas leak
1984	Leminsk-Kuznetsly, USSR	300	Defense plant accident
1985	Durango, Spain	148	Aircraft crash
1985	Off Ireland	329	Plane demolished in midair (sabotage?)
1985	USSR	150	Aircraft crash
1985	Near Dallas, Texas	137	Aircraft crash
1985	Central Japan	520	Aircraft hit mountain
1985	Central Chile	151	Earthquake
1985	Central Mexico & Mexico City	7,000	Earthquake
1985	Eastern Brazil	150	Floods
1985	Northeastern Brazil	100	Floods
1985	Northern Italy	361	Dam burst
1985	Sichuan Province, China	275	Flood & landslides
1985	Northeastern China	180	Floods
1985	Eastern & Central U.S.	123	Storms & coldwave
1985	Pabna, Bangladesh	300	Hailstones
1985	Southern Bangladesh	10,000	Cyclone & tidal wave
1985	Chekiang Province, China	177	Typhoon
1985	Ponce, Puerto Rico	150	Mudslide
1985	Vietnam	600	Typhoon
1985	Southeastern Nigeria	100	Ferry capsized
1985	Dhaka, Bangladesh	450	Ferry capsized
1985	Chittagong, Bangladesh	100	Ferry broke in half
1985	Harbin, China	174	Ferry capsized
1985	Mindoro Strait, Philippines	115	Ferry sank
1985	Awash, Ethiopia	450	Train derailed into ravine
1985	Nelas, Portugal	100	Two trains collide head on
1985	Europe	320	Freezing weather
1985	Colcabamba, Peru	120	Mudslide
1985	Behar State, India	100	Heatwave
1985	Colombia	25,000	Volcanic eruption & mudslides
1985	Gander, Newfoundland	256	Airplane crash
1986	Chernobyl, USSR	31 (but up to 40,000 may die of cancer)	Nuclear power plant fire, explosion & partial meltdown

1990	Cove Neck, New York		Avianca Airliner crash, Flight 52
1990	Bronx, New York	87	Happy Land Social Club fire
1990	Qinghai Province, Central China	115	Earthquake
1990	Kilauea, Hawaii		Volcanic eruption destroys 116th home since 1983
1990	Bay of Bengal, India	85	Cyclone
1990	Philippines	350	Earthquake
1991	Mount Pinatubo, Philippines	310	Volcanic eruption
1991	Ohio River Valley	35	Flood
1991	Persian Gulf War	100,000	War victims
1991	Oklahoma & Midwestern U.S.	24	Tornadoes
1991	Colorado Springs, Colorado	25	Air crash
1991	Bangladesh	140,000	Cyclone
1991	Shigaraki, Japan	42	Train crash
1991	Oakland-Berkeley, California	24	Conflagration
1991	Philippines	3,455	Typhoon
1992 1993 1994 1995 1996 1997 1998 1999 2000	History is expected to repeat itself as disasters similar to the actual ones recorded above recur, with only the names and places changing	1,000s	Air disasters, Weather disasters, Transportation disasters, Hazmat disasters, . . . continuing

Figure 1-1. Disasters in Perspective.

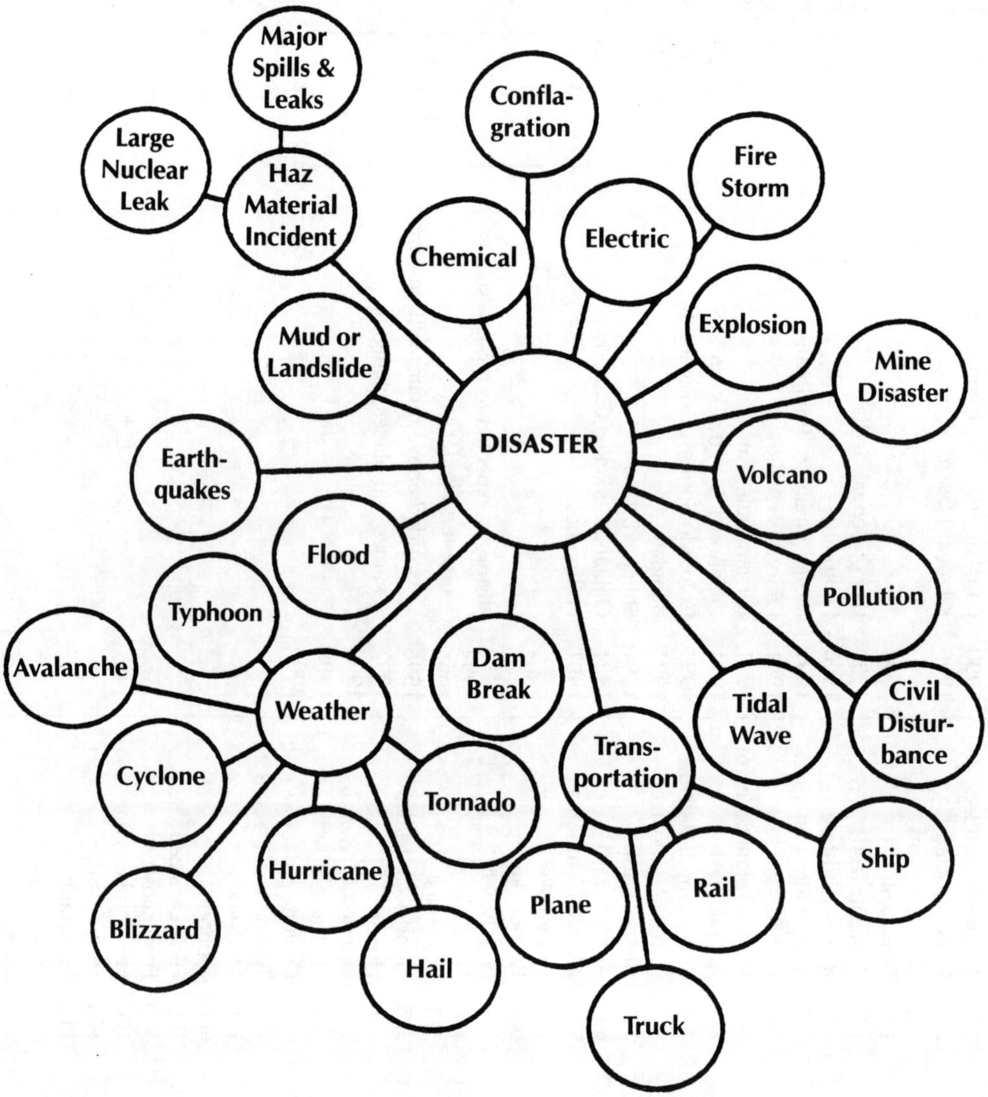

by the responding fire department and was listed in the *Year Book* of the *World Book Encyclopedia* under the heading Disasters.

Though a study of the disasters listed in Table 1-1 might be both depressing and impressive, it is certain to be revealing, for it discloses how ubiquitous disasters really are. Figure 1-1 shows the interrelationships among the various types of disasters.(8) This book is designed to provide fire officials and others involved with emergency response with practical suggestions for minimizing the adverse effects of a disaster once it occurs. As Table 1-1 shows, disasters are inevitable. The responsibility of fire officials and emergency officials, then, is to anticipate disasters of all sorts and to prepare for them.

SIMILARITIES AND DIFFERENCES BETWEEN FIRE DEFENSE CONTROL AND DISASTER CONTROL

It is obvious that the fire department has a role in both disaster and fire defense control. Detailed information and examples show how the role of the fire department in disaster control is consistent with its traditional role in fire suppression.

Although the fire department more commonly is perceived as a fire suppression and fire protection force, it is nonetheless the logical organization to respond to other disasters as well. Many similarities exist among disasters, fire situations, and other emergency responses of the fire department. Lets consider similarities and differences.

Similarities Between Disasters and Fire Situations

IMMEDIATE RESPONSE

In any community the fire department is poised and ready to swing into immediate action at the sound of an alarm. Its continual readiness makes it a logical organization to respond to emergencies other than fires.

MISSION OF THE FIRE DEPARTMENT

Whether a fire is raging or another kind of major mishap is in the making, human lives and property usually are endangered. An important mission of the fire department is to save lives and property. Therefore it is appropriate for the fire department to respond to other events that also threaten lives and property.

PERSONNEL REQUIREMENTS

A disaster requires large numbers of persons who are immediately available and trained to operate efficiently in extreme stress situations. Fire fighters who routinely encounter personal hazards in fire suppression also can be expected to manage the risks that occur with other types of disasters.

INCIDENT COMMAND

The Incident Command System, which gained wide acceptance by the nation's fire departments by the early 1990s and is promoted by the Federal Emergency Management Agency (FEMA) and the National Fire Academy, lends itself naturally to most emergency incidents, from a simple structure fire to major calamities affecting thousands of persons.

PRE-INCIDENT PLANNING

Progressive fire departments have long recognized the value of planning to foresee dangers and plan actions in response to fire potential in conventional commercial and industrial establishments. Advance planning also is necessary in an effective disaster control program.

MUTUAL AID

In fire departments the term mutual aid has long represented a second line of defense that can be called on when needed. Through mutual aid agreements, fire departments in different jurisdictions agree to come to the aid of others as requested. Since this activity traditionally involves large-scale or multiple fire emergencies, it is useful in disasters as well.

Differences Between Disasters and Fire Situations

The following factors differentiate disasters from the more common fire situations.

SCOPE OF EMERGENCIES

Usually a disaster emergency covers a larger geographical area, affects more lives, and is generally broader in scope than the building fire or other emergency routinely encountered by the fire service.

TYPES OF EMERGENCIES

As shown in Table 1-1 and Figure 1-1, disasters take many forms. They can be man-made or natural, deliberate or accidental, and include a diverse range of disaster types. There is more categorization of disasters than fires, and a broader range of variation among the disasters.

PLANNING EFFORT

Although most fire department planning is local, and to a certain extent regional, the scope of disaster planning is of necessity regional, statewide, and national. Likewise disaster planning includes a much larger cross section of political, social, and governmental agencies. Later in this chapter we will cover these various levels of planning in greater detail.

The Fire Department Response to Disasters

If we weigh the similarities and differences between fire situations and disasters the similarity factor does seem to dominate. See Figure 1-2 for a pictorial view of this weighting. Fires and disasters, although separate events, are related because the fire department is the logical agency to respond to both.

In recent years, as the hazardous materials problem has grown more complex, many fire and disaster situations have become interwoven. This provides yet another reason why the fire department is the logical agency to respond first to disasters.

Disasters often include fires, hazardous materials, and other components

Figure 1-2. Fires and Disasters: Similarities and Differences.

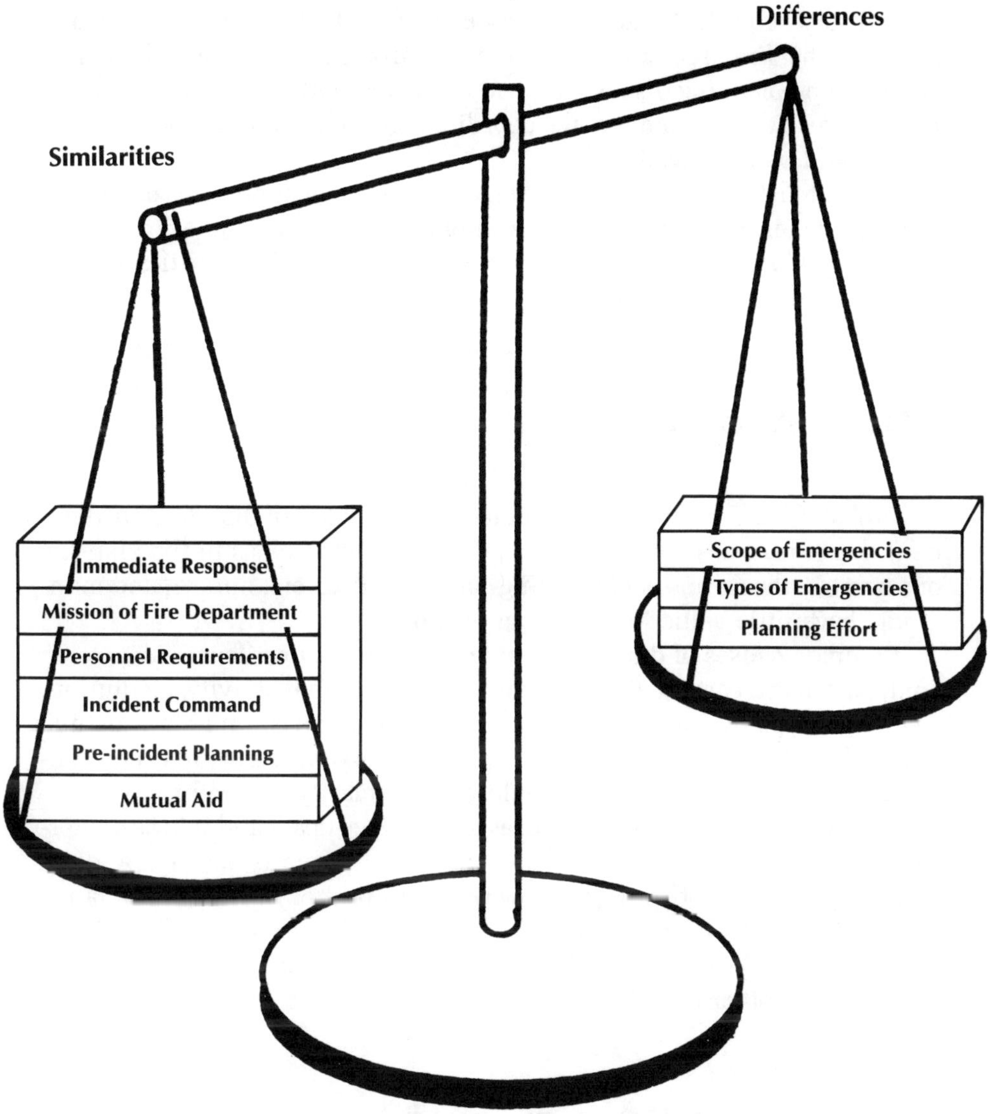

of the fire fighter's job description. Sometimes, fires and explosions escalate to include all of the standard components of a "disaster." Consider what the Cincinnati Fire Division faced when a fire and explosion suddenly became a hazmat incident, a mass-casualty incident, a major EMS triage operation, and a widespread evacuation operation: "The CFD's high degree of professionalism would be put to the test on July 19, 1990, when an explosion and fire at the BASF plant (in the Evanston residential neighborhood) reverberated throughout the City."(9) The source of the explosion was a reactor vessel that was being cleaned with a solvent. When a pressure relief disc ruptured, releasing a cloud of flammable vapors, it took only a source of ignition to produce an explosion that could be seen over 25 miles away, and heard over 10 miles distant. Although 83 persons were injured, only two were killed. A four-story building was demolished, and some of the other 17 were damaged on the 9.5 acre site where resins were manufactured for can and paper coatings. Exploding drums in the storage yard and collapsed stairways complicated the rescue of injured employees, but the hospital disaster plan proved very helpful. The area was evacuated for a half mile downwind, and some of the fire fighters required hospitalization for heat exhaustion. Although preplanning discussions had been held with most of the responding agencies, it was found that the coordination and communications among them could be improved.

It is not just luck or accident when a fire department operates efficiently during a major disaster. It is the result of effective planning, training, and cooperative coordination with supporting agencies. These activities are prerequisites of a successful disaster response.

CAUSES AND EFFECTS OF DISASTERS

Probably the simplest terms for classifying causes of disasters are *accidental*, as in the case of falling aircraft, and *deliberate*, as in the planting of a bomb. Other terms of classifying causes of disasters, *natural* and *manmade*, have little value for planning purposes.

Common causes of disaster include fear, famine, pestilence, fire, water, cold, heat, blast, earthquakes, contamination, criminal activity, revolution, aggression, apathy, and panic. Some persons may question how "apathy" or indifference could cause a disaster, but discourse with a serious environmentalist or a safety engineer should quickly clarify the point.

Regardless of the cause, most disasters produce death and injuries, interrupt communication systems, cause extensive property damage and, in addition to cutting off power and water supplies, may result in contamination of air, food, or water sources. As pointed out in an early U.S. Navy publication on disaster control, one nuclear detonation can produce all of the effects in other disasters.(10)

CIVILIAN AND MILITARY ROLES IN DISASTER CONTROL

What is meant by the term "disaster control"? It may have various meanings in the different emergency services, but for this book we will assume that "disaster control" is the use of all available measures to prevent casualties and damage, to minimize the effects of disaster, and to recover from it as soon as possible.

What is the role of the emergency services in disaster control? Ordinarily, the departments of fire, law enforcement, engineering, public health, plus medical people, are the principal "first-in" response forces in a major emergency or disaster. They are sometimes assisted by similarly trained personnel from the state or nearby communities, but they are the "front-line troops" who must make the rescues and save lives and property.

Cincinnati fire fighters had to contend with hazardous materials, toxic air, mass casualties, evacuation, and mass destruction at this BASF plant explosion. (Photo courtesy of Gary Auffart.)

Massive property destruction confronts emergency responders handling natural disasters. (Photo courtesy of Ron Buckingham.)

"Civil defense" refers to the collective activities of *all* emergency forces, professional and volunteer, and those who help to evacuate the casualties, care for the wounded and dispossessed, bury the dead, feed the hungry, shelter the refugees, and restore things to normalcy.

What is the military role in disaster control? If the disaster happens to be a war, then military force will repulse an enemy attack and provide military defense; in other disasters it may also have limited responsibility. Military forces are often called to assist in coping with major disasters, in deploying helicopters and boats in search and rescue operations, but the overall directions and coordination of disaster control and relief in peace time is a civilian responsibility.

We turn now to a consideration of actual disaster planning.

WHAT IS A DISASTER PLAN?

A disaster plan is an attempt to anticipate the effects of a particular disaster and develop responses to alleviate the effects on the community. From the fire department's perspective, disaster planning serves as a discipline for initiating prompt and effective actions that may be unusual in comparison to the normal response mode (large scale vs. ordinary).

The completion of a fire department disaster plan is not an end in itself; it is really a beginning. Having a plan does not guarantee that the actual disaster response will be effective. This concept was exemplified by President Eisenhower when he stated, "Plans are worthless, but planning is everything . . . keep yourself steeped in the character of the problem you may one day be called upon to solve, or help solve."

Fire departments have too frequently applied the converse of this philosophy to their disaster planning responsibilities, thinking that getting a plan on paper fulfilled their responsibility.

There are about nine good reasons why you need to draw up a plan of action for disaster control. The obvious benefit of having a plan is to reduce the time required to go into action by eliminating doubt and confusion. Here are some of the motives for putting the plan in written form:

1. Fulfill the responsibility of command.
2. Support the plans of higher authority.
3. Point up the gaps which still need study; an idea that seems complete is often found to lack important details when the first attempt is made to write it down.
4. Define the objectives of the plan.
5. Define lines or scope of authority and fix responsibility.
6. Avoid the weakness inherent in a "one-man show." The man who has the plan in his head is extremely valuable as long as he is around. If he should keel over from a heart attack when the signal for action is sounded, it may be rather difficult to locate a psychic medium who can obtain the plan from the dear departed chief.
7. Reveal the need for coordination of collaborative units.

The military provides support resources. Here, assistance arrives from the Coast Guard. (Photo courtesy of U.S. Coast Guard.)

Disasters cause disruptions in transportation arteries. Here, a critical bridge has collapsed during the Great Alaskan Earthquake. (Photo courtesy of Ron Buckingham.)

8. Accurately determine available forces and equipment.
9. Bring together in one document all relevant data necessary for a comprehensive picture of a disaster control operation, from the smallest to the largest unit.

A written plan is essential to explain the duties and responsibilities of those expected to respond. However, the planning process may yield far greater value than the actual plan. The planning process should include integration of local officials and involvement of all agencies expected to respond during a disaster. If this interaction is properly guided and motivated, it will continue as an ongoing process of participation. Disaster planning is a dynamic process and with modern wordprocessing and desktop publishing capabilities, the planning document can also be dynamic.

It is not necessary for municipalities, fire departments, and other governmental agencies to start from scratch in the development of a disaster plan. Much information already exists and, where applicable, should be used. Large amounts of federal funding have in recent years gone into the field of disaster planning, and while some might question the cost effectiveness of many federal programs, including those of FEMA, it is nonetheless incumbent upon fire departments and government agencies to take advantage of the federal tools that can be truly helpful.

FEDERAL EMERGENCY MANAGEMENT AGENCY

FEMA can provide the foundation for a comprehensive national emergency management system wherein federal, state, and local emergency management organizations become partners. The federal emergency organizational development is depicted in Figure 1-3. A local disaster plan evaluation is enhanced by an understanding of how a specific plan relates to functions on the chart.(11)

THE INTEGRATED EMERGENCY MANAGEMENT SYSTEM (IEMS)

The IEMS Concept

The Integrated Emergency Management System (IEMS) concept was developed by FEMA to facilitate the development of comprehensive emergency management capability at federal, state, and local levels of government.

IEMS is a method of needs assessment and program planning. This chapter will present the IEMS concept and explore other concepts that are part of a fire department's planning and preparation for catastrophic disasters.

FEMA has high expectations for IEMS. Although IEMS is not an imme-

Figure 1-3. Federal Emergency Organizational Development.

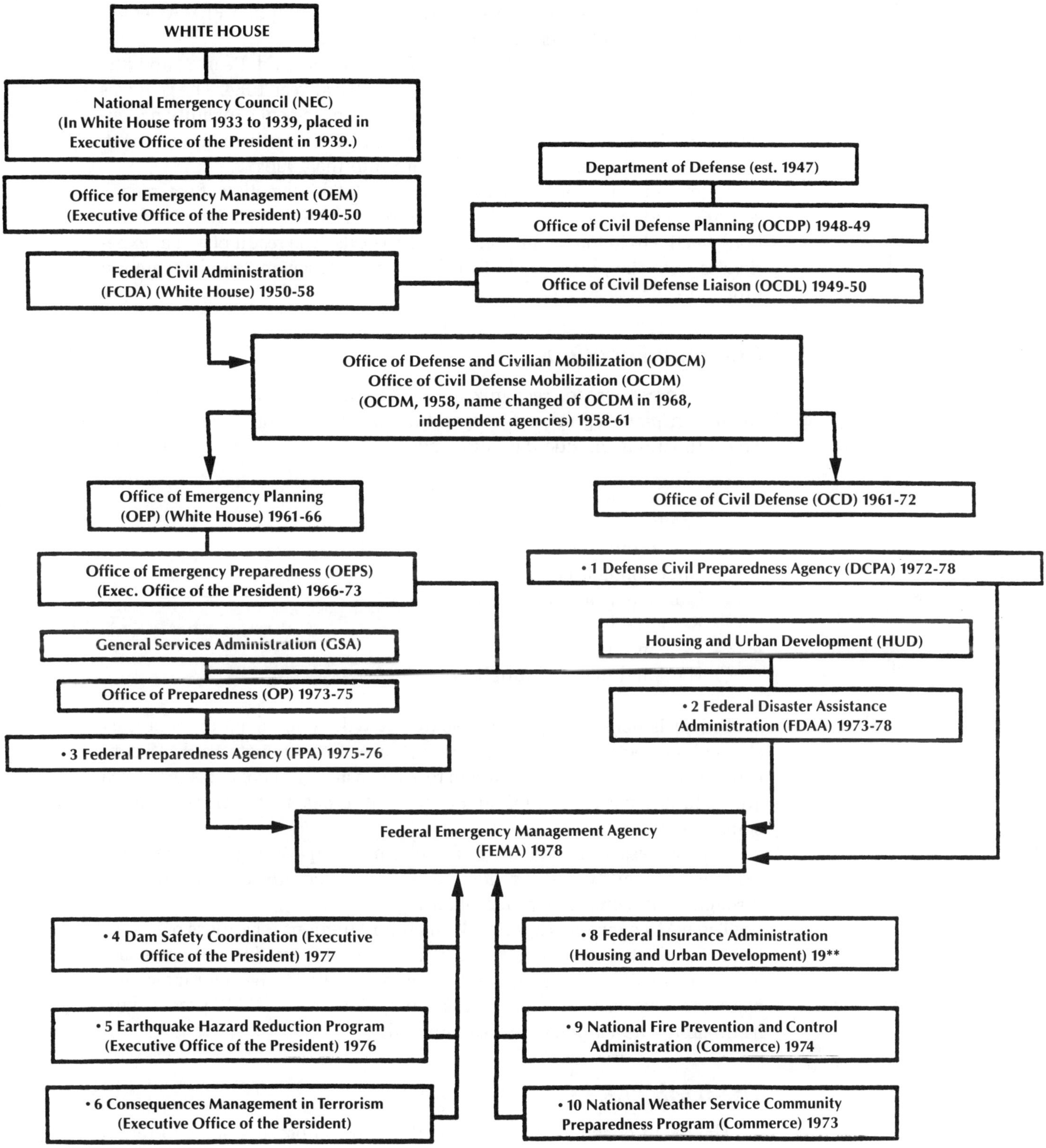

diate panacea for all the problems facing the emergency management community, an integrated approach, encompassing all the hazards confronting our communities, is the most effective way to accomplish FEMA's emergency management mission.(12)

The IEMS concept, endorsed by President Ronald Reagan and the Congress, was first implemented by FEMA in fiscal year 1984. The goal was to develop and maintain credible emergency management capability nationwide, for all types of large-scale emergencies, at all levels of government. It has remained a viable federal program into the 1990s.

Federal initiatives are designed to achieve total integration of emergency activities and to develop the capabilities required by governments to function effectively in the broad spectrum of hazards. The result of actual experiences with the process help to refine the guidance and implementation procedures.

Common Functions

The IEMS approach recognizes that common functions need to be considered in planning for and responding to all catastrophic emergencies. These functions include the following:

- Evacuation
- Sheltering
- Communications
- Direction and control
- Continuity of government
- Resource management
- Law and order
- Food and medical supplies

Fire jurisdictions of all sizes eventually will be involved in, or at least affected indirectly by, large-scale emergencies. Even though the IEMS concept stresses the functions common to all disasters, it recognizes that there are unique aspects of individual incident types that warrant special consideration.

The process is aimed toward a continual upgrading of emergency response capability and a reduction of duplicated efforts and resources through joint, communitywide planning.

The IEMS concept is applicable to all jurisdictions regardless of size or level of sophistication, even though not all are confronted by the same hazards, and not all have or require the same capability. The need for integrated emergency management capability is nationwide, and the IEMS process will facilitate the development of that capability.

The IEMS Process

The result of actual experience with the IEMS process helps to refine the guidelines and implementation procedures. FEMA has developed the Hazard Identification, Capability Assessment, and Multi-Year Development

Plan (HICA/MYDP) as tools for sorting out where we are today in disaster planning and where we should be going tomorrow and beyond.(13)

As a planning tool, this process can guide a local jurisdiction through a logical sequence for identifying hazards, assessing capability, setting priorities, and scheduling activities to improve capabilities within a given time frame.

STEP 1: HAZARD IDENTIFICATION (HI)

Knowing the types of large-scale emergencies likely to occur in one's locale, the probability of their happening, and the magnitude of potential problems are essential ingredients of emergency planning. All potential hazards affecting the jurisdiction must be identified. Hazard identification is a structured approach for determining hazards that pose a significant threat to local jurisdictions. Four questions guide the evaluation of each hazard threatening a community. It is the responsibility of the fire department and other planning agencies to determine the severity of the threat.

There is no use in gathering a mass of data about disasters that are not likely to happen in your community. If you live in Ohio you will not need data about volcanos or tidal waves. But do not rule out earthquakes in any part of our country, or flash floods, cyclonic windstorms, or blackouts. It is better to give these possibilities (marginal for some areas) some thought

Advanced planning—always the key to effective disaster mitigation. (Photo courtesy of the National Fire Academy.)

rather than none at all. The first thing to do is to make an appraisal of your community's susceptibility to disaster.

You can informally jot down the types of major disasters most likely to occur in your community. If your response district is near an airport then a plane crash is possible; if near a railroad yard or highway, fire or explosion in cargoes of hazardous materials or leaks of toxic or flammable gases from derailed tank cars or overturned tank trucks might occur. If you are in or near a forested or brush area, then woodland conflagration could be your greatest disaster potential.

Next, completion of a formal HI is a giant step toward a credible emergency management system. Determining the hazards and the impact they present to the community is only a small effort for a planning group, but the knowledge of those problems and the concerns generated provide the impetus to follow through the process and determine solutions to those problems. The complete exercise is found in Appendix A.

STEP 2: CAPABILITY ASSESSMENT (CA)

After the potential hazards to a community have been identified and analyzed, and preparedness and response priorities have been established, the next step is to assess current capabilities of the local jurisdiction in handling a large-scale incident. The current capabilities assessment should include an in-depth examination of emergency management staffing, Emergency Operation Center (EOC) criteria, warning systems, public information and education, communications, evacuation capability, operational plans, emergency resources, mutual aid, training, and exercises. Most of these capability concerns are common to all hazards. A fire department should also examine its capability to effectively respond to each of the particular hazards identified.

In addition to identifying the strengths of a jurisdiction, the CA will also identify weaknesses. Capability shortfalls that are identified should be the basis for preparing a Multi-Year Development Plan (MYDP).

STEP 3: MULTI-YEAR DEVELOPMENT PLAN (MYDP)

Based on the capability shortfall identified in Step 2, the MYDP should outline what needs to be done to reach the desired level of capability. A step-by-step procedure should be set up to include the following.

- Total resource requirements for alleviating shortfalls that the jurisdiction plans to address
- Additional capability development not necessarily identified in the CA
- Major maintenance efforts to sustain the current capability

The MYDP, used with the HI and CA becomes a convincing tool when approaching your budget director for establishing priorities for future funding. The problems and the solutions are both well defined.

HICA/MYDP Summary

The IEMS-HICA/MYDP is a long-term concept aimed at improving emergency management capability in all jurisdictions. The improvement that is envisioned by FEMA is a gradual, continual upgrading of capability year by year.

The goal is to develop and maintain credible emergency response capability nationwide, but to accomplish this, HICA/MYDP must be accepted by local fire departments.

This explanation of the IEMS program of HICA/MYDP has been but a brief overview of the process. To fully implement the process on a formal basis, FEMA provides indepth instruction and the format for completion of HICA/MYDP. All local jurisdictions receiving Emergency Management Assistance (EMA) financial support from FEMA (through state emergency management agencies) are required to complete the process.

This overview, however, should provide a progressive fire department with the conceptual tools necessary to approach the problem of disaster planning and preparedness in its community so that it is prepared for large-scale incidents.

LEVELS OF PLANNING

There are five different levels for disaster planning: national, state, regional, local, and fire department level. In the larger fire departments, a sixth may be desirable at the division or bureau level.

National Plan

Among other things, the national plans prepared by FEMA, Office of Civil Defense, EPA, and other agencies provide for a nationwide warning system: federal control of fire fighting on federal land; state control of fire fighting on state and private land (only when outside of organized urban areas); advice and guidance on training matters; and some financial assistance to lower echelons (usually with "strings" attached) for all types of disasters. National plans also provide that in the case of conflagrations, primary reliance is to be placed on local fire departments. The jurisdictional guidelines based on fire control are extended to include other types of national emergencies or disasters.

State Plan

State plans generally provide for fire departments to operate as they normally would at large fires; mutual assistance to other fire departments is voluntary except when a state of extreme emergency has been declared. Such plans usually provide for integration of state and federal forestry service operations. They also establish a chain of command for requesting mutual aid; for example, the city fire chief will request assistance from the state's area command, and it can relay the request to the regional command, and if necessary, to the state director. Such plans almost invariably provide that the local fire chief will be in charge of a fire in his own jurisdiction, and usually provide for a statewide emergency communications network.

Regional Plan

An example of a plan more regional than state or local is the Firescope program, an outgrowth of the disastrous California brush fires in 1970. "Firescope" is the acronym for FIre fighting RESources of California Organized for Potential Emergencies, and the plan is intended to be a coordinated effort of all fire fighting agencies in California joining in one organization to fight wildland brush fires. Administrative complications can arise as a commander attempts to weld many large agencies into a cohesive organization: U.S. Forest Service, California Division of Forestry, California Office of Emergency Service, and numerous fire departments. The project utilizes computers programmed to tell where fire crews and equipment are, how long it will take to get them somewhere else, their individual capabilities, and other operational conditions. The computer can even tell the last time the crews were fed and the availability and need for fuel for fire apparatus (which run out of gasoline about every two hours).

The plan was designed to reduce the cost of fighting fires as well as losses in structures and watershed, because it enables the fire fighters to know quickly where the fire will spread, how fast it will travel (based on weather data and fuel load input), where structures are located that need immediate protection, and what resources are available at any time or place. Firescope was a predecessor to the incident command system, which is becoming a standard operating procedure for the departments throughout the entire North American continent.

Local Plan

Local plans, though often at variance in minor details and titles from one community to another, generally provide for the local fire chief to be under the direction of the city manager, mayor, or disaster coordinator of the city or the chairman of the county's board of supervisors (as head of the civil defense organizations in incorporated and unincorporated areas, respectively). The civil defense coordinator (or disaster preparedness coordinator, or whatever title he assumes for this position) is usually assigned the duty for keeping the plan up to date. In smaller cities, this officer may hold this position as a collateral duty to his regular job as fire chief, safety director, or some other official position.

Local plans often establish about a dozen emergency divisions, such as fire, police, health, utilities, public works, housing and shelter, transportation and communications, property and procurement, personnel and recruitment. These divisions would be represented at a convening of the disaster control organization in the event of a catastrophe. Other features of a local plan usually provide for succession of command, commandeering of services and property of private citizens at the direction of the Disaster

Control Commander, when deemed necessary to protect lives and property of the people as a whole, and for fining and imprisoning anyone who obstructs members of the Disaster Corps in the performance of their duties. The local plan also sets forth the conditions required for authorizing declaration of a disaster and activating the disaster organization on a duty status.

As part of the IEMS process the local plan is officially called The Community Emergency Operations Plans (EOP). It has three components: (1) The Plan, (2) Functional Annexes, and (3) Appendices.

THE BASIC PLAN

The basic plan is a component of a community's EOP that provides an overview of the emergency organization, operational concepts, situations to which the plan is applicable, and the legal authority for the plan.

The plan identifies the missions of the individual agencies participating in the plan and assigns the authority for specialized decision making to the appropriate agency. Responsibility and authority to perform specific tasks are delegated to the agency best able to cope with a given problem. Thus, all fire and rescue problems and any related decision making is handled by the fire department. A prime concern of fire department planning then is to ensure that the responsibilities assigned by the EOP are covered in the annex to the plan. One very important aspect of the basic plan is that it has the authority of local law. A disaster plan, duly approved by law, gives all agencies within a jurisdiction the legal authority to perform the assigned duties.

FUNCTIONAL ANNEXES

Functional annexes are components of the EOP that develop specific information and direction with general application to all incidents. Each operational section (fire, police, public works, etc.) develops an annex providing information on that section's operations plan. Only information that is pertinent and useful to its participants in the plan is included. Cognizance of each section's operational concepts is essential to instill cooperation and coordination of efforts. It is helpful if each annex follows the same general format to facilitate usage. A suggested format is shown below.

- Authority–enabling laws and ordinance
- Responsibility–as defined in the charter and EOP
- Objectives–missions assigned to the section
- Operational Concepts–outlined in broad terms
- Organization–chain of command and the emergency organization
- Resources–personnel and types of equipment
- Mutual Aid and Special Needs
- Liaison–methods for interdepartment and interagency interface
- Recovery Operations

APPENDICES

The more specific intradepartment operational information may be addressed in appendices, which may or may not be part of the planning document. The overall EOP addresses common functions associated with the management of all disasters; appendices are developed to address unique aspects of particular types of disasters. See Figure 1-4 for a schematic breakdown of the component parts of a standard disaster plan.

Figure 1-4. Components of a Standard Disaster Plan.

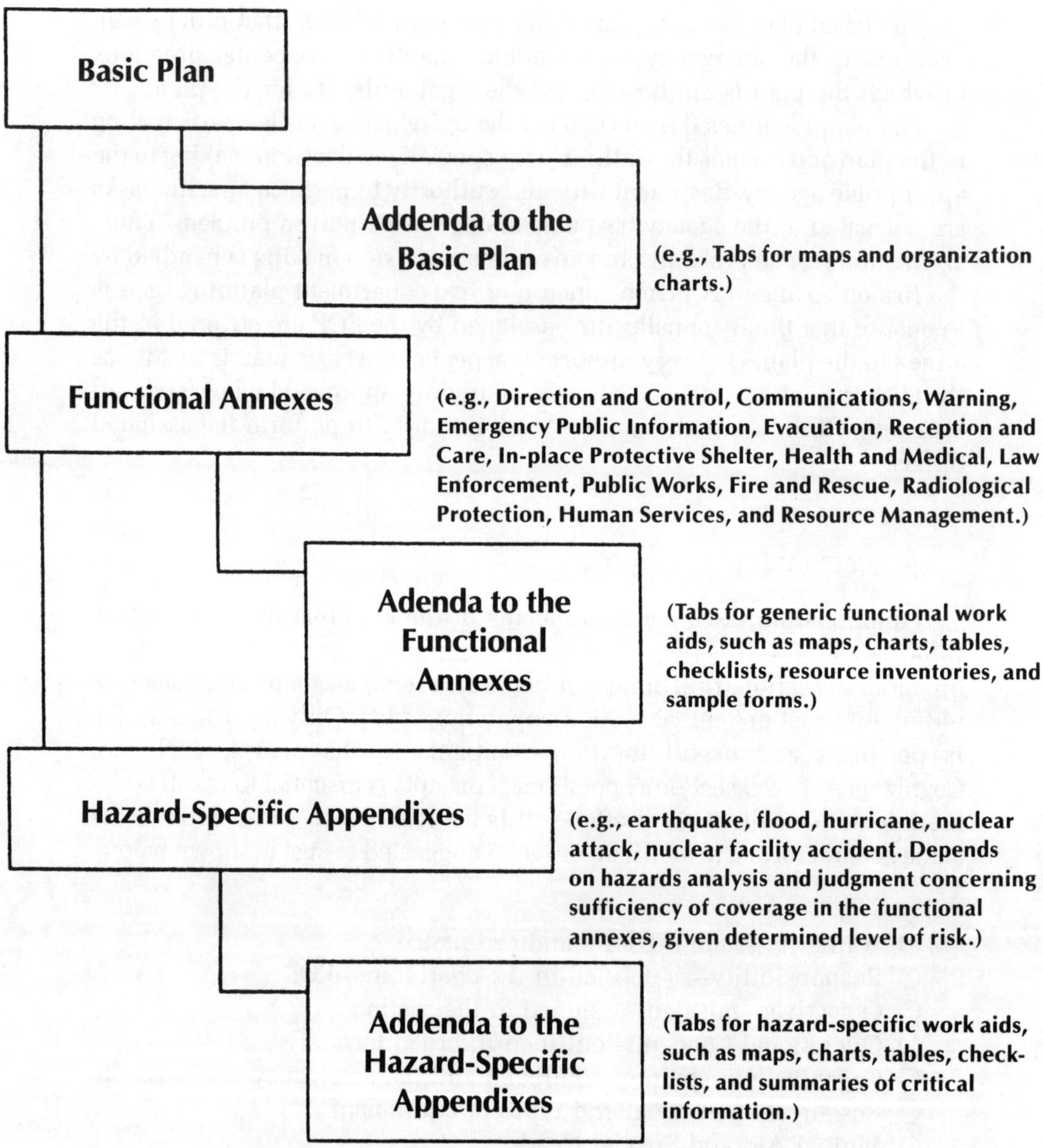

THE FIRE DEPARTMENT DISASTER PLAN

Hazard Identification, Capability Assessment, and Multi-Year Development Plan

The motivation and basis for developing or revising a fire department disaster plan could be the result of an HICA/MYDP. Completion of this process gives the planner knowledge of the hazards to consider and the capabilities of the department to respond.

At the fire department level, plans usually include a standard operating procedure for implementation of incident command and for augmenting normal first alarm response assignments with additional companies, either singly, in task force increments, in predesignated second, third, fourth, and other alarm assignments to box (or phantom box) locations, with provision for moveup (or transfer) of other companies or fire departments to cover for the displaced companies.

Fire department plans might also include a Fire Division Operations Plan as an annex to the community's civil defense and disaster plan. This usually identifies the fire department's authority, such as may be found in state law and local statute; the fire department's responsibility under local ordinance, such as, control of fires, explosives (demolition of dangerous structures), control of oil, chemicals, and ruptures in pipe lines conveying same; training of auxiliary fire fighters; the fire department organization (functionally and geographically); and annexes and appendices as deemed appropriate. (See Appendix B for a sample basic Disaster Plan furnished by Maplewood, Minnesota.)

The fire department's Disaster Plan might well provide for two plans of operation either of which could be used from the beginning of the major emergency, depending upon the seriousness of the prevalent condition or, in a rapidly developing danger, might be used in succession.

A "Plan One" operation could be designated for handling major fires, such as highrise or watershed fires, and could go into operation automatically any time a request for more than a certain number of companies is requested, e.g., five (a typical first alarm assignment to a highrise fire or a brush fire in an inhabited area); this would be useful in a community where large-scale operations are expected. Plan One would utilize on-duty personnel, and would provide extra back-up companies for covering spot fires, to fill in for apparatus that might break down, and respond to sudden ignitions from flying brands. It would also provide for automatic dispatch of needed special apparatus, such as helicopters, air units, medic units, fuel trucks, or water tankers.

The organization called for in Plan One would be flexible, since the billets could be filled in the sequence needed as help arrives. For example, the incident commander could assign the next officer to respond to the plan's functions, with responsibility for formulating manpower assignments and crew operations. The next to arrive could be designated as the assistant field commander for *Fire Suppression*, to coordinate the activities of the sector commanders; the third could be designated as the assistant field commander for *Staging*, and so forth. The Incident Command System is designed to assign functions in a modular fashion, including command post communications, fuel supplies, and other functions.

In larger departments, the field headquarters commander might be a deputy chief, assistant field commanders might be assistant chiefs, and sector commanders, battalion chiefs. However, the rank is not important. The important thing is getting the needed functions performed, and if the fire marshal or the training officer happens to show up before the chief from the adjacent battalion, he could be made an assistant field commander and put in charge of whatever billet needed filling most urgently at the time.

A "Plan Two" operation might be used for very large brush fires, airport or harbor fires, structural conflagrations, or where more than one major fire is burning at the same time. This plan would have to be invoked by the department field commander. It would not have to follow a Plan One, but could be designated from the start. Nor would it be limited to fires, but could be called for a dam break, earthquake, or any large-scale emergency for which no special plan has been formulated.

In addition to regular response assignments, a Plan Two would recall off-duty personnel normally assigned to department bureaus, such as personnel and training, supply and maintenance, maps and drafting, hydrants, medical liaison, training, and fire prevention, to provide a staff for the field commander (or to augment the one already obtained under a previously called Plan One). Plan Two also would provide a flexible organization, the billets of which could be filled as personnel arrive, in accordance with the importance and imminent need of the tasks to be carried out. Plan Two augments the staff of the incident commander; if a Plan One had been called previously and a command post set up initially with sector commanders for the Plans, Fire Suppression, and Staging, then under Plan Two an additional assistant field commander for Liaison would be provided to handle coordination of representatives from civil defense, Red Cross, law enforcement, and other fire departments. As more assistant chiefs arrive, they can be assigned to replace any battalion chiefs who might be filling field billets, and the battalion chiefs could be used to replace any captains who had been assigned as sector commanders.

As mentioned at the outset of the discussion, the planning level for the larger fire departments might well go to the division or bureau level. In fire departments having divisions or bureaus for handling such functions as supply, maintenance, personnel, training, and fire prevention, subordinate commanders in charge of such bureaus or divisions could be required to submit a detailed plan identifying the support responsibilities assigned to members within their respective commands; these become appendages to the fire department's Disaster Plan.

Preparing the Disaster Plan

Many formats are acceptable for an operational plan. The essence lies with the plan's content and its applicability to the disaster situation.

A fire department disaster plan may take the broad approach that deals with the common characteristics of all disaster responses. If this concept

This rigging equipment is typical of resources beyond those within even large fire departments. (Photo courtesy of the Federal Emergency Management Agency.)

is utilized, then annexes and appendices need to be developed to detail those specific actions to be taken during particular types of disasters.

Some departments have developed specific plans for specific hazards, with each plan standing separate and complete. The Los Angeles City Fire Department utilizes a separate Earthquake Operational Plan. The basic plan with accompanying information and worksheets is enclosed in a packet and carried in every emergency vehicle.

It should be obvious that it would be unwise to wait until a possible disaster is imminent before preparing a plan to cope with it. Yet that is what happened in the Los Angeles fire department when the Cuban missile crisis was announced. Author Chuck Bahme was at the time a Naval Reserve Officer. He received a call from the Navy Department in Washington, D.C., alerting him that he might be recalled to active duty in a matter of hours. Then the fire chief asked him to prepare a plan for response to an enemy nuclear attack and requested that it be ready for presentation at the special fire commission meeting the next day. Though the city had plans for an enemy attack with incendiary bombs that were prepared in World War II, plans for brush fires of varying magnitudes, and plans for earthquakes, it had no plan for fire department action in the event of a nuclear attack on the city, notwithstanding the fact that Los Angeles was considered a number one target in the early sixties. A few years later, in 1965 when the Watts riots broke out, there was no special plan for coping with an attack by armed criminals. A task force plan had to be developed and tested during the four days that "all hell broke loose."

So perhaps it is not so obvious that the time to start planning is *now*,

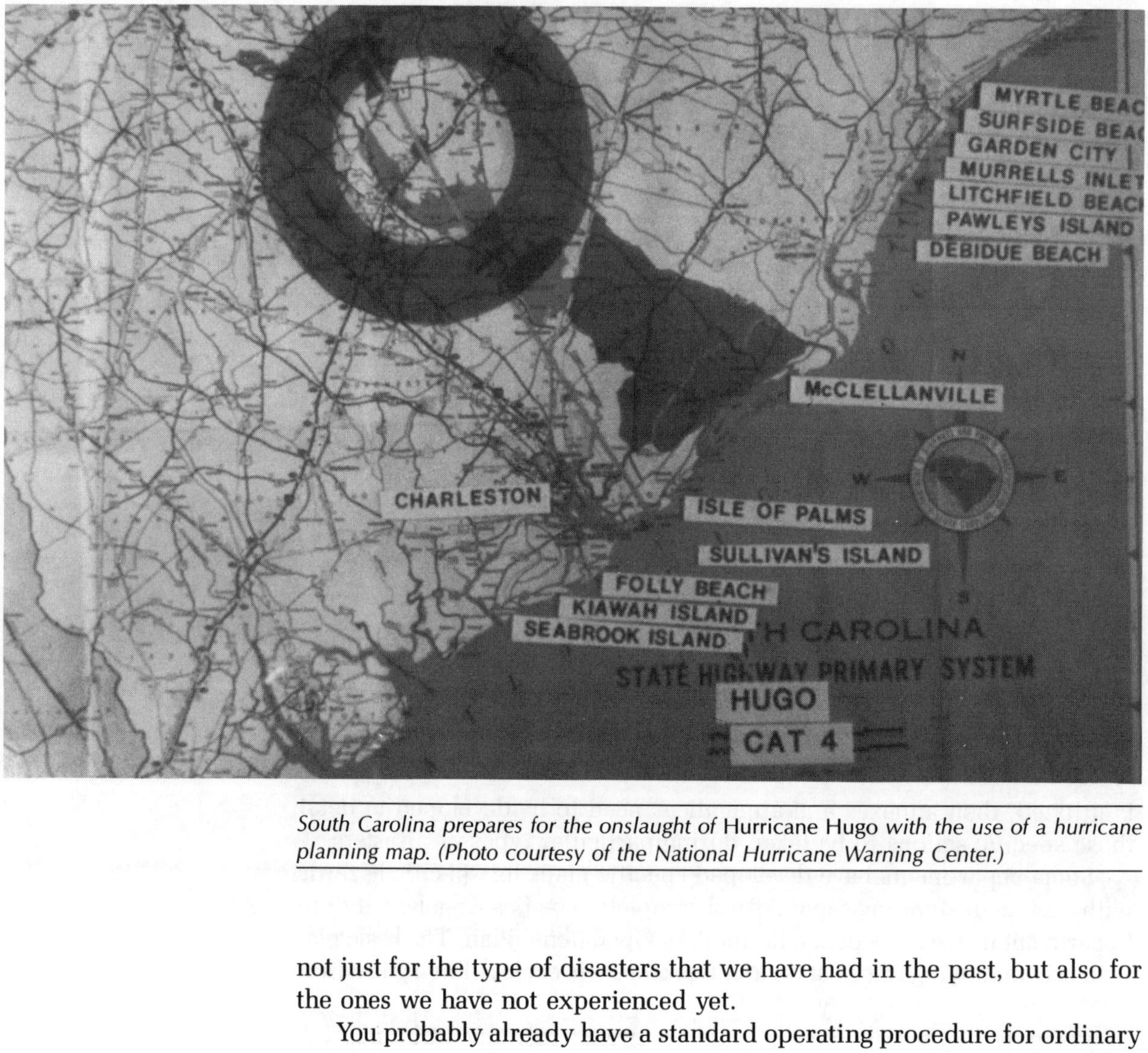

South Carolina prepares for the onslaught of Hurricane Hugo *with the use of a hurricane planning map. (Photo courtesy of the National Hurricane Warning Center.)*

not just for the type of disasters that we have had in the past, but also for the ones we have not experienced yet.

You probably already have a standard operating procedure for ordinary emergencies, such as a grass fire, auto fire, or structural fire, and perhaps some "odd-ball" emergencies, such as acid spills, bomb threats, cave-ins, a child locked in a bathroom, an elevator stuck, or a person stuck in a bath-tub. You should take certain actions to develop a broad plan for major disasters. We already have discussed the theoretical need for planning. There are some practical steps to begin the planning process: (1) assemble the data you need, (2) make an inventory of resources, (3) study available plans and organization charges, and (4) begin to compose a plan to suit your needs.

Anticipate the Unexpected

Regardless of where you are located, an earthquake could knock out water and power systems as well as cause other damage.

In the San Francisco earthquake that occurred late in 1989, water pressure was reduced to almost zero in the older marina district. A major fire erupted which threatened to become a conflagration encompassing many blocks of the city. Because San Francisco had a special unit including 3,000 feet of five-inch hose, it was able to compensate for a loss of water mains and prevent the major fire from spreading.(14)

Those fire departments that cope best in emergency disaster situations are those that have anticipated them in advance. Tornadoes are more of a probability in some states than others, but windstorms can happen anywhere. If you are in a hurricane belt, high winds and flooding could be your primary concern. Coastal cities are more apt to be hurricane targets than others. Almost any community could be in the path of radioactive fallout in event of a nuclear accident.

Inventory Your Available Resources

With a list prepared of your most likely disaster probabilities, a second listing will be an inventory of available resources within a reasonable distance of your community. A list will need to be developed of the people in other departments and agencies who you can call to provide assistance in handling disasters you are likely to experience.

Most disaster situations will require interaction and mutual reliance from many different supporting agencies or organizations. This support could come from federal, state, or local governments as well as private business, industry, or volunteer organizations.

If the fire department expects to use all of the available resources during a disaster, it must be knowledgeable about the individual resource, plan of operation, response capability, and procedure for obtaining service during an emergency. There are advantages to establishing rapport with those agencies beforehand or including them in the planning process.

- The fire department becomes familiar with the service or function available for support.
- Mutual concerns of authority, command, and expected participation can be resolved.
- The fire department gains a friend instead of an adversary, and the department plan, in effect, becomes their plan as well.

The necessary resources to respond to major disasters are out there and available, but without interagency planning, organization, and cooperation clearly established before the fact, those resources cannot be used most effectively.

Study Available Plans and Organization Charts

If possible, before changing your existing plan or formulating a new one, take a look at plans of other fire departments of comparable size and hazard potential.

After you have studied other plans, you might select the features and nomenclature most suitable for your circumstances and include them in the draft of your proposed plan. Then check with your nearest civil defense agency to ascertain how your proposed organization and plan for deployment of forces fits in with applicable federal, state, local, industrial, and military plans. Make needed modifications.

Some plans are designed to cope with enemy attack, some for handling wildland fires, some for multiple company operations at large structural fires, lumberyards, piers, and wharves. Some plans merely call for sounding of more alarms to obtain additional manpower and equipment, letting the incident commander decide whether to summon special equipment, call for mutual aid, request recall of off-duty personnel, or other logistical support.

Some plans do not contemplate sounding additional alarms under a typical running card system, but the incident commander calls for the exact number of each type of fire company he wants, as he wants them, and at the same time he directs them to the sector of the emergency where he wants them to report. Covering companies' notifications (arson squad, police department, water department, power company, Red Cross, etc.) are made by the dispatching staff at the predetermined stage of the operations, without special request of the incident commander.

In fire departments that already have companies operating regularly in task force formation (e.g., two engines and a ladder company), the incident commander at a major emergency may simply call for the number of task forces needed under a plan that would automatically bring response of additional chiefs and administrative officers to staff a command post at the scene of the action.

Some plans have annexes that specify operational procedures for special emergencies, such as floods, hurricanes, widespread arson fires and riots, harbor fires, airplane crashes, and earthquake.

All existing emergency response plans, even outdated plans, should be reviewed. The review should include any relevant information such as mutual aid agreements, action guides, standard operating procedures, and other documents that reflect previous thinking on the subject. As mentioned above, it might be helpful to review plans of other departments of similar size and disaster potential; however, it is not a good practice to rely exclusively on plans developed by another department. Each department has unique characteristics and specific problems that should be addressed from a local perspective.

Plan Content

After the planning format has been determined, it is time to consider the types of information that should be included in the plan. Listed below are some content considerations for a fire department disaster plan utilizing the broad approach.

- Emergency Notification List—Identifies personnel who shall be notified and methods of notification.
- Amendment Procedures—Establishes a method for notification and recording of changes to the plan.
- Plans Distribution—Accounts for all organizations, sections, and fire stations receiving copies of the plan.

- Assumptions—Describes the background considerations and intent of the plan.
- Table of Contents—Provides quick access to pertinent sections, as well as maps, charts, checklists, and definitions of responsibilities.
- Supporting Agencies—Establishes methodology for obtaining outside resources, identifying their response capability, and predetermining arrangements.
- Hazard Identification—Identifies and analyzes the specific hazards to be anticipated. Frequency and probability of occurrence as well as areas of high risk or special concerns are discussed.
- Specific Plans (Probably in the form of Appendices)—Prescribes detailed plans for responding to each of the particular hazards identified in the previous step. A list of specific actions to be considered by the fire department in each event is suggested. Each level of command must be fully aware of the operational changes and unique responsibilities that may evolve with each hazard.
- Training and Testing—Trains fire department personnel and supporting agencies in all aspects of the plan. This not only makes the response personnel aware of their assigned duties, but also tests the validity of the plan. A frequency for training and testing should be established so that the plan does not just sit on the shelf and collect dust.

Get Your Plan Accepted

The best plan in the world is of little value if it gets no farther than the drawing board. If you go about it right, your plan will have a good chance of being approved.

A chief officer should discuss the disaster planning proposal with agencies affected outside the fire department. After acting upon all suggestions received from his or her own staff, the next step is to lay the groundwork for obtaining approval of the fire commission, board of directors, or other governing board. He or she should get on the agenda early, so as not to feel rushed in the presentation, as would be the case if scheduled just before noon or closing time. A good, but not too lengthy, introductory statement of the need for the new or revised plan is important. It is helpful in making a presentation to use a chart to show the deployment of forces, organizationally (bureaus, divisions, battalions, companies) and geographically (location of division headquarters, battalions, and special equipment, such as fire boats, crash companies, rescue companies, mountain patrols, dispatching centers, helicopter hangars, and communications vans).

The chief might also show charts depicting successive steps of deployment of additional companies and staff officers as they arrive to assist in handling a greater alarm emergency. This could be done with simple flip charts or in a more sophisticated fashion using computers and color graphics. As the assumed emergency requires buildup of forces, printed colored cards, representing units responding, can be placed on a flannel board or other large board, together with the additional staff officers, command posts, apparatus pools, and other special elements needed.

Planning continues during a major disaster with the use of tracking equipment at the Emergency Operations Center (EOC). (Photo courtesy of the National Fire Academy.)

Issuance of Directive

After approval of a governing board or commission has been obtained, it is necessary to get the signature of the proper authority on the directive, which will authorize the plan to be put into effect. This directive should provide for succession to command in your organization, and because of the possibility of several high ranking officers being incapacitated in a single explosion or other mishap, succession should be specified at least three deep, preferably from department chief to company commander in smaller fire departments where only a few chief officers might be available to take over command. In an explosion of a truck load of dynamite and blasting agent in 1959 in Roseburg, Oregon, the highest ranking officer surviving after the blast to command the department was a lieutenant.(15)

This directive should also set forth the circumstances under which the officer-in-charge at the scene of the incident may (or must, preferably) declare a "Major Emergency" to activate the first steps in the disaster control plan. For example, the directive might provide that whenever the situation requires summoning of mutual aid or recall of off-duty personnel, then a Major Emergency must be declared.

A third matter that this directive might cover is designation of an Operations Officer (and successors or alternative, in the event of his unavailability or need for relief) who will be required to report to the EOC when a major emergency is declared, to carry out the provisions of the plan. The directive should specify that he has the authority, under designated circumstances, to recall off-duty platoons, singly or all at once, man reserve equipment with special duty personnel, off-duty members, personnel from other departments, and volunteers trained for such emergencies; request air tankers from the forestry service; request food for emergency personnel, including the EOC and dispatching staff; commit the fire department to pay for additional logistical support needed to cope with the emergency, such as private contractors having bulldozers, tank trucks, cranes, backhoes, and related equipment; and commit the department to pay overtime for mutual aid forces called to help in accordance with the plan.

A fourth provision in this directive might be the designation of a liaison officer (or equivalent title) to have authority for coordinating with agencies other than the emergency service in time of disaster.

In some departments it might even be desirable for the directive to set forth powers and duties of other subordinate officers who have special assignments in the disaster plan, e.g., apparatus control, field communications, air operations, and public relations.

A fifth provision in your plan, in an absolute necessity, is the procedure by which your standard incident command system for routine fires is to be utilized and expanded as necessary to accommodate all of the forces in your disaster plan. In Chapter 3 we will show how the primary incident command system includes the four general functions of planning, operations, logistics, and finance. In a disaster plan many of the subordinate functions in these four major categories will be identified in advance and titled. In the time of disaster they merely will need to have personnel assigned.

Disseminate the Plan

A plan is of no value until officers become thoroughly familiar with it. Using flip charts, a flannel board, or similar graphic aids, the operations officer (in larger departments) or other suitable officer could present the plan to highest echelon officers (deputy, assistant, battalion chiefs, bureau commanders). Succeeding presentations could be made to company commanders, and periodic review should be included in the training sessions of all officers.

Try Out the Plan

A plan is no better than the paper it is written upon unless it works when put to the test. Don't wait until disaster strikes to try it out. As a fire official, you should cooperate in all civil defense drills where fire department participation is desired.

It would also be advisable to hold your own test exercises. The conducting of a simulated emergency for each platoon will help to familiarize officers and fire fighters with nomenclature, procedures for reporting in at a command post, setting up the communications van, field headquarters, personnel and apparatus pools, sector commands, air operations, feeding on the line, reproducing needed maps, liaison with the press, police, public works, utility companies, Red Cross, civil defense, and other agencies.

You are almost sure to find gaps in the plan that need to be filled in, and some provisions that do not work as expected. So try to eliminate the bugs that show up on dry runs and make necessary revisions or additions to the plan.

Administer the Plan

It is not enough to draw up a plan, get it adopted, disseminate it to the troops, run a few tests, and then file it away in a drawer. A number of things must be done to make sure that the plan does not become obsolete. It must be reviewed constantly for revision. A high-ranking officer should be responsible for keeping it up to date. Even in the largest departments this task might be a collateral duty, with a title such as Emergency Liaison Officer, for someone who has been designated as the Operations Officer. Modern word processing technology is insufficient to guarantee revision. The human leadership element is required.

Make sure that disaster plan billets are kept filled. As personnel are transferred or retired, replacements should be designated promptly, and given copies of the directives.

Since conditions and assignments of officers are bound to change, all key personnel should be kept informed of latest developments affecting the plan. Instruction should be provided on a continuing basis (at least three deep) for each position in the plan.

Execution of the Plan

Plans are not self-executing; even the best plan is of little value if the officers are reluctant to put it into effect at the scene of an emergency. For example, suppose you monitor a neighboring fire department's frequency on the radio and hear the first-in officer-in-charge call for six more companies, at a going brush fire. Then ten minutes later you hear his superior officer call for six more (bringing the response to eighteen engine companies) without declaring a major emergency. You might assume they are operating under a plan that provides for automatic declaration of a major emergency after a certain number of companies are called or they have some reluctance to invoke their plan. Or their fire department might not have a plan.

While it is true that no one wants to be considered a "shaky chief," the importance of invoking the plan at the earliest sign that destructive forces are getting out of control cannot be stressed enough. If you wait for the mayor to declare a local major emergency, or for the governor of your state to declare an extreme emergency, you could have a long wait. In the Watts riots when such a declaration was needed, Chief Bahme found that the mayor was out of town, the governor was out of the country, and the deputy mayor and lieutenant governor were hesitant to act on their own authority. While a fire chief cannot put the city's plan into effect, he or she should have no hesitancy invoking the department's plan.

Even if the fire chief lacks authority to declare a local disaster (so as to activate resources outside the fire department such as the state's mandatory mutual aid provision), upon activating the local fire department plan he or she should simultaneously advise the community's administrative officer or civil defense director that a situation exists warranting such a declaration.

For example, if the earth shakes and it is suspected serious damage may have occurred in your own or some neighboring community, the chief might wish to make a helicopter reconnaissance for possible fires, dam breaks, or destroyed buildings. Critical observations can be made quickly from the air. If while flying over an area outside of the city, the chief sees people trying to rescue victims from a collapsed building, he might ask a dispatcher by radio to notify that jurisdiction of the conditions observed, and assist in requesting a response. Otherwise, it may be a couple of hours before other agencies, such as the public works department, water and power departments, gas company, health department, and sewer maintenance division, check this particular location.

CONCLUSION

This chapter has given us an overview of the similarities and differences between conventional fire defense planning and disaster planning. We have considered all of the primary ingredients that must go into effective disaster planning, including the need for a strong incident command system, an effective network for notifying other disaster agencies, and a working operational plan to mobilize your own fire defense forces. Succeeding chapters will provide additional information on incident command, on handling varying types of specific disasters, and on gaining the ability to effectively utilize fire department resources for disaster control.

CHAPTER 1 REFERENCES

1. Scawthorn, Charles. "First Report on the Quake" in *Firehouse*, Vol. 14, No. 12 (December 1989), p. 31.
2. "Desert Victory" in *Newsweek*, Vol. CXVII, No. 10 (March 11, 1991), p. 20.
3. Kelley, Jack. "Capping the Fires in Kuwait," *USA Today*, July 31, 1991, p. 6-A.
4. Bahme, Charles W. *Fire Officer's Guide to Disaster Control*, 1st ed., Boston, MA: NFPA, 1978, p. 2.
5. *USA Today*, April 10, 1991, p. 1.
6. *World Book Encyclopedia*, 1991.
7. *World Book Encyclopedia*, Vol. 5, p. 181, 1969.
8. Kramer, William M. "Introduction to Disaster and Fire Defense Planning" in *Disaster and Fire Defense Planning*, Emmitsburg, MD: National Fire Academy, 1989, pp. 1–8.
9. Uebelacker, Carl. "Explosion and Fire in Cincinnati Kills Two, Injures 83" in *Firehouse*, October 1990, pp. 43–51.
10. Disaster Control (Ashore and Afloat), NAVPERS 10899-A (1964) p. 287.
11. Kramer, William M. "Training and Preparedness" in *Disaster and Fire Defense Planning*, Emmitsburg, MD: National Fire Academy, 1989, pp. 5–12.
12. Federal Emergency Management Agency. *Hazard Identification, Capability*

Assessment, and Multi-Year Development Plan: Overview (CPG 1–34), Washington, D.C. January 1985.

13. Federal Emergency Management Agency. *Hazard Identification, Capability Assessment, and Multi-Year Development Plan: Overview* (CPG 1–35), Washington, D.C.: January 1985.
14. Kramer, William M. "Questions on San Francisco Quake" in *Heat Sheet*, American Heat Video, January 1990.
15. Carlson, Gene, and Warren Isman. *Hazardous Materials*, New York: Macmillan Publishing Co., 1980, p. 10.

CHAPTER 2

HISTORICAL LESSONS

SAVING LIVES and protecting property—that is what the fire service is all about. Members of the fire service are charged with the responsibility of safeguarding our communities from devastation. Yet, year in and year out, countless lives are lost and millions of dollars in property damage result from floods, tornadoes, earthquakes, hurricanes, and other catastrophic disasters, despite the presence of the firefighting forces.

There is a need for all levels of government to limit the impact of these occurrences on the people that the fire service is charged to protect. Many activities can be undertaken to improve our preparedness, but the available resources are limited.

The problems connected with a catastrophic disaster can be overwhelming to a fire department and to the community. However, Chapter 1 showed how many of the problems can be anticipated. Appropriate preparation and planning are responsibilities of fire service leaders. A properly developed community emergency operation plan (EOP) and a viable fire department disaster plan may be the keys to preventing mismanagement of an incident.

We learn from history. Plans should not merely follow the theoretical outline produced by the Federal Emergency Management Agency (FEMA) and other governmental agencies but should rely on information gained from actual previous incidents. In this chapter we look at the historical background of disaster planning and blend actual lessons learned with planning for large-scale operations.

New large diameter hose has permitted application of massive volumes of water to control major fires and mitigate major disasters. Compare the hose used by the Dayton Fire Department with that used by their predecessors. (Photos courtesy of Charles Steinbrunner- (TOP); Jim Lanagan (BOTTOM)

GEOGRAPHICAL DIFFERENCES

In the first chapter we saw how disasters can be expected to strike anywhere, anytime. Despite the fact that many of the negative aspects of disasters are foreseen and are rightfully anticipated in a disaster plan, there are always new complexities that are unforeseen.

Of all the types of disasters we will study, the ones most closely related to the ordinary or routine emergencies faced by the fire service are large-scale fires—conflagrations, high-rise fires, and wildfires, for example. With an historical perspective, we see how unusual weather or climate affects these various major fires differently in different parts of North America.

The $5 billion inferno that razed 2,500 of California's most expensive homes and snuffed out two dozen lives in the hills of Berkeley and Oakland late in 1991 provided ample evidence of the hazards of the combination of wind, weather, wood shingles, and water deficits. (As reported in the November 4, 1991, issue of *U.S. News and World Report*, p. 22.) Also consider the drought stricken summer of 1990. On May 8, 1990, the state of Michigan received an unusual share of devastation from wildfire when 86 homes and 126 other structures were destroyed. Conditions became ominously right with temperatures in the low 80s, relative humidity at 26%,

This advancing wall of flame in a west Texas wildfire presents different tactical needs than those in an urban setting. (Courtesy of Ron Buckingham.)

and winds out of the southwest at about 20 miles per hour, gusting to 40 miles per hour. Although unusual for this area of the country, during the afternoon of May 8, there were 15 different wildfires, one of which came within 200 yards of several large flammable storage tanks.(1)

Meanwhile, in parched southern California where wildfire is not unusual, some additional but different complexities arose. In the coastal community of Santa Barbara, wind blown flames erupted and burned across thousands of acres. More than 600 fire fighters battled the flames in the initial stages and many more were added later in the operations. Besides forcing unprecedented mass evacuations throughout southern California, the blazes injured scores of fire fighters and closed an unexpectedly large number of highways.(2) In Chapter 7 of this book we will cover not only disastrous wildfires, but catastrophic fires of all sorts, and illustrate not only geographic disaster planning considerations, but many other factors as well.

Although we have chosen wildfires to illustrate geographic differences, the same could be shown for any of the many different types of disasters covered in this book. Also, different geographic regions often have different resources dictating alternate tactical considerations. Just as fires in urban or metropolitan areas with large crews of on-duty fire fighters and an extensive water supply, are fought differently, from identical fires in rural areas relying on natural water supplies, tankers, and all volunteer forces, so too will identical disasters require different approaches based on different resources.(3)

PLANS DEVELOPMENT AND DISASTER PREPAREDNESS

The fire service has always led the field in concern for public safety; consequently the fire department could be the lead agency in developing or revising a community's plan for disaster preparedness.

A survey of fire chiefs conducted by the International Association of Fire Chiefs (IAFC) indicated problems that confront chiefs in plans development and disaster preparedness. Among the most frequent responses were these:(4)

- Lack of interest by other agencies and city officials
- Limited resources—personnel, money, and time—for planning
- Inadequate cooperation with other organizations
- Lack of a common mutual aid radio frequency with fire and rescue agencies.

In the 1990s, a quality disaster plan for the community developed in accordance with Chapter 1 can serve as the motivational tool necessary to overcome those problems.

The Benefit of Disaster Planning

In response to fireground action, or for any other emergency, the fire officer must make quick, important decisions, based on individual judgment, past experience, and training. Most will be correct for the situations; some will be wrong, and corrective actions must be accomplished immediately. But a fire officer has the duty of making decisions quickly, without much help from anyone else. Perhaps the easiest of these relate to the tactics of fighting fire; usually there is time enough to correct wrong judgments, or poor placement of apparatus and equipment. The toughest decisions concern the expedient ways of removing injured and endangered people, while others are equally at risk.

"Triage" is a term applied to military and political decision making when many casualties occur and a choice must be made quickly on which persons to save and which to let die. Actually, "triage" means simply a sorting, a process of selection and rejection, but for fire officers in a time of disaster, "triage" can create high-stress moments of personal decision making.

When confronting the first shock of a disaster situation, every fire officer has a responsibility and a personal obligation to remain calm and use the best judgment for the situation. Military officers are trained to analyze a situation quickly and to make expedient decisions. What is the situation (immediate surroundings and broad-scale)? What is the strength of the enemy? (What is the "disaster"?) What is the strength of our command? (Who and what do *you* have under immediate command?) How do we resolve this situation?

In the length of time that passed as you read those questions, you might have to decide some day whether to save ten endangered people in a building, or to rescue people screaming for help in rushing flood waters. That is the responsibility of a fire officer. Disaster control calls for careful planning. Although every fire department has found the value of adopting a standard operating procedure for handling "run-of-the-mill" emergencies, this SOP usually will not suffice for handling major disasters, and there is no substitute for having a carefully thought out and tested plan when catastrophe strikes. But even the best of plans incorporating all of the key components from the previous chapter cannot cover all possible contingencies. Unforeseen circumstances that arise call upon every officer to use his or her best judgment in improvising a suitable course of action under conditions where a plan is vague or impossible to follow. The fact that no plan is likely to be perfect for all conditions does not excuse the failure to chart a course of action for known types of disasters that can reasonably be anticipated to occur within your command. Remember the old saying, that even if you are on the right track, you will get run over if you just sit there! Just as a written plan remains static if not continually revised, a field command plan will also falter if not updated constantly.

The need for planning is not limited to large cities. Major disasters can occur anywhere. Any town located in, or adjacent to, a forest or brush area could be wiped out in a conflagration of the type that hit Berkeley, California, in 1928 and 1991, or Peshtigo, Wisconsin, on October 8, 1871,

Hydrants obstructed by debris from the Xenia, Ohio, tornado, left, and by heavy snow and ice during the Buffalo blizzard of 1977, right. (Photos courtesy of Dennis McCloskey.)

the same day the great Chicago Fire broke out. The loss of life in the Peshtigo fire was more than four times that in Chicago.(5)

Any town located on the banks of a river could be washed away in minutes if a dam breaks, as happened in Johnstown, Pennsylvania, on May 31, 1889. About 2,200 of the city's residents perished in the flood and fire that ensued.(6)

Any town located near a major highway, served by a railroad, in the flight path of an airport, located in a "tornado alley," hurricane belt, or earthquake zone, could be the scene of a disaster. Even when the town is "just missed" it could be called upon to help its neighbors, and in the absence of a plan defining the amount and conditions of response, problems are likely to arise.

Then, too, disasters can strike in totally unexpected ways, and suddenly involve seemingly impervious structures. Consider what happened in the nation's capitol on November 20, 1990, when a construction site cave-in spread to an adjoining street, causing massive structural damages and life-threatening damage to electrical cables. Power was cut in buildings in the heart of Washington, D.C., but fortunately no one was hurt. A three-block area was cordoned off.(7)

Scope of Plan

To assure efficiency of response a plan must be more than a mere inventory of equipment and personnel. It is not enough to compile a list of volunteers and a list of equipment that might be used in disaster control; a definite plan for their utilization is a "must" for effective action. Lists soon become obsolete, people change jobs or move away, and equipment is transferred, substituted, or junked. A *team* is required, and teamwork is impossible without a continually updated plan of action, an organization, and leadership to keep it in training.

A plan must have sufficient scope and breadth. Knowing what to do is not enough; even a mosquito in a nudist camp knows what to do, but without a plan it doesn't know where to start! A good plan encompasses the *what, when, where,* and *how*! For example, you roll up to an oil refinery in which a high-pressure gasoline pipe near the cracking tower has burst and ignited; assuming that you know *how* to extinguish a flammable liquid fire, your plan for that target hazard would cover such basic matters as *where* to attack, *how* to control (use of available valves and fixed extinguishing systems), and *when* to move in for the "kill" (after the flow has been stopped). If careful prefire planning is so invaluable for a single target hazard, it is easy to see how important it is to plan for an earthquake, hurricane, or other major disaster.

A plan is essentially a guide. Allowances must be made for the fact that no two emergencies are exactly alike, though disasters of the same type are apt to require the same general approach. Where certain details may be helpful, such as the particular method for operating master release locks on the county jail (in the event that evacuation has to be carried out), these details should be included in a suitable appendix to the plan.

HISTORICAL BACKGROUND

The history of disasters and their control parallels the history of mankind. Prehistoric man lived in a cave and carried a club, prepared in a sense for emergencies that might be encountered. We have all heard of the great biblical flood—Noah's famous Ark was used to circumvent the disastrous effects of the flood. Medieval castles had moats, drawbridges, defensive forces, and supplies in store for anticipated sieges. Early American colonial settlements organized and took protective measures against conflagrations and other threats that confronted their societies. Although rudimentary, according to our contemporary thinking, these are examples of disaster preparedness. This book is primarily concerned, however, with more recent history since this is the era that has provided the organizations, technology, and management expertise necessary to effectively control disasters.

Trends in Disaster Planning

Developments and changes in the perception of disaster planning by various agencies involved have permitted better control over disaster situations. Also, the nature of assistance that can be brought to bear on a disaster situation has been broadened. Hence it is important to take an especially close look at relatively recent times as we attempt to determine those factors that have brought disaster control to its current state.

Basic Characteristics of Disaster Planning During the 1960s

In a government report entitled *The Role of Local Civil Defense in Disaster Planning*(8) the basic characteristics of disaster planning were identified and studied from the 1960s through the 1970s. Some of its conclusions identify significant events, forces, and developments from earlier historical periods that have affected disaster planning through the 1980s, and are shaping the direction for the 1990s and beyond.

In the government report, the decade 1960–1970 is described as a period of transition in disaster planning. Certain common attributes among cities were identified and some basic trends were listed. The following paragraphs identify the characteristics of that period.

In the 1960s disaster planning was located in three different segments of the community with minimum contact among the segments. These three segments were the local civil defense office, private health and welfare agencies such as the Red Cross and the Salvation Army, and municipal emergency organizations, such as police and fire departments. While the civil defense agencies were concentrating on nuclear attack, the fire department and other agencies with stronger ties to serving the local community became much more concerned with conflagrations, floods, hurricanes, and tornadoes, which they felt presented more imminent types of threats.

As a consequence, most community organizations tended to engage in planning in isolation from other community organizations. Because of the various contradictory pulls that many community organizations felt, each organization tended to develop its own planning, often with little relationship to other organizations within the community.

Planning throughout the period tended to be cyclical and spasmodic. At the beginning of the 1960s, the Berlin crisis and more important, the Cuban missile crisis provided an increased interest in civil defense activities. These crises also led to dramatically increased federal funding for certain types of civil defense related programs, in particular, fallout shelter construction, location, and supply. The massive Alaskan earthquake in 1964 provided a different impetus to disaster planning; in particular localities, specific localized disaster often acted as stimulus to increased and improved planning. In the late 1960s riots and civil disturbances suddenly shifted the focus toward urban concerns.

Direction of Disaster Planning in the 1970s

Although the previously mentioned trends seem to have dominated disaster planning during the 1960s, a number of crosscurrents were creating change within American communities and the American Fire Service. These began to shape disaster planning in the 1970s.

The scope of disaster planning was broadened to include a wide range of disaster agents. While natural disaster agents became increasingly integrated into the planning of various community organizations, other types of emergency situations that required attention and action appeared within local communities. For example, the combination of a superhighway system throughout the nation and new manufacturing processes accentuated the problems of handling dangerous materials in transit. Such problems had previously been confined to port cities or to major rail lines. Increased air travel in planes with larger passenger capacities created new dimensions to be considered in emergency planning.

There was a shift in the focus of disaster planning from emphasis on the nation's security to concern with the viability of the local community. Since most of the planning for civil defense in the 1960s had started with assumptions predicated on national security, communities were perceived as smaller governmental units that would act collectively to support national goals. In the 1970s this assumption began to have less appeal. Local communities had their own problems that might or might not be related to national security. For example, persistent local flooding would necessitate greater attention than some more distant problem. Although the concern over national security did not disappear, greater amounts of time and energy were expended on local problems.

The number of community organizations involved in disaster planning increased. In the 1970s fire departments and other organizations became more interested in disaster planning. There were many different reasons for this. Sometimes, experience on the local level with a disaster situation revealed the lack of previous planning. Other times, the impetus came from national organizations and was translated down to the local level. Some local organizations were also prompted to make their own plans because of ongoing planning in other organizations on which they would depend. All of these diverse reasons combined to increase the number of organizations involved in disaster planning.

Trends of Disaster Planning for the 1980s and 1990s

NEW FEDERAL FOCUS

In reading about IEMS in the first chapter we saw how the federal government is now putting more federal resources into disaster planning. At the National Fire Academy, fire chiefs from throughout the United States are taking the necessary courses to bring a unified body of knowledge regarding disaster planning back to their own departments. A standardized body of knowledge has been developed through the efforts of the National Fire Academy. Also, the Superfund Amendment and Reauthorization Act (SARA) Title III, is providing numerous new opportunities for the elimination and control of disasters in the hazardous materials arena.(9)

NEW TECHNOLOGY

Technological advances in electronic data storage have facilitated the development of sophisticated computer-aided dispatch and enhance such systems as Computer Assisted Management of Emergency Operations (CAMEO).(10) In addition, developments in the medical field have been responsible for the placement of hundreds of new paramedic units in the ranks of the fire service, greatly increasing the capacity of the fire service in handling mass casualties. As a result many more lives are being saved in disaster situations. The increasing role of the fire service in providing emergency medical services during disasters has become well documented in recent incidents. During the May 1991, EMS Conference sponsored by the International Association of Fighters (IAFF) workshops were provided on many issues relevant to disaster mitigation: the future of EMS, training needs and assessment, EMS legal issues and liability, trauma systems, quality assurance.(11)

INTERAGENCY COOPERATION

In the 1960s agencies essentially planned independently. In the 1970s some integration occurred, but in the 1980s a genuine attempt for harmonious integrated planning among all agencies became evident. In the 1990s, the fire department is no longer merely planning for its own activities in the event of disaster, but also planning its role in a larger scenario involving multiagency response to disaster situations. New mutual aid and mutual response pacts also ensure that when disasters cross jurisdictional boundaries or when disasters are of such magnitude that they cannot be handled adequately by resources within one jurisdiction, then other agencies are not only available but capable of being dispatched immediately.

PLANNING IS CONTINUOUS AND CUMULATIVE

In the 1960s planning tended to be cyclical and spasmodic. In the 1990s we reached a point when planning became continuous and cumulative in the major cities of the United States, and although much remains to be done, the master plans for disaster control in the major cities are far superior and far more functional than their counterparts of three decades ago. Disaster plans in most metropolitan areas now include the functional components outlined in Chapter 1 and are quite comprehensive.

GENERIC INFLUENCE

One other important new trend should be identified. A great deal of early planning tended to center around the roles and activities of particular individuals. In other words, in many organizations, planning was the concern of a particular person within the organization, e.g., Battalion Chief John Jones, rather than the concern of a particular position within the organization, e.g., Director of Safety. Because of this tendency, there was little continuity of planning. When the particular person left the organization the knowledge and the skills went with him or her. By the 1990s this problem had been identified. Plans were designed around organizational priorities rather than individuals.

RESCUE DOGS

One familiar companion being used as a new resource in the aftermath of many disasters is man's best friend. Dogs trained by the U.S. Disaster Team Canine Search and Rescue Unit are welcome sights to trapped victims in the aftermath of a disaster. Numerous success stories already have been recorded in which the specially trained dogs located missing persons, victims in demolished structures, mud slides, earthquakes, tornadoes, and hurricanes. Due to their keen sense of smell, they can accomplish many things that humans cannot, such as searching quickly through collapsed structures.

In early 1991 representatives of the country's Canine Search and Rescue (SAR) teams met to form "the National Confederation of Search Dogs." The group was formed to emphasize training and share ideas, so that by the end of 1992 an international rescue force could be available at a moment's notice. The U.S. Disaster Team Canine Search and Rescue Unit is sanctioned by the U.S. government's Office of Foreign Disaster Assistance for response in international emergencies.(12)

Search and Rescue (SAR) dogs are becoming a valuable resource. Left to right: "Stormy," Handler Bill Kramer, and "Gus." (Photo courtesy of Dianne Para.)

THAWING THE COLD WAR—KINDLING THE HOT WAR

With the destruction of the Berlin Wall much more than bricks and mortar came crashing down. As relations appeared to improve between President Gorbachev and President Bush, massive cuts in defense spending were contemplated in both the Soviet Union and the United States. This temporarily translated into less emphasis on civil defense needs and a reduction of planning for enemy attack. However, the expected "Peace Dividend" was soon derailed with the advent of the 1991 war in the Persian Gulf where over a million members of multinational military forces successfully conducted a six-week war to restore the legitimate government of Kuwait, which had been invaded by Iraq.(13) Since the United States provided the major component of the allied forces in order to carry out the United Nations' resolution affecting Arab countries, there is genuine concern that it may become the future target of terrorist activities on the part of some Middle Eastern nationals. In Chapter 13 we will discuss the possible effects of the independence of the former Soviet republics, as well as other global tensions, which might cause us to reconsider any lowering of our guard and compel us to include some provision for enemy attack in our disaster planning.

USING PLANS—LESSONS LEARNED

Every disaster has its lessons and the problems presented to the Fire Service through time have enough in common that some generalization can be made from past experience. First, it must be recognized that not every action taken to cope with a disaster has been successful; much can be learned from mistakes of others. So let us proceed with an analysis of situations for which alternative courses of action might produce better results in the future.

General Actions

No matter what the disaster or its magnitude, there are certain actions the first arriving officer-in-charge probably takes. He or she should establish command as soon as possible and establish communications with some central dispatch center, giving a description of conditions as they appear. As units arrive they are deployed to tentative positions, pending further reconnaissance for a better understanding of the problem.

The incident commander then reports this revised sizeup to an arriving superior officer, stating the nature and magnitude of the disaster, who and what are threatened, and specialized information such as "the best place to cut off fire spread," considering forces at hand and those that can be made available quickly.

Upon the completion of this report, additional help is requested as the situation demands, perhaps activating a plan used by the department for emergencies of this kind and obtaining a planned response of certain units.

At this point the difference between having a plan or not may well mean the difference between an orderly operation or chaos.

Early in the operation, the incident commander surveys the situation to direct rescue of persons in danger; those in immediate peril receive attention first, and special rescue teams that have responded are used for this purpose. The saving of lives in immediate jeopardy is the first task of importance to be accomplished.

Of course, it is essential to confine destruction as much as possible; sometimes it is necessary to decide what to sacrifice and what to save. Such might be the case in a conflagration where forces at hand or the water supply may be insufficient to save buildings already fully involved, making it necessary to direct all efforts to saving uninvolved structures and protect exposures.

Finally, in any type of disaster, the commanding officer continues to use all available forces to effect control until the emergency has ceased, and retains command until relieved by higher authority.

Declaring a Major Emergency

Problems can arise if there is a lack of understanding of what constitutes a major emergency that warrants invoking an existing disaster plan. Admittedly, it is difficult to spell out exactly, under every circumstance, when a disaster is deemed to exist. A situation that commences as a fairly limited emergency may develop suddenly into a full-blown disaster.

In judgmental matters where lives are endangered, it is better to declare a "Major Emergency" or "Disaster" prematurely, than to wait too long. As with routine fire situations, it is better to call for help and not need it than to wait too long, in the vain hope in vain that the situation will improve.

Many officers put off making a disaster announcement in hope that conditions will change for the better, thus avoiding inconveniencing personnel who have to be summoned for duty. The fact that the declaration, even if it turns out to be unnecessary, will provide a good drill is often overlooked. Also forgotten are the facts that it may require several hours to bring in some types of equipment; that evacuation cannot be effected instantaneously; that it takes considerable time to obtain military assistance; and that every delay in utilizing forces summoned through a formal declaration may mean larger losses that will be difficult to explain in the postdisaster critique of your operations. Moreover, there may be financial compensation in such a declaration; operational expenses may be reimbursed under governmental assistance laws, such as helicopter rentals, air tankers employed, extra services hired, and meals purchased. Low interest loans may be available for victims of the disaster.

Communications

It is vital for the first-arriving officer to give an accurate initial report. The dispatcher and superior officers may be listening for a description of the emergency, its extent, and the need for additional resources such as ambulances, heavy equipment, hazmat teams, and heavy rescue units.

Heavy equipment was summoned by fire fighters when a chemical tanker careened off a bridge and landed in a river north of Dayton, Ohio. (Photo courtesy of Dennis McCloskey.)

Sometimes dispatch centers fail to make the planned notifications after receiving a report of a disaster. Envision a helicopter pilot on a routine reconnaissance who observes a collapsed structure in which a hundred or more victims may be trapped; he requests the dispatcher to send companies to assist in rescue. Even though the structure is in another jurisdiction, the dispatcher should, without special request, notify the fire department in the area where the structure is located.

Another problem arises when the dispatch center is trying to cope with all the dispatches and notifications required during a major emergency prior to the time when the fire department can set up an Emergency Operations Center (EOC). Other officers in the department may not be aware that the dispatchers are so busy and may attempt to carry on routine business as usual. Once again, this points up the need for the officer-in-charge to make the declaration required for activating an EOC as soon as it becomes apparent that the emergency is one that will require an augmented staff.

A similar problem is likely to arise when simultaneous major emergencies occur in the same jurisdiction. When this happens the need is obvious for more radio frequencies, more patience, and more assistance at the communications center. Under such conditions, the thoughtlessness of the "air hog" becomes pronounced, and the need for more training in radio courtesy is magnified.

As a contrast, there is the officer who fails to use the radio enough. It is important that field commanders give progress reports at reasonable intervals so the department commander and operations officer can anticipate further needs of the combat forces.

Interagency communications problems are created by lack of a common radio frequency for use by mutual aid companies. This is not merely a problem of facilitating deployment of assisting units, but also of their safety.

Color-Coded Radio Frequencies

A common problem at major emergencies is on scene communication among units with portable radios. Any major incident with complexities will generate numerous radio transmissions cluttering air waves. On one hand an argument can be made to increase the number of frequencies but at the same time this may cause some units to miss critical messages from other units who are operating on different frequencies. And too, an oft-mentioned complaint among mutual aid responders is that they cannot communicate to the host fire department because they do not enjoy the benefit of a common frequency. The ideal solution is yet to be developed but one alternative that has merit was reported by Rick Donahue, Fire Marshal for the City of Madison Heights, Michigan. At an educational workshop sponsored by the University of Cincinnati, Rick presented his concept to several dozen fire chiefs from throughout the United States. According to the plan, originated in the city of Troy, Michigan, and now used by the participants in the Mutual Aid Pact to which the city of Madison Heights belongs, all cities in the Mutual Aid Pact have a separate primary frequency plus several common ones. Different types of radios and different frequency numbers by the Mutual Aid Pact are resolved by the use of color coding. For example, all responding companies for a given sector could be instructed to switch to "orange" when frequency numbers do not line up (F2 in Madison Heights might be F3 in Royal Oaks). Accompanying this plan could be a procedure for use both in larger cities where many units are assembled for a major incident and in regions where numerous small fire departments are pooled to handle a disaster; namely, the use of overlapping frequencies for adjacent levels of command. This allows all units to ultimately reach the incident commander by simultaneously monitoring only two frequencies. Modern radios have scan features that make this feasible. See Figure 2-1 which explains how the overlapping frequencies parallel closely the overlapping levels of command.(14) All units have all channels, and can switch to Channel 1 for emergency transmission, which can be rebroadcast through the incident command chain.

Envision a major brush fire in which aerial surveillance is used. In the absence of identifying symbols on top of fire apparatus, an aerial recon officer will find it difficult to notify an endangered unit that the fire is about to sweep around toward its location. Even without a common radio frequency, the air officer can notify the field headquarters commander of the situation, and if the unit's fire department can be identified, ask that the notification be made through that department's dispatcher, or by means of portable or vehicular radio from that unit's command at the nearest command post.

Figure 2-1. Overlapping Frequencies for Incident Command.

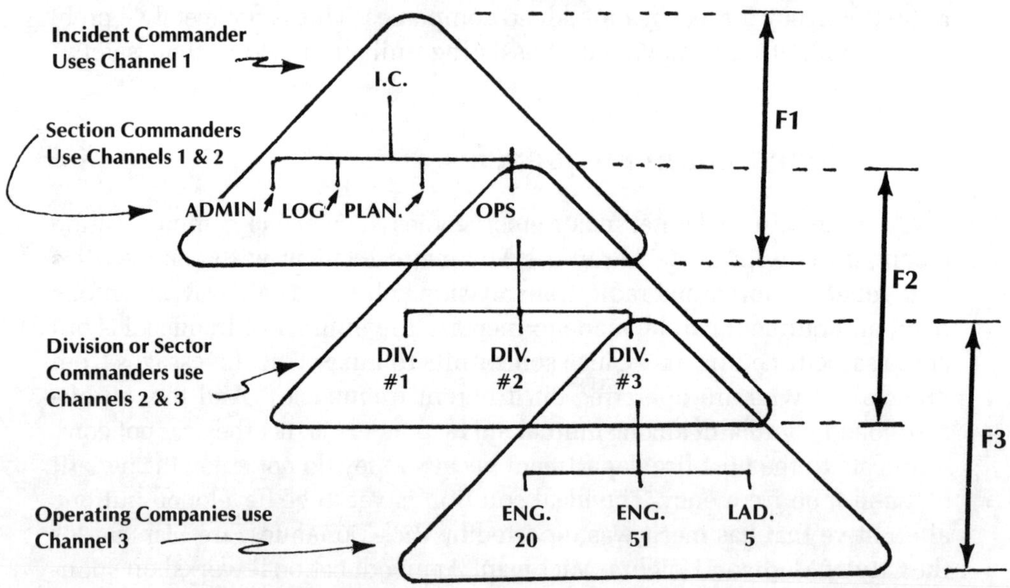

When an unidentified unit in a remote location cannot be described sufficiently, it may be necessary for the pilot to land at the nearest command post to permit the air recon officer or someone else to drive a vehicle to the unit and give the warning. But it is difficult to identify at ground level the fire roads and locations of units seen from an aircraft even using a sketch drawn while airborne.

Deployment of Forces

One lesson learned in deployment of forces at a rapidly spreading disaster is the need to visualize how the situation will look ten minutes after your arrival, and how your apparatus and manpower will be deployed by then. Failure to anticipate possibilities may result in their being placed where they will have to be moved quickly to escape destruction.

The spotting of fire apparatus at an emergency provides a good example. If you have seen a fire truck stop in front of a building upon responding to a reported "gas leak" in that building, you would have no doubt that the truck would have been stopped elsewhere if the officer anticipated that gas would explode shortly after his arrival. The same is true of parking a rig downhill from a flowing gasoline spill or a liquid oxygen leak; experience has shown that these actions can be serious mistakes.

Examples of other mistakes:

- ○ Sending companies into dead-end canyons where fire can make an end-around sweep and trap them.
- ○ Failing to anticipate the path of the fire, flood, tornado, or other destructive force.

- Using helicopters in hazardous areas or after dark; failing to anticipate thermals, downdrafts, power line obstructions, and poor visibility. (Here is an example of what happened to one helicopter. Both occupants were seasoned veterans of the Los Angeles Fire Department, but neither spotted a power line stretching across Big Tuhunga Canyon. Both were killed in the crash after the helicopter struck the line.)

Deployment of mutual aid companies can be a problem when time comes for relieving their members who have been working for a long time. If their home community is a considerable distance away, there may be difficulty in arranging for bringing in relief crews and returning working crews to their home stations. Sending them to empty fire stations for covering assignment may be better than utilizing them in the disaster area.

When using camp crews from correction centers, rules may require sending them back to their quarters after not more than eight hours of work. It matters not whether they are busily engaged in building levees, dikes, or firebreaks; when the eight hours have elapsed you will need replacements ready to carry on the work.

Another problem is the necessity for having forces available to handle a second emergency. If you recall all off-duty personnel to handle the first situation, you must rely on mutual aid companies to respond to the second incident and you must relieve your own members if the operation extends over a period of several days.

Relieving

From practical experience it has been found best to make reliefs in daytime, while there is sufficient light to locate companies where access may be difficult. Normal shift changes usually occur during daylight hours. It is less complicated to leave apparatus being used in critical or front-line operations, and merely change the crews; changing fire trucks especially where hose has been laid out along narrow roads or through complex buildings is time consuming and usually unnecessary. By bringing in fresh crews on a transportation truck, hose streams can be kept in operation and relieved members can return for food and rest.

Another advantage of arranging reliefs in daylight hours is that the on-coming crew can obtain a clear view of the fire situation. Burned-out stairways and unsafe structural conditions are more easily seen in daylight. In brush areas, if an acre or two of unburned brush is within the burn, it could flare up during the night; however there would be no great concern or hasty calling for more help, because the crew would know that the fire could not go anywhere. Because a fire usually looks worse at night, a better perspective is possible if you have seen it in the daylight.

It should be understood in advance of relief time the designation relieving crews and chief officers will assume. Ordinarily, the replacement crew for Engine 34's apparatus will be designated as "Engine 34" regardless of the company to which it was assigned before recall to active duty.

For chief officers who respond from home or from regular assignments after putting an acting commander in charge of their battalion or division, a new designation must be adopted. For example, suppose that chiefs from Battalions 6, 8, 10, and 12 have been directed by the operations officer to leave their commands in the charge of acting battalion chiefs and report to the field headquarters for assignment to specific sectors; there, they might be directed to establish sector command posts and take charge of all companies operating in that area. The acting chief left in charge of the battalion will now be addressed by the dispatcher by his battalion number, and the chief who was formerly "Battalion 6" will become "Sector 3" (or whatever his crew command is designated).

The same should be true of division commanders who have been directed to turn over their commands to acting assistant chiefs and to report to the field commander at the disaster scene to take charge of air reconnaissance or other assignment. The battalion chief in charge of Battalion 5 might be directed to assume an assistant chief's duty as commander of Division 2, to permit the latter to report to the field commander for duty at the disaster scene; the former officer will thereafter identify himself on the radio as "Division 2" and the assistant chief will use the new title assigned to him by the field commander in making radio reports at the disaster scene.

In these circumstances, some fire departments require chief officers to include their last names when making radio reports, for it helps the operations officer to keep the officer status board up to date. As chiefs in the disaster area are relieved, or given new assignments, the field commander knows "who is in charge of what" if chiefs use their names when making radio reports. If two chiefs have the same last name, or as in the Los Angeles Fire Department where two chiefs operated in the Bel Air conflagration with the same first and almost the same last name (Bill Johnson and Bill Johnston), the latter used his initials "W.W." rather than "Big Bill," commonly used by his friends to distinguish him from "Little Bill."

When the emergency is over, and the tapes of the radio messages are furnished by the communications center to the department commander to review in preparation for a critique of the entire operation, it is helpful to be able to identify who called for what units at what time, rather than to guess the identity of voices coming over the tape, or to try to trace records over a period of several days.

Another thing that must be understood: the meaning of the expression used by many officers and dispatchers, "Go Home." In some departments, after companies have worked several hours at a real "worker," the officer-in-charge will bring in fresh companies to do the overhauling and tell the tired company officers and their crews to "Go Home," meaning to go back to their quarters. However, during one conflagration that persisted for several days, when this command was given to fire companies, instead of returning to the apparatus pool at the base camp, where the commanding officer meant them to go, they went back to their own quarters fifty miles away. Some recalled personnel, learning that the recall order had been cancelled, instead of returning to the personnel pool, left their rigs in the road and went to their residences!

Be specific if you suspect that your directions may be misunderstood.

Don't rely on code numbers or informal expressions that may be misinterpreted.

Feeding

Many disasters require that forces remain in the field for long periods of time. Every fire officer usually is trained to look out for the welfare of his personnel, and one consideration is the need for periodic rest and food. The company commander on the front line who has been advised by his superior officer that arrangements have been made for feeding the fire fighters at an approximate time will not clog the airways with periodic requests such as, "When do we eat?" However, if the food doesn't arrive within a few hours of when it is due, a respectful inquiry might be in order.

Some departments find it easier to feed on the line, rather than to take out crews and return them to a staging area or camp for meals. Canteen vehicles are used for hauling hot food to remote sectors. This is satisfactory unless food runs out before the last company is reached or unless a company is in such an inaccessible location that it is overlooked by the food wagon's driver. In such cases, a return trip with more food must be made.

Police, utility crews, press members, and others assisting get hungry too, and it is well to order enough extra rations to feed these people.

Some departments have their own field kitchens and can prepare meals very quickly. Others have contracts with catering companies that deliver hot meals within four hours after request, in any number from 100 to 1,000. This service generally is satisfactory, the food is tasty, and in sufficient quantity to provide generous servings. However it is costly.

The use of meal tickets has been tried in some large-scale emergencies where the cost of the food (several dollars per person) delivered by catering services is relatively high and the number of onlookers so great that the discouragement of "freeloaders" seems warranted. However, meal tickets are not satisfactory for members who want a second helping, and the use of tickets is considered unfriendly by visitors who are not issued tickets.

By locating the field kitchen a hundred yards or more from the field headquarters and personnel pools, the priority status for feeding crews brought in from the fire front can more easily be maintained.

Whenever the military are called upon to assist in a disaster operation, their officers generally request that the fire department not attempt to feed military personnel. The reason for this is that they usually make provision to take care of their own independently of civilian food supplies.

Pending the arrival of the food or its preparation, sandwiches, doughnuts, hot coffee, and milk may be provided by the Red Cross and Salvation Army. Also, neighbors and fire buff organizations often tender refreshments to members on the line, and such thoughtfulness is appreciated. However, some departments forbid their members to accept food from strangers not known to be affiliated with a reputable organization, particularly under circumstances where there is a possibility that food or drink may be contaminated.

Medical Care

Fire fighters injured at major disasters often are reluctant to report anything of a minor nature until they are relieved and return to base camp or a fire station. Some fire departments have their own doctors who can treat minor injuries or order such injured personnel to a hospital.

To handle serious injuries, ambulance personnel usually will be stationed at command posts and first aid stations, ready for transportation to a nearby hospital or university where medical services are available.

Record-keeping

Another lesson learned at major disasters is the importance of keeping adequate records of matters that will be subject to inquiry when the emergency is over. This should include the record of units called to the scene and to apparatus pools. In addition to designating units by company number and department (in case of mutual aid response), records should show the time of arrival, time of deployment from the apparatus pool, and time returned to pool or place of origin.

Similar records should be kept for personnel who work at the emergency; recalled members should be logged in, so should the names of persons who relieve crews of mutual aid companies. Time of release from duty at the conclusion of the emergency should also be shown, for this information will be required in computing overtime pay. The same type of records should be kept for personnel who report to other posts of duty on a recall order, such as supply and maintenance shops and the communications center.

To assist in acting upon financial claims, records should be kept of contractual services used in the emergency, such as air tankers, tank trucks, cranes, and others. It is not unusual for contractors to bring in equipment voluntarily to assist in a disaster, but records should be kept of the time "donated" at the scene. If federal disaster relief money becomes available later, among those seeking reimbursement might be the companies and individuals who originally made their services available without any thought of financial gain.

Water Information

In many large emergencies it is very important to have information concerning location of water mains, shut-off valves, storage tanks, reservoirs, and other essentials. For a small community this data may be in a command post kit, carried in the trunk of the chief's car, but in a disaster affecting hundreds of square miles, it probably will be necessary to call upon personnel of water companies to supply maps of systems in the area affected.

Responding to Other Communities

Whether mutual aid agreements specify it or not, it is the general practice for the officers-in-charge of forces responding to another jurisdiction to retain command over their own members even though the overall incident commander generally should be from the jurisdiction of origin. If the necessity arises to withdraw mutual aid forces to handle a sudden emergency in their own community, generally it is understood that this may be done upon notifying the overall incident commander.

If the need arises for calling nearby communities to send help, the request should be explicit as to type of equipment needed, whether it is to be fully manned or merely brought by a driver to the apparatus pool, the probable duration for which they will be needed, and whether the assisting community or the assisted community will provide personnel to relieve initial crews and take care of logistical support, such as food and fuel for apparatus. These matters may be covered in a mutual aid agreement, but disaster conditions may be such that modifications have to be explained; if a question is asked later concerning the exact request and agreed upon response, a tape recording of the verbal agreement (made by either party) will help resolve ambiguities.

Legal Problems

Explicit authority must be assigned by the governing body of your community, such as that found in mutual aid agreements, before you can legally send your forces out of your jurisdiction, unless, of course, your state law provides otherwise. Some state laws require that a response *must* be made to assist communities that have suffered a disaster, after a formal proclamation by the designated official has been made. However, such laws usually permit retention of adequate forces to protect the home community.

There is a possible liability of a community and its public officers for failure to take reasonable steps to safeguard the lives and property endangered in the disaster, and there are related legal questions. For example, to what extent can you strip your own community of protection in order to respond great distances to help another? What is your liability for accepting services of private contractors whom you erroneously assumed were donating their time and use of their equipment in rescue operations? If federal funds are made available with which to reimburse claims for lost and damaged fire equipment, special contractual expenses, and other allowable items, will you have complied with conditions precedent for eligibility to the file such claims? What is the liability to nonfire department volunteers who are injured while assisting you in your operations? Does your state's worker's compensation law specifically provide that all volunteers or personnel requested to assist in an emergency are covered in the protection afforded for compensable injuries to the same extent as your paid personnel or registered volunteers?(15)

Public Information

Lessons have been learned regarding the necessity for considering the news media during and after disaster control operations. Information must be made available as soon as reasonably practical. No doubt your officers have been instructed to refer requests for data to the officer you have designated to handle this information in your disaster plan. Specific kinds of data might be number of possible dead or injured, dollar loss estimate, possible duration of the emergency, and the number of companies engaged in operations, Undoubtedly, your officers have been instructed not to give names of victims until relatives and next of kin have been notified; unless this policy is followed, a traumatic experience can result. For example, a young wife was shocked to learn from a televised newscast that her husband had just been killed while fighting a fire; the fire fighter had been author Charles Bahme's aide, and he had planned to tell her the sad news in a personal visit. The death occurred while the Los Angeles Fire Department was assisting in another jurisdiction, and somehow, perhaps by observing the name on the man's crushed helmet, the news media were able to identify the deceased member.

The subject of fire department public relations has been extensively discussed in other publications.(16) Obviously, on questions for which you can only guess the answers, it may be advisable not to hazard any opinion; this was the situation in Los Angeles that night in 1941 when an "enemy bomb attack" was presumed to have taken place; while antiaircraft artillery was filling the sky with flak, off-duty fire fighters were recalled to duty and the chief was requested to tell the press how many planes the Army had shot down, and where they had fallen in flames.

No doubt the experts on public relations will be able to help you provide the right answers in situations where it may appear to you that some other department has been negligent in creating or permitting conditions to exist that may have caused or contributed to the disaster; such might be the case in a flood, where you have reason to believe that a dam and reservoir was constructed over an earthquake fault; or in an area where oil wells have been pumping for many years with resulting weakening of the subterranean support for the dam. Nevertheless, it would not be a good idea to point the finger of blame at another government agency based on your suppositions for the disaster, nor wise to volunteer such opinions to the news media. Better to reserve them until requested to comment in a proper forum.

There is no harm in painting a rosy picture to news representatives of the great work your forces are performing, sometimes at considerable personal danger, to bring the situation under control. Nor is there any objection to giving ample credit to the foresight of your mayor, city council, and fire commission for all the support they have given to your budget requests for personnel and equipment that enabled you to better cope with the disaster conditions. You might add that you are fairly sure that these forward-looking officials will see the wisdom of providing your department with more needed equipment, such as portable radios, helicopters, training centers, personnel, or whatever else may be needed to handle this type of disaster more effectively in the future.

Evacuation Order

The question of responsibility for issuing an order to residents of an area threatened by wind-swept fire, dam rupture, hazardous material spill or other disaster, should be settled long before such an emergency arises. That this is not always so is illustrated by an article in the daily newspaper of a large metropolitan city: ". . . Mayor __ disclosed Tuesday that neither the city's fire department nor its police department is quite clear which department has the authority to order evacuation of a disaster area. Both departments thought the other had the final authority. . . . We will find out just who has the legal authority and have the authority widely disseminated."(17) The mayor pointed out in this article, however, that both departments had always proceeded to carry out an evacuation once one was ordered *no matter who issued the order*!

When such an order has been issued to residents in the path of a conflagration, fire department personnel are apt to be so busy with fire control that the burden is likely to be on law enforcement personnel to spread the warning and carry it out. From past experience it has been found unwise to force anyone to leave home premises if they object; assault and battery, even though committed with a worthy purpose, could result in legal consequences that put the ejecting officers in a defensive and possibly untenable position in later court action. Chapter 14 will treat the subject of evacuation in greater detail.

Maps and Photographs

It has been found very useful to obtain maps and aerial photographs for evaluating the disaster and deployment of forces in the area of devastation. Maps reproduced at field headquarters with copy machines and duplicators are very helpful. Portable Fax machines are now available for field use and can be used for rapid receipt of helpful charts, and map sections.

Aerial photographs, television and videotape cameras (e.g., in helicopters) provide a continuing size up of conditions, resulting in the better utilization of resources, personnel, and material. Photographers, cartographers, and draftsmen can be of great assistance.

Compound Threats

Throughout these first two chapters we have learned about varying types of disasters, and the many different components of disaster planning. It is now quite obvious that disasters can take many forms, but we have yet to discuss how they may take on several forms simultaneously. They may or may not be independent of one another. Figure 2-2 shows how one form of disaster can trigger other forms. This dependent relationship may occur in immediate succession or gradually, giving rise to compound threat situations.(18)

As you continue to read successive chapters, remember that various types of disasters, all of which are discussed separately, can often occur in pairs or multiples.

Figure 2-2. Compound Threat Situations.

Major Event	Possible Ensuing Events											
	Flash Floods	Slow Rise Flood	Stormtide	Thunderstorm	Tornado	Torrential Rains	Tsunami	Landslide, Dam or Levee Break	Transportation Accident	Fire	Explosion	Hazardous Material
Hurricane			●	●	●	●			●	●	●	●
Flood								●	●			
Thunderstorm					●	●				●	●	
Tornado				●		●			●	●	●	●
Torrential Rains	●	●						●	●			
Winter Storm		●						●	●			
Earthquake							●	●		●	●	●
Tsunami	●							●				●
Landslide, Dam or Levee Break	●	●							●			
Transportation Accident										●	●	●
Fire											●	●
Explosion										●		●
Hazardous Material										●	●	

Compound hazards occur during times of emergencies as seen in this photo of an interwoven tanker truck and ship. (Courtesy of Ron Buckingham.)

CONCLUSION

In the second chapter we have traced disaster planning through history and discussed numerous lessons learned from actual large-scale operations. We have seen how effective planning can mean the difference between success and failure in a major disaster operation and have detailed the many considerations that must temper actual operations in the field. The fire officer who has the ambition to tackle the tedious job of preparation and planning in advance of an emergency is the officer who will have the best chance of rising to the occasion as an effective field commander during the actual emergency. Detailed planning, a comprehensive organizational structure, and skillful leadership are all needed in the modern fire service as it is called upon to fulfill the vital role of disaster mitigation.

CHAPTER 2 REFERENCES

1. "Michigan Wild Fire Destroys 86 Homes and 126 Out Buildings" in *Wild Fire News and Notes*, Vol. 4, No. 3 (May/June 1990), p. 1.
2. *Cincinnati Post*, June 28, 1990, p. 1-A.
3. Kramer, William M. "Meeting the Community Fire Threat" in *Community and the Fire Threat*, Emmitsburg, MD: National Fire Academy, 1991.
4. International Association of Fire Chiefs. *Disaster Planning Guidelines for Fire Chiefs*, Washington, D.C.: Federal Emergency Management Agency, July 1980.
5. Bahme, Charles W. *Fire Officer's Guide to Disaster Control*, 1st ed., Boston, MA: NFPA, 1978, p. 11.
6. Ibid.
7. *USA Today*, November 21, 1990, p. 3A.
8. Dynes, Russell R., and E.L. Quarantelli. *The Role of Local Civil Defense in Disaster Planning*, Washington, D.C.: Government Printing Office, 1975.
9. Isman, Warren E. "Wanted: Comments on Response Standards" in *NFPA Journal*, March/April 1991, p. 110.
10. McCloskey, Dennis L. "Emergency Scene Reception of Computer Data" in *Hazardous Materials* (Course Guide by William M. Kramer), Cincinnati, OH: University of Cincinnati Press, 1991.
11. *International Fire Fighter*, Vol. 73, No. 5 (November/December 1991) p. 7.
12. "Rescue Dogs Find Survival Scent," *Atlantic City Press*, April 2, 1990, p. C4.
13. Weldon, Curt. "The Fight for Fire Protection" in *Firehouse*, Vol. 16, No. 4 (April 1991), p. 20.
14. Donahue, Rick (Fire Marshal, Madison Heights, Michigan). "Color Coded Radio Frequencies," unpublished report prepared for University of Cincinnati, July 1991.
15. Bahme, Charles W. *Fire Service and The Law*, NFPA No. FSP-3A, 1976.
16. *Successful Public Relations*, NFPA No. FSD-8SP, 1974.
17. "Disaster Study Finds Weak Link in Procedures," *Los Angeles Times*, June 23, 1976.
18. Strope, Walter E. *Disaster Response Planning*, Menlo Park, CA: Stanford Research Institute, 1974, pp. 1–8.

CHAPTER 3

ORGANIZATIONAL STRUCTURE AND INCIDENT COMMAND

A FIRE DEPARTMENT IS organized like any other business or commercial organization with a pyramid type structure. While we may have a chief executive officer at the head of a business with numerous vice presidents in charge of various functional operations, we have a fire chief at the head of the fire department with assistant chiefs in charge of various functional operations. The analogy is rather complete as we go through the varying ranks below the assistant chief level in a fire department and compare ranks with those in a typical private industrial or business organization. Harry Carter pointed out similarities in organizational structuring and leadership between the public and private sectors.(1)

The fire service organizational structure is established to handle both administrative duties on an on-going basis and emergency operations—from routine fires requiring a minimal number of resources to major disasters taxing the full organizational structure. In *Fire Engineering* David M. Lesak showed how relying solely on the Incident Command System is insufficient for today's wide ranging large scale incidents. He shows how a management decision process with roots in the standard organization can combine with a functional field Incident Command System to spell success at the scene of a major emergency.(2)

GENERAL REQUIREMENTS OF ORGANIZATIONAL STRUCTURE

A good organizational structure is indispensable to effective command at an emergency situation; it is the only way you can be sure that someone is always in charge and that the "right hand knows what the left hand is doing." The Incident Command System (ICS) has become synonymous with emergency organizational structuring. In this chapter we will consider the broad approach to organizational structuring prior to the occurrence of a disaster, and incident command procedures that are used at an actual disaster incident, keeping in mind that they are somewhat interdependent in a quality organization.

Broad in Scope

The organizational structure of a fire department should be such that it can handle a broad scope of operational situations. It should be able to function in any type of disaster, whether arising from an accident, natural catastrophe, sabotage, or enemy attack; annexes to the plan may call for some augmentation to meet special situations, but the basic organization and chain of command will stay the same.

Standard Units

The organizational structure should provide for the standardization of units; for example, engine companies, ladder companies, task force units, paramedic units, rescue companies, and other universally identifiable specialty units. This permits greater simplicity of organization, more rapid alerting and mobility, and greater interchangeability with, or integration into, other fire-rescue organizations.

Practical

The organizational structure should be practical. It should be such that units already in use in everyday operations can be deployed with the least amount of delay, and under commanders with whom they are accustomed to respond. It would be impractical to require them to wait for decisions from numerous layers of command before being authorized to go into action at a disaster. It is not logical to require that company commanders of fire forces arriving at the scene of an ordinary fire stand by until the battalion chief gets there. Nor should the operations officer be required to defer activating the emergency operations center in a major disaster until the department commander gives him permission.

Easily Understood

The organizational structure should be easily understood. It should be such that it can be plotted on a chart, clearly showing the chain of command and the relationships of one unit to another, with their respective spheres of activities clearly defined.

Flexible

The organizational structure should be flexible. It should be such that it is possible to use all or any part of it in an emergency. Members should be trained to perform alternate jobs if necessary.

Mobile

The organizational structure should be mobile. It should be such that immediate support can be given to surrounding areas in event of widespread disaster. Units designated for outside aid should be such that no area of your own command is left without some minimal protection.

Coordinated

The fire department should be properly coordinated with other organizations, whether civil or military, private or governmental, which also are prepared to engage in disaster control activities.

Staffed

Adequate staffing is essential. There often is a great difference between "paper" strength and manpower actually available, especially in volunteer fire departments at certain times of the day. The only way to be sure of maintaining the minimum number of personnel for key positions, such as fire chief, dispatcher, apparatus operator, ambulance operator, company commander, and fire fighter, in sufficient depth to guarantee around-the-clock availability, is to employ such persons for these duties exclusively. The direction of fire fighting forces should not be assigned as a collateral duty except in the very smallest community where there is no other alternative. It is just as unfair to expect the local gas station attendant to take charge at an earthquake disaster as it is to expect the local pharmacist to perform an appendectomy. Many smaller fire departments are beginning to employ a full-time permanent, paid nucleus. It is then possible in small communities having an ample number of energetic, dedicated volunteers, to fill in the ranks in case of disaster.

Well-Trained

Fire fighters should be well-trained and the personnel should consist of more than just a list of names of "bodies" who have been assigned certain disaster duties. It is not enough, for example, to designate a mythical person to each battalion, on each platoon, to be "radef" (radiological defense) officer for that command, with the duty of monitoring the assigned district in case of nuclear accident or radiation hazard arising from spills or leaks. The effectiveness of any operation will depend upon the skill and teamwork developed in drills and instruction given to designated personnel before the disaster.

Well-Equipped

The fire department should be well-equipped. Standardized equipment should be available in sufficient quantity to enable trained personnel to be dispatched immediately with minimum encumbrances and delay. For handling disasters created by natural phenomena, the anticipated requirements for that area, such as floods in some regions and earthquakes in others, should be given due consideration.

Properly Directed

Finally, the fire department should be properly directed. The respective units, from the smallest to the largest, should be commanded by officers who are well informed, and who have leadership abilities necessary for coping with emergency situations. Since any officer may be off duty, on vacation, on leave of absence, or otherwise unavailable in time of disaster, it is important that successors in the line of command also be competent leaders.

Principles of Organization

The general principles of organization upon which an effective Incident Command System are built, usually include Unity of Command, Span of Control, and Delegation of Authority.(3)

Unity of Command is a principle that applies to the person commanded.

The "team concept" is vitally important for fire fighters operating under an incident command system. (Photo courtesy of the National Fire Academy.)

The Buffalo blizzard of 1977. Extreme cold or heat situations mandate that a rehab sector be employed in incident command systems. (Photo courtesy of Ron Buckingham.)

It requires that he or she have only one immediate superior, and that all tasks be assigned through that superior.

Span of Control is a principle that requires that the number of persons reporting directly to an officer should not exceed that which can be supervised effectively. The number will vary with the type of work involved, but the authorities generally agree that it is approximately five.

Delegation of Authority is a principle that recognizes the fact that, while a person can never absolve himself of responsibility for performing a mission by requesting others for assistance, he or she should delegate certain tasks to subordinates and at the same time grant them sufficient authority to carry out such tasks effectively. Work should be delegated down to the lowest level at which it can be done, but delegation of work must always be accompanied by authority to accomplish it.

Special Principles Applicable to Disaster Control Organizations

The *General Build-Up* principle provides for augmenting forces as the size of the emergency increases, or the number or emergencies occurring in succession increases. In departments using running cards in dispatching, the forces can be increased by calling for second, third, fourth, and other alarm responses; owing to the move-up feature, whereby other companies are directed to fill in at other stations, a wide displacement of companies can occur as a result of a major emergency. If another greater alarm emergency develops while the first is in progress, the running card system presents some difficulties. For example, Chief Bahme responded on one occasion to a greater alarm brush fire on Mt. Washington in Los Angeles, and it was the third major emergency in progress. The nearest first-in engine company had to come from San Pedro, about twenty-five miles away, because the nearest move-up companies were already responding to the second major brush fire in the San Fernando Valley.(4)

The *Triplication* principle decrees that if you need "x" number of officers and men to cope with a major disaster on the first day of operations, you should have three times that number in order to provide reliefs when the emergency extends into the second and third day of continuous operation, or when the second and third emergency occurs while the first one is still in progress. Fire departments operating on the three-platoon system with fully manned companies can meet this requirement, but those depending upon small departments with no assistant chiefs to relieve the chief, nor additional volunteers to relieve the ones on the line, are almost solely dependent upon help from neighboring departments to provide needed reliefs. If these departments have their own emergencies to contend with, as was the case in late 1989 in the area south of San Francisco, when the large earthquake struck, outside help may be unavailable. This could also occur in tornadoes, conflagrations, and other widespread emergencies.(5)

Line and Staff

The terms "line" and "staff" sometimes are applied to officers and sometimes to functions, and there seems to be no unanimity of opinion among authorities as to the exact scope of these terms. The confusion is further compounded by the fact that in the military as well as in the fire service there are staff corps officers, such as medical, dental, or supply performing so-called line functions, and line officers performing so-called staff functions.

It has been said that line officers exercise command authority, and that line functions are those for which the organization was primarily created to carry out; that staff officers act in an advisory capacity, do the planning, training, selling of ideas and public relations, in support of the primary functions. It has been recognized, however, that practically every organization, from the smallest integral unit to the largest overall structure, has line and staff functions, and that practically every officer, excepting perhaps the person at the top, can be considered line and staff. It is more correct to refer to duties or individuals as "essentially staff" or "essentially line."

INCIDENT COMMAND SYSTEM (ICS)

Even if your fire department has met all the organizational requirements just discussed, there is an additional dimension of organization structuring required at the time of emergency. We have shown how a fire department should be broad in scope, include standardized units, be practical, easily understood, flexible, mobile, coordinated, well-staffed, well-trained, well-equipped, and properly directed. All of these criteria, however, are insufficient if at the time of emergency this organizational structure cannot be used effectively. This is where a functioning incident command system comes into play and serves the needs of the field commanders and operating forces on the front line.(6)

Throughout the history of the fire service, it has been observed that progressive changes are often brought about as the result of a disastrous loss of life or property. The Incident Command System (ICS) was no exception.

In 1970, southern California experienced two weeks of devastating wildland fires. Five hundred and eighty thousand acres of watershed area were destroyed. The combined cost and loss figures totaled $18 million per day or $750 thousand per hour. This multijurisdictional disaster was the impetus for the development of an improved interagency incident management capability.

We saw in Chapter 1 how ICS was one of the beneficial results of a federally funded project called FIRESCOPE. (The acronym stands for Fire Resources of California Organized for Potential Emergencies.) All the involved fire service organizations recognized the advantage of combining resources under a common organizational structure during multiagency major emergencies. They also recognized that the system should not be designed as a multiagency, major emergency system only, but should be a day-to-day operational system for all the participating agencies applicable to any and all emergency situations. Each participating agency, of course, had a unique emergency organization, staffing, terminology, and training; and as is characteristic of the fire service, each agency ran strong on pride and egotism.

The provincial barriers eventually took a back seat for the good of the service. A lot of mutual give-and-take was required by each agency in order to develop a command system that was realistic, workable, and acceptable to all. The process of coming together as federal, state, county, and city fire service agencies, working out differences of opinion, working together for the common good of all the concerned agencies and ultimately the common good of all the publics served, resulted in the Incident Command System (ICS).(7)

The ICS required mutual agreement and acceptance of four things:

- the organizational structure
- common operational procedures
- common terminology
- personnel qualifications.

The ICS, as it is used today, is the result of much input from participating agencies nationwide, including the National Interagency Management System (NIMS). The ICS is a viable system and will continue to be subject to modification as experience is gained through operational usage.

One problem already encountered in even the most progressive of fire jurisdictions is a situation that suddenly develops when seldom-used mutual aid is called. Even though the participating fire departments may all be familiar with, and use, an incident command system, the terminology, procedures, and the "rules of conduct" may vary from one jurisdiction to the next. Hence, even though all departments would like to unify under a joint command, these differences prevent a smooth operation.

At a GLS Chemical Warehouse fire in Broward County, Florida, in July 1988, three different fire departments converged simultaneously on the scene of a potentially disastrous hazmat fire. The departments were Broward County, Hollywood, and Port Everglades. A postincident critique indicated that while all three were familiar with incident command, all three did not use the same brand, and therefore were unable to cooperate with the greatest of harmony. The Port Everglades chief, Huey Manges, stated that "one of the outcomes of this fire was that all fire departments in the jurisdiction will begin to use the same incident command system, and conduct joint exercises."(8)

Characteristics of the ICS

The most salient characteristic of the ICS is its ability to expand from a routine incident to a complex multiagency incident in a logical progression. Other characteristics are listed below:

- The ICS is an all-hazards system and is adaptable to all types of emergency incidents.
- Jurisdictional authority is determined by the geographic location of the incident.
- Span of control is the basis for modular expansion of the system.
- Unity of command is clearly established by the organizational format.
- The incident commander may be an engine company captain or, if conditions warrant, may be the chief of the department.
- The ICS utilizes commonly understood organizational structure, terminology, operational procedures, and qualifications for positions.
- Everyday operational usage provides a smooth transition in a large-scale emergency situation.

Functions of the ICS

There are five major functions utilized in the Incident Command System.(9) They are identified in Figure 3-1.

- Command
- Operations
- Planning
- Logistics
- Administration

Concurrent with the development of the federal focus on incident command, Fire Chief Alan V. Brunacini of Phoenix, Arizona, developed a fire ground incident command system based on a quite similar model, which became widely accepted throughout the United States.(10) While both the

Figure 3-1.

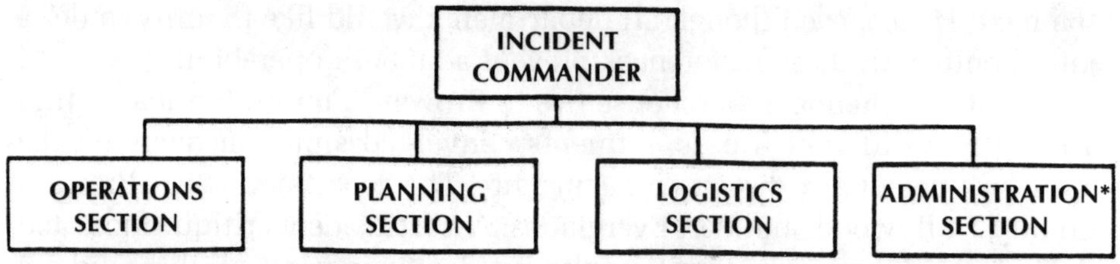

*Formerly called "Finance Section"

FEMA and the Brunacini versions of incident command were capable of being broadened or reduced so as to handle either large-scale or small-scale incidents, the Brunacini model was more effective for routine day-to-day emergencies, whereas the federal model lent itself more readily to large-scale incidents such as wildland fire emergencies which extend over numerous weeks or even months.

The Brunacini model, for example, used the term both functional and geographical *sectors* while the National Fire Academy model originally used divisions and groups. It is the opinion of the authors, and many other professionals in the fire service, that the word *sector* is much more specific and permits the use of one term rather than two, simplifying command procedures. Originally, the Brunacini model had a fireground commander whereas the FEMA model had an incident commander. Both, now use the term *incident commander* and both use the term *sector* in systems accepted by fire departments coast to coast.

This national standardization is the outcome of a meeting among interested parties that took place in Cheshire County, Virginia, on August 15, 1991, followed by another meeting a month later in Phoenix, Arizona in which plans were finalized. It was at these planning sessions that new terms such as *critical factors* and *task analysis* were first popularized in the Incident Command System. Additional refinements and improvements will be made in the Incident Command System in the years ahead but it no doubt will remain the vanguard of effective fireground planning and management.

COMMAND

The Command function is managed by the incident commander (IC). The IC commands the incident, develops the strategic decisions, and is responsible for the results. The IC implements additional functions that may be required to support the incident and retains the responsibility for any functions that are not implemented. The IC retains the approval for ordering or releasing resources.

Multijurisdictional incidents will involve a unified command element that will bring together jurisdictional incident commanders to develop a common and consistent action plan to make the best use of all available resources.

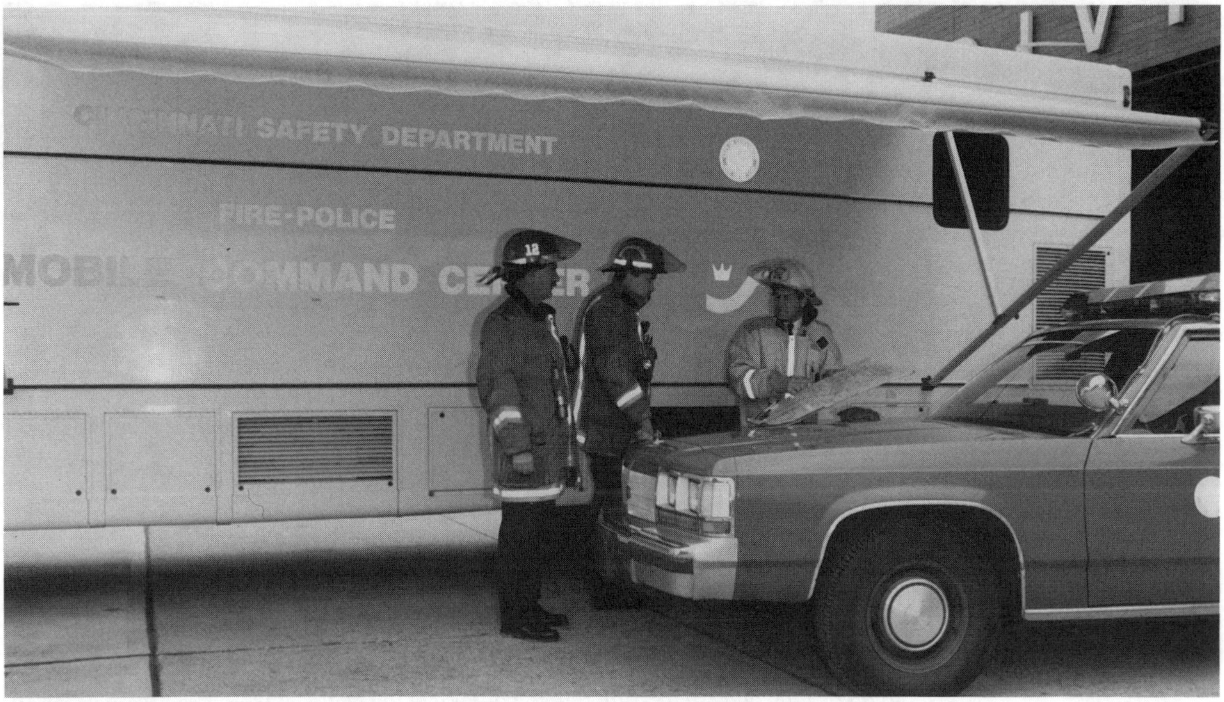

Incident command: District Fire Chief William Kramer of Cincinnati confers with two company officers at a command post. (Photo courtesy of Gary Auffart.)

OPERATIONS

The Operations Section is managed by the operations chief or operations officer, a member of the general staff who is responsible for all operations directly applicable to the primary mission of the incident (fire suppression, rescue, emergency medical services, etc.). The operations chief allocates and assigns resources to accomplish control of the incident.

PLANNING

The Planning Section is managed by the planning chief or planning officer, a member of the general staff who collects and evaluates incident status information needed to fulfill these responsibilities:

- understanding the current situation
- predicting the course of the incident
- preparing control objectives and alternate strategies
- conducting planning meetings and preparing the Incident Action plan.

LOGISTICS

The Logistics Section is managed by the logistics chief or logistics officer, a member of the general staff, who is responsible for providing all the facilities, services, and materials required to support the system.

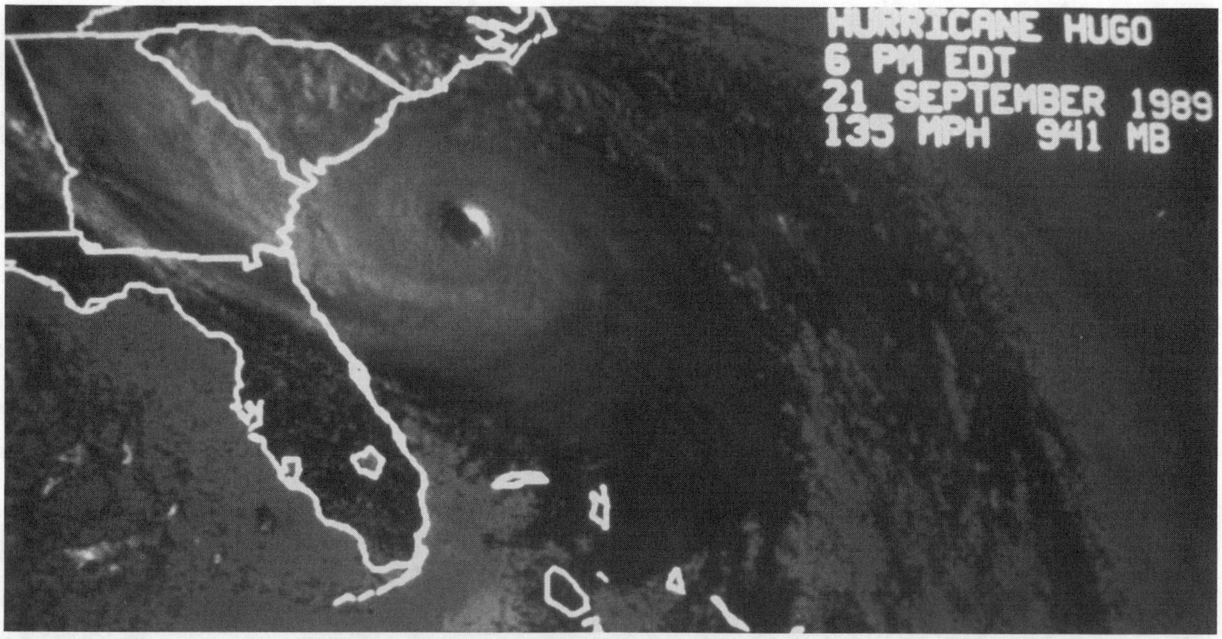

ABOVE: *A planning section uses this aerial photo of Hurricane Hugo, delivered by satellite, to predict the hurricane's path. (Photo courtesy of the National Hurricane Warning Center.)*

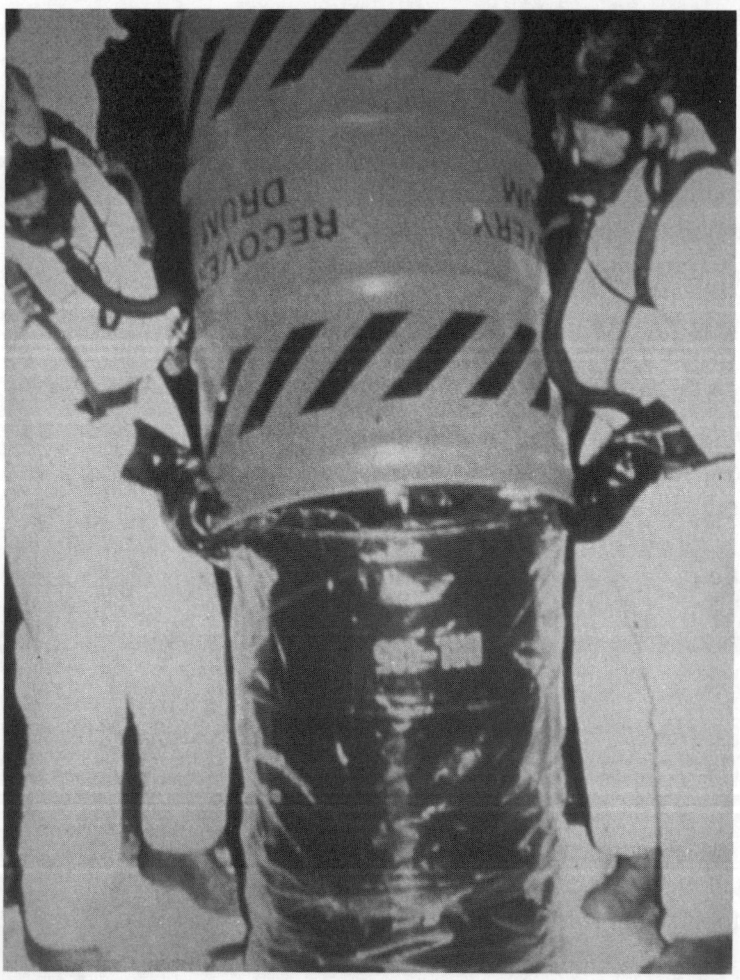

The logistics section ensures timely arrival of recovery drums to control a hazardous materials incident. (Photo courtesy of the National Fire Academy.)

ADMINISTRATION

The Administration Section is managed by an administrator or finance officer, a member of the general staff who is responsible for administrative and financial aspects of the incident.

ICS Functional Sections Expansion

As an incident grows in complexity, the functional sections may be expanded to accommodate the responsibilities of the section and to ensure a proper span of control.

COMMAND

The Command Staff function may be provided initially by a chief's adjutant, but early on in a complex incident consideration should be given to providing sufficient personnel to this important aspect of the command function.

The three Command Staff functions, shown in Figure 3-2, are as follows.

1. The liaison officer provides an interface for all assisting agencies and identifies current or potential interagency problems.
2. The information officer provides information and media releases within the constraints of the IC.
3. The safety officer monitors hazardous and unsafe situations and initiates actions to prevent unsafe acts by incident personnel.

Figure 3-2. Command Staff Functions.

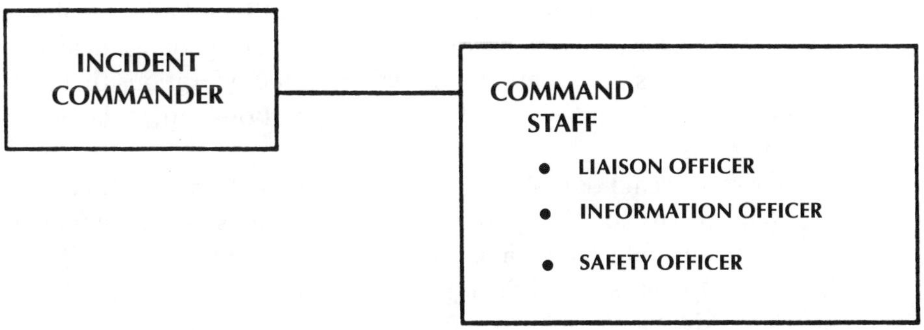

OPERATIONS

The expanded version of the Operations Section, shown in Figure 3-3, consists of the following.

- The Staging Area, supervised by the staging area manager, manages all the resources at the location that are ready and available for assignment. The complexity of an incident may require more than one staging area.

Dangerous working conditions mandate a safety officer to ensure that fire fighters are given maximum protection. (Photo courtesy of the Federal Emergency Management Agency.)

- Sectors are managed by supervisors who direct the operations of strike teams or other resources within a specified geographic area of the incident. Fire suppression and search and rescue are typical section commander responsibilities.

  NOTE: "Divisions" and "Groups" were terms originally used in lieu of "Sectors" in the first version of the federal model, and they still are used in some areas, including major FIRESCOPE operations on the West Coast. Divisions are used to designate geographical sectors, and groups are functional or task-oriented operations that may cross division boundaries. EMS and air operations typify the group designations.

  For large-scale operations the number of sectors (divisions and groups) may exceed the span of control guidelines—usually five. In this case an intermediate level of command may be instituted between Operations and Sectors, called "Branches." These ideally are established for specific disciplines (e.g., Law Branch, Fire Branch, Medical Branch, etc.)
- Groups are managed by group supervisors. They supervise functional or task-oriented operations that may cross division boundaries. EMS and air operations typify the group designations.

Since 80 percent of the nation's fire departments are totally volunteer and another 10 percent are staffed with combination career/volunteer personnel, some fire departments do not lend themselves to idealized, regimented company response. Fire Chief Steve Meyer of Garrison, Iowa, has developed a more practical and understandable concept, simply designating

Figure 3-3. The Operations Section.

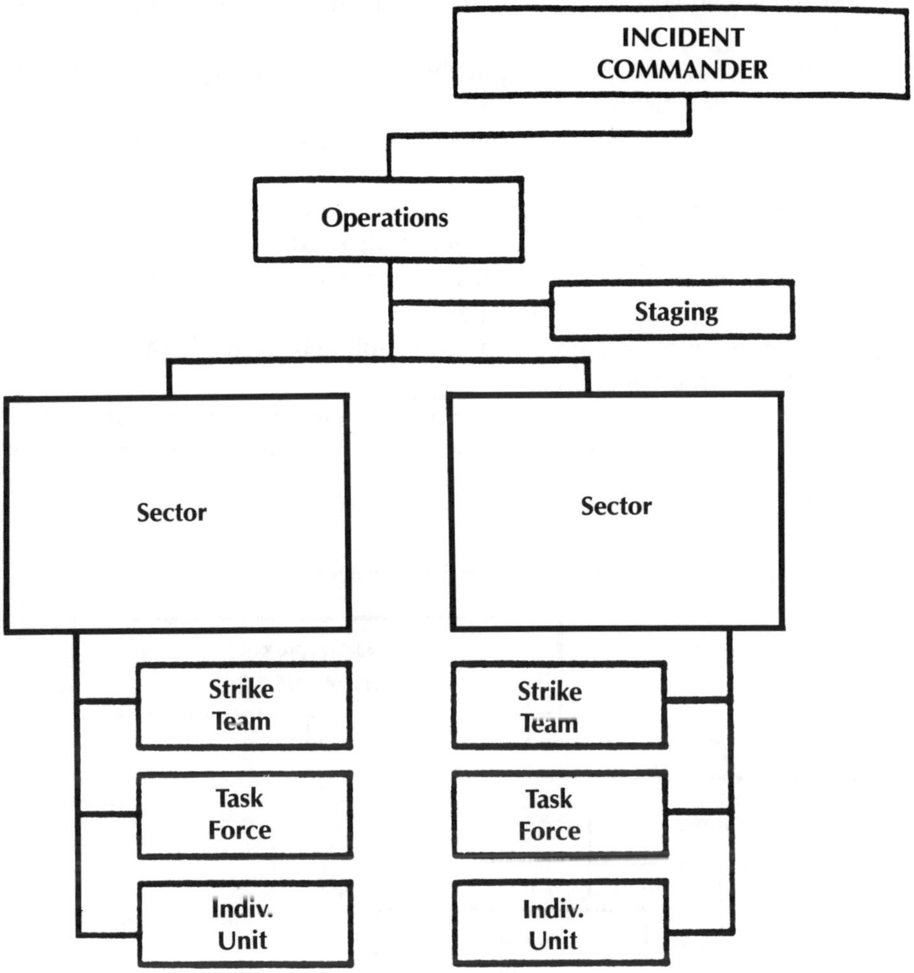

tactical assignments to make-shift groups, called "Crews." Most often divided by members responding with pieces of apparatus, the Crew concept is nothing more than a contingent of fire fighters assigned to a specific foreground operation such as suppression, ventilation, or search and rescue.(11) In large-scale disaster situations, the fire departments accustomed to using the CREW concept can adapt it to the larger role they are called upon to fulfill.

Most fires and other routine emergencies do lend themselves naturally to division of work into component parts. The most common division is by geographical area. A workman's torch caused a blaze in the historic Church of the Holy Apostles, in Manhattan, on April 9, 1990, causing $500,000 damage. Chief Devin Byrnes, Incident Commander, faced fire on three fronts: the cockloft, the roof, and the church proper. This lent itself naturally to three primary sectors, each of which could be subdivided by subordinate commanders.(12)

PLANNING

The expanded version of the Planning Section is shown in Figure 3-4.

- ○ The Resource Unit is responsible for maintaining Resource Status information (RESTAT).
- ○ The Situation Unit is responsible for maintaining incident status information (SITSTAT).
- ○ The Documentation Unit maintains accurate and complete incident information records for legal, analytical, and historical purposes.
- ○ Technical specialists are advisors who have special skills to support incident operations. Special skills might include expertise in fire behavior, water resources, environmental concerns, hazardous materials, weather prediction.
- ○ The Demobilization Unit develops the demobilization plan, which includes methodology for de-escalation and closing down of incident operations.

Figure 3-4. The Planning Section.

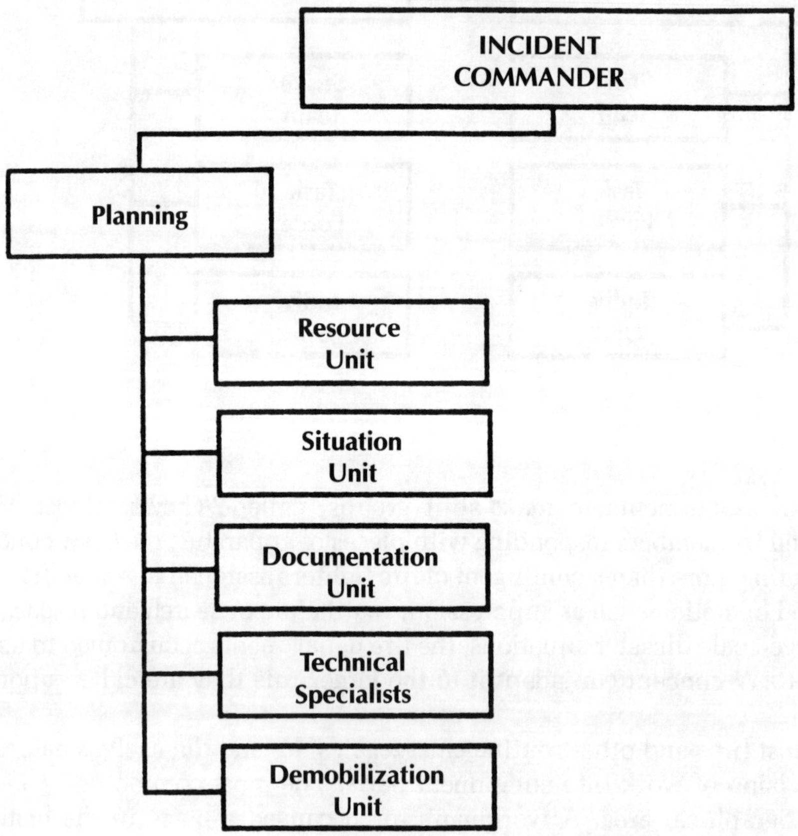

LOGISTICS

The expanded version of the Logistics Section is shown in Figure 3-5.

- The Communications Unit is responsible for the effective use of all incident communications equipment and facilities, including distribution, maintenance, and repair.
- The Medical Unit is responsible for providing medical aid and transportation for injured and ill incident personnel. (Note that a medical group under the Operations Section would be developed to handle large-scale civilian casualties.)
- The Food Unit is responsible for feeding at all incident facilities, including menu planning and food.
- The Supply Unit is responsible for ordering personnel, equipment and supplies, and maintenance of proper inventories.
- The Facilities Unit is responsible for the layout, activation, and maintenance of all facilities required to support the operation (e.g.,

Figure 3-5. The Logistics Section.

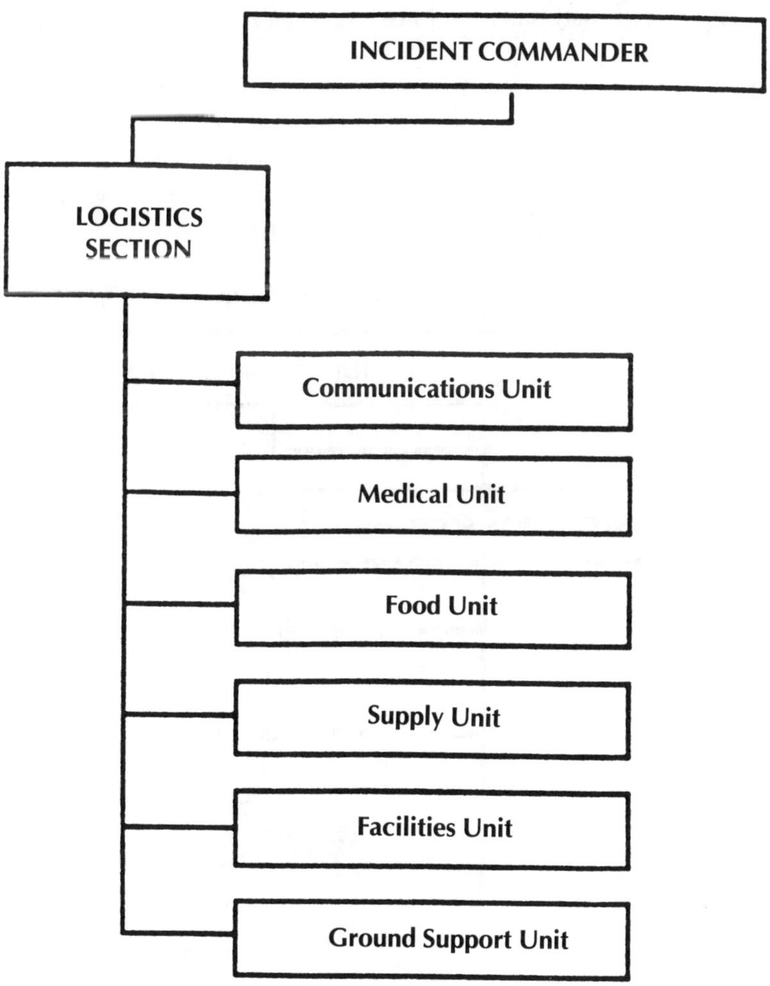

Incident Command Post, Base Camps). This unit provides sleeping and sanitary facilities for all incident personnel.

- The Ground Support Unit is responsible for the service and repair of vehicles, transportation of personnel, supplies, food and equipment. It also implements a traffic plan for the incident.
- Sector directors for Service and Support Units may be implemented as the span of control is exceeded.

ADMINISTRATION

The expanded version of the Administration Section is shown in Figure 3-6.

- The Time Unit is responsible for maintenance of personnel time records.
- The Procurement Unit is responsible for all vendor contracts utilized during the incident.
- The Claims Unit is responsible for processing all claims arising at the incident.
- The Cost Unit is responsible for collecting all pertinent cost data and providing cost estimates and cost savings recommendations.

The Administration Section originally was called the Finance Section. The finance function is usually operable at large incidents where costs, billings, and other monetary issues become important. Financial elements involved in smaller incidents might be so trivial that the finance function was usually combined with some other function.

Figure 3-6. The Administrations Section.

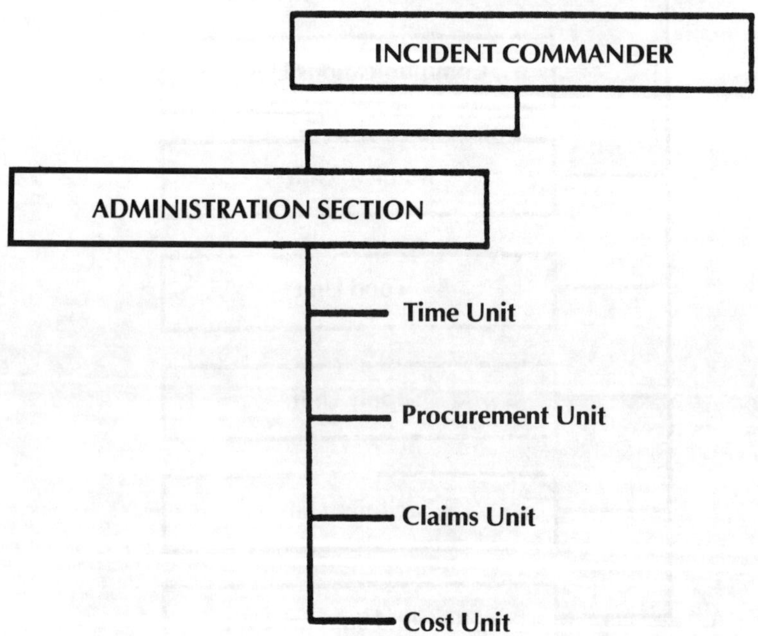

One reason for the change is that for most routine fires the finance function does not become relevant, and often is not even activated. It does become relevant, however, when major expenditures of forces need to be tracked and numerous supplies are purchased and administered through the Logistics Section. Such would be the case for an ongoing emergency that lasts for days or weeks.

Another occasion when a finance officer might prove helpful would be the tracking of major losses at a fire. Consider the costs when Universal Studios were involved in a massive fire in late 1990. Although Steven Spielberg praised the fire fighters for saving the vast majority of the studios, those lost included priceless theatrical props and movie memorabilia that were valued at millions of dollars.(13)

ICS Expanded Organization

The ICS expanded organizational chart shown in Figure 3-7 depicts the relationships of the functional subdivisions. A large-scale ICS may not need to incorporate all the subdivisions shown; however, the relationship and responsibilities should be understood.

The five major functional responsibilities have been separated, and the three command staff functions identified. The proper span of control is maintained and unity of command is clearly defined by the lines depicting who is in charge and to whom each subdivision is responsible.(14)

ICS Modular Expansion

As previously stated, the ICS is designed for every day, every incident usage. It can be expanded from a routine incident to a complex multiagency incident in a logical, sequential progression. The incident commander may initially be an engine company captain performing all the required command functions or, if conditions warrant, command may evolve to the chief of the department with all functions established.

As an incident grows in complexity, the incident commander can expand the ICS to meet the needs created by the incident or reduce the span of control to a reasonable level.

Figure 3-8 depicts a typical first-alarm assignment. the incident commander (battalion chief 1) and his adjutant can handle all the required command and staff functions. Span of control is manageable.

The incident illustrated above has now expanded to a greater alarm. Five additional engines (two are mutual aid) are on the scene. Police and TV news have arrived. Assistant chief 1 is on scene and in charge. The incident has become more complex. Reasonable span of control is exceeded. The IC can no longer manage all the required functions. The command staff functions of liaison and information should be established to interface with other agencies and handle media information. Additional chief officers are

Figure 3-7. ICS Expanded Organization.

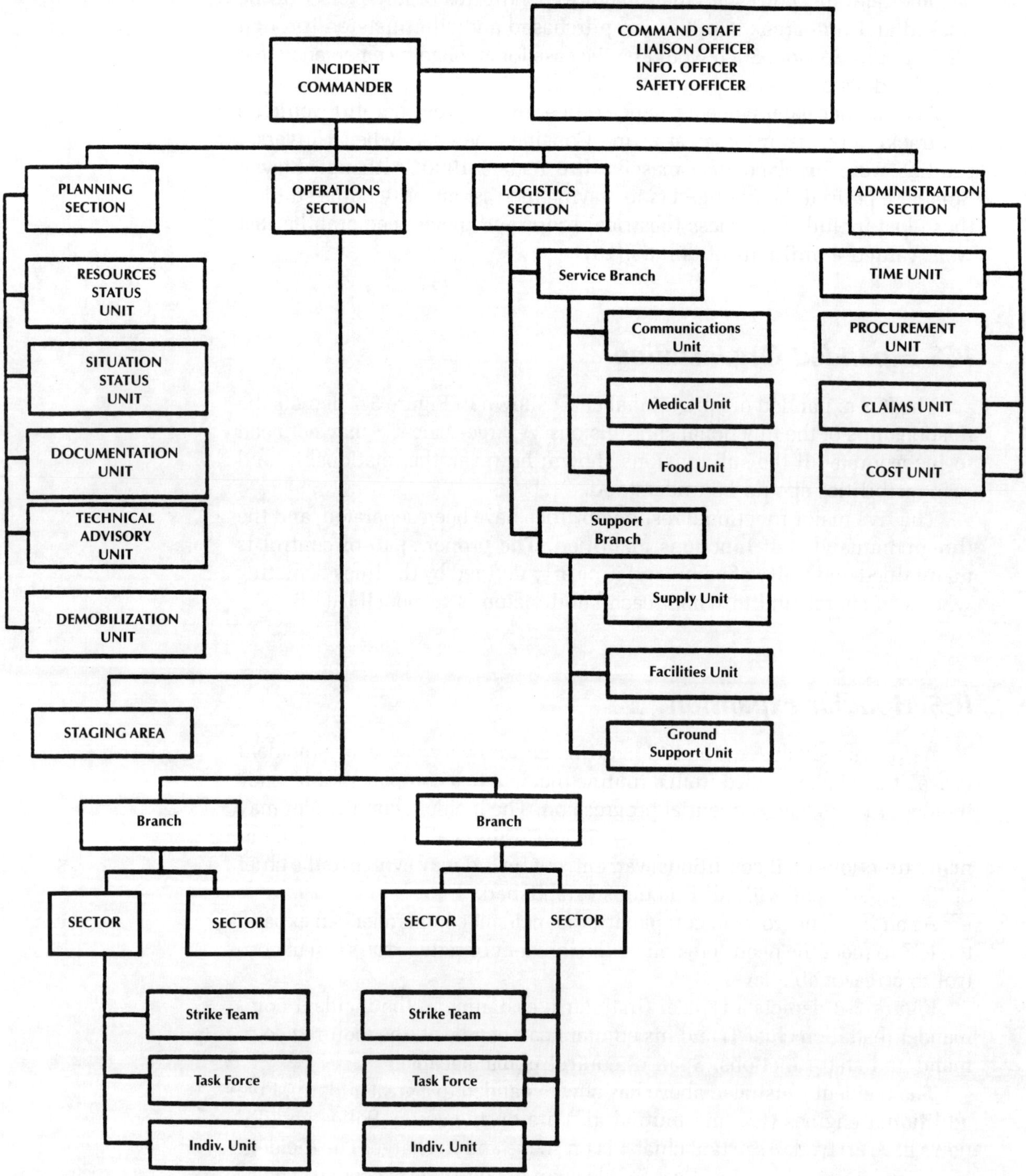

Figure 3-8. A Typical First-Alarm Assignment.

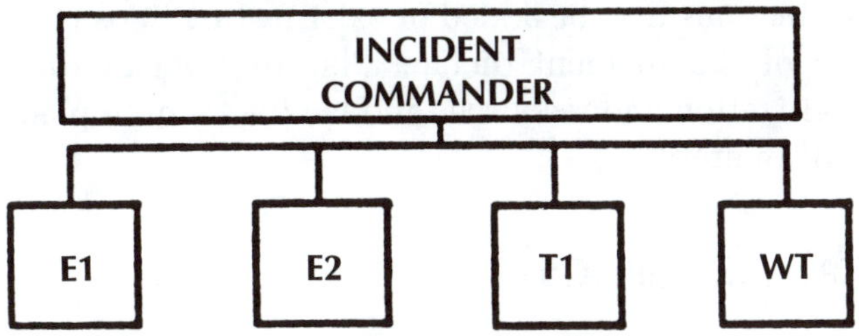

Figure 3-9. A Greater Alarm Assignment.

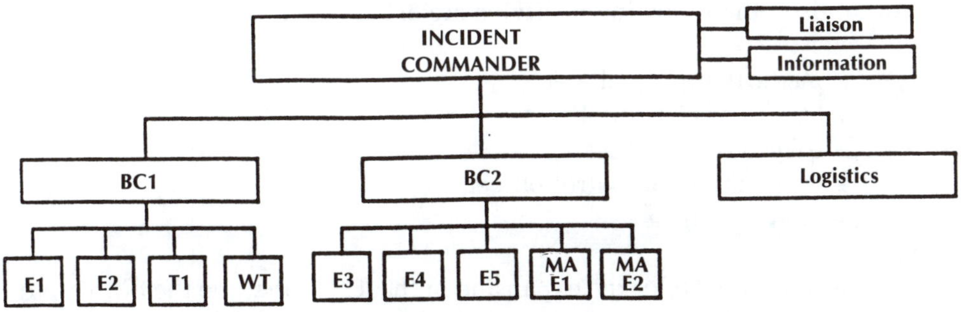

Figure 3-10. Medical Group Organizational Chart.

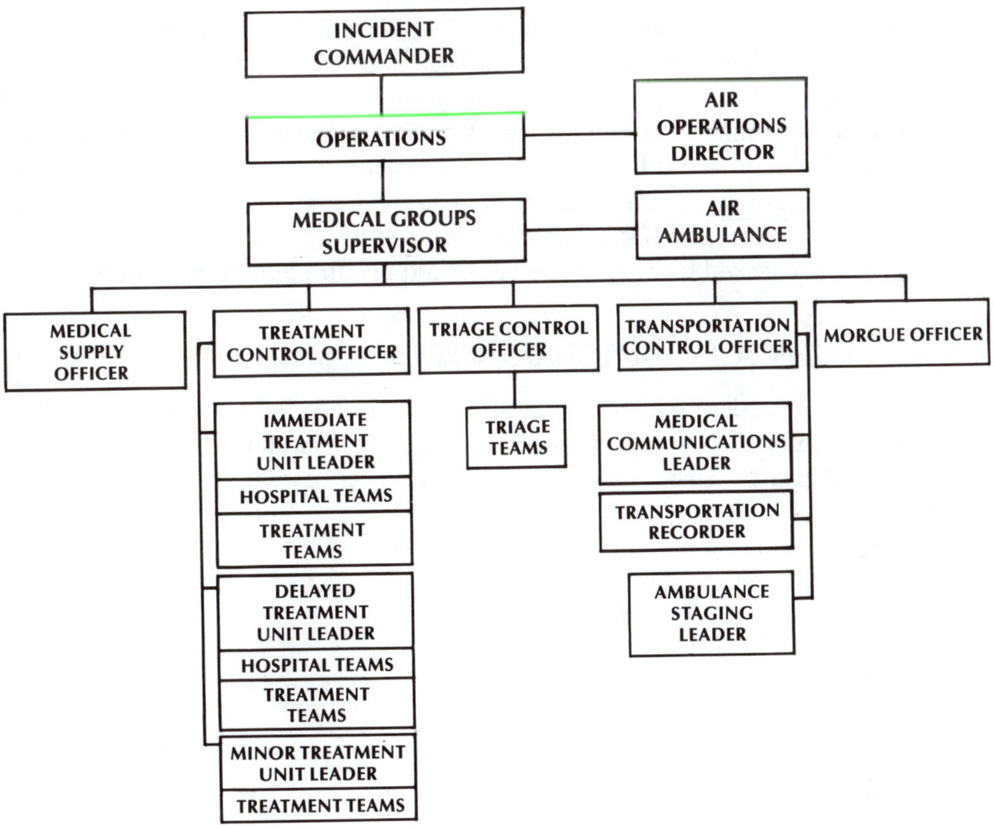

required for the command function. Logistics is set up to provide support for the operation (Figure 3-9).

Support functions may be added or existing functions may be expanded as the needs of the incident dictate. Planning ahead to anticipate the required organization and resources and having them in place is the key to a a successful operation.

MEDICAL GROUP ICS

Any large-scale disaster could involve a mass casualty or EMS incident. The Medical Group is a subsection of Operations and would be implemented in the event of a large-scale EMS incident. Under the medical group supervisor, the following positions are established as necessary:

- medical supply officer
- treatment control officer
- triage control officer
- transportation control officer
- morgue officer

See Figure 3-10 for the expansion of the ICS to include the EMS group, and refer back to Figure 3-7 to see how the expansion of the Medical Group relates to the overall ICS chart.

The tremendous value of the Incident Command System where triage is needed at a major emergency was illustrated in Newington, Connecticut, on January 2, 1991. Here, a fire at the U.S. Veteran's Hospital presented what was called "the most significant challenge ever faced by the Newington Fire Department." In all, 63 persons were rescued, 15 of which were removed by ground ladders or aerial towers. A patient triage and control area established in an adjacent building allowed the fire department to account for all victims. Eight patients were treated, seven for smoke inhalation and one for both smoke inhalation and second degree burns. Fifteen hospital persons also were checked for smoke inhalation and one fire fighter was injured.(15)

HAZARDOUS MATERIALS GROUP ICS

The Hazardous Materials Group, shown in Figure 3-11, illustrates the expansion of another group function to meet a specific need. Under the hazmat group supervisor, the following positions are established as necessary.

- Evacuation Team
- Save Refuge Area
- Perimeter Control
- Site Control

Site Control consists of the following:

- Entry Team
- Decontamination

Figure 3-11. Hazardous Materials Organizational Chart.

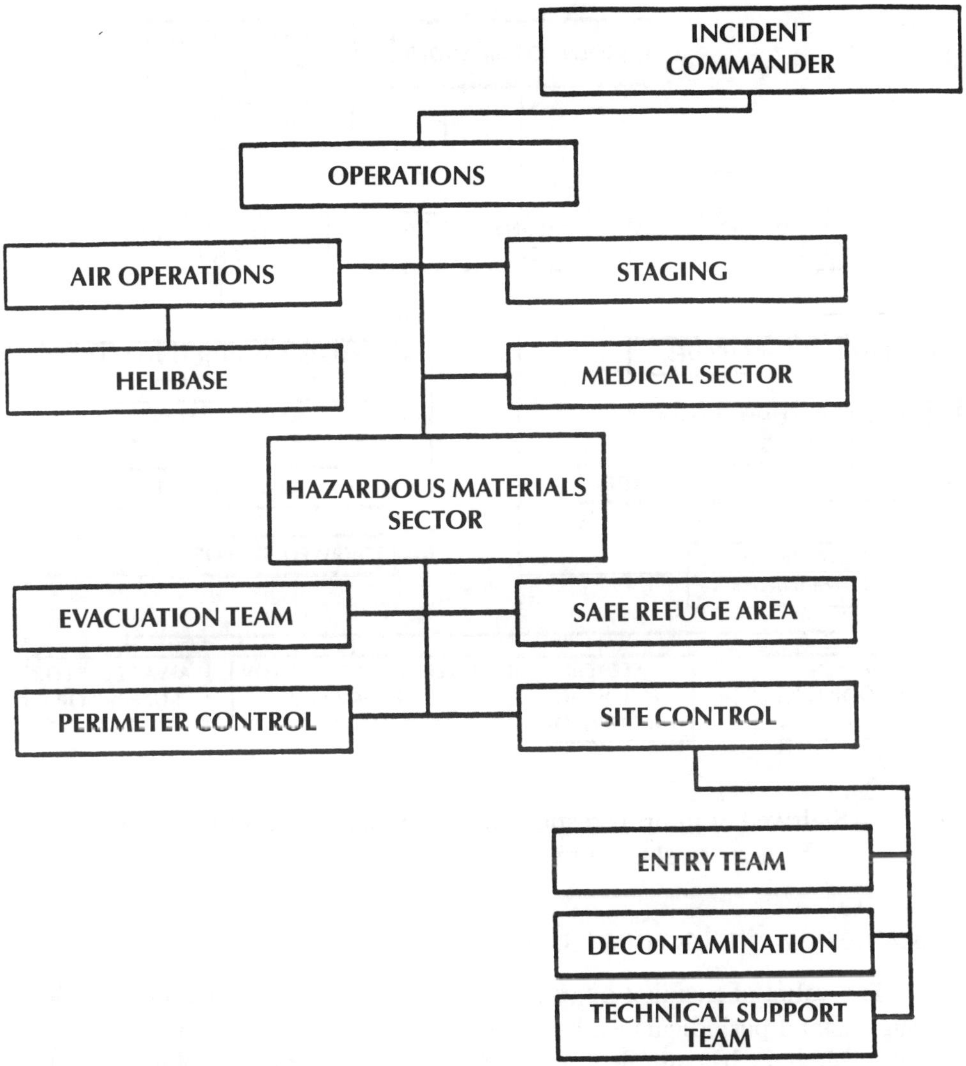

○ Technical Support Team

Refer back to Figure 3-7 to see how the expansion of the Hazardous Material Group relates to the overall ICS Chart.

HIGHRISE ICS

The following Highrise ICS is a typical ICS format with slight modifications. See Figure 3-12.

○ Systems Control coordinates movement of resources between base (outside resource pool) and staging (inside the building resource pool). Systems informs incoming companies of the availability of elevators or stairways. It is responsible for controlling the air-handling system.

Figure 3-12. High Rise ICS.

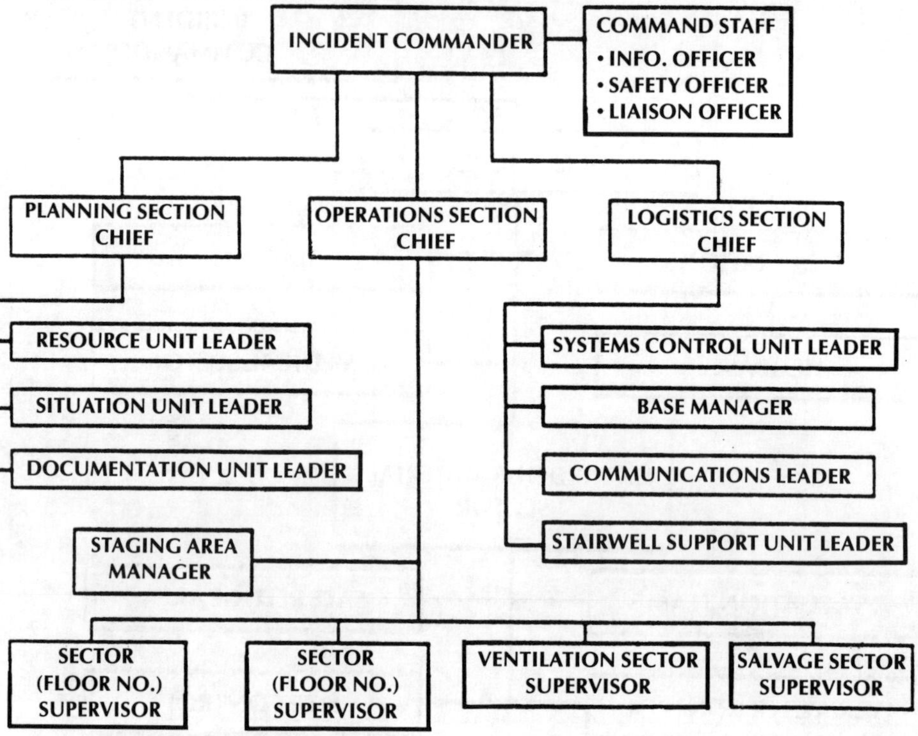

- Stairwell Support is responsible for transporting equipment from the lobby level to staging via the stairway(s).

Success with Incident Command

The value of incident command often has been documented in video journals and periodicals in the fire service since its inception in the early 1980s. One such example was the Great Bend, Kansas, explosion at the Wiretech Services Company on October 30, 1989. Assistant Chief Marion Root took command and established a command post northeast of the oil producing complex. The one-square-mile area surrounding the blast area was divided into six sectors, with a team assigned to each. The Great Bend forces were augmented by mutual aid and in the end, "all responding agencies realized the effectiveness of the Incident Command System as a scene management tool."(16)

Although many systems exist throughout the nation for the command and control of resources at emergency incidents, the National Fire Academy has adopted the ICS as its base for teaching the concepts of incident command. This chapter has presented a skeletal view of the system. Intensive preparation and training are necessary to properly equip a department, a community, or a region to apply the ICS on an expanded basis. ICS is a day-to-day, every incident command system. Yet, it is expandable in the logical steps to provide a smooth transition into a large-scale, multiagency command system.

Think of the ICS as a box full of tools available to help the Incident Commander on a daily basis. The ICS should

- use only those tools that are needed for the job
- leave unnecessary tools in the box
- but remember, the tools are there in the box to help, however complex the job may become.

ICS Terms

Knowledge of the terminology used in the ICS is extremely important if communications within the system are to be clearly understood. Refer to the list of definitions presented in the Glossary following the final chapter of this book which contains basic disaster and fire defense planning terminology. Incident Command System terms you should know are:

Base	Group
Branch	Incident Action Plan
Camp	Incident Commander
Clear Text	Incident Command Post (ICP)
Command Staff	Incident Command System (ICS)
Company	Sector
Division	Staging Area
Engine Company	Strike Team
General Staff	Task Force

ICS Focus

The establishment of the following required functions, early in the incident will relieve the incident commander of those many important concerns and allow focus on the main responsibility of managing the incident.

- Strike teams and division supervisors (5 to 1 ratio)
- Staging area for incoming units and reserve units
- Operations chief to supervise the above
- Resource Status, Situation Status, Documentation and Technical Unit (weather) due to complexity of incident
- Planning chief to supervise the above
- Communications, Medical, Food, and Facilities to accommodate personal needs and ensure radio contact
- Logistics chief to supervise the above
- A Time Unit to insure accurate timekeeping for mutual aid resources (perhaps this function could be assigned to Liaison)
- Liaison, Information, Safety are all essential functions in the described incident.

Command Headquarters

As we followed the incident command system through its various hypothetical phases, we saw how the system could grow in terms of resources and command personnel to accommodate emergencies as they escalate in size. As the complexity of the emergency grows, so too does the command headquarters. In relatively small incidents, the command post is nothing more than a staff car and a status board with the incident commander in charge of planning, operations, logistics, and administration. As the emergency grows, or at the onset on a larger emergency, the command post or field headquarters must expand in size and complexity incorporating not only the incident commander but perhaps several other subordinate commanders in charge of the various sectors or divisions. If the emergency covers a large part of the community, or is expected to last over several days, or both, then there is a need for an emergency operations center (EOC) where more durable record-keeping is possible.

The Emergency Operations Center, often referred to as the EOC, is the location from which the commander of the organization issues orders to his forces as a whole. If you are speaking of the community's EOC, it might be called the Disaster Control Center, and the mayor might be the commander of the organization. For this book, we will refer to the fire department's center for emergency operations control with the letters EOC. Most large cities and other large governmental units designate a fixed structure that could be used as an emergency operations center

Radio communication becomes essential between adjacent levels of command when visibility deteriorates. (Photo courtesy of The National Fire Academy.)

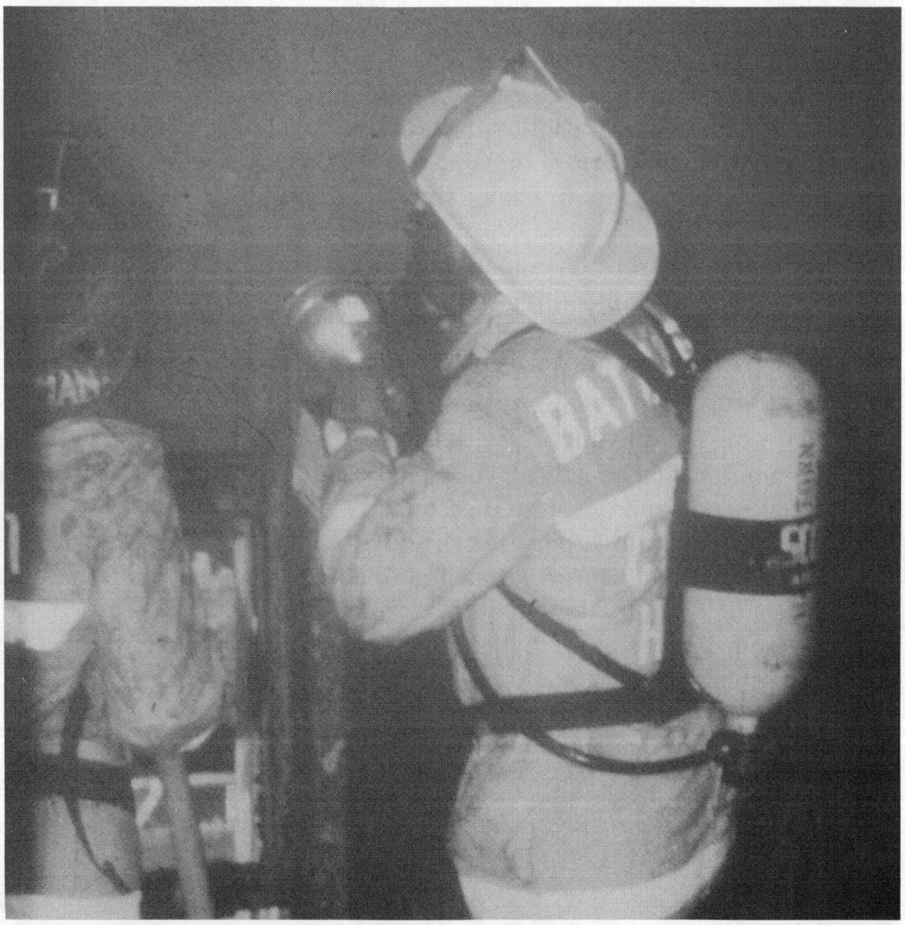

(EOC) during a protracted emergency or disaster. This could be located in the same building with a dispatch center or other fixed structure. It also could be improvised in a selected structure in proximity to the heart of the disaster at hand. Where an EOC has been designated, provisions can be made in advance for logistical supplies, communication networks, overnight lodging, and other conveniences that will be helpful to sustain a command staff during a long period of time.

In addition to the emergency operations center there will be at least one field command post. If the incident commander is located at the emergency operations center it is likely that the operations officer will be found at a primary field command post. If the disaster covers a wide geographical area such as an earthquake or large-scale brush fire, subordinate or sector command posts might be necessary for subdivisions or sector commanders operating under the overall operations officer. The key requirement for effective organizational structuring is that the combination of command post and emergency operations center follow the same clean lines of command outlined on the incident command organizational chart.

FUNCTIONS OF EOC

The emergency operations center is a place where accurate reports of conditions at the scene of the disaster can be received, recorded, and evaluated. Some reports received will be far from accurate, for example when the source is an anonymous telephone call, but data supplied by the emergency forces is likely to be trustworthy. Though seemingly conflicting reports may be received, the discrepancies often can be explained by the fact that the observers were witnessing the events from different vantage points.

Based upon the evaluated data, command decisions can be made and issued to your forces—decisions that the field commander would be unprepared to make without reports of the overall picture. For example, one evening during the Watts riots of 1965, the incident commander at the EOC located in the dispatch center was receiving reports from helicopter observers, police liaison officers, and command post fire officers, that heavy rifle fire was being experienced at a number of intersections where fire fighters were engaged in knocking down fires. There were no National Guard troops mobilized at that time, and law enforcement personnel were insufficient to provide adequate protection for fire department members. After conferring with the deputy chief who was directing field operations, the incident commander used the media to warn the public that fire fighting forces were pulling out of a two-square-mile area, and anyone choosing to stay within that area should be prepared to fight his own fires. Later this evacuated area was increased to four square miles.(17)

The EOC is not only a place where command decisions can be issued but is also the place where orders can be given to heads of subordinate forces, such as the logistics division, which are necessary to support operating forces carrying out command decisions. Requests for food, fuel, battery and tire replacements, and reconnaissance can be relayed to the appropriate bureau representative.

The EOC is also a place where a graphic plot of the situation can be maintained as events transpire, providing they do not happen too fast to be neatly plotted. During the Watts riots just mentioned, it was necessary to record dispatching of apparatus from other parts of the city to the various "charcoal alleys" with a large felt-tip pen on blank flip-chart paper. As the charts became filled, they were torn from the pad and attached to the wall with masking tape. This was not such a problem after the command posts became activated, for it then became necessary to record only the apparatus dispatched to apparatus pools of the command posts; pools officers kept track of their own dispatching.

Status boards also are maintained at the EOC to provide instant information about who is in command of what elements of the organization at any given moment, which units have been recalled to duty, which have been relieved, what mutual aid forces have been summoned, and other details.

The size of the emergency will dictate the composition of the personnel in the emergency operations center. If the operations chief has numerous sector or division commanders he or she may also be found in the emergency operations center. During an extended emergency it is quite logical for the incident commander and the chiefs of operations, planning, financing, and logistics to all be located in the emergency operations center, with the subordinate commanders in charge of field command centers. In relatively small- and medium-scale emergencies this also is the location of the overall incident commander. In large-scale operations the primary field command post will be the location of the operations officer in charge of all field forces. Where separate subordinate sector command centers are used, there remains one central command post in the field which is manned by the operations officer.

The incident commander probably would use the EOC as headquarters during the disaster operations; this officer may or may not be the fire chief. In the absence of the chief, one of the deputies would serve as incident commander. In any case, this commander probably would remain at the EOC at least until an evaluation of the situation has been made, decisions rendered, forces deployed and supported.

The operations officer (probably the executive officer to the chief or a bureau commander) is responsible directly to the incident commander, and usually has overall responsibility for the movement of all apparatus, equipment, and personnel. This officer also has responsibility for coordinating department activities during a major emergency and for keeping the incident commander informed of the status of operations, calling attention to the need for calling mutual aid or for recalling off-duty personnel. If the operations officer is unable to communicate with the chief, then actions that are necessary must be performed to maintain basic requirements for the community's protection.

Miscellaneous assistants may be needed at the EOC for special emergencies: chemical engineers, radiological safety personnel, clerks, messengers, and draftsmen may be needed in any sizable emergency.

SELECTING AN EOC

Among features that should be considered when selecting a designated EOC are the location, construction, and design. The location should preferably be central in the community it serves, and in an area not subject to inundation, landslides, or serious fire exposure. This ensures the probability that it can be used in a multitude of disaster types.

A structure housing an EOC should ideally meet the same requirements as for a communications center. It should be resistive to fire, flood, earthquake, thermonuclear bombs, radiation, fall-out, sabotage, or any other disruptive forces. Many of these characteristics are, however, compromised where it is necessary to establish an alternate EOC in the vicinity of a disaster.

Where possible the EOC should be equipped with an automatic fire protection system, emergency power sources, back-up communications system, and be self-contained to the extend that, when activated, emergency personnel will be able to remain in the center for at least a week. This means that there must be adequate provision for feeding, sleeping, and sewage disposal, notwithstanding any failure of public utilities systems which normally serve the EOC.

If located in the same facility as the dispatching center, the EOC should be separated from the room in which alarms are received and equipment dispatched. Ideally there should be see-through, sound-proof partitions, which permit the staff to view the apparatus status board being maintained by the dispatcher, and remote speakers, capable of modulation, which can transmit sounds from the dispatcher's area. Likewise, the chief dispatcher should be able to see such items as a situation map and officer status board being maintained in the EOC area.

A gallery, separated by see-through partitions and having a remote speaker from the dispatching office, may be used to accommodate other city officials not directly involved in command. One corner of this gallery could be used by representatives of the news media and the department's public affairs officer. In addition to monitoring the main radio frequency used in the disaster control operation and television broadcasts from department and news helicopters, persons in all three of the locations would be able to see the status boards and situation map and be able to determine at a glance the operational status and the command status. The operational status could be any of the following: Plan I, Plan II; recall status, civil defense status (major emergency, local peril, disaster, extreme emergency, etc.). The command status could give names and locations of various commanding officers as posted (incident commander, operations officer, sector one commander, etc.).

FIELD COMMAND POSTS

At the scene of every emergency, large or small, someone in the field must be in command. The location selected for issuing orders is deemed to be the command post.

At a small emergency the command post (CP) may be a department vehicle, and a chief's driver or adjutant may be the staff. If the chief wishes to investigate the incident or make a personal sizeup, he or she should use the adjutant or other member of the department to stand by for radio reports and to relay orders to the arriving units. Should the situation develop to the point where the chief wants to call for a greater alarm assignment of apparatus, he or she could ask one of the later arriving chiefs to establish a division or sector command post at the opposite side of the emergency. A superior officer may arrive and use this initial location for a field headquarters and use the original responding chief as an assistant field commander. The new assistant can standby to relay orders the commander may issue while making a reconnaissance of the situation.

Other command posts may be set up at other points, where sector commanders may be assigned. In the Bel Air fire the Los Angeles Fire Department had three command posts at intervals along the south perimeter, in addition to field headquarters. Should the emergency develop further requirements, the field commander might declare a major emergency, and thereby put the machinery in motion for obtaining additional support and activating an EOC.

FUNCTIONS OF COMMAND POST

The command post, usually called the CP, is actually a sector office of the fire department headquarters, where the officer-in-charge first arrives on the scene. In almost all cases this is where the incident commander sets up shop and where he or she should remain. The exception is for a major disaster that results in the activation of an Emergency Operations Center with primary command transferred to the EOC. If another CP is established, the first one probably will become the field headquarters, temporarily at least, in which case the second CP is a sector office for the field headquarters. A CP provides a designated spot for the officer-in-charge to receive information by radio and personal reports from officers at the scene, and to issue orders to his forces already at, or arriving upon, the scene. It also provides a location where someone is always present to answer inquiries from the news media, coordinate the efforts of law enforcement officers on the scene, confer with utility company representatives who may have reported to assist, Red Cross personnel, and others.

At large-scale operations, the CP may become the location of a sector command, where forces sent in to handle that part of the disaster may report in, receive orders, and be deployed by the sector commander. The sector commander will be guided by instructions from the field headquarters commander, if one has been established, or from the EOC, if it has been activated.

COMMAND POST AND EOC

The main difference between a CP and an EOC relates to scope of functions and sometimes to location. Whereas the EOC provides the "What," "When," and "Where" of command, the CP provides the "How"; and whereas the EOC may be a fixed location, remote from the disaster, a command post may be located closer to the scene. And, finally, whereas the EOC

A 2,500 gallon propane tanker overturned on the Seaford/Oyster Bay Expressway in 1988. Fire fighters used unmanned monitors. (Photos courtesy of Harry Loud.)

will have to handle all of the problems involving reliefs, feeding, fuel for apparatus, supplies, mutual aid, and liaison with other agencies, the CP will put greater emphasis upon fire fighting, rescue, and operational tasks.

REQUIREMENTS FOR CPS

The two requirements for a CP are equipment and personnel. With respect to equipment, ideally, a CP should have about the same facilities as a control center, but in more mobile form. It will seldom achieve that goal. Although mobile command posts are becoming popular, and will be explained in greater detail in Chapter 6 dealing with "Communications," a command post is often nothing more than a clipboard in the hands of the initial incident commander. Sometimes the trunk of the chief's car, which responds to greater alarms, is equipped with a fold-out table and folding chair, maps of the area, electric lighting, speaker extension from car radios, pennant for the radio antenna (with "CP" on it), and forms for use in setting up an apparatus pool. If the engine is kept running to turn the alternator, the exhaust fumes could be a problem unless diverted to the side. A portable radio and a public address system, either portable or fixed, are also useful.

With respect to personnel, the officer-in-charge at a CP may only have his driver-adjutant to assist him or her only at the beginning, but if a greater alarm or major emergency develops, the plan should provide staff to help with planning, logistical, and operational problems.

CONCLUSION

This chapter has focussed on resources used by fire departments in the handling of all incidents from mere routine to major disaster scenarios. We have both outlined the general requirements for effective organizational structure in the 1990s and beyond, and we have shown how the incident command system allows for modular building of resources to handle escalating demands as a major disaster unfolds. In the next chapter we will look more directly at fire department resources and discuss ways in which human resources can be augmented to allow successful mitigation of disasters.

CHAPTER 3 REFERENCES

1. Carter, Harry. "Command Post—The Basics of Leadership" in *Firehouse*, Vol. 16, No. 1 (January 1991) p. 26.
2. Lesak, David M. "Operational Decision Making" in *Fire Engineering*, Vol. 142, No. 6 (June 1989), p. 63.
3. Carter, Harry. "Command Post—The Roots of Management" in *Firehouse*, Vol. 16, No. 3 (March 1991) p. 84.

4. Bahme, Charles W. *Fire Officer's Guide to Disaster Control*, 1st ed., Boston, MA: NFPA, 1978, p. 33.
5. CNN Broadcast, January 10, 1990.
6. *Command and Control of Fire Department Operations at Catastrophic Disasters: Student Manual*, Emmitsburg, MD: National Fire Academy.
7. *Field Operations Guide* (ICS 420-1), Incident Command System Publication, Stillwater, OK: Fire Protection Publications, 1983.
8. Kramer, William M. *American Heat* video journal, St. Louis, MO, August 1988.
9. National Interagency Incident Management System. *Incident Command System: Operational System Description* (ICS 120-1). Stanford, CA: FIRESCOPE Program, 1981.
10. Brunacini, Alan V. *Fire Command*, Quincy, MA: NFPA, 1985.
11. Meyer, Steve. "Organizing Crews in Incident Management" in *Fire Engineering*, Vol. 144, No. 3 (March 1991), p. 113.
12. Hashagen, Paul. "Workman's Torch May Have Caused Blaze in Historic Manhattan Church" in *Firehouse*, Vol. 15, No. 8 (August 1990) p. 51.
13. *American Heat* video journal, St. Louis, MO, December 1990.
14. *Incident Command System*, 1st ed. Stillwater, OK: Fire Protection Publication, 1983.
15. Scialabba, Richard. "On the Job—Connecticut" in *Firehouse*, Vol. 16, No. 4 (April 1991), pp. 52–55
16. Meisinger, Richard A. "The Great Bend Explosion" in *Fire Engineering*, Vol. 144, No. 1 (January 1991), pp. 59–62.
17. Bahme, Charles W. *Fire Officer's Guide to Disaster Control*, 1st ed., Boston, MA: NFPA, 1978, p. 35.

CHAPTER 4

RESOURCE MANAGEMENT AND AUGMENTATION

DISASTERS OFTEN DEVASTATE widespread areas that reach far beyond the confines of a local community. It would be erroneous to regard an earthquake, flood, or nuclear detonation as occurring within city limits when the city is but a small segment of the disaster area. John Carr and Lesley Omans pointed out that even a typical aircraft incident—which we tend to think of as being isolated—is usually a multijurisdictional incident "requiring a wide range of emergency response and management services."(1)

Experience with tornadoes in all parts of the United States and Canada illustrates how damage may spread across more than one community, even into other states or provinces. Thus, it has been accurately stated that "disaster control planning must be geared to an 'area concept.'"(2) Not only must such planning provide for an effective system of collective assistance among all political subdivisions within a defined area, but, because of the vast potential effects of large scale disasters, coordination of planning becomes a factor of great importance. Without a high degree of coordination, effective mutual aid is impossible; and the need for coordination extends down the chain of command to the smallest unit.

"A major disaster could result in temporary paralysis of extensive areas. In such an event, the problem of debris might create serious, although temporary, transportation problems at a time when food, clothing, medicine, materials for repair and construction, and other essential supplies are urgently needed. In addition, contamination from chemical, biological, and radiological sources may prevent transportation facilities from operating in or through certain areas."(3)

As disasters grow in complexity additional resources need to be summoned. In this chapter we will consider how to build resources to meet the growing demands of an emergency and how to manage expanded forces in a major operation.

Resources needed for wildland fire fighting include bulldozers and heavy equipment. (Photo courtesy of Jeff Whelan.)

AGENCIES INVOLVED IN DISASTER CONTROL

No fire department, regardless of how amply supplied with apparatus and manpower, will ever be capable of handling the largest emergency that might occur within its response district. The fire department in the City of Los Angeles is large by any standard, and yet on numerous occasions it has needed assistance from Los Angeles County, Beverly Hills, Santa Monica, and other departments when urban wildland interface fires have suddenly become overwhelming. On December 23, 1989, Los Angeles City used one of the largest turnouts of its own personnel and equipment in history plus mutual aid to combat numerous massive simultaneous structure fires, including one high-rise fire in a conflagration in its Westwood area. Fire fighters fought blazes in 28 structures all told with a dollar loss of approximately $40 million.(4)

In addition to mutual aid the fire department commander usually has a number of agencies that can be of great assistance in a disaster situation. Not all these agencies may have representatives within jurisdictions but they should be taken into consideration, commencing with those "closest to home." Following is a discussion of some that could be included among those requiring coordinated efforts in effective disaster control operations.

Other Departments of Local Political Entities

Even though a fire department may not be affiliated with a city or county government, as in the case of a fire district, advance planning can make available the services of the local health department and its chemists in emergencies involving dangerous chemicals and biological agents. The public works department has engineers and heavy equipment operators whose efforts can be coordinated at the scene of cave-ins, landslides, and building collapse. The police and sheriff's departments can provide coordinated assistance in traffic control and other public safety actions.

Red Cross, Salvation Army

The American National Red Cross has many community chapters that respond to disasters when notified of conditions qualifying for their assistance. At one time the Red Cross considered anything a disaster where five or more persons were killed. Even though reimbursement might not be obtained from the national organization, local Red Cross personnel often respond to fires having less than five casualties where displaced families need clothing or shelter.

One function that must become a part of incident command at major emergencies is coordination among the Red Cross, the Salvation Army, and other volunteer relief agencies, all of which may be attempting to solicit food, distribute blankets, and other supplies, as well as arranging shelter for refugees.

Civil Defense Agencies

In larger departments, a fire officer may be assigned the collateral duty of responding to major emergencies for the express purpose of providing coordinated liaison to the local civil defense agency. People often ask, "Where does 'civil defense' come into the action?" The collective activities of *all* the emergency forces, professional and volunteer, and those who help evacuate casualties, care for the wounded and dispossessed, bury the dead, feed the hungry, shelter the refugees, and restore things to normalcy are "civil defense." If the military forces are repulsing an enemy attack, that is "military defense." If civilians are coping with an emergency, that is civil defense. If the National Civil Defense organization activates a local headquarters at the place of a disaster, this can be a powerful resource.

Civil Preparedness Agency

It is true that military forces often are called upon to assist in major disasters, but the overall direction and coordination of the disaster control and relief is a civilian operation. As long as an emergency can be handled by "front-line" forces within a single department, there is seldom need to call the community's civil defense staff. However, if conditions develop where it becomes necessary to call outside help or mutual aid for an extended period, creating logistics problems, e.g., fuel, food, pay for overtime, pay for contracted helicopters and air tankers, then it becomes helpful to use the services of the civil defense coordinator.

Industrial Representatives

The Disaster Planning Guide for Business and Industry urges every industrial facility to become prepared to cope with the hazards and disasters of today's complex world.(5) Industrial managers who follow the excellent advice offered in this DCPA publication will be able to provide a readily available and highly trained source of volunteer leaders and advisors, program planners, emergency operations personnel, instructors, technicians, and other workers needed for disaster control operations.

Fire chiefs are familiar with the services rendered by public utility employees in the water, power, gas, oil, and chemical companies, but the availability and need for coordinating services by other industrial organizations at an emergency may not be fully understood without thorough investigation.

Military Organizations

While primary efforts of the armed forces are to carry out their assigned missions, with second and third priorities being given to assisting other military activities and other federal (nonmilitary) agencies, the fourth priority is to render maximum practical aid to civil authorities in domestic catastrophe. In a major flood, for example, the military often is called upon to provide helicopters and boats for use in search and rescue operations. During the 1960s and early 1970s, the National Guard was called upon to quell civil disturbances. Military personnel assigned to protect fire fighters have sometimes included fire department officers called out by their National Guard units, and such officers proved to be invaluable in coordinating the activities of both fire and riot control.

The Corps of Engineers has various capabilities in disaster operations. These are outlined in publication ER-500-1-1, Emergency Employment of Army and Other Resources, which discusses advance preparation, flood reconnaissance, disaster reporting, flood fighting and rescue work, rehabilitation of flood control, works damaged or destroyed by flood, and emergency protection and rehabilitation of federally authorized shore protection damaged or destroyed by coastal storms. Such work results from the statutory authority of the Corps of Engineers.(6)

Disaster activities of the Corps of Engineers involve conducting surveys of disaster areas, furnishing recommendations and reports, performing specific assignments of an emergency and recovery nature covering protective work, clearance of debris and wreckage, emergency work for the protection of life and property, permanent repair restoration, reconstruction or replacement of public facilities, advance preparation during and after natural disasters, other than floods and coastal storms.(7)

Naval forces render assistance to civil authorities when established civilian agencies are overwhelmed during a domestic emergency. Such assistance may involve the saving of human life and property; the furnishing of relief

Aerial surveillance and vertical evacuation can be provided by helicopters. Here, a U.S. Air Force helicopter hovers during a disaster drill. (Photo courtesy of U.S. Air Force.)

in the form of food, shelter, and clothing; and the preservation of law and order in a specified area. "Assisting the public in time of need has long been traditional with the United States Navy. In time of hurricane, tornado, fire, earthquake, flood, and other disaster, the Navy has been on hand to lend assistance with men, materials, and supplies. Naval officers may well take pride in serving with a Navy which has earned the gratitude of countless disaster victims both in this country and in foreign lands."(8) When a series of devastating tornadoes hit the southern Ohio Valley in the early 1970s, it was the Marine Corps Reserve that responded in force during a coinciding drill weekend to help alleviate the loss and suffering of the tornado victims.

Government Employees

One resource that can be mobilized by cities, counties, and other large governmental jurisdictions is the group of employees who are normally at work and who quite likely would be at their work stations when a disaster would strike. Christopher Burton of the Los Angeles Fire Department documented how the City of Los Angeles has a plan in place that will mobilize city employees by making each department general manager responsible for his or her employees during an emergency. Included are employees in the 51 floors of the five Civic Center buildings, outlying office buildings, and maintenance yards. The fire department took a leadership role and showed other city agencies how to employ the same principles used on the fireground—unity of command, span of control, and modular management.

Now, city employees respond to emergencies in their own buildings or facilities and operate under a building emergency coordinator.(9)

Transportation

In a disaster that require deployment of all the fire department's apparatus, plus helicopters, patrol boats, buses, and passenger vehicles, it may be necessary to call upon outside resources, such as the civil defense organization's transportation division, to provide additional equipment to set up motor pools, convert trucks into ambulances, convert tank trucks into water trucks, set up fuel depots and keep them supplied.

Property and Procurement

If the fire department needs additional bulldozers, cranes, skip loaders, back hoes, towing vehicles, and tank trucks the property and procurement division of the civil defense organization could be requested to obtain them from contractors and industry, and their use could be coordinated with that of fire department equipment. Equipment operators may be needed also, with provision made for their relief after a period of service.

Utilities Control

In the event of a widespread disaster causing the toppling of high-voltage power lines, or rupturing of gas mains, it may be necessary for the safety of fire fighters and rescue teams to call upon the utility officials to obtain the services of public utility personnel to operate switches and valves for de-energizing and controlling such hazards. The utility division of the civil defense office can be helpful in making requests. Direct requests normally made to these utility companies by the fire dispatcher for assistance in day-to-day emergencies may not get the attention that a request coordinated with the civil defense agency would receive, once that organization's control center is activated.

In many metropolitan areas the electric and natural gas suppliers are members of the same utility, creating a "Gas & Electric Company." If a disaster such as a tornado or hurricane causes widespread power outages due to downed power lines it may occasionally be a serious life-threatening electrical emergency. Due to jammed or nonexistent phone lines into electrical trouble centers, there may be no feasible way to summon an emergency electrical crew. In these cases if a gas trouble crew can be summoned, and this will often be much easier to do, the gas crew will often have an emergency radio link with the electrical crews and can then summon an emergency electrical crew.

Traffic Control

Fire and rescue operations are greatly impeded in the absence of coordinated traffic control on the part of local law enforcement agencies, state highway patrol, air traffic control (where fire department operations involve

Inclement weather creates a strain on operating forces, necessitating abundant manpower reserves. (Photo courtesy of Ron Buckingham.)

helicopters, bird-dog planes, and air tankers), and harbor traffic control (where fire boats are summoned to handle pier and ship fires, assist in oil spill control, or dive for persons trapped in sunken vehicles). In major disasters in which the civil defense organization is activated, usually a police division can be called to assist. In cities having large ports, there may be a harbor division that can coordinate problems in that sector.

Health and Sanitation

Although it is sometimes possible to set up peripheral command posts near a public building or park where sanitation facilities are available, if this is not feasible, emergency toilet facilities can be rented or constructed. If the control center of the civil defense organization is activated, its health and sanitation division can assist in securing these facilities.

Food and Shelter

There could be situations where it may not be possible to obtain food supplies or send relieved members of the fire department back to their quarters, which may have collapsed in an earthquake, become contaminated from radioactive fallout, or been seized by hostile rioters. Under such conditions it may be desirable to call the Red Cross, Salvation Army, and other agencies to furnish food and shelter for the members affected. The Housing and Shelter Division of the civil defense agency can be helpful.

Special Services

In a major disaster, such as a large earthquake, it may be necessary to call the civil defense organization for assistance in getting the streets sufficiently cleared of debris to enable fire and rescue apparatus to gain access to fires in hospitals, nursing homes, and occupied schools, where helicopter response would not be adequate to cope with the life hazard involved. The public works division can be called upon to assist in this work, and in extricating the injured and the dead. Services of personnel trained in first aid, emergency medical technicians, paramedics, nurses, doctors, ambulance drivers, and coroner's deputies might also be needed.

In a nuclear incident, the services of personnel trained in monitoring radiation might be needed to supplement work by fire fighters trained in such monitoring. Their efforts would need to be coordinated with those of trained personnel in monitoring centers in police stations, airports, hospitals, and field offices of federal agencies.

In a criminal disturbance of widespread proportion it might be necessary to request the National Guard to supplement efforts of local law enforcement agencies to provide physical protection to fire fighters in their quarters and while responding to and operating at emergencies.

Personnel Augmentation

The problem of augmenting your own fire department personnel at the scene of a disaster is not simple. You may need to supplement on-duty forces by recalling off-duty members, or at least one of your off-duty platoons, volunteers, trained auxiliaries, mutual aid assistance, or specialists. You may have to call outside assistance.

It may not be necessary or advisable to recall all off-duty members at one time. In larger fire departments the need may require notification of only off-duty members of part of one platoon. If that does not provide enough manpower, then all members of the platoon might be recalled. To relieve the dispatching staff of the task of making such notification, the file of members' telephone numbers could be kept at the fire prevention bureau. Its members could use automatic dialers to make calls, and public radio and television announcements could supplement the telephone calls.

In smaller fire departments, where members are furnished with radio-activated alerters, some or all of the off-duty members and registered volunteers can be notified to report for duty.

Activities of specialist personnel supplied by the federal government for your assistance have to be coordinated. The city's civil service department usually can provide assistance. For example, in a disaster involving dangerous chemicals, or ruptured liquefied petroleum gas lines, the city or

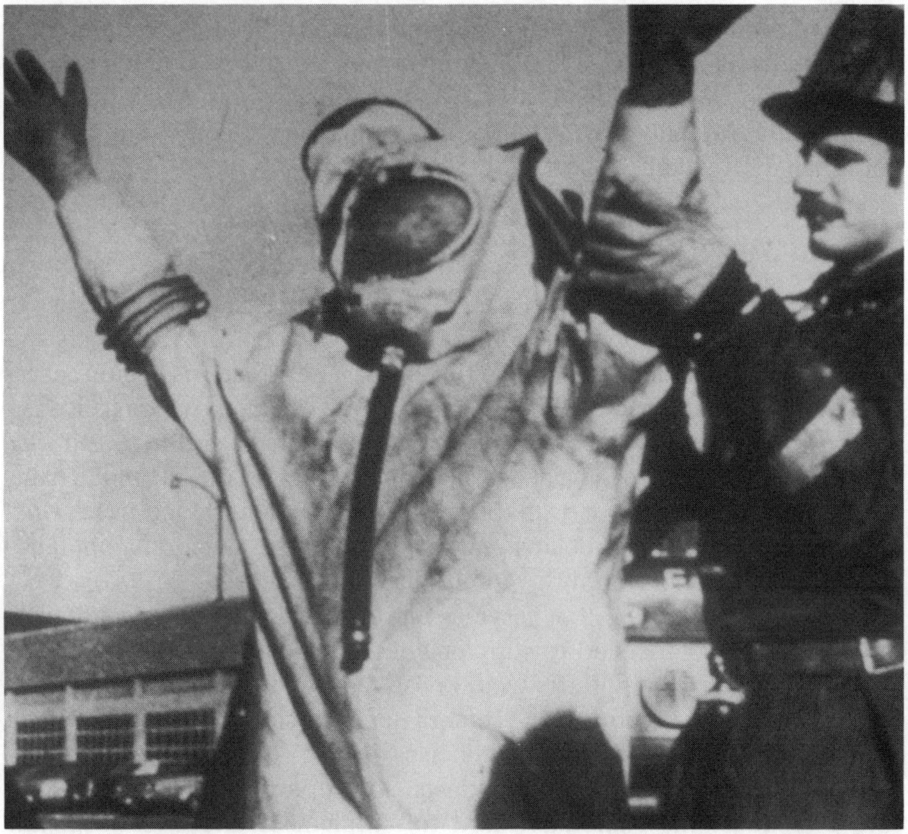

Support personnel must be available to assist with entry suits and new specialized equipment used by the fire service. (Photo courtesy of the National Fire Academy.)

county personnel division may be able to obtain the services of chemical engineers and other petroleum specialists to help supply and evaluate the information needed to handle the emergency. They can help coordinate the information that the fire department obtains from its own reference materials and from CHEMTREC.(10)

MUTUAL AID IN DISASTER CONTROL

When resources within a fire department are insufficient to handle a disaster or other major emergency, the incident commander turns to Mutual Aid. In the broadest sense this is the concept of one fire department being assisted by other fire departments during a time of particular need.

Many different arrangements currently exist throughout the country to fulfill this essential concept of supplementing a community's emergency response capability. The arrangements may vary from verbal understandings to formal signed agreements. Participation may involve only neighboring communities, may be regional, or may be statewide.

The concept has been refined in many adjoining communities to include automatic first-alarm response into each other's jurisdiction based upon established criteria.

Long distance mutual aid. Here, all-wheel-drive pumpers are being loaded aboard a C5A transport for delivery to assist Buffalo during the blizzard of 1977. (Photos courtesy of Ron Buckingham.)

History of Mutual Aid

The concept of mutual aid, that is, one department coming to the aid of another, is as old as the fire service. On an informal basis, mutual aid has prevailed for many years. There has been a tendency in recent years, however, to establish it on a more formal and organized basis.

There is good reasoning behind this move. Many mutual aid operations in the past have proven to be almost as disastrous as the original emergency. Companies have arrived uninvited and have undertaken tasks that were not assigned to them. The chief officers did not know who was present or what they were doing. Orders were often disregarded or countermanded and units worked at cross-purposes. In recent years, progressive thinking has prevailed. The result has been better planning, organization, and training in mutual aid, and a dedicated commitment through a formal agreement.

The City of Freehold, New Jersey, for example, used mutual aid from a number of surrounding communities to successfully limit a blaze to one floor of a major condominium/commercial warehouse in December 1989. This was just one of many successes recorded by the "Western Monmouth Mutual Aid Association." This association had conducted extensive prefire drills on the same "A&M Karaghuesian Rug Company" site. That spelled success when the second more extensive fire erupted in the complex on April 8, 1990. Here the stubborn fire was held to a single floor in part of the three-block-long facility. In all, 41 different fire agencies in three counties had been involved either in direct mutual aid or in move-up operations.(11) Many small fire departments combined can equal the resources of a large fire department and in fact are necessary for the protection of major occupancies in small jurisdictions.

Meaning of Mutual Aid

"Mutual" by definition denotes a sharing between partners. Thus, mutual aid is not a one-way street but, according to agreement, a two-way street with aid flowing in either direction as the particular need indicates. Of course, some communities have much more to share than others, but any mutual aid agreement must involve a degree of reciprocity. **Mutual aid, then, is an agreement in which two or more parties agree to furnish resources or facilities to render service to each and every other party of the agreement, based on established criteria.**

Legal Considerations with Mutual Aid

In some states, the emergency organization may be identified in statutory form, with an established chain of command, communications net, and provision for mutual aid agreements between political entities. As pointed out in *Fire Service and the Law,*(12) the general rule is that a fire department may not operate outside its jurisdiction unless expressly authorized to do so by statute. Several cases have held that although it is proper to obtain fire protection from outside aid, a city has no authority to enter into a contract to help another jurisdiction unless a law specifically allows. In many states, mutual aid contracts are authorized that may provide for reimbursing the assisting agency by the assisted agency for any judgments that may have to be paid as the result of negligent actions of fire fighters, loss of equipment, and other mishaps. The legal effects upon the rights of fire fighters relating to worker's compensation and pensions, when operating outside their own jurisdictions in the absence of a valid aid agreement can be detrimental. It is not sufficient that such an agreement be entered into solely by the heads of the emergency forces; it requires confirmation by the governing bodies of the parties participating in the contract.

Where a state law specifically provides that a government agency may go outside its jurisdiction to help another agency extinguish a fire, and provides that a charge may be made for rendering such assistance, then even in the absence of a written contract, such billing action would be legal.

Some cities and fire districts help each other without an agreement if it appears that the fire is likely to spread into their jurisdictions, and such help is always permissible, whether there is a law on the subject or not. However, where the arrangement is not mutual, but only involves outside aid to be given by one of the parties, a contract is required under a law authorizing this arrangement. Such a contract should provide for the contingent liability that could result in furnishing fire protection to another entity on a fee basis. Even if there is no fee involved, sending fire apparatus out of the city to fight a fire could remove the city's immunity from liability and deprive the fire fighters of rights and privileges they would otherwise have in event of injuries sustained in a collision or at the fire.

See Appendix C of this book for an example of a Mutual Aid Pact between a municipality and a university, and Appendix D for an example of a countywide system (Santa Cruz County).

Types of Mutual Aid

Agreements may involve any of a variety of mutual aid arrangements. Three prevalent concepts are

- automatic aid with neighboring communities and industry
- mutual aid with neighboring communities and industry
- mutual aid on a regional or statewide basis.

AUTOMATIC AID WITH NEIGHBORING ADJACENT COMMUNITIES AND INDUSTRY

Existing jurisdictional boundaries often pose a quandary to those charged with planning fire protection for cities, counties, and other political subdivisions. Oftentimes it is just not cost effective or practical to build fire stations in remote areas or geographic fingers that protrude through another jurisdiction. Frequently, another agency could respond to those problem areas more quickly. In addition, smaller jurisdictions with major hazards, such as chemical processing plants, may wish to expand their initial response to first alarms in such facilities. Such problems can be resolved by negotiating an agreement of automatic aid.

Figure 4-1*. Automatic Aid *summary for the Los Angeles City Fire Department.

PROVIDED TO LOS ANGELES CITY—BY:		PROVIDED BY LOS ANGELES CITY—TO:	
Los Angeles County			
Watts/Willowbrook	= 1 Engine	Baldwin Hills	= 1 Engine
Strip	= 1 Engine (fire incidents) 1 P/M Squad [EMS incidents] (Squad provided under oral agreement)	West Hollywood	= 1 Engine
Median	= 1 Engine	Universal City	= 1 Rescue Ambulance (provided under oral agreement)
Initial Action Zones	= Full Brush or Structure Assignments	Initial Action Zones	= Full Brush or Structure Assignment
Incidents Exceeding L.A. City Capabilities	= Resources upon = request as per M. of O	Unincorporated Areas	= Resources upon request as per M of O
		Fire Protection Districts	= Resources upon requests as per M or O
		Incidents Exceeding County Capabilities	= Resources upon requests as per M or O
Beverly Hills			
Fire and EMS incidents in lower Franklin Canyon	= 1 Engine	Incidents Exceeding Capabilites of BHFD	5 Cos. not more than 2 trucks and 1 B/C/
Santa Monica City			
Santa Monica Canyon	= 1 Engine or 1 P/M Engine	To Fill 1 Fire Station vacatee SMFD dispatched into L.A.	= 1 Engine
Venice	= 1 Engine	Incidents Exceeding SMFD Capabilities	= 5 Cos. not more than 2 trucks and 1 B/C
Bristol Ave. (Brentwood)	= 1 Engine		
101 Ocean Avenue	= 1 Task Force and 1 B/C		
Area south of Sunset Adelaide to Sunset at PCH	= 1 P/M Engine		
Burbank City			
Equestrian Center EMS	= 1 Engine + 1 P/M Squad capable of transportation	Working High Rise Fire	= 5 Cos. not more than 2 trucks, 1 B/C, 1H?
Fire	= up to 1 Task Force, 1 Squad, and 1 B/C	Working Brush Fire	= 5 Cos. not more than 2 trucks, 1 B/C, 1H?
		Unusual Medical Transports	= 1 Helicopter (helicopter support upon request of L.A. B/C or pilot

Long Beach Naval Station

Fire Incidents (Dispatch upon request of L.A. City I.C.)		Fire Incidents	= 1 Task Force, 1 Engine, 1 B/C
Ferry St. to Dock St. to Henry Ford	= 1 Engine		
Fire in L.A. Harbor	= 1 2,000 GPM Tug with 8-member crew		

Ventura County

Cash Compensation based upon Established and Formula	Fire Incidents in Bell Canyon	= 1 Task Force 1 B/C
Ambulance	EMS Incidents	= 1 Engine (No Rescue Dispatch)

United States Forest Service

Brush Incidents Involving L.A. River Upstream Control Project I.A.Z. Assignmnt	= Air Tankers + other special equipment upon request	Brush Incidents Involving L.A. River Upstream Control Project I.A.Z.	= Brush

City of San Fernando

Cash Compensation	Full Fire and EMS Service

City of San Segundo

Area 1 and	Imperial Blvd on the north 116th St. on the south Aviation Blvd on the west Inglewood Blvd. on the east	up to 1 Task Force	Incidents Exceeding El Segundo Capabilities	= 5 Cos. Not More Than 2 Trucks 1 B/C
Area 2	All portions of the City of L.A. south of Imperial Blvd. bordering the ocean	up to 1 Task Force &/or 1 Paramedic Squad Capable of Transport		

12/84 0730F

The career Middletown, Ohio, Fire Department (LEFT) and the LeSourdsville Volunteer Fire Department (RIGHT) participate together in the city of Middletown's Bicentennial Parade in July 1991. (Photos courtesy of Captain James Hall.)

Automatic aid agreements vary according to specific situation. One arrangement might stipulate that the closest station is automatically included in the first-alarm dispatch to certain areas even though it is in an adjacent jurisdiction. Many response patterns can be utilized to achieve optimum protection in those areas that otherwise might be deficient. Again, this is a two-way street with responsibilities and benefits to both departments and the end result being better fire protection to the citizens on both sides of the boundary.

Figure 4-1 illustrates the automatic aid arrangements between the City of Los Angeles and adjoining jurisdictions. Notice that the exchanges involved are not exactly the same, but in the long run, the value received is viewed as comparable.

Each arrangement is formalized with a letter of agreement and a contract outlining the specific understandings. They are then signed by the appropriate officials, establishing legal authority.

Another variation of automatic aid occurs when a group of adjacent smaller cities, each with a separate fire department, enters into agreements whereby the resources of all the departments are utilized for common response within the combined areas. The concept is much like separate divisions in a larger department, except that the political integrity of each department is maintained. The agreements may provide for a common dispatch center, common training facility, and standardized training procedures.

Political autonomy is maintained, yet total resource availability on a routine basis is increased. The concept is cost effective and provides vastly improved fire protection for each of the participating cities.

Many examples of automatic response between adjacent jurisdictions are in evidence throughout the North American continent. As fire departments become more synchronized with one another due to the fact that they are using common incident command procedures and are using standardized tactics and strategies as taught by such national centers as the National Fire Academy and *American Heat* videos, it is logical that more fire departments will work harmoniously together. One midwest example involves the cities of Dayton and Kettering, Ohio, which share a rather lengthy common border. For years these two fire departments stopped at each other's border on first alarm responses and summoned additional help from deep within their own territories. That changed when Fire Chief James Frey from Kettering and Fire Chief Glen Alexander from Dayton decided to form an AMAR agreement in 1984.(13) This extension of their already existing Montgomery County Mutual Aid pact established a procedure whereby in border areas the closest fire companies from each of the two jurisdictions would respond regardless of the territory in which the fire originated. As a result many border line fires witness the swift arrival of both Dayton and Kettering units and both sides realize a net positive gain. In Appendix E you will find a copy of the Montgomery Mutual Aid Pact and the specific AMAR agreement between the cities of Dayton and Kettering, Ohio.

Mutual aid agreements may be limited to response arrangements on the periphery of the jurisdiction, such as called for by the memorandum of understanding that the City of Los Angeles had with the County of Los Angeles with respect to mountain regions along their borders. It was never felt necessary by the chiefs of these departments to enter into a formal contract, for it was understood that neither department would ever bill the other for its services or for equipment lost or damaged while operating together, or for reimbursement of medical costs or pensions arising out of injuries sustained at such joint operations. The memorandum of understanding, legal under California law, provided for predetermined response in this "twilight zone" without necessity of obtaining special approval of higher authority in each instance. Requests for response of equipment to more remote areas within the other's jurisdiction required such approval.

In an editorial of the *California Firemen*,(14) Dwight Glass, then president of the California State Firemen's Association, commented upon the fact that many areas in California have what is called "automatic aid," such as that described in the preceding paragraph, and that it is also common among smaller communities elsewhere. He pointed out that Shasta County fire departments have long been hampered by calls from outside their districts from people who do not pay for the protection. They now have solved the problem through automatic aid to these areas, financed by the county taxing the areas not covered by the departments. "Aside from the obvious advantages, such as reduced insurance costs to county residents, the plan also presents possible or partial solutions to the FLSA (Fair Labor Standards Act) requirements and the 911 system. Shasta County was also able to provide revenue-sharing funds with all local departments receiving some benefits. The added cost of mutual aid incurred by the FLSA amendments may be circumvented due to having these areas merely added to your response district. The county pays a fee for calls, which also helps offset those costs involved in providing such fire protection.

"Emergency reporting under P.S.A.P. (Public Safety Answering Point) 911 requires that your department respond if called. The location of the call may be very remote from your department and in an area which pays for no services at all. Since the County of Shasta will be providing funds for answering these calls and almost every area where a call might originate is paying its share, departments as well as citizens will benefit."

Based upon the experience obtained while operating at many disasters, it has been found that mutual aid arrangements will only be successful if there are definite provisions for the assisting department to retain command over its own forces while in another jurisdiction, although deployment of forces will usually be made in accordance with the overall plan of the official directing the disaster control operation. It is also best that formulated plans allow the emergency departments to function within the scope of their normal organizational and command procedures.

EXPANDED USE OF OUTSIDE RESOURCES

Both mutual aid and automatic aid are activated upon specific or predesignated requests and represent a cooperative agreement between the agencies. They are designed to provide additional firefighting resources in specified situations, usually those beyond the control of the services, personnel, equipment, and facilities of a given jurisdiction. In addition, other specific situations, along with the anticipated Mutual Aid Resources, often are described in the written agreement. In the past many mutual aid agreements were strictly verbal. In recent years, however, the trend has been toward formalized written agreements.

MUTUAL AID WITH NEIGHBORING COMMUNITIES AND INDUSTRY

The authors of a mutual aid study in the Los Angeles area had this comment: "Mutual aid in the Los Angeles area, as in all of California, is a child of necessity, sound in principle and exceptionally well adapted to the local problem. Both its strengths and weaknesses lie in the vast number of fire services involved in the program, but the organization is being perfected by continual practice. . . . Adequate protection of any jurisdiction would not be possible without mutual aid."(15) They also concluded that the success of the program is due in a large degree to the fact that assistance is rendered automatically and not subsequently billed for by the assisting agencies. Finding that mutual aid has achieved results beyond expectations under exceedingly complicated conditions, they felt that the mechanics of the program were worthy of consideration for adaptation to mutual aid programs elsewhere and in national planning.

MUTUAL AID WITH INDUSTRY

When we think of mutual aid, we usually think of political jurisdictions such as cities, towns, townships, or other such political subdivisions. In many locations, however, this concept has been expanded to include various industrial brigades. On October 23, 1989, when a massive explosion rocked a Phillip's 66 Plant in Pasadena, Texas, a mutual aid system brought immediate first responding units from industrial fire brigades. As soon as the alarm was sounded Phillip's notified CIMA (The Channel Industries Mutual-Aid Organization) which summoned a general alarm mutual aid complement. CIMA is composed of 96 industrial members located on the Houston ship channel, and it also includes six cities and six governmental agencies. Its prearranged mutual aid plan is triggered the minute an alarm is sounded. A common communications network is manned on a 24-hour basis with teams at various brigades assigned specified emergency duties. When a stricken company or facility requests assistance it retains authority over its own emergency while CIMA puts out the alarm for mutual aid. At the Phillip's explosion, industrial complexes from up and down the entire ship channel responded. Although only 15 pieces of fire apparatus were used, 65 were available. A stream of more than 50 ambulances shuttled the 113 injured employees (most of whom were suffering from broken bones, cuts, and bruises from flying debris) to area hospitals.(16)

MUTUAL AID WITHIN THE REGION OR THE STATE

In many locations it may be necessary at times to supplement firefighting resources with additional resources that far exceed those available in the neighboring communities. In such cases, it is beneficial to have an established mutual aid relationship on a statewide or regional basis.

Let us take California's plan, for example: "Although Mutual Aid plans

In disasters, damaged highways can often hinder the arrival of mutual aid reinforcement and block movement of a fire department's own resources. (Photo courtesy of Ron Buckingham.)

and agreements have existed in California for many years, the present State Fire and Rescue Plan has combined the State Fire Disaster Plan and the State Rescue Plan, dating back to 1950. The original plans and all revisions were prepared and adopted after careful consideration by the Governor's Fire Advisory Committee.

"The Plan is actually very simple and uncomplicated, based on the concept of "Self-Help" and "Mutual Aid." The State of California and nearly all of the County and City governments are signatories to a Master Mutual Aid Agreement. Mutual aid extended under this agreement and the operational plans adopted pursuant thereto shall be available and furnished in all cases of "Local Emergency," "State of Emergency," and "State of War Emergency" as defined in the Emergency Services Act. However, the "Act" also provides the basis for exchange of mutual aid under any and all other circumstances."(17)

The plan provides for the following:

1. Mobilization, organization, and operation of the state government's fire fighting resources.
2. Uniform procedures for the most effective mobilization, dispatch, and coordination of *all* fire fighting resources on a local area, regional or statewide basis.
3. Provisions for recruiting and training auxiliary or volunteer personnel to augment full-time forces.
4. Inventory and cataloging of all fire and rescue resources in the state from local and private levels to federal level.
5. Receiving and disseminating emergency information, data, and directives affecting the fire and rescue services.
6. Coordination and implementation at the state level of government (by State Fire and Rescue Coordinator).

CALIFORNIA—THE TRENDSETTER

Throughout this chapter we have on numerous occasions cited examples from the west coast. This was not without good reason, as the State of California has been a clear leader in innovative response patterns among adjacent jurisdictions and has been a trendsetter in the development of mutual aid and mutual response agreements. We have seen already how current incident command procedures were developed from west coast initiatives. Because of the tremendous wildland fires that erupted suddenly, and threatened thousands of acres of residential and wildland area, fire departments in California realized at an early date how new concepts of mutual aid could be vital and necessary.

Many locations within the state are subjected to seasonal threats of uncontrollable wildland fires that may burn for days or even weeks. The following excerpt was taken from a news report on September 12, 1987, about California. "About 1,240 timber and brush fires had burned about 549,810 acres and more than 170 fires remained uncontained. Of nearly 13,000 people fighting the fires, three were killed and 99 injured.

"So far, firefighters have contained 203,350 acres since the lightning-started fires began August 28, and 346,000 acres remained uncontained. The largest fire had blackened 132,000 acres of the Klamath National Forest. Most of the 15,000 people who had been evacuated have returned home. Nearly 100 structures were destroyed, including 39 homes."(18)

A fire situation such as that demonstrates the value of regional or statewide mutual aid. A similar situation several years ago in southern California brought about the then new concept in regional mutual aid resource availability known as FIRESCOPE.

FIRESCOPE

In the second chapter you were introduced to FIRESCOPE, an acronym created from FIrefighting RESources of California Organized for Potential Emergencies. (Originally FIrefighting REsources of Southern California.)

The Firescope project was chartered by the United States Congress in 1972. It was developed as the result of the tragic 1970 wildland fire season in southern California in which more than 500,000 acres burned, 700 homes were destroyed, and 16 lives were lost in a 13-day period. The devastating fire siege clearly pointed out some major problems relating to mutual aid, incident management, communications, and multiagency coordination. These problems had long been recognized by the involved agencies, but as is so often the case, it took a disaster to initiate change.

MultiAgency Coordination Across Jurisdictions

The congressional action charged the U.S. Forest Service (USFS) with the mission of assisting the southern California fire services in making significant improvements in multiagency coordination on multijurisdictional emergencies.

The initial funding (1972–1977) came from the Forest Service Research Funds with five million dollars being appropriated for system design. The Forest Service then established a program and management for the project. The program concept included direct participation and guidance from the other agencies of the fire service within the particular area of concern. A decision process was set in place, utilizing representatives of seven key fire service agencies who were then referred to as Partner Agencies. The partner agencies provided experienced personnel to staff committees for lesson plan development, decision-making, and interagency training.

The original concern had focused on major wildland fire problems so it is not coincidental that all of the partner agencies had wildland fire responsibilities. Those agencies were the U.S. Forest Service, California Department of Forestry and Fire Protection, Los Angeles County Fire Department, Los Angeles City Fire Department, Santa Barbara County Fire Department, Ventura County Fire Department, and the California Office of Emergency Services Fire and Rescue Division.

After the establishment of the FIRESCOPE project, an initial problem analysis identified five major components to be pursued in meeting the mission of the program.

- Develop standardized terminology and incident organization structure
- Provide for multiagency communications
- Develop improved methods for status-keeping
- Develop improved methods of forecasting fire behavior
- Provide multiagency training in the systems developed under FIRESCOPE

Two specific systems were developed in the effort to achieve those components.

- Incident Command System (ICS): A system for on-site management of all emergencies (discussed in detail in the previous chapter)
- Multiagency Coordination System (MACS): That element of FIRESCOPE which coordinates multiagency resources, commitments, and availability(19)

The implementation of the system began in 1977 when the Forest Service with the partner agencies began to train the fire agencies of southern California on the concepts and application of ICS. Then, in 1978, the Operations Coordination Center (OCC) was opened in Riverside, California. The OCC was the operational base for the MACS for all of the FIRESCOPE region. By 1982 the partner agencies had fully implemented the ICS and were using the standardized lesson plans and training materials.

By the end of 1982 FIRESCOPE had many major achievements. In addition to the ICS and the MACS/OCC mentioned above, the system included the following:

- Automated weather stations—a system of remote weather stations for electronically transmitted weather data
- Common mapping system—orthophoto maps, urban-interface use
- Communication systems management
- Fire behavior modeling—development of computer models for forecasting fire behavior
- Infrared monitoring—monitoring fire perimeters for accurate mapping, intelligence by aircraft
- Interagency communications—better multichannel communications, radio caches, and communications vehicles, and interagency agreement on frequency assignments.

All-Risk Application

In 1984, the partner agencies realized that an optimal system should be operational statewide with more emphasis on all-risk application.

The Office of Emergency Services Fire and Rescue Division (OES) coordinated a major effort to spread FIRESCOPE throughout California.

In 1987, another disastrous fire season gave the fire services of the northern part of California an excellent opportunity to utilize and evaluate the ICS and MACS element over a period of several weeks. The ICS helped to standardize all operations, minimize costs, reduce injuries, and provide allocation of qualified personnel and resources on many different fires. The MACS element provided mobilization and coordination in moving over 20,000 fire fighters from all parts of the nation to meet the varied operational needs. The systems were effective.

The components of FIRESCOPE had again proven their worth and this lead to statewide acceptance of those operational concepts. Another major plateau for FIRESCOPE came about in November 1987, when the operations teams met for the first time using a new definition for the same acronym FIRESCOPE, namely FIrefighting RESources of California (formerly southern California) Organized for Potential Emergencies. FIRESCOPE has expanded to a statewide operation.

The systems that have been developed through FIRESCOPE have been tested and proven in multiagency incidents over and over again. The Incident Command System (ICS) is being used daily by many departments throughout the nation and it is the model command system being taught at the National Fire Academy. The concepts and application of ICS were presented in Chapter 3.

MACS Modes of Operation

The MACS function occurs at the OCC and is staffed by participating agencies. There are four operational modes established within MACS. Each mode is based on definition of the current hazard threat within a region.

Modes 1 and 2 reflect a noncritical regional situation. In these modes there are no major multiple incidents that would require extended use of multiagency resources. The basic difference between Modes 1 and 2 is seasonal.

Mode 3 reflects a serious situation or the potential for a serious situation that involves the use of multiagency resources. An example might be severe weather conditions or a tsunami warning. A major commitment of fire suppression resources to out-of-region incidents would also initiate Mode 3.

Mode 4 is an all-out regional effort where resource use priorities require multiagency coordination. In Mode 4, a regional GHQ will be established at the OCC. Agency representatives to the MACS/OCC/GHQ under Mode 4 should be fully authorized to speak and commit for their agency. The agencies that will be represented within the MACS/OCC/GHQ include all the aforementioned partner agencies plus the Orange County Fire Department, a later addition to the partner agencies. If, because of the character or intensity of a particular situation, it is felt that the inclusion of some other agencies on the MACS/GHQ would improve the operation, those agencies may be invited to participate in Modes 3 and 4.

Functions of the MACS/GHQ in Mode 4 are as follows.

- Evaluate new incidents
- Assign the following priority to incidents

Life-threatening situations
Real property threatened
High damage potential
Incident complexity

- ○ Ensure that agency resource situation status is current
- ○ Determine specific agency resources requirements
- ○ Determine agency resources availability (available for out-of-jurisdiction assignment at this time)
- ○ Determine need and designate regional mobilization centers
- ○ Allocate resources to incidents based on priorities
- ○ Anticipate future agency/regional resource needs
- ○ Communicate GHQ decisions back to agencies/incidents
- ○ Review policies/agreements for regional resource allocations
- ○ Review need for other agencies involvement in GHQ
- ○ Provide necessary liaison with out-of-region facilities and agencies as appropriate
- ○ Critique and recommend improvements to GHQ operation

SPECIFIC ACTIONS REQUIRED—MODE 3 ACTIONS

1. OCC must make required notifications to all agencies of Mode 3 declaration.
2. All agencies must be prepared to respond to a conference call initiated by MACS Support Services Manager to be held within one hour of declaration of Mode 3.
3. All agencies must provide MACS Information Management System with a committed and available resource report within one hour of Mode 3.

SPECIFIC ACTIONS REQUIRED—MODE 4 ACTIONS

1. All actions within Mode 3
2. All agencies must provide representatives to the OCC/MACS/GHQ.

Reports into the MACS Information Management System

MACS Information Management System deals with both resource commitments and resource availability during major incidents.

There are direct hot lines to all partner agencies from the OCC in addition to intercom circuits and computer terminals.

APPLICATION OF THE SYSTEM

With the system in a state of full implementation (Mode 4), the following scenario depicts a typical operation of MACS:

- ○ All on-going regional high-priority incidents are monitored.

- All available resources are identified and categorized.
- Resources are allocated on incident priority basis upon request from the in-charge jurisdiction.
- Upon notification, resources are dispatched by the agencies directly to the incident or to a mobilization center for that incident.
- Responsibility for relief of fire fighting crews lies with the participating agency. At a recent FIRESCOPE incident, Los Angeles City relieved their crews via helicopter thus saving much time involved in travel.

The MACS component of FIRESCOPE represents a concept for coordinating the availability and allocation of emergency resources on a regional, areawide, or statewide basis.

It emphasized efficiency in distribution of available resources based on incident priority determination. The partner agency representation on the MACS/OCC/GHQ assures mature judgment in making those decisions.

Although initially directed toward the wildland fire problem in southern California, the MACS concept of FIRESCOPE can be used for coordination of many emergency services in any disastrous situation that overtaxes local response capability. For fires, floods, earthquakes, etc., the system has been tested and it works.(20)

PROCEDURES—MUTUAL AID REQUESTS

Fire and rescue mutual aid procedures are planned and conducted in accordance with an accepted standard operating procedure to insure the most effective results. The following procedure has been tried and proven to be adaptable to any community where emergencies are of such magnitude as to require combined efforts of personnel and equipment from two or more jurisdictions. It is readily and economically established and productive of maximum benefits to all participants and should be reflected in operational plans produced by local people on the ground at local and regional levels.

1. Local. All requests for outside assistance shall be made by the appropriate fire official or his duly authorized representative, and shall be made to the Operational Area Fire and Rescue Dispatch Center for action.

2. Operational Area. The operational area fire and rescue coordinator or his representative, upon receipt of a call for assistance, evaluates the request, and determines where such assistance should come from according to prearranged plans. He then issues the necessary instructions for appropriate action.

3. Region. When an area fire and rescue coordinator needs outside assistance, requests are made to the regional fire and rescue coordinator or his representative at the Regional Dispatch Center, who follows the same system as an area fire and rescue coordinator. If the emergency is of such magnitude that two or more regions are involved, requests from the regional fire and rescue coordinator shall be directed to the state fire and rescue coordinator for appropriate action. In all cases, the dispatch center at the next higher level of jurisdiction will be advised immediately.

MUTUAL AID—DRAWBACKS

Mutual aid is not without problems, especially in matters of communications and training, but when a disaster strikes, it is invaluable. It may not always be available, for afflicted communities may be so occupied with their own problems that little or no aid can be extended at the time when it is most needed. This was the situation in Watsonville and other small communities around Santa Cruz during the 1989 earthquake.(21) Detroit was more fortunate during the time of the riots in 1967. Detroit had never deemed it necessary to enter into any mutual aid agreements with the surrounding communities, for no one had ever visualized a time when a fire department as large as Detroit's would have to call for help. When the riots broke out, however, the outlying fire departments, including the one from Windsor, Canada, voluntarily responded and proved to be of inestimable value in preventing greater destruction, possibly total destruction, of the vulnerable areas of the city. As a result of this experience, a plan was quickly drawn up by the Detroit Fire Department.

Concerning mutual aid and major emergencies, the NFPA states that "It should be recognized that mutual aid tends to have deficiencies. The parties to the plan may only give help to the extent that they feel they can do so without seriously reducing the local protection, although the better plans provide effective coverage for all districts dispatching apparatus. Each local department may have its own operating methods, and types of equipment, so that maximum coordination may be hampered. This is far less serious than formerly, due to state fire training programs, large-scale exercises, and fire officer command schools and conferences."(22)

COORDINATION AT LARGE-SCALE EMERGENCIES

There are three places where it is extremely important that activities be coordinated in disaster control: at the scene of the emergency, at the emergency operations center of the fire department, and at the control center of the civil defense organization.

At an emergency that does not require response of aid from other fire departments or agencies, the fire chief's job largely involves direction. Coordination between fire forces, police forces for traffic control, utility company representatives, and news media can be handled by fire department personnel assigned this task in the major emergency plan. If multiple company response expands into a large-scale operation that requires mutual aid, then the fire department's plan should provide for designation of a liaison officer to coordinate activities of outside companies and assist forces from other agencies.

The next place where coordination is highly essential is in the emergency operations center of the fire department. Forces dispatched to the major emergency have to be monitored, augmented, relieved, fed, and rested. Other greater alarm fires may occur at the same time and must be given the proper attention. During the Bel Air conflagration in Los Angeles, while practically all of the city's fire fighters were engaged for four days in trying to prevent spread of the wind-swept brush fires throughout the entire region, sixteen greater alarm fires occurred elsewhere in the city and had to be suppressed.

Obviously, the regular dispatching staff has to be augmented at the department's emergency operations center. Under such conditions, the plan should provide for representatives of various bureaus in the department to report to the operations officer in the staff room set aside in the EOC for consultation with these members and representatives from other agencies.

In a really large disaster involving several communities, the command of the overall fire fighting efforts, under a plan promulgated by the state, might be the responsibility of an area fire coordinator, operating in the control center of the civil defense organization. The fire chief would still command the fire fighting within his own area, but if several fire departments are overwhelmed simultaneously, it would become necessary for the next higher echelon to allocate forces from uninvolved areas within the state to those jurisdictions in greatest need. There has to be a high degree of coordination between the city, county, fire district, and forestry fire services called upon to render assistance in such large-scale operations. The greatest problem probably relates to communications at the control center.

CONCLUSION

The effective control of disasters depends upon the mobilization and deployment of adequate resources. In this chapter we have seen how fire departments must somehow be able to augment their existing complement of manpower and machinery if they are to be able to cope with the worst-case scenarios that might befall their communities. A national trend toward standardization and harmony seems to permit expanded use of mutual aid and mutual response pacts for the future. It is becoming obvious to fire service leaders everywhere that as we fight fires and fend off disasters in the 1990s we must do our job from a regional perspective and consider our organizations, regardless of how large they are, to be parts of a much larger system of manpower and machinery that can be mobilized against disaster forces.

CHAPTER 4 REFERENCES

1. Carr, John N., and Leslie P. Omans. "Aircraft Incident Management Systems: A Team Approach" in *Fire Engineering*, Vol. 144, No. 3 (March 1991), p. 143.
2. *Disaster Control* (Ashore and Afloat), NAVPERS 10899-A (1964), p. 288.
3. *Disaster Control*, NAVPERS (1964).

4. *Firehouse*, June 1990, pp. 28–32.
5. *Disaster Planning Guide for Business and Industry*, Defense Civil Preparedness Agency, CPG-25, 1975, p. 39.
6. Primarily pursuant to Public Law 93-288, Disaster Relief Act of 1974. (42 U.S. C. 4401 et seq.) (84 Stat. 1744).
7. *Ibid.*
8. *Disaster Control* (Ashore and Afloat), NAVPERS 10899-A (1964) p. 43.
9. "Disaster Plan for City Employees" in *Fire Engineering*, Vol. 144, No. 3 (March 1991), pp. 127–130.
10. See Chapter 12 for a further discussion of CHEMTREC and chemical disasters.
11. Carter, Harry R. "On the Job—New Jersey" in *Firehouse*, Vol. 16, No. 1 (January 1991), p. 43.
12. Callahan, Timothy, and Charles W. Bahme. *Fire Service and the Law*, Boston, MA: NFPA, 1987, pp. 102–105.
13. AMAR—"Automatic Mutual Aid Response"—An agreement between the Ohio cities of Dayton and Kettering, enacted in 1984.
14. Glass, Dwight. "Mutual Aid?" in *California Firemen*, Vol. 21, No. 5 (May 1975), p. 3.
15. Storey, T.G., and C.L. Harrison. *California-Los Angeles Mutual Aid Study*, Project OCD-PS-64-229, U.S. Forest Service, Department of Agriculture, March 1964, p. 88.
16. "On the Job—Texas, Explosion Rocks Phillips 66 Plant in Pasadena; Mutual Aid System Activated" in *Firehouse*, Vol. 15, No. 3 (March 1990), pp. 59–61.
17. State of California, Office of Emergency Service Division. "History and Organization," R.G. Barrows, p. 2, 1972.
18. Petterson, Roger. "Fires at a Glance," *Daily News*, September 12, 1987, p. 12.
19. *Multi-Agency Coordination System (MACS) Procedures Guide* 410-1, April 1, 1988.
20. Federal Emergency Management Agency. *Command and Control of Fire Department Operations at Earthquakes and Other Catastrophic Disasters: Student Guide.* Emmitsburg, MD.: National Fire Academy, March 1986.
21. "San Francisco Earthquake," in *American Heat* video journal, St. Louis, MO, December 1989.
22. *Handbook of Fire Protection*, 14th ed., Boston, MA: NFPA, 1976, pp. 9–29.

CHAPTER 5

TRAINING AND PREPARATION

IN THE FIRST FOUR CHAPTERS you saw how man's efforts to cope with disaster situations are complex, sophisticated, and include many agencies besides the fire department. All persons involved in disaster planning must operate under the philosophy that since disasters will occur, they must prepare themselves adequately for the inevitable.

This preparation takes many forms of education, training, rehearsals, meetings and conferences, and a host of other techniques, all of which are necessary if persons involved will really be able to cope with the situation. In 1989 when a massive earthquake struck the San Francisco bay area, a genuine disaster occurred. Over 40 lives were lost; property damage was extensive. There was, however, warning of this occurrence, if perhaps not of its magnitude, and many agencies used prior training in the efforts to mitigate losses. Training is an essential ingredient of any disaster control program if the program is to be effective. For example, the Incident Command System, introduced in Chapter 3, is useless if a fire department has not been trained in its proper use and application.

Training made the difference on October 16, 1990, when the Denver Fire Department in Colorado received a call for a rollover tanker on Interstate 70. Initial response consisted of an engine, a truck, the hazmat response team, and a district chief. Shipping papers indicated that the product in the tanker was 7300 gallons of lethal unleaded gasoline. Two weeks prior to this incident, Denver Fire Department personnel had attended a training session on "hot tapping," a technique using pneumatically driven holes into tank compartments for off loading the product. At the scene, the Denver Fire Department used the assistance of Amoco technicians and successfully used hot tapping to off load the product, eliminating the danger. Although this incident caused the largest traffic jam in Denver's history, it was successfully concluded largely due to proper training.(1)

Safety Systems, Inc., of Jacksonville, Florida, conducts a one-day hazardous materials seminar for fire fighters in suburban New York. (Photo courtesy of Harry Loud.)

NEED FOR DISASTER CONTROL TRAINING

Few fire officers will ever respond to a sufficient number of disasters to gain the experience really needed for handling large scale emergencies. Most fire departments can train their members in mechanical skills and in the ability to direct the operations of engine, truck, salvage, rescue, and other single companies. Fire departments have an obligation to use their organization and facilities to provide the broader training necessary for coping with a major disaster.

The ability to plan, organize, command, and coordinate activities of all companies in a fire department, plus companies that might respond under a mutual aid plan, cannot be learned in typical exercises conducted at the drill tower. It is not enough that you learn how to work with your own resources. You must learn to work with other agencies—not just other fire departments—but also the police, sheriff, health, public works, and other local departments, as well as the state and federal agencies, civil defense, private ambulance services, mortuaries, Red Cross, military, and others.

Training is most beneficial if used in conjunction with the potential disasters in a given area. Hence, we begin this chapter with a discussion of disaster preparation evaluations. Next we discuss training and rehearsal procedures and show how they should be used in a modern fire department. This approach allows those responsible for training to create realistic training scenarios based on the actual types of disasters most likely to befall a community.

In Chapter 1 we introduced the concept of hazard identification (HI). This provides an excellent prelude to training, in that it identifies the types of disasters toward which a fire department should orient its disaster training. Completion of the HI exercise in Appendix A provides a meaningful launching point for effective training. In this chapter we provide a shortened approach, a four step "needs assessment" which can be similarly effective and can be used with or without the accompanying HI exercise.

CONDUCTING A NEEDS ASSESSMENT— "WHAT COULD HAPPEN?"

As a prelude to Disaster Training, it is well for fire officials to conduct a needs assessment to determine the types of disasters toward which training should be directed. This assessment contains four elements:

1. problem identification
2. population analysis
3. potential disasters ratings
4. objective setting

Problem Identification

Often the need for a plan is perceived only after some calamitous event. For example, a hazardous chemical spill that kills a number of people may dramatically illustrate the requirement for hazmat training. One of the goals of this book is to make you aware of the need for comprehensive planning and proper training before a disaster occurs. Even without some calamitous event in a given peaceful community, the absence of any disaster plan should suffice to indicate the need for one.

Population Analysis

The second step of a needs assessment identifies the characteristics of your population. Census bureaus, building departments, and your own fire department records are all sources for identifying the nature and extent of your community's vulnerability to various kinds of threats. Later chapters will identify the different kinds of threats. You should compare your community's disaster potential with each of the disaster types listed in Figure 5-1. Then ask yourself: What other problems are unique to my situation that we should prepare for and how likely are these events to occur? If our plan is to be effective, what are our unique training needs?

Figure 5-1. Potential Disasters.

	Probability of Occurrence			Disaster Factor			Disaster Rating (probability × danger)
	Likely (3)	Possible (2)	Unlikely (1)	High (3)	Average (2)	Low (1)	
Wildland Fire			X			X	1
Conflagration		X		X			6
Civil Disturbance	X			X			9
Tornado	X				X		6
Flood		X			X		4
Hurricane			X		X	X	1
Earthquake		X		X			6
Volcano			X			X	1
Chemical Spill	X			X			9
Industrial Disaster		X			X		4
Nuclear Attack		X		X			6
Terrorist Attack		X			X		4
Aircraft		X			X		4
Rail		X		X			6
Highway	X				X		6
Marine			X			X	1
Other							P.O. × D.F.

Potential Disasters Rating

After you rate each disaster according to its likelihood of occurrence, you rate each according to how dangerous it is. In other words, how great a threat to lives and property would this event present if it did occur? The disaster rating, then, would be the product of likelihood times danger. For example, if wildland fires are unlikely to occur in your area and would present a low danger if they did (meaning there was very little wildland in your area) that would receive a rating of $1 \times 1 = 1$. A likely tornado of average danger (relative to the other possible disasters) would receive a disaster rating of $3 \times 2 = 6$. You will want to prepare a disaster plan for all of these events, even unlikely ones. Those events most likely to occur and of greatest danger should be weighted more heavily.

Objective Setting

At this point we are ready to take our disaster training and test its mettle for actual field practice. If your own fire department or city disaster plan is not yet in place, internal fire department analysis, or existing disaster plans borrowed from federal, state, and county agencies will allow you to proceed with training.

Envision a feedback loop where field testing will lead to refinements in plans and generate a succession of improvements as the training mirrors preplan strategies. The following questions will allow you to plan for training and provide improvements for successive interactions.

- *How* has the training been developed?
- *Who* are the participants?
- *Who* knows about it?
- Is training primarily by simulation?
- Will we use a mock emergency?
- Will training be based on an actual occurrence?

The answers to these questions allow you to establish training objectives in advance rather than begin random training which may or may not optimize your return on time invested.

A similar four-step exercise led the Philadelphia Fire Department to realize that a fire at the Chevron U.S.A. tank farm near Philadelphia's International Airport was a likely disastrous situation. The simulated disaster drill included a chain of events including explosions, fires, hazardous material release, and injuries. As part of the drill extra pumpers, ladder companies, and chiefs were assigned.(2) Fifteen days later the drill paid off when an actual fire at the refinery challenged the Philadelphia fire fighters to the limit but resulted in successful extinguishment without a loss of life. At 10:30 p.m., on August 9, 1987, the explosion and fire at the Chevron U.S.A. refinery was reported. Alternate plans had been developed because of painful lessons from the past that had resulted in a fire at the same tank

Contemporary fire fighters must be well versed in the placards and markings of transportation modes such as railroad cars. (Photo courtesy of Ron Buckingham.)

farm which had killed eight Philadelphia fire fighters 12 years previously. The tank was 140 feet in diameter and 48 feet high and contained gasoline. Surrounding it were additional tanks containing gasoline, naphtha, and gasoline-like products.(3)

TRAINING FOR DISASTER CONTROL

Few fire departments have the organization and facilities to directly provide the broad training required for coping with a major disaster. Most departments therefore are required to expand their training capabilities as much as possible to include disaster training, or to take advantage of broad training elsewhere. This chapter will provide information relative to both components of preparation. Training is defined in a multitude of ways depending on where and when it is used. In this chapter we differentiate training from education and show the interrelationship between these two terms. An adequate disaster plan of necessity incorporates elements of both. Figure 5-2 shows this contrast.

Figure 5-2.

TRAINING	*VS.*	*EDUCATION*
Task Oriented		Theory-based
Memorization		Reasoning
Skills		Knowledge
Practice		Judgment
• "How"		• "How" and "Why"
Learning is		Learning is
• more predictable		• less predictable
• more measurable		• less measureable

This comparison of training and education is important for city and fire officials, because both types of preparation are needed by those who must deal with disaster situations. Certainly there is a need for education. Most curricula and bachelor degree fire science programs include full-scale courses in disaster planning. The University of Cincinnati holds week-long accelerated workshops each summer in Cincinnati, each winter in Florida, and each spring at the Alberta Fire Training School in Canada. The Open Learning Course "Disaster and Fire Defense Planning" is always in high demand. Education alone, however, is insufficient. Training is even more crucial so that persons responsible for action on the front line in disaster situations can swiftly and methodically carry out their expected obligations.

Very often the trainer and trainee both have various types of experience. We should take advantage of prior knowledge and experience. Favreau and Gillespie(4) have provided us with valuable information on how learning concepts can be applied to delivery of effective training. See Figure 5-3.

According to Favreau and Gillespie, people learn from their five senses as follows.

1.	Sight	75%
2.	Hearing	13
3.	Touch	6
4.	Smell	3
5.	Taste	3
	Total	100%

A trainer is therefore called upon to make maximum use of visual devices if training is to be effective. To a certain extent, hearing is important as well. This underscores the need for realistic facilities, models, and modern training tools that capitalize on these senses.

Figure 5-3.

Learning Principles	*Application*
People learn best:	To fill need for disaster control training:
• By doing →	• Provide different types of physical facilities for actual rehearsal/hands-on learning
• When there is a need for learning →	• Recall and relate prior knowledge and experience
• New things in terms of old →	• Expand small emergencies and fire defense training into larger experiences and disaster control training

High Tech Training Facilities

Throughout North America high tech training academies are being constructed which are proving valuable in training fire fighters within lifelike yet safe environments. For example, as we went to press the Bergen County, New Jersey, police and fire academy was constructing a complex scheduled for completion in 1992. It included the capability of creating fire scenarios in a wide variety of structures using a computerized propane fire simulator to control and monitor the fire environment. The extensive funding required for the project was provided through a bonding ordinance, and the description, sketches, and architect's rendering were all sent to fire instructors, fire chiefs, advisors, and fire academies in New Jersey and other states. After a questionnaire solicited input from this cross section of the fire service community, a state-of-the-art facility was designed. The completed structure contains 18 model occupancies commonly found in communities throughout the country: hair salon, retail store, townhouse, dwelling, warehouse, grocery store, gas station, auto body shop, tenement, restaurant, luncheonette, a forward command or staging area, professional office, hotel, highrise office, auditorium, movie theater, and basement space designed for nonhostile rescue situations.

Within seven minutes the complex's smoke system provides nontoxic smoke throughout the building thereby forcing fire fighters to use a kneeling search technique. Then a sophisticated ventilation system is capable of removing three floors of smoke within 60 seconds.(5)

Meanwhile, out west, Nevada's fire training academy also provides an

Fire fighters learn differences in various types of high-pressure cylinders at a training session. (Photo courtesy of Ron Buckingham.)

excellent example of realistic training. Drivers on U.S. Highway 395 north of Reno are often startled to see flames and dense black smoke rising off the end of a runway where championship air races are held each year. This is typical of the problems facing students and instructors at the University of Nevada's fire protection training academy. More than 3,000 students come from all over the world to attend a one-week course each year in production and pipeline fires, hazardous materials, manufacturing fires, and emergency management. One of the unique factors of this academy is that the training staff works closely with the oil producers in processors who employ many of the students. As these students learn alongside municipal fire fighters, in a facility mutually supported by private industry and the fire service, a valuable link between the industrial private sector and the fire service is formed. In all, 17 props are available on the campus, including an overhead range which includes both LPG (liquefied petroleum gas) and liquid fuel loading racks. One challenge is a process unit simulating a refinery distillery with leaking valves and gaskets. "In this hurricane of flame, students learn how to isolate and shut off valves, send in three or more teams together, and coordinate an incident command system."(6)

Multicity Training Centers

Many fire departments cannot afford expensive training centers, but by pooling resources, a number of communities are constructing facilities for joint operation. Typical is the multicity fire training center operated in Inglewood, California. Serving Inglewood, Redondo Beach, Manhattan Beach, Hermosa Beach, and El Segundo, the facility was built in 1972 for about $800,000 through issuing bonds under a joint powers arrangement with the five participating communities. The cost today would be at least twice as great. Financial support for the center is shared by users on an equitable basis, and while it was planned originally to provide recruit training, the program has been broadened.(7)

Another example of multicompany drills in large-scale operations for mutual aid companies can be found in the "Tri-Borough Area" which includes three townships along the Delaware River northeast of Philadelphia: Cinnaminson Township, Palmyra, and Riverton. Fire companies within these townships conduct four multicompany drills yearly on a rotating basis.(8)

An Inter-Agency Fire Center operated by the U.S. Department of Interior's Bureau of Land Management in Boise, Idaho, provides an advanced fire management course and training in fire fighting techniques through the cooperation of four federal agencies. The Bureau, the Forest Service, the Department of Agriculture and NOAA Weather Service, and the Department of Commerce combine plans, communications, and operations at this location.

In the early 1990s, it was the Idaho State Fire School that did much of the research on Class A foams. At this facility, the relatively new concept of CAFS was first demonstrated in 1991 showing how this high energy foam generation system used an air compressor to accelerate the distance from which the foam can be propelled from the nozzle. The information from this training school soon spread throughout the United States showing how this new technology would be able to push the ability of the fire service to new limits. This was especially valuable for rural fire departments working with limited water and limited resources.(9)

Multidepartment fire training can also be achieved in test exercises involving buildings slated for demolition and removal. The California State Fire Training instructors invited a large number of fire departments to participate in the "Willow Burn" project.

This plan called for the total operation to be directed by an incident commander; under him would be the operations chief who would supervise tactical officers and crew chiefs. A safety officer would be responsible for overall safety. These positions would be rotated among the participating fire chiefs. Because blocks of multistory buildings were burned in this project, the training extended over considerable time. Using this basic plan, a number of cities and the California Division of Forestry later participated in a similar multidepartment training program involving the burning of a large number of buildings near Pacifica, California.(10)

In the Midwest often a dozen or more volunteer fire departments will participate in the control burning of a grain elevator. These old structures, once abandoned, often pose a significant fire threat to a rural community. A controlled burn provides the opportunity for many rural volunteer fire departments to gain live burning experience and at the same time practice on what can be one of the most deadly types of rural fires due to the explosiveness and high fire load associated with the older wooden grain elevators.(11)

With forethought and careful planning, buildings designed for demolition can be used to provide realistic disaster control training that requires forcible entry, ladder rescues, and command post operations.

While EPA and air pollution authorities are reluctant today to sanction live burns, they will often cooperate when convinced of the value to the fire service. After all, a poorly trained fire department could be responsible for unchecked fires which create considerable air pollution.

All across the American continent more and more realistic disasters drills are being conducted. These are taking on added degrees of realism as they begin to expand across several days and several geographical jurisdictions. In the April 1987, issue of *Fire Engineering*, Mr. David A. Norris of the Portland, Oregon, Bureau of Fire documented two such drills that took place in the Portland area. This was the beginning of a series of drills that have been conducted annually ever since. He explained the complexity of

Proper training can prevent fire fighters from applying water to deadly metal fires, which can erupt violently when exposed to water. (Photo courtesy of Ron Buckingham.)

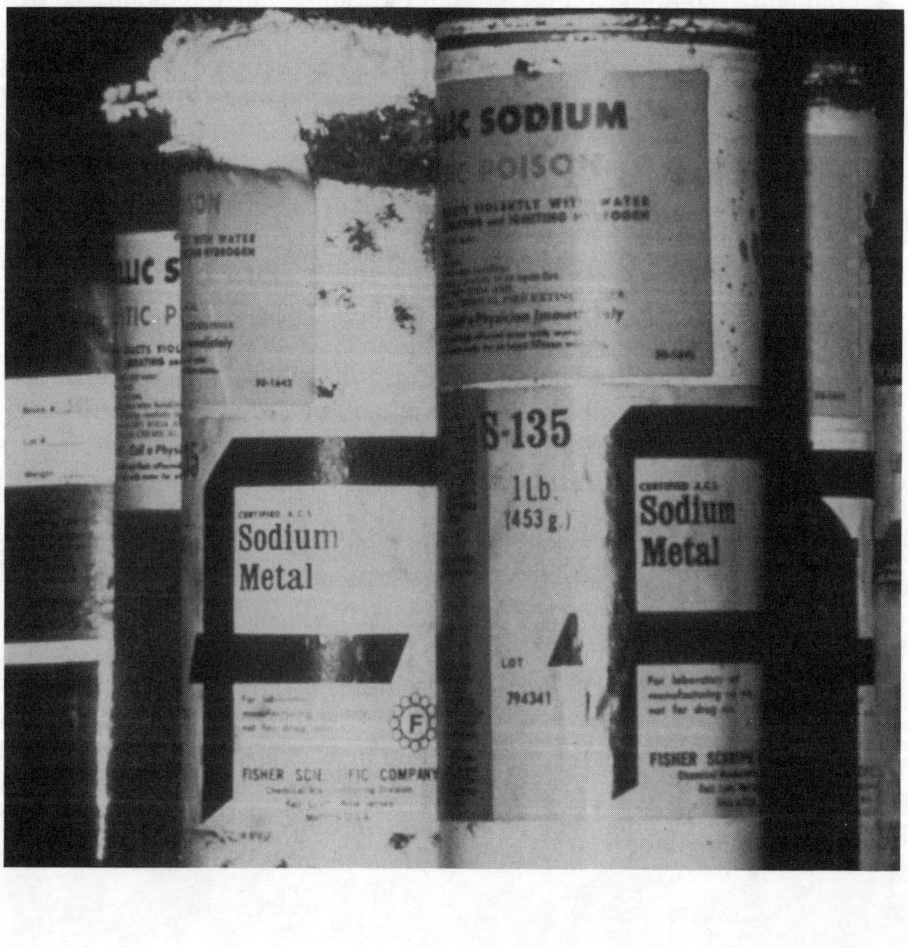

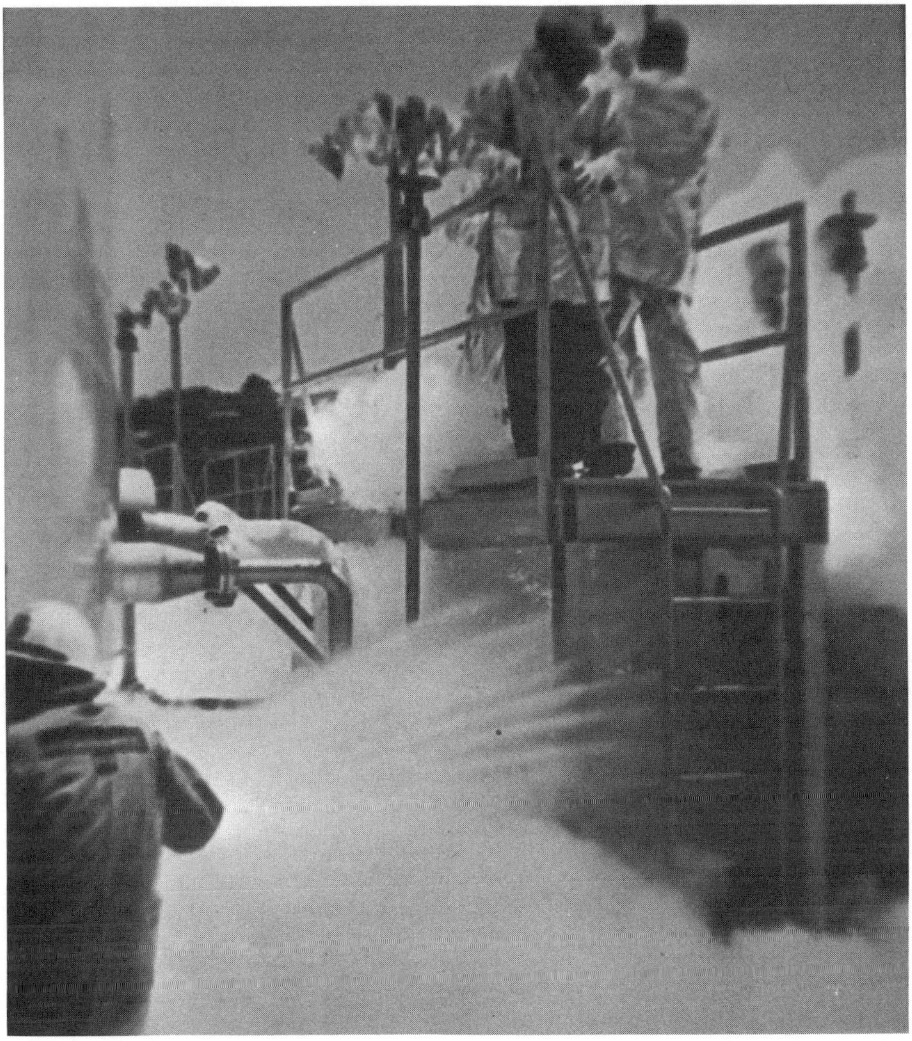

Fire fighters use a roller towel and water spray to form an ice plug in a leaking tank. (Photo courtesy of Charles Bahme.)

working with both internal and external forces in a major disaster scenario. While the fire chief can cope with internal problems through SOPs, he or she has a more difficult time in dealing with the external forces. He indicates that today the fire service is faced with coordinating the efforts of up to 30 separate agencies, from volunteer groups to federal organizations. The time to decide how these various agencies will be used is prior to an actual disaster. Questions such as "what services will be needed, who can supply them, will the type of help needed be forthcoming?" should be answered in a predetermined plan. Also critical, according to Norris, is the issue of "Who is in charge?" These and other issues are best worked out prior to an actual drill and then refined during the drill scenario.(12)

In Portland some of the participating agencies in the disaster drill were the Willbridge Mutual Aid Association; the ECHO (Emergency Communication Hazard Obligation), consisting of the safety officers of the major chemical and industrial corporations, utilities, railroads; and CAER

(Community Awareness and Emergency Response), a 94-member association from six counties and two states. When invitations went out to all members of the Willbridge, ECHO, and CAER associations, a planning meeting was heavily attended. A problem arose in limiting the number of participants. When the multiagency disaster exercise was actually conducted, 21 separate organizations played a role. The actual exercise involved the following components:

a. An acid spill with victim at McKesson Chemical Company
b. A railcar rupture with victim involvement
c. A diesel trailer rupture
d. A fuel spill into the Willamette River.

In addition to the participating agencies, more than 120 top city officials attended. This exercise provided several lessons.

1. More press packets for the media were needed.
2. The public address system was inadequate both for the orientation and the viewing areas.
3. Some agencies experienced notification failures of their personnel in an overzealousness to create surprise.
4. Hearing was difficult because of the wide area.
5. Objectives for all participating agencies should have been prepared in writing and authorized.
6. An official announcement that the exercise had been completed was needed to keep participants from being "left hanging."
7. Fire alarm dispatch needed check lists in order to make contact with all participants.

These lessons were applied to a second exercise, an earthquake scenario, which included twenty-five private companies, nine city bureaus, six state departments, three federal agencies, two volunteer organizations, and three regional districts, for a total of 48 organizations and 500 actual participants. The second exercise was virtually flawless and proved the value of learning lessons ahead of time. It was decided to make the disaster exercise a yearly event.

Special Facilities

In many areas of the world there are facilities for conducting drills for special hazards, such as the mockup for natural gas and petroleum fires in Nevada. The chemical industry also can give assistance in conducting training in the techniques of handling chemical disasters, such as was done in the Linden, New Jersey, Industrial Mutual Aid Council.(13) In conjunction with this type of training facility, many departments, both volunteer and career, are using videos provided by *Fire Engineering*, *American Heat*, the NFPA, and others in preparing for various types of disasters.

The U.S. Navy has been conducting training in crash fire fighting at its air stations, which fire officers from nearby civilian fire departments have been invited to attend. Full-scale mockups are used in U.S. Air Force fire training.(14) The Navy has also permitted fire fighters from nearby cities to complete courses in shipboard fire fighting. The Los Angeles Fire Department sent several hundred of its members from the harbor area to the San Diego Naval Fire School. Such training privileges are often reciprocated: Naval reservists from New Haven and New London, Connecticut, were invited to attend the Eastern Connecticut Fire Fighting School in Willimantic, to learn techniques of fighting oil fires.(15)

Facilities

As a means of obtaining information for preparing plans and a scale model of a large training facility in Los Angeles, author Charles Bahme visited fire academies all over the United States and in many other countries. Most were suited for training recruits in mechanical and tactical skills; some had facilities for hot fire problems out-of-doors and within structures. Some had classrooms, libraries, equipment display rooms, pump and hose testing facilities; several had equipment for simulating fires in energized electrical equipment; a few had laboratories for study of electrical, chemical, and hydraulic problems; a few had gymnasiums, swimming pools, and athletic fields. In some foreign academies were quarters for faculty members, a hospital, and a mortuary. Only one had a mockup of a bomb-damaged street with buildings for training to rescue "victims" under varying conditions of fire, smoke, broken gas mains, interrupted water supply, and partly collapsed buildings. ("Disaster streets" were visited that had been constructed by civil defense agencies for general instruction.) In 1990, *American Heat* video productions documented excellent realistic flashover training provided by the Swedish fire service.(16)

Many articles have been written about U.S. and Canadian fire academies which might be of interest to fire department administrators. These include facilities in New Jersey (17), Nevada (18), Memphis (19), Toronto (20), Washington (21), Wisconsin (22), Louisiana (23), Delaware (24), New York (25), Massachusetts (26), California (27), Pennsylvania (28), Texas (29), and Hong Kong.(30) Many articles also document the founding of the National Fire Academy and show how its programs have been a source of valuable information for disaster and fire defense planning.

As the Fire Service entered the decade of the 1990s the National Emergency Training Center (NETC) housed the National Fire Academy and the Emergency Training Institute. For information about current training opportunities offered by the NETC, write the National Emergency Training Center, 16825 South Seton Avenue, Emmitsburg, Maryland 21727.(31)

Table 5-1 shows a listing of various disaster training aids available from the Learning Resource Center at the National Fire Academy. Fire chiefs and training officers can write or phone the Learning Resource Center, National Fire Academy, Emmitsburg, Maryland 21727; Phone 800 638-1821.

Table 5-1. Training Aids From The Emergency Management Information Center

	Site	*Date*	*Events*	*Contents***
1.	Three Mile Island, PA	March 28, 1979	Nuclear Plant Incident	M, V, OT
2.	Livingston, LA	September 28, 1982	Hazardous Materials Spill	M, V, OT
*3.	Long Beach, CA	December 1, 1980	Naptha Pipeline Explosion	M, S, A, OT
4.	Somerville, MA	April 3-4, 1980	Hazardous Materials Spill	M, V. OT
5.	Kenner, LA	July 9, 1982	Plane Crash	M, OT
6.	Washington, DC	January 13, 1982	Plane/Subway Crash	M, V, OT
7.	San Diego, CA	September 25, 1978	Mid-Air Collision	M, V, S, A, OT
8.	Waverly, TN	February 22, 1978	Hazardous Materials Blast	M, V, OT
9.	Kansas City, MO	July 17, 1981	Hyatt Regency Collapse	M, V, OT
10.	Cocoa Beach, FL	March 27, 1981	Condominium Collapse	M, V, OT
11.	Anaheim, CA	April 21, 1982	Urban Fire	M, V, OT
12.	Las Vegas, NV	November 21, 1980	MGM Grand Hotel Fire	M, V, OT
13.	Lynn, MA	November 28, 1981	Urban Conflagration	M, V, S, OT
14.	Colorado	July 15, 1982	Lawn Lake Dam Collapse	M, V, OT
15.	California	January 2, 1982	Floods	M, V, OT
16.	Arizona	December 17-20, 1978	Floods	M, OT
17.	Hawaii	November 23, 1982	Hurricane *Ida*	M, V, OT
18.	Brownsville, TX	August 10, 1980	Hurricane *Allen*	M, V, OT
19.	Mobile, AL	September 12, 1979	Hurricane *Frederic*	M, V, OT
20.	Kalamazoo, MI	May 13, 1980	Tornado	M, S, OT
21.	Grand Island, NE	June 3, 1980	Tornado	M, V, P, OT
22.	Wichita Falls, TX	April 10, 1979	Tornado	M, V, OT
23.	Los Angeles County, CA	February 9, 1971	Earthquake	M, V, OT
*24.	Jackson/Hinds Co., MS	January 12, 1982	Winter Storm	M, V, OT
25.	Illinois	March 24, 1978	Ice Storm	M, V, OT
26.	Massachusetts	February 6-7, 1978	Blizzard	M, V, S, OT
27.	Mt. St. Helens	May 18, 1980	Volcanic Eruption	M, V, OT
28.	California	1978–1981	Brush Fires	M, V, OT
29.	Coalinga, CA	May 2, 1983	Earthquake	M, V, OT
30.	Galveston, TX	August 18, 1983	Hurricane *Alicia*	M, V, OT
*31.	Philadelphia, PA	May 3, 1984	Center City Fire	M, P, OT
32.	Williston, VT	July 7, 1984	Amtrak Derailment	M, V, OT
33.	Barneveld, WI	June 7, 1984	Tornado	M, V, OT
34.	Roseburg, OR	August 7, 1959	Fire, Explosion, Conflagration	M, V, OT
*35.	Jackson Township, NJ	May 11, 1984	Haunted Castle Fire	M, V, P
36.	Dundalk, Baltimore Co., MD	October 22, 1984	Furniture Store Fire	M, V, P

37.	Shreveport, LA	September 17, 1984	Cold Storage Building Fire	M, V, P
38.	United States of America		Toxic/Hazardous Waste Dumps	M, V, S, A, OT
39.	Honolulu, HI	August 15, 1980	Chevron Tank Farm Fire	M, V
40.	North Carolina	April 1985	Wildland Fires	M
*41.	Fairfax County, VA	April 4, 1985	Wolf Trap Farm Park Filene Center Fire	M
42.	Mississauga, Canada	November 10, 1979	Train Derailment	M
43.	Benicia, CA	January 19, 1985	Molten Sulfur Spill	M, V, S, OT
44.	Frederick County, MD	August 25, 1985	Bus Crash	M, V, OT
*45.	Hallet, OK	June 25, 1985	Fireworks Factory Explosion	M, OT
46.	Addison, TX	June 19, 1985	Gas Line Rupture Fire	M, V, OT
47.	Dallas-Ft. Worth Airport, TX	August 2, 1985	Air Crash	M, V, A, OT
48.	Bradford, England	May 11, 1985	Stadium Fire	M, V
49.	Phoenix, AZ	November 15, 1984	Flammable Liquid Tank Explosion	M, OT, S, A
50.	Buffalo Creek, WV	February 26, 1972	Slag Dam Break/Flood	M, V, OT
51.	Los Angeles, CA	April 29, 1986	Central Library Fire	M, V, OT
52.	San Ysidro, CA	July 19, 1984	Mass Casualty Incident/McDonalds Restaurant	M, OT
53.	Mexico City, MX	September 19–20, 1985	Earthquake	M, V, A
54.	Pine Bluff, AR	June 10, 1985	Train Derailment	M, V, OT
55.	Chase, MD	January 4, 1987	Amtrak/Conrail Collision	M, V
56.	San Juan, Puerto Rico	December 31, 1986	Dupont Plaza Hotel Fire	M, V
57.	Miamisburg, OH	July 8, 1986	Train Derailment	M, V, A
58.	Cerritos, CA	August 31, 1986	Mid-Air Collision	M, V, S, A
59.	Bridgeport, CT	April 23, 1987	L'Ambiance Plaza Building Collapse	M, V, OT
60.	Miami, Fl	March 27, 1987	Cyclo Chemical Fire	M, V, S
61.	Wayne Township, IN	October 20, 1987	Ramada Inn: Military Air Crash/Fire	M, V, P, OT
62.	United States	1974–1989	Wildland Fires	M, V, S, A
63.	Pampa, TX	November 14, 1987	Hoechst/Celanese Chemical Plant Explosion	M, V
64.	Loma Prieta, CA	October 17, 1989	Earthquake	M, V, OT
65.	San Bernardino, CA	May 12, 1989 May 25, 1989	Train Derailment/Pipeline Explosion	M, V, OT
66.	South Carolina	September 21, 1989	Hurricane *Hugo*	M, V
67.	Sioux City, IA	July 19, 1989	Air Crash/Gateway Airport	M, V, OT

This listing represents all case studies active at the time we went to press.

*NOT AVAILABLE FOR OFF-CAMPUS LOAN
**M = MANUAL (s), V = VIDEO CASSETTE (s), P = PHOTOGRAPHS, S = SLIDES, A = AUDIO CASSETTE (s), OT = OVERHEAD TRANSPARENCIES
Available from EMIC. LIBRARIAN, LEARNING RESOURCE CENTER, NATIONAL EMERGENCY TRAINING CENTER, 16825 SOUTH SETON AVENUE, EMMITSBURG, MARYLAND, 21727. 1-800-638-1821 (OUTSIDE MARYLAND). 301-447-1032 (MARYLAND)

WHAT DOES A FIRE ACADEMY NEED FOR DISASTER CONTROL TRAINING?

An all-around regional, state, or provincial fire academy should permit recruit training, advanced training in fire prevention, arson investigation, apparatus operation, aerial reconnaissance, fire suppression, prefire planning, communications, strategy and tactics, command and staff, rescue practices, EMS and paramedic skills training, and disaster control. Mockup structures such as autos, tank trucks, aircraft, electrical vault, ship's engine room, dwelling, oil pits, and drafting sump should be in a separate area of the drill site, where extinguishment, rescue, and command training can be conducted under conditions of heat, smoke, carbon monoxide, monitored from a control tower overlooking the area. There are restrictions on open-air burning and environmental regulations that must be followed in all training.

Recall New Jersey's high tech center, referenced early in this chapter. It is desirable to have such facilities for teaching "sizeup" incident command, and joint company operations. A "disaster village" might consist of intersecting streets, one with adjoining intact buildings, the other with buildings partially destroyed as though hit by falling aircraft, bomb, or earthquake. Similar to a movie set, "disaster village" may contain a hotel, garage, drug store, theater, restaurant, beauty parlor, dry cleaning shop, and other occupancies filled with props which can be inspected by students in prefire planning and fire inspection courses.

A hotel can be on one corner of intersecting streets and be designed to serve as a drill tower, with a fire escape on one side, overhead power lines running down the street, autos parked near fire hydrants, and typical conditions of a congested business district.

Fire and rescue problems can be simulated with built-in gas lines to supply fire and smoke-producing devices in a single room, a series of rooms, or floors in succession. Such problems can be developed to test decision-making ability of student officers who would be rotated through command assignments of companies, battalions, divisions, and departments. To add realism, tape-recorded cries of help could emanate from loud speakers on the upper floors, accompanied by mechanical ejection of dummy "jumpers" from the windows; these could be activated by the drill master in the control tower. He could also wait until engine companies placed hose lines, then cut off water supply by operating a control valve in one of the water mains supplying pumpers. This should test the ability of student commanders to issue commands that could remedy this situation, such as calling for relay from more remote mains that might still be in service. If a simulated power failure blacks out the communications center, another test can be created for officers who depend on radio for issuing orders at an emergency.

The "disaster village" can be used for practice in rescuing persons on upper floors, inside mockup elevators, or in gas-filled basements (such as in a leak of a refrigerator system).

Officers and fire fighters might be summoned day or night to respond to unannounced problems. Some can be assigned to the communications

Back to basics. Fire fighters advance hose lines on simulated fire to practice realistic flame and heat conditions during a training session. (Photo courtesy of Charles Bahme.)

center to learn skills needed for receipt of alarms, dispatching of apparatus, recording movement of equipment and chief officers, making proper notifications, and other essentials. There should be a driver training course with narrow, winding roads, dead-end streets, steep hills, obstacles, and a skid area. An overturned school bus can be placed at the bottom of a ravine from which dummies (or students) could be rescued.

Many of the disasters that we will study in this text, such as hurricanes, floods, and similar water emergencies, will find the fire service involved in water rescue. Hence, for regions of the country where the fire service is involved heavily in water rescue, a training academy might logically include a pond, with perhaps a car in the bottom. There students can learn how to make underwater rescues. Since in these rescue conditions water is usually murky and visibility often limited to one or two feet at best, scuba diving should simulate these conditions to be realistic. Technical assistance in teaching realistic diver training with proper attention to both safety and efficiency is available from professional consulting firms. The May 1990, issue of *Fire Engineering* showed the key ingredients of successful diver training.(32)

An area can be designated for radiological training, where isotopes are planted for students training in monitoring. A decontamination center can be set up for radiological and other hazmat simulations.

In a room designed at the disaster control center, students can act as the staff handling messages sent by radio (intercom) from a command post in another room. A large sand table can portray a mockup disaster condition, with messages sent by the command post based on conditions presented.

If the fire academy is located in hilly terrain, large operations can involve deployment of companies in two or more sectors at a simulated brush fire. A mimeographed fire problem can be handed out to the first officer to arrive on the scene. He would then radio orders to arriving companies to take positions he thinks best for the conditions stipulated. The first chief to arrive can take over and announce location of his command post; establish communications; determine the nature, magnitude, and probable spread; determine the forces required; deploy available forces; and request additional help.

The instructor can then hand out additional written descriptions of the conditions affecting fire spread, change of wind velocity and direction, location of spot fires, relocation of some companies, call for additional support, and invoke the department's plans for major emergencies.

WHAT SHOULD DISASTER TRAINING ACCOMPLISH?

In addition to a thorough knowledge of the disaster plan, fire officers and fire fighters must have fundamental knowledge of disasters and the ability to think and act quickly and decisively. Field exercises should test the plan and disclose any weaknesses, so that changes can be made. Such exercises can indoctrinate officers in control center and command post techniques through a series of prepared incidents, and at the same time they may indicate gaps in the knowledge of members who are supposed to be familiar with details of their own units.

Exercises in command decisions should include practice in (1) evaluating the fire or other emergency, (2) ordering resources, (3) planning an attack and control strategy, (4) assigning resources, (5) making decisions as the situation develops, (6) dealing with emergency control problems, and (7) monitoring and evaluating decisions.

What Should Be Done First?

In analyzing your training needs take stock of abilities of your fire department and of mutual aid companies. Determine how well they can handle ordinary emergency situations and estimate the amount and type of training that will be required to expand these abilities for handling large-scale emergencies. How much can you safely assume that your members and members of neighboring fire departments already know about handling potentially disastrous situations?

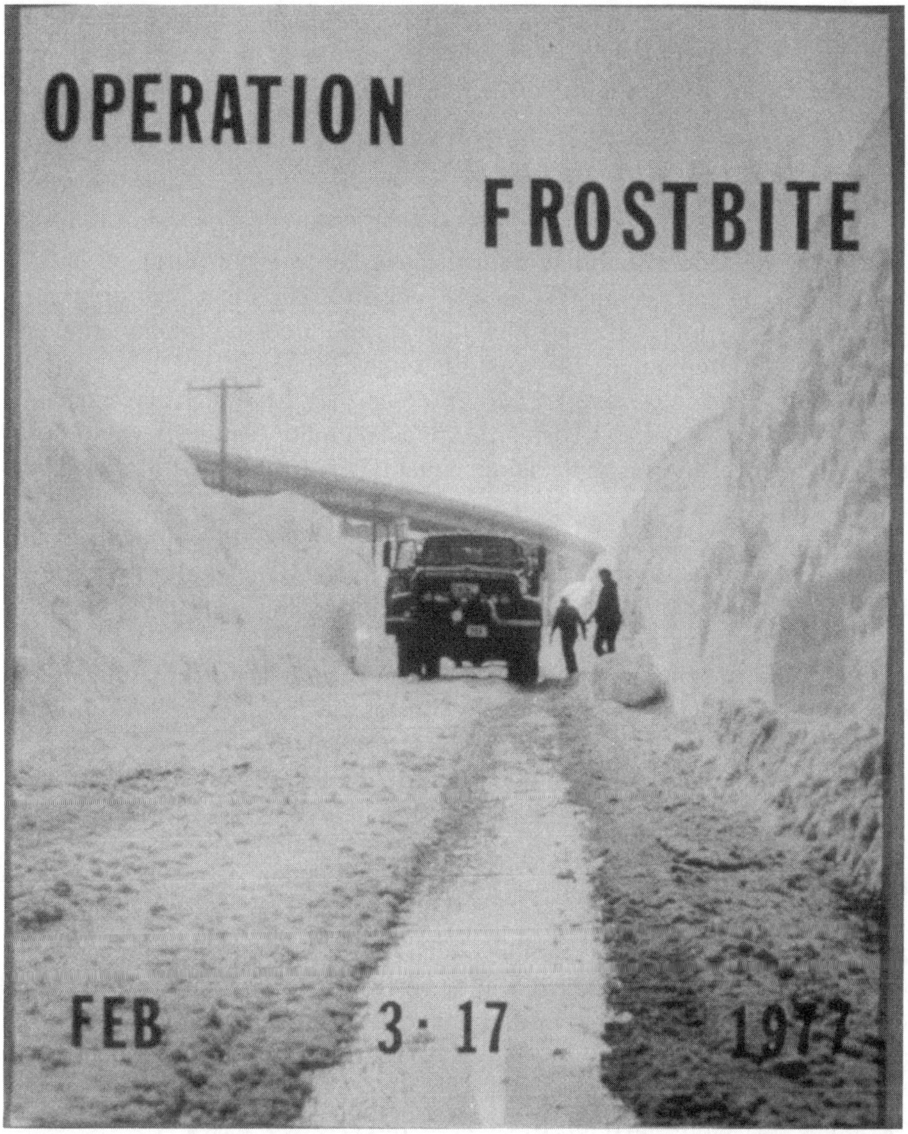

Documentaries which show how a historical disaster was handled can often provide excellent training resources for fire departments. (Photo courtesy of the Buffalo, New York, Fire Department.)

Have the Basic Lessons Been Mastered for Small Emergencies?

The old adage "Before you can learn to run, you must learn to walk" applies here. Before you are qualified to extricate fifty victims from an overturned bus, you must learn to extract one victim from a crushed vehicle. Until you are competent to extinguish a one-acre grass fire, you cannot expect to learn to cope with a 1,000-acre brush fire. When you have learned how to rescue one person from the rubble of a collapsed building, you are then ready for a test exercise involving uncovering hundreds of victims in a simulated earthquake disaster. Before you can learn to manage a battalion of companies, you have to learn how to manage a single company, and before you can direct operations of several fire departments, you must master the management of your own under difficult conditions.

What About Assumptions?

Basic skills, including that of management, must be mastered in order to acquire the proper foundation for disaster control. For purposes of this discussion, can you assume that each of your fire companies, including company, battalion, and division officers, have mastered the fundamentals of rendering the emergency services customarily carried out by your fire department? You may have to, and you also may have to make the same assumption for your neighboring fire departments. Assumptions are necessary when life and death decisions must be made quickly, but they carry a risk.

Some assumptions you cannot make safely. You cannot assume that your officers, or those of mutual aid companies, will be fully aware of the hazards involved in a potentially catastrophic situation until you supply them with the information. For example, at a large train wreck, one of the fires might involve a box car of ammonium nitrate; you cannot assume that incoming companies realize the lethal characteristics of nitric oxides in the billowing smoke, or the possibility of detonation of the entire mass.

On the other hand, if you request a mutual aid company to put a master stream into operation at a given location, you have the right to assume that its crews know how to do so; and its officer has the right to assume you know how many other companies are pumping from that water main.

When you request a forestry crew to cut a firebreak up a ridge, you have the right to assume that its members have practiced this skill often enough to carry out the task. These fire fighters have the right to assume that you have aerial reconnaissance and fire patrols informing you of the fire's advance, so if the fire "fishhooks" around the base of the ridge on which they are working, you will warn them in time to escape the upward sweep of flames that could envelop them very quickly.

Thus, one of the first steps in acquiring management skills for handling a large-scale operation is to become thoroughly familiar with the ability of companies and department officers you will direct to carry out your plan of action. Start with making sure that your own companies and officer personnel know how to carry out basic operations you have prescribed as part of regular training, including engine, ladder, salvage, first aid, and rescue practices. Next, be sure that your officers fully understand your disaster plan and the specific roles they are expected to carry out in multiple-company operations.

Since you may be required to direct deployment of mutual aid companies, you should know the general extent of their competence to carry out certain tasks. If they come from the "flat-land" and have never had brush-fire training, you would not direct them to take the flank of a rapidly advancing, wind-driven fire; you would not even ask them to carry out "cold-trailing" unless accompanied by experienced members of your command who would provide instruction and direction.

If you have your own forces trained to handle any emergency that may arise within your command, and you anticipate that arriving mutual aid units may be unfamiliar with potential fire problems, then it would be wise to arrange for test exercises to familiarize these outside departments with what is expected of them in extraordinary circumstances.

Similarly you must assume your equipment will work as planned, knowing full well there is likely to be a breakdown that will have to be repaired in a sudden "Plan B maneuver." If a ½" static kernmantle rope is to be used in a high angle rescue maneuver, we would like to feel that it will withstand a 2,000-pound load if so rated. Yet tests at the National Association of Search and Rescue (NASAR) in October, 1990, showed that a 19 foot drop of a 450-pound rescue load was enough to sever such a rope.(33)

WHAT METHODS CAN BE USED FOR DISASTER CONTROL TRAINING?

First to be considered are the methods and devices that can be used in your existing training. If your local community college offers a good campus-based course in disaster control, it would be practical to encourage your key personnel to take this course. The Open Learning Fire Service Program, from its inception in 1978, decided that a course in Disaster

Ben Klaene of the University of Cincinnati teaches in a modern classroom employing overhead projector, personal computer, VCR, and television monitor. (Photo courtesy of Brian Barkley.)

Planning and Preparedness was essential as a component part of its core baccalaureate curriculum. Many persons read this text in conjunction with a required college course entitled Disaster and Fire Defense Planning.(34) It is the opinion of both authors that disaster planning should be a required course in any upper level college curriculum for a fire-related degree.

Briefing sessions can be held within the regular officer training program, using publications as the basis for discussion. Application of your department's major disaster plan, mutual aid agreements, and large-scale operations would logically follow this discussion. Problems can describe a disaster of the type that might occur in your area, and participants in training can write their answers to questions relating to actions required to cope with the hypothetical situation.

Various training aids can be employed in problem-solving and decision-making exercises, such as the tabletop mockup and the fire simulator. To test the ability of an officer to manage a large number of companies under realistic conditions, multicity training centers often are used, with pooled resources. In addition to the local and regional training, an increasing number of states are conducting command and staff schools for fire officers.

Some fire departments conduct realistic multicompany drills, using sites where large structures or whole blocks of buildings have been slated for removal.

Tabletop Training

Tabletop training in multiple company operations can be useful. The State of Virginia's Fire Service Training personnel, finding the typical seminar "too big" for small departments or "too small" for large departments, began to offer two-day courses in four locations throughout the state in decision-making techniques and fireground tactics employing a tabletop exercise. The tabletop provides a measure of realism with three-dimensional structures and an audio system to simulate transmission of radio messages. Miniature buildings and facilities are placed upon a large table covered with heavy paper on which streets are drawn with a marking pen. Miniatures include main street buildings, elevated water storage tanks, horizontal LP-gas tanks, electric substations, railroad tracks, freight train, highway vehicles, fire hydrants, fire stations, and fire apparatus.(35)

A potential disaster situation, such as might occur if a number of LP-gas tank cars were to be derailed in the center of the village, can be simulated, with a problem posed to officers undergoing training. A problem involving ruptured tank cars and ensuing fires would have to be solved through the use of microphones in simulated command posts and in the dispatcher's office, by orders to departmental and mutual aid units to be deployed at the scene of the emergency. To achieve the aim of the course—forcing the commanding officer to make actual decisions under tension and stress—the instructor can create complications, such as loss of water, collapsed buildings, power failure, injured fire fighters, trapped occupants, breakdown of equipment, or the presence of explosives.

The U.S. Navy has been making effective use of the tabletop method of training in its disaster relief courses conducted at the Sea Bee (Construction

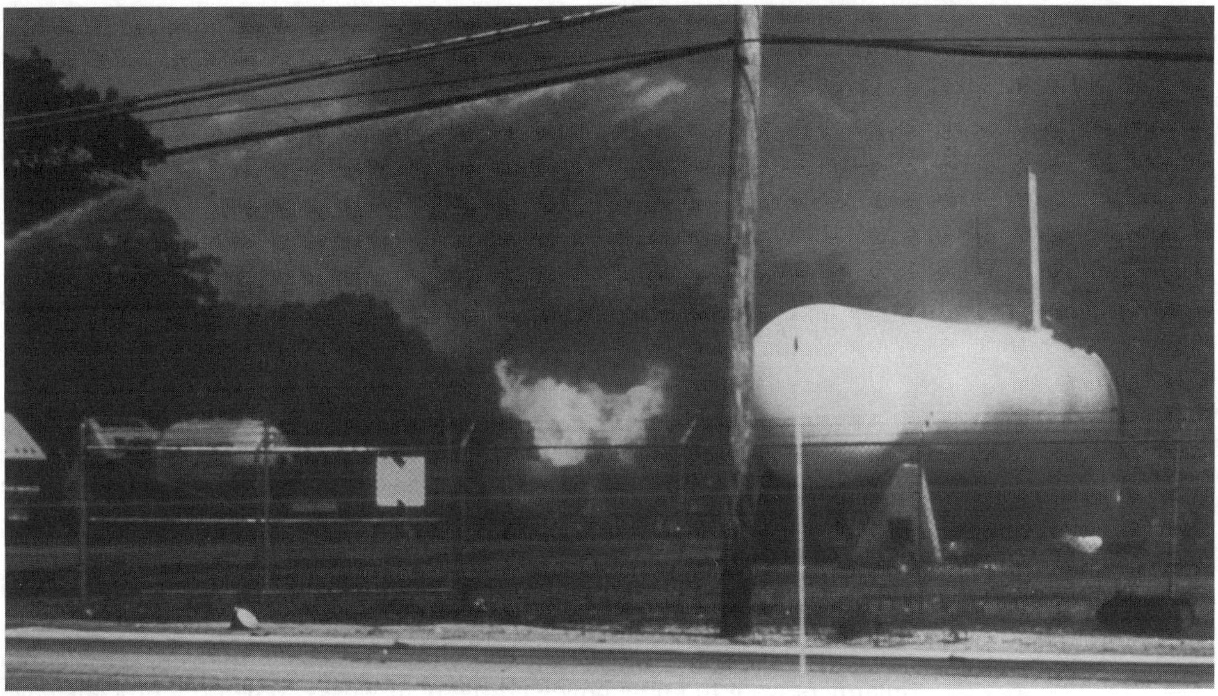

Proper training has led fire fighters to apply a protective stream to the uninvolved LP Tank at the Glover Gas Facility in Long Island, New York. (Photo courtesy of Harry Loud.)

Battalion) training center at Port Hueneme, California. On a gigantic scale, this method is used also by the Naval Amphibious Training Center in San Diego. In this facility the "tabletop" is the entire floor of a large building having a grandstand from which students can witness the mechanical deployment of large numbers of model ships, planes, helicopters, and amphibious boats, accompanied with the noise and flashing of guns, and rockets flying from ship to shore.

Simulator

Similar to tabletop training is the device known as the "simulator." The Elgin Air Force Base in Florida has a 28-foot van trailer that permits fire fighters to "handle" multialarm fires and cope with problems that could arise anywhere within the base's 722 square miles, including docks, runways, tank farms, and structures. It utilizes a 35-mm slide projector, tape recorders, an intercom, and overhead projectors that superimpose transparencies on the slide screen. Scenes are projected from the back of the van. The students sit in front and receive the general situations by radio (intercom) while viewing the slide of the problem area or structures. As the student assuming the role of the chief deploys fire apparatus, the trucks are shown on the screen, along with hose lays, special equipment and whatever else the student thinks the situation demands. The director can indicate a broken hose line, shift in wind direction, and similar variables that affect the handling of the emergency. If the director feels the deployment is not correct, or too slow, or inadequate to cut off the spread of the fire, the director can show its spread to other structures. At the critique following the session, the mistakes can be pointed out and the exercise rerun to show how it should have been handled.(36)

In the years that have elapsed since the U.S. Forest Service developed a simulator for training fire officers in forest fire fighting others have been constructed and put to good use. Simulators have several advantages over conventional classroom training: (1) they permit participation of trainees; (2) they provide decision-making in stress situations; (3) they put an emphasis upon teamwork; (4) they relate the exercise experience to job experience; and (5) they provide rapid feedback of the results of decisions.

"Macro-training" for Disasters

So far in this chapter we have discussed training facilities and training techniques. While these all are necessary, they all are smaller or "micro-pieces" of the training requirement. There also is need for large-scale training exercises which contain characteristics of actual disasters and provide a broad or "macro" perspective. Persons involved in disaster planning usually will advocate a large-scale exercise. The reason, of course, is that the planners want to know how well the plan works. When fire officials have spent considerable time and effort to develop a disaster plan, they too will want to know how well it works.

Disaster Simulation

As a method for overall evaluation of a disaster plan, simulation is the least expensive and can provide valuable insight into the operation of the plan. The plan can be tested by developing a scenario of a disaster occurring in the jurisdiction covered. Think big—one of the things that makes an event a disaster is that greater destruction occurs than anyone anticipated. Some of the examples in this book can be used as a basis for establishing the scenario. Gather the people involved in the plan together with a map of your community and have each person respond as if this were an actual disaster.

In a simulation each player has a specific role. By pretending to react to a disaster, participants can discover role conflicts, necessary channels of communication, how resources can be monitored, and other information valuable to the successful implementation of the plan. Don't forget to start the simulation with everyone dispersed. It is not likely that a disaster will occur with everyone in place at the command post. Your goal with the simulation is to make sure that all responsibilities and communications run smoothly—at least on paper. Again, make sure everyone understands his or her responsibilities.

The major advantage of simulation is that it involves few people and little expense. In the formative evaluation stage the plan can be tested without deploying personnel and resources, and it can be repeated many times. The disadvantage is that these "paper" simulations are unlikely to turn up any deployment problems. For example, during the simulation the fire chief may say that at this point he deploys three companies to preassigned points. It may not occur to anyone that the earthquake has just demolished one company's apparatus and made passage for another impossible.

Dry Run

The best way to test all phases of your training before a disaster is to conduct a dry run. This is very similar to a simulation, but in this case you actually deploy people and pretend that you are responding to a disaster. Though this is an expensive operation, you can also use your dry run as a way to increase community awareness and support of your preparedness efforts. The Portland exercises described earlier are excellent examples of the benefits gained from actual practice.

Next to the real thing, the dry run has the greatest potential of preparing emergency responders for actual disasters. In March, 1991, *Firehouse* magazine contained an article entitled "High-Rise Fire Training."(37) This article described how 2,250 fire fighters, using approximately 650 pieces of apparatus, participated in two separate highrise training exercises in Sacramento, California. Thirty northern California and Nevada fire departments fought simulated blazes in the tenth and eleventh floors of the Renaissance Tower, Sacramento's newest highrise. As this article was being distributed, three Philadelphia fire fighters lost their lives in the highrise fire at One Meridian Plaza, not having the opportunity to even see the article.(38)

GATHERING DATA IN ACTUAL DISASTER

Of course, the most accurate way to evaluate your disaster plan is to see how well it works in an actual disaster. No matter how good your plan is, you will find features that can be improved. The importance of keeping accurate records during a disaster should be stressed. These records can be of great help in evaluating your plan. Perhaps the greatest amount of information, though, comes from those who have participated in the disaster. Plan on interviewing participants and victims to determine what happened and what could have been done better.

For any evaluation technique, data collection should be planned early. Ask yourself, "What do I want to know about our plan?" and then, "What kind of information do I need?" Always plan your data collection to answer a specific set of questions. One golden rule of data collection is never collect data that you do not need. In your questionnaires, interview schedules and other data collection efforts, you should know why you are asking each question, and how the answer will be used in the decision-making process. Finally, ask "How can I collect the information needed?" The answer to this question will tell you how you will collect your data. It is possible to receive assistance from your local college or university and from other agencies within the local government for help in defining your data collection needs and activities. Nationally recognized fire science institutions such as The

University of Cincinnati or the University of Maryland often can provide consultants.

CONCLUSION

This chapter has provided a variety of information on disaster preparedness, and various types of training that are required by contemporary fire organizations and other emergency responders if a municipality is to successfully cope with whatever type of disaster may befall it. The chapter also has pointed out the interrelationship between preparedness and training and provided tools for fire service leaders to analyze the potential disasters most likely to strike a given community.

It is important for fire service leaders to expand conventional fire service training to include the broader more comprehensive realm of disaster preparedness. Expansions in the types, scopes, and sizes of training facilities and training academies are evident today. Likewise, there is a corresponding increase in the spectrum of coursework offered at the National Fire Academy, through local training programs, and in the educational realm at institutions of higher learning.

By taking advantage of the modern tools and opportunities available, a fire service leader today can insure that his or her community will not be caught unprepared when a disaster of any sort materializes. We will move into the next chapter with a discussion of modern communications and information management—the lifeblood for any disaster defense organization.

CHAPTER 5 REFERENCES

1. "Rollover Tanker Puts Training to the Test" in *Fire Engineering*, Vol. 144, No. 3 (March 1991), pp. 105–106.
2. Janda, Gerald. "From Drill to Real in 15 Days" in *Fire Engineering*, Vol. 141, No. 3 (March 1988), pp. 32–33.
3. McLaughlin, Philip J. and James Smith, "Plan B" in *Fire Engineering*, Vol. 141, No. 3 (March 1988), pp. 21–24.
4. Favreau, Donald F., and Joseph E. Gillespie. *Modern Police Administration*, Englewood Cliffs, NJ: Prentice-Hall, 1978, pp. 125–129.
5. Evans, John C. "High Tech Training" in *Fire Engineering*, Vol. 144, No. 3 (March 1991), pp. 118–125.
6. Ross, Christopher. "Nevada's Fire Training Academy" in *Firehouse*, Vol. 16, No. 1 (January 1991), pp. 50–51.
7. Smith, Joseph G. "Multi-City Fire Training Center" in *Fire Command!*, December 1971, p. 20.
8. Burns, Robert. "Multicompany Drill Gives Experience in Big Operations" in *Fire Engineering*, September 1973, p. 140.
9. Davis, Larry. "Class A Foams" in *Firehouse*, Vol. 16, No. 4 (April 1991), p. 75.

10. Evans, James R. "Multi-Department Fire Training" in *Fire Command!* March 1972, p. 26.
11. Fitzwilliam, John F. "A New Resource for North Dakota Firefighters" in *Firehouse*, Vol. 16, No. 4 (April 1991), p. 28.
12. *Fire Engineering*, April 1987.
13. McGary, Roger A. "Hazardous Materials Training" in *Fire Engineering*, May 1972, p. 39.
14. "Full Scale Mockups Used in Air Force Training" in *Fire Engineering*, May 1971, p. 29.
15. "Fire Fighters Learn Techniques, Dangers of Controlling Oil Fires" in *Naval Reservist News*, June 1976, p. 8.
16. "Swedish Flashover Training"—Two part series in *American Heat* video productions, St. Louis, MO, 1990.
17. Evans, John C. See No. 5.
18. Ross, Christopher. See No. 6.
19. "Training, Equipment Stressed in Memphis" in *Fire Engineering*, December 1974, p. 52.
20. Baird, Donald. "Toronto Opens Fire Academy" in *Fire Engineering*, May 1971, p. 25.
21. Wilson, James. "Four-City Use of Fire Academy" in *Fire Engineering*, August 1972, p. 98.
22. Benkert, Joseph. "Regional Fire School Ingredients" in *Fire Engineering*, August 1972, p. 83.
23. "Louisiana Dedicates 25-Acre Center for Firemen Training" in *Fire Engineering*, August 1973, p. 42.
24. Sylvia, Richard. "Students Get Feel of Command in Delaware All-Day Exercise" in *Fire Engineering*, January 1974, p. 49.
25. Sylvia, Richard. "Education Tradition Continues at Site of New York State Fire Academy" in *Fire Engineering*, April 1974, p. 5.
26. Chapman, Edward. "Training Academy Developed for Massachusetts Firemen" in *Fire Engineering*, February 1972, p. 30.
27. Cockerton, Ward L. "Training, Education Plan Developed in California" in *Fire Engineering*, p. 50.
28. Burns, Robert. "Training Facility Developed by Volunteer Fire Company" in *Fire Engineering*, December 1973 p. 56.
29. Rauch, John R. "3,000 Go To Texas Fire School For All Levels of Training" in *Fire Engineering*, August 1972, p. 78.
30. Kwok-Choi, Ki. "The New Training Complex in Hong Kong" in *Firemen*, June 1969, p. 29.
31. *Annual Catalogue of Courses*, 1992: Published by the National Fire Academy, Emmitsburg, MD, 1991.
32. Hendrick, Walt. "Training Divers for Minimal or Zero Visibility" in *Fire Engineering*, Vol. 143, No. 5 (May 1990), p. 67.
33. Pedrotti, Dean R. "High Angle Rescue" in *Firehouse*, Vol. 16, No. 2 (February 1991), p. 20.
34. Kramer, William M. (ed.) *Disaster and Fire Defense Planning* (an Open Learning course), Emmitsburg, MD: National Fire Academy, 1992.
35. James, Paul E. "Tabletop Firefighting Training" in *Fire Engineering*, April 1973, p. 49.
36. Barrow, B. F. "Firemen Get Training on Simulator At Elgin AFB" in *Fire Engineering*, October 1972, p. 55.
37. Barkdull, Walter L. "High-Rise Fire Training" in *Firehouse*, Vol. 16, No. 3 (March 1991), p.
38. Caption for Cover Photo, *Fire Engineering*, Vol. 144, No. 4 (April 1991), p. 1.

CHAPTER 6

COMMUNICATIONS AND INFORMATION MANAGEMENT

TO HAVE SUCCESSFUL COMMUNICATIONS and to manage information for disaster control there must be a good organization, properly staffed, equipped, and trained for emergency communications services. Fire scene officers and fire service dispatchers are usually capable of handling the normal emergencies. However, as Warren Kimball pointed out in one of the first guides published on fire service communications, in a major emergency caused by a natural disaster, it is likely that telephone lines may be down, other alarm and communications facilities adversely affected, and a vast increase in the work load imposed upon personnel and equipment in the control center. "Quite commonly, most of these problems are encountered simultaneously in major emergency situations. The fire service, and particularly its alarm service and operations control personnel, must be prepared to operate effectively under extremely adverse situations. This requires planning and preparation for emergency communications."(1)

While communications problems are as old as mankind, the technology available to help fire departments deal with communications needs, both in the field and in administrative offices, continues to expand at a rapid pace. As we went to press, fire departments throughout North America, both large and small, were using personal and portable computers extensively, were communicating over fiber optic cable, portable fax machines, multichannel high-power portable radios with priority scan features, and other sophisticated equipment unknown in the past.(2) By the time you read these lines, additional advances will have been made in communications technology.

COMMUNICATIONS PERSONNEL

A fundamental prerequisite for emergency communications is redundancy—backup personnel, backup systems, backup energy supply, and even alternate facilities from which to operate in the event the main control center is completely disabled. With respect to personnel, there must be a sufficient number of trained dispatchers who can be called upon to respond quickly to the emergency operations center. These may be uniformed or civilian members of the fire department, but terms of their employment should make them readily available at all times. They should be given pagers or radio receivers that can be activated when recall of the off-duty communications personnel is required. Reliable means for recalling the off-duty communications repair personnel should also be established, in the event that equipment requires servicing during the emergency.

All personnel who may be called upon in a major emergency should receive regular training in the use of alternative systems and equipment and should become thoroughly familiar with the department's emergency operations plans, recall procedures, and notifications.

As the use of desktop and portable computers and other automated electronic equipment increases, regular periods should be scheduled when the communications personnel revert to manual methods to stay familiar with alternate equipment that may have to be used in an emergency.

One distinguishing characteristic of major emergencies or disasters is the dramatic increase in the number of decisions that must be made and the number of people who must make them. High-quality decisions require the communication of timely, valid, and useable information among a very large number of individuals and agencies. Communications needs during disaster are unique and often critical. "We in the fire/rescue service are especially in need of good communication skills on many levels. We operate in situ-

Author William Kramer serves as Incident Commander while adjutant James Miller plots progress at a simulated highrise fire. (Photo courtesy of Gary Auffart.)

ations that are hazardous, where we must depend on others to understand our instructions and to carry them out as we intend."(3)

After-action critiques of disasters almost always identify communications as a major problem area. Often the problem is not inadequacy of communication hardware but lack of an adequate incident management structure that delineates how information should flow. In the absence of such structure, key players frequently develop the notion that everyone needs to talk to everyone else. One of the greatest improvements in fire service communications in the last decade is not the further sophistication of equipment but the development of the Incident Command System (ICS) for more effective use of existing technology in multiagency operations at major emergencies. In Chapters 3 and 4, you were introduced to the application of ICS and other systems in large-scale operations, often involving mutual aid.

This chapter addresses the nature of communications needs during major emergencies and looks at some recent developments in communications technology. Incident management structures and enhanced technological tools, such as the computer, designed to improve the availability and management of information, will also be addressed.

COMMUNICATIONS NEEDS IN FIRE DEFENSE AND DISASTER OPERATIONS

A study of basic communications in fire defense and disaster operations requires an analysis of *who* needs to communicate *what to whom.* This chapter begins with an overview of these simple concepts and then goes on to discuss individuals and agencies involved in effective communications.

Who Needs to Communicate?

The set of people who constitute the communicators and those who "need to know" can be extraordinarily large. Certainly the officers, members, and representatives of the agencies having direct functional responsibilities need to communicate. Additional communicators are the various units of command, senior policy officials (including elected officials), dozens of support agencies, the news media, and the general public. Let's take a closer look at this group of communicators by examining the various layers or levels of organization that typically develop during a major emergency.

Beginning with agencies at the national level and progressing through the state to the local level, the following links are necessary.

- Between national and regional offices of various federal agencies including FEMA, the Department of Energy, Nuclear Regulatory Agency, EPA, Coast Guard, FBI, and the Department of Defense. In many cases, these agencies primarily provide support services, but in some cases, such as major hazardous materials incidents or acts or nuclear terrorism, federal agencies will be in the lead role.

- Between regional offices of FEMA and state offices of emergency management or emergency services. FEMA prefers to deal with state offices rather than directly with local office.
- Among state, area, county, and local offices of emergency management and the local emergency operating center (EOC).
- Among senior elected and appointed officials including department heads and police and fire chiefs.
- Among agency chiefs, field commanders, and dispatch or control centers.
- Among incident commanders at different command posts.
- Among field commanders and operating companies, task forces, strike teams, tactical teams, air attack forces, and other special units such as medical, police tactical (SWAT), or evacuation teams.
- Among company and team leaders.
- Among operations and planning section personnel and technical advisors.
- Between members of the public and agency dispatch centers (primarily the public requesting services or reporting damage).
- Between senior policy officials and members of the public with policy officials giving the public information on self-protection and on specific actions requested of them). The relationship of these entities and their communications links are depicted in Figure 6-1.(4)

What Information Needs to Be Communicated?

Information that needs to be communicated generally involves guidance, direction, requests for assistance, status reports on the incident, and updates on resources and operations. The following are some specific types of information to be communicated.

DISASTER ALERTING AND WARNING

This alerts the public to existing or potential emergency conditions, such as hurricane, flash flood, tornado, chemical spills, evacuations, or enemy attack. Warning involves the provision of additional information on protective actions and usually relies heavily on the electronic media. Awareness is now developing of the need to communicate with hearing-impaired people, which involves the use of interpreters, message "crawlers," or closed captioning on television broadcasts.

ASSESSMENT OF THE NATURE OF THE DISASTER PROBLEM

Information must be obtained from responders, the public, or reconnaissance concerning the type of disaster, geographical area affected, extent of damage, deaths and injuries, disruption of essential services such as utilities, and exposures requiring protection.

Figure 6-1. Agencies and their Communications Links.

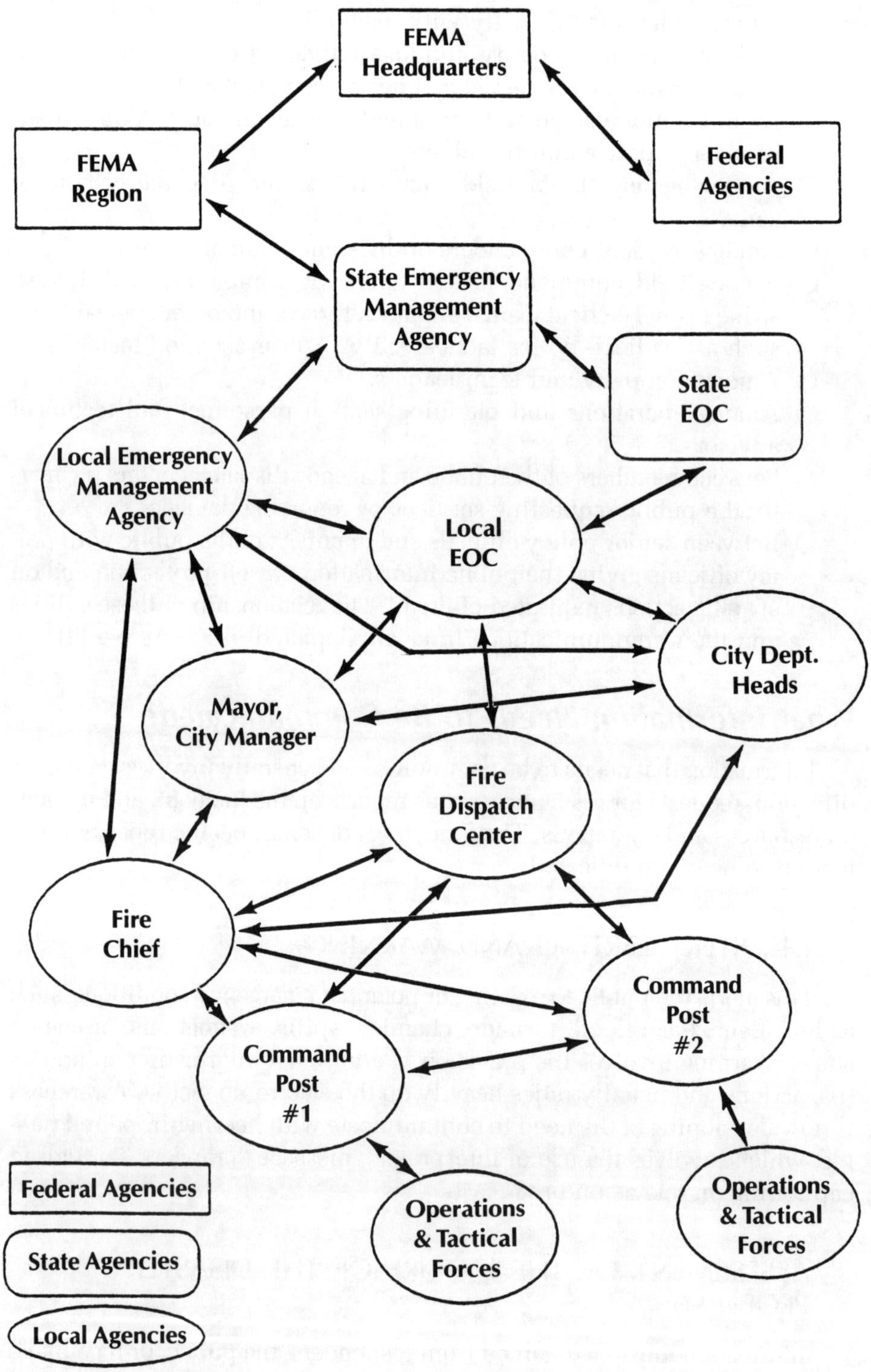

POLICY GUIDANCE

These are actions that involve complex or controversial economic, sociological, or legal issues and may require policy decisions from senior elected or appointed officials. Large-scale evacuations or crowd control actions during civil disturbances are examples of such actions.

TACTICAL INFORMATION

Command and control information is required for deployment of forces. Following the size-up, this information includes an estimate of the types and numbers of forces needed to initiate rescue and fire defense operations, protect exposures, manage casualties, and take appropriate measures for disaster control.

SITUATIONS STATUS (SITSTAT)

This is information concerning the current extent of the problem and progress being made in controlling it.

RESOURCE STATUS (RESTAT)

RESTAT includes information concerning the initial status of personnel, apparatus, equipment, facilities, and response capability immediately following the onset of a disaster. RESTAT during response operations includes ongoing assessments of resources committed, resources out-of-service, and resources available for assignment.

PLAN ACTIVATION

Orders to activate emergency operations plans appropriate for the type of emergency may come from authorized field commanders, emergency management directors, or senior elected or appointed officials. Activation also involves notification of all agencies required to send representatives.

EXTERNAL SUPPORT NEEDS

Information must be communicated concerning the type and number of external resources needed such as mutual aid, police, medical, military, communications, disaster relief services, radiological or hazardous materials teams, and technical advisors.

TECHNICAL DATA AND ADVICE

Reports must contain technical information including chemical or radiological data, meteorological data, fire spread or toxic plume modelling, medical data, and legal advice.

ENVIRONMENTAL CONDITIONS

The presence of traffic hazards, flooded areas, adverse weather conditions, toxic gas, or hostile criminal activities must be noted.

REQUESTS FOR SERVICE

These are individual requests for service from the public. If telephones are overloaded or out-of-service, requests for assistance may not be received from the most heavily affected areas.

PUBLIC INFORMATION

Information must be provided to the public through the news media concerning the general nature of the incident, its cause, what is being done to resolve it, who are the key decision makers, and who, if anyone, is responsible for causing the incident.

ORGANIZING FOR EFFECTIVE DISASTER AND FIRE DEFENSE COMMUNICATIONS

Chief Steve Meyer of Garrison, Iowa, studied command systems in detail and, like many others, concludes that the solution to many of the communications problems encountered in disasters is the rapid establishment of a comprehensive incident management system.(5) We next consider various levels of this management system, the responsibilities of the communications unit in the incident command system (ICS), and organized methods for communicating with the public.

The overall incident management organization at the local level can be viewed as consisting of three levels—policy, coordination, and operations. Each of these levels is discussed below along with a look at the types of communications needs associated with them.

Policy

Those people normally associated with establishing policy guidelines during disasters include senior elected and appointed officials, such as the mayor, city manager, director of public safety, county commissioners, city council members, fire chief, police chief, health commissioner, EMS chief, and others. In many disasters, particularly long-term situations such as hurricane, flood, or chemical spill, this group will convene for meetings at an off-site location such as an EOC.

The policy group is responsible for providing policy guidance to incident commanders concerning such areas as large-scale evacuations, resolution of agency jurisdictional disputes, operations involving large financial commitments, operations involving sociological or political sensitivities, and legal concerns.

Policy-level officials should not become overly involved in controlling field operations, which should be left to incident commanders. Policy officials normally communicate with field commanders, technical advisors, or dispatch centers, and they rely most heavily on land line or mobile telephone.

Coordination

Coordination involves gathering, analyzing, and sharing of information among agencies about the disaster and the identification and availability of resources. Coordination activities normally do not involve command decision making, although some decisions, such as mobilization of resources, may be made according to pre-established policies.

Coordination can occur at a variety of locations both on and off site, including the following.

DISPATCH OR OPERATIONS CONTROL CENTERS

In day-to-day operations and many major emergencies, a great deal of coordination occurs among agency dispatch or operation control centers. Much of the associated communication utilizes direct telephone lines, regular telephone, dedicated mutual aid radio frequencies, or, in some cases, data communications between computers.

NINE ONE ONE EMERGENCY NUMBER

Many local telephone companies have made available the number "911" to assist the public in reporting fires and other emergencies and to avoid the confusion of having an assortment of telephone numbers to call. Enhanced systems visually display for the dispatcher the telephone number and address from which the call is made.

Most major cities and counties have adopted this concept, and some states have set a deadline date for this number to be put into use in every community. Communications centers have difficulty in handling the volume of traffic that results from major disasters.

The adoption of 911 or 911 Enhanced requires establishment of a central communications center to receive all calls for fire, police, and other emergency services. "Where only one municipality is involved and the municipal and telephone central office services boundaries coincide, there is no particular problem in adopting 911, provided the fire, police, and other municipal departments concerned can reach agreement as to planning and operational procedures. However, problems do arise when two or more communities decide to implement 911 service, and the telephone central offices serve a portion of each community. Under these conditions, it is desirable to establish a single communications center to handle emergency calls from the entire area. Even so, the communications operator will have to ascertain from which community or district a call originates and the correct address or location of the emergency. Similar or identical street names in the communities served are apt to cause confusion.(6)

Fire departments that want to continue operation of alarm centers where street fire alarm box circuits terminate must retain around-the-clock personnel to relay alarms received from this source to the central communications center, or else the alarm equipment will have to be relocated to the central facilities.

Few street alarm box systems using telegraphic communication remain in service, however, since the accessibility of public phones and coinless 911 systems are almost universal.

As 911 and Enhanced 911 systems become more prevalent, communities must guard against a tendency to describe these systems in an abbreviated form, either written or spoken. Terms such as E911 or "Nine-eleven" should be avoided since there are documented cases of adults attempting to dial E-9-1-1 or children searching for an "eleven" on a telephone button. It is just as easy to emphasize 9-1-1, "nine-one-one" or say "nine-one-one enhanced."(7) Enhanced versions of 911 display the caller's phone number and address on a terminal while the information is being received by emergency call takers.

EMERGENCY OPERATIONS CENTER

In Chapter 3 we saw how an emergency operations center (EOC) is activated when a major emergency is of a magnitude or duration that requires close coordination among a large number agencies. EOCs may be activated at the municipal, county, regional, or state level.

Typical agency representation includes one or more police and fire departments: EMS and medical agencies, Red Cross and other disaster relief agencies; streets, public works, and parks departments; utilities; public transit agencies; financial, technical, and legal advisors; senior elected and appointed officials; and representatives of the news media.

These representatives must communicate with their policy officials, dispatch or control centers, and field operation people. A variety of communications modes are used including radio, telephone, amateur radio, teletype, fax machines, and computer data modems.

SPECIAL COORDINATION CENTERS

A variety of special coordination centers, normally utilized to coordinate day-to-day activities, can take on increased importance during major emergencies.

- Regional fire coordination centers such as those utilized as part of Southern California's Multi-Agency Coordination System (MACS) keep track of all fire resources of local, state, and federal agencies and can immediately assemble special teams and resources.
- Regional medical coordination centers have been established throughout most of the country and are responsible for coordinating the use of ten nationally designated UHF medical radio frequencies.

These centers also coordinate use of an older interhospital system known as the Hospital Emergency Administrative Radio (HEAR) network.
Many of these centers also track available paramedic, EMS, and helicopter ambulance and rescue units throughout the region and maintain inventories of available hospital beds and capabilities.

- Joint public information centers normally are assembled after disaster onset at a location in the vicinity of the incident or in some cases at the EOC. They are staffed by public information officers from the principal agencies involved and are established to closely coordinate the public information that is released. This can be crucial if specific protective actions are being requested of the public. If these centers effectively meet the needs of the news media they can prevent news media from interfering with commanders or seeking out conflicting information from different agencies.
- Field incident command posts are the location for coordination of on-site activities. Coordination activities principally occur in the planning section, but may also take place in operations, logistics, and command.

Operations

Operations refers to direction and control of field operational activities associated with bringing emergency conditions under control. Until recently, standard strategies for organizing field activities during disasters, which typically involved dozens of agencies, were usually absent. One exception was the project called FIRESCOPE (introduced in Chapter 3) which involved a number of local, state, and federal fire agencies in California. As you recall, the Incident Command System evolved from FIRESCOPE. This was a major breakthrough in incident management because of several key features.

- The system employs a very high degree of standardization in organizational structure and terminology. This is a key feature in its utility during large, multiple-agency operations.
- The system establishes a comprehensive and detailed organizational structure in which everyone at the site knows to whom they report and who reports to them. This formalized information flow drastically reduces confusion and minimizes the amount of necessary communications.
- The system maintains jurisdictional authority by placing the senior officer of the jurisdiction in the role of incident commander.

The communications unit within the ICS structure is part of the logistics section (see Figure 6-2). This unit is responsible for developing the field communications plan, which allocates radio frequencies to be used by various sections, branches, and groups. It is also responsible for acquiring field communications facilities, such as a special van or trailers, and for acquiring additional mobile and portable communication hardware.(8)

Communications at the operations level are heavily dependent on radios

Figure 6-2. Communications Unit in Incident Command System.

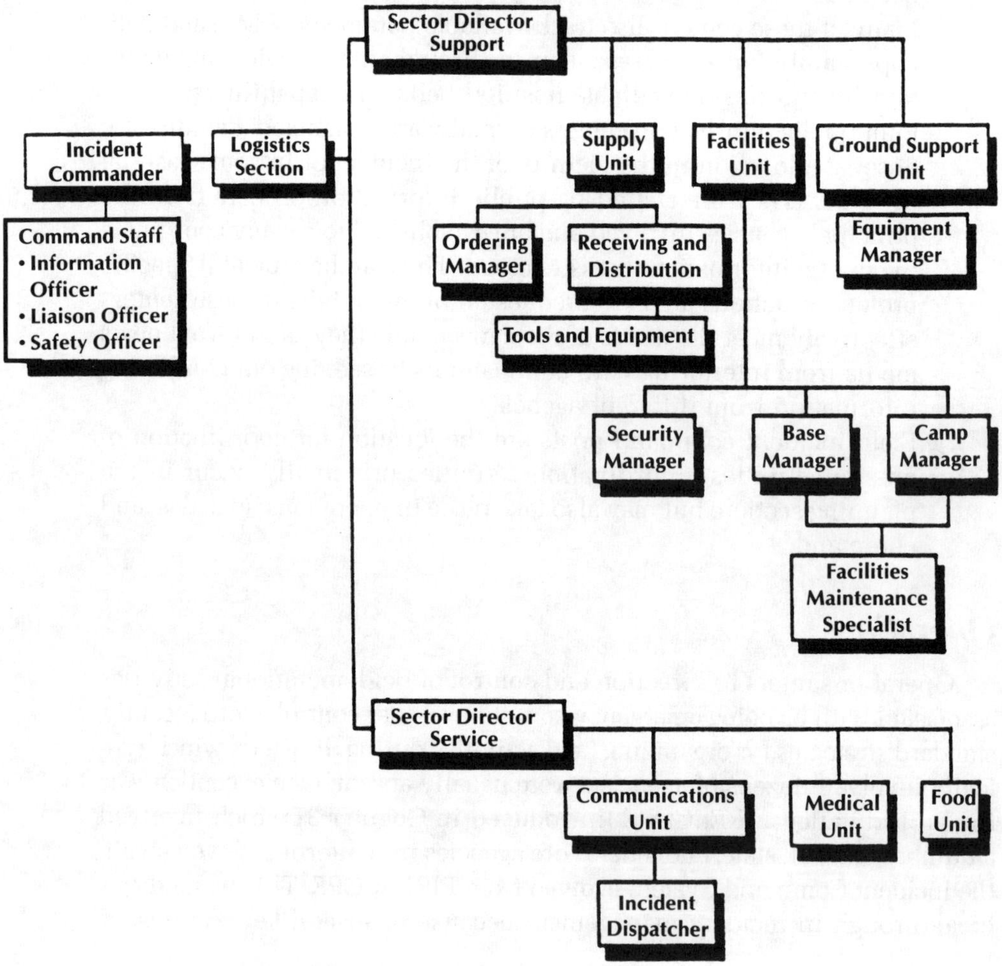

because of the need to communicate with mobile personnel. However, recent advancements in the areas of mobile telephones and portable computers are having a major impact on field operational communications. These hardware developments are described later.

Communicating with the Public

The importance of communicating with the public during disasters is often overlooked by agencies such as the fire department. Yet a variety of crucial problems can develop during disasters.

- The public may not be able to request public safety services because the telephone system is damaged or overloaded. If no calls for help are coming from these areas; this may lead to failure to respond to the most heavily damaged areas of the community.

- Specific protective actions may be indicated for the public. The public needs coherent, consistent, and appropriate guidance.
- Specific actions by the public may be needed to facilitate government response actions. These might include conservation of water for fire fighting, removal of cars from snow-covered streets, special closing of businesses, and minimizing use of the telephone.

Effective communications with the public involves the establishment of special organizational elements.

- Establishment of a joint public information center, described earlier, assures that information going out is properly authorized, accurate, relevant, and not contradictory.
- Fire stations or other government facilities can be utilized as places where the public can access the system in person.
- Amateur (HAM) radio and C.B. networks can be organized so that information can be gathered and funnelled into an EOC or other appropriate point. Many state and local emergency management organizations have formal associations with amateur radio groups known as the Radio Amateur Civil Emergency Service (RACES).
- Public access channels in a cable TV system can be used. The St. Petersburg, Florida, Fire Department was one of the first fire departments to communicate with the public using cable TV.(9)

During disasters, one form of communication is that provided by amateur radio operators, such as these in Dayton, Ohio. (Photo courtesy of Dayton Amateur Radio Association.)

FREQUENCIES USED IN COMMUNICATIONS

The number of frequencies available for radio communications in a major disaster is very important. When only two were available in the Bel Air conflagration, it was necessary to reserve one for handling emergencies in other parts of Los Angeles. The major operation involving hundreds of units over a 50-square-mile area had to be managed with the other radio frequency. Consequently, it was almost impossible to transmit and receive all messages needed to handle dispatching, placing, and reassignment of the 500 pieces of fire apparatus operating in various sectors of the fire. Mutual aid companies from 23 nearby cities could not be effectively controlled, because they were operating on their own frequencies, all different. Fortunately, the radio frequency for ground control and bird dog planes directing air tankers was separate from the rest of the air traffic. It would have been desirable for the fire suppression officer on the front line to have access to that frequency to describe specific locations where water drops were needed. Instead, he had to relay radio requests via the field headquarter's communications truck.(10)

Following this fire, additional frequencies were obtained for the Los Angeles Department. Since that time (1961) many more have been assigned to the fire service, but the number is not unlimited. The frequency spectrum is being used up at a rate that is sure to result in a serious shortage. For example, one article on this subject points out that there were in northern California only 27 VHF frequencies available for assignment to local agencies and, by 1976, there were 433 licensees on these frequencies.(11) The crowding continued in the 1980s but has been alleviated somewhat in the 1990s by the introduction of the 800 MHz Frequency Band and Trunked Radio.

800 MHz Frequency Band

Most fire and public safety agencies, until recently, have utilized radio frequencies in the Very High Frequency (VHF) band in both the low (30–50 MHz) and high (155 MHz) ranges. Some agencies, mostly police, have moved into the Ultra High Frequency (UHF) band in the 450 MHz range. But dramatic increases in public safety communications needs and the limited availability of frequencies have forced agencies to share channels. This can create mayhem when widespread disasters such as hurricanes, earthquakes, tornado outbreaks, or floods force all of these agencies to communicate at one time.

The situation has been relieved somewhat with the opening of a new 800 MHz frequency band. The new band is most effective when coupled with new computerized trunking technology.

Trunked Radio

Most public safety agencies are assigned one or more radio frequencies which they may share with a fixed number of other agencies. Although the

Many natural disasters result in downed power lines and disrupted communication as seen in this photo from the Xenia, Ohio, tornado of 1974. (Photo courtesy of the Dayton, Ohio, Fire Department.)

amount of radio traffic on these frequencies varies, usually most are not busy. Trunked radio technology originally was developed in the telephone communications field to take advantage of this unused time on these radio frequencies.

Trunked radio systems utilize a bank of 20 to 50 radio frequency pairs all in the same radio band (VHF, UHF, or 800 MHz). A relatively large consortium of agencies usually participates in the system and all utilize identical portable or mobile radios that can communicate on any of the radio frequencies.

Certainly improvements have been made in radio communications. Two decades ago, communication specialists were already worried about the crowding and depletion of the older VHF frequencies. Some were clear channel, having transmitters on high mountains which send loud signals for a hundred miles; these would be difficult to reassign. But VHF clear channels are not available today, as they were twenty or thirty years ago; and present day applicants for frequencies must take into consideration the need for down-tilt antennas, constant tone-coded squelch systems, and low-level sites when developing their new systems. "It is essential that those responsible for communications operations and engineering in the public safety sector begin to conserve available frequency allocations; to begin to re-evaluate their needs; to re-engineer their systems; to provide adequate, but not excessive, coverage; to look to coordination (not necessarily consolidation) with like agencies in adjoining jurisdictions; and to recognize that the possibilities of clear channel assignment of fire frequencies are as remote as the return of the nickel beer."(12)

Since this was quoted back in 1973, we possibly could state today that clear channel assignments are as remote as the "25 cent beer," and add that in light of 800 MHz and trunk radio technology, they are no longer necessary.

The key to the system is a two-way communications link between each radio and a central computer that takes place on a special data frequency. Each radio transmits a unique unit identifier code. The computer is programmed to group units (using the identified code) so they can communicate much as they did with dedicated frequencies. The difference is that any particular communications exchange can take place on any available frequency. To accomplish this the computer tells each radio which frequency is available to transmit on and which to receive on with each transmission and reception.

The importance of this technology is that it permits a very large number of agencies to communicate on a relatively small number of frequencies by maximizing the use of each frequency. Users need not share on-air time with anyone unless desired. The system also allows any individual to be linked or regrouped with any other simply by entering a computer command. Entire groups of units can be regrouped according to predetermined communications plans in the event of a major emergency. The major disadvan-

Modern satellite communications are capable of transmitting technically sophisticated graphic messages which can be color coded and used for detailed disaster planning. (Photo courtesy of the University of Cincinnati.)

tage to the system is the extensive cost associated with the primary and backup computer systems.

Satellites

The use of satellites for public safety communications has not advanced as quickly as some observers predicted when the technology was first introduced. Mobile satellite communications have been used successfully in large mountainous states with small populations such as Wyoming. The federal government has also made use of satellites for long-distance communications.

TELEPHONE COMMUNICATIONS

Cellular mobile telephone has had a major impact on emergency communications by permitting field personnel to communicate with essentially any off-site individual. Cellular telephone is based on the trunked radio technology described above. It employs a much larger number of frequencies than does the old mobile telephone system thus eliminating waiting time for an open channel. Disadvantages of the system are the monthly system access costs and its restrictions to the large urban areas.

Facsimile or "fax" technology allows hard copies of documents to be transmitted between two points using special machines linked by regular telephone lines. This can be very useful in accurately communicating technical information such as hazardous chemical data or casualty lists. At this time fax machines can be used with cellular mobile telephone if "inverters" are installed on the vehicle to provide 110 volt AC power to the fax machines. Portable battery-powered fax machines have just recently become available that eliminate the need for 110 volt power.

The Harrisburg rural fire protection district (RFPD), which protects approximately 64 square miles of western Oregon's Willamatte Valley, reported in 1990 that it had successfully used movable fax machines mounted in fire vehicles. A fire department member staffs these stations during incidents and transmits information from its own computer and from the Oregon state fire marshal data base via the fax machine.(13)

MOBILE COMMAND CENTER

One of the most useful facilities for improving operations during major emergencies is the mobile command center. Early versions were used to coordinate communications at incidents. However, more recent configurations also focus on providing conference and planning space for incident commanders and their staffs. Shown here is a mobile command center used in Cincinnati. It incorporates a forward communications section and a rear conference area with cellular mobile phone, weather monitoring instru-

Pictured here is Cincinnati's Fire-Police Mobile Command Center. Because of its size, driver Jerome Carter views the rear of the van through a television monitor, rearview mirrors being inadequate. (Photos courtesy of Gary Auffart.)

ments, portable computer, portable copying machine, and a complete hazardous materials library.

The Office of Emergency Services in California uses a fire communications and status-keeping vehicle, called COMSTAT, designed in accordance with technology recommended for the FIRESCOPE program. To accomplish its mission of aiding fire managers in tactical planning, COMSTAT has an extensive radio communications system, infrared imagery equipment, a remote weather system, a minicomputer for status keeping, and closed-circuit television. It will house the ground terminal for fire perimeter information telemetered from an aircraft, making photographs almost immediately available to fire managers. Its minicomputer, programmed with data as to the location of every man and piece of equipment assigned to a fire, will provide information to fire officers at five remote terminals. Two portable weather recording stations feed wind, temperature, and humidity information to the computer base, making the current weather situation available to COMSTAT at five terminals.(14)

Oregon's forestry radio section uses a fast-moving mobile radio net to assure optimum communications at fires where more than a thousand acres are involved. The net consists of a jeep with a propane-powered solid-state mobile relay station and a crank-up antenna; a pool of other radio equipment, including two control stations, six mobile and thirty-five portable radios, is also reserved for the mobile fire net. Oregon's 16 million acres of forest and range land is interlaced with powerful repeaters, or relay and lookout stations, and more than 1,300 radios.(15)

RADIO PROCEDURES

A flagrant abuse of good operating procedure, not necessarily associated with a disaster, is the overabundance of testing. In most areas you can hear fire departments testing mobile radios and home receivers daily, or even several times each day, dawn to dusk. Such tests and needless reports interfere with emergency transmissions of neighboring departments and may interfere with operations in other states, due to "skip." One Christmas morning, Chief Bahme had an annoying experience while trying to receive reports from his first-in companies at a going brush fire in Hollywood's Laurel Canyon. He was unable to receive reports because the mayor of a large city in Tennessee was sending greetings to all his fire fighters. A fire captain in Washington announcing the work detail and a fire fighter in Pennsylvania radioing his wife that he would be home late for lunch on frequencies that also "skipped" to Los Angeles also interfered, although they probably did not realize it. Such needless testing and "chit-chat" is not only a violation of FCC rules but can hinder or endanger fire fighters at emergency operations.

Radios used daily need no special tests. If a radio is malfunctioning, wait until all departments with emergency communications are "clear" or back in quarters before making your test.

Some fire departments use the "ten code"; some overuse it. If a driver gets the location of a fire over a pager or telephone, there is no need of his getting a block away from the station and then giving a "10-4" or "10-8" and asking for the location. The dispatcher can assume the driver will respond to the call upon hanging up the telephone, without a radio code number. One large city found, as later confirmed by recent scientific studies, that plain language is more effective and takes less transmission time with fewer errors than the ten code.(16)

Members within the same department should consider others who may need to use the radio for important messages and avoid "hogging" the airways. As long as an emergency can be handled entirely by one company, there is little problem, but when more than one company is working, especially at a major emergency where a hundred companies may be engaged, there should be established radio procedure to avoid interference with officers who may be trying to give or receive instructions, call for help, or keep the dispatcher and the officer-in-charge informed of the situation.

As the magnitude of emergency grows, communications problems increase, especially with arrival of companies called under mutual aid agreements that have different radio frequencies. Even this difficulty can be overcome by lending portable radios on your frequency to outside commanders, or by parking one of their chief's cars next to your communications van at field headquarters.

CIVIL DEFENSE COMMUNICATIONS PROCEDURES

In a large disaster, extending over several days, the commanders of fire departments in the affected area will probably meet to confer periodically on overall strategy. They will leave their operations officers to carry out agreed plans and liaison officers at the civil defense center to receive and send information that may be of assistance or to request help in problems of traffic control, physical attack, and transportation for refugees. In summary, then, the more direct the method, the greater the clarity of communications and the less chance for misunderstanding and delay.

FIELD COMMUNICATIONS

In large disasters, the civil defense organization usually has a communications division that can be called upon to obtain additional telephone services, such as field microwave relays to portable switchboards, additional radio network links for emergency services, and, the services of amateur radio operators.

DISASTER WARNING DISSEMINATION

The sooner and more accurately that emergency services can be notified of an impending disaster, the more effectively they can operate. Levels of government are supposed to maintain a nationwide warning system, but the only way to learn if it is working is to have test exercises. Tests have been witnessed that did not result in the fire department getting the proper bell and light signals in the dispatcher's office from the next higher echelon, as provided in the rules of the national warning system. Electronic failures and human error had to be corrected to make the system function as designed.

Warning and Alert Signals

Everyone is supposed to know that the Defense Civil Preparedness Administration has established an *Attack Warning Signal* consisting of a three- to five-minute wavering sound on sirens, or a series of short blasts on whistles or horns. The public is also supposed to know that an attention or *Alert Signal* consists of a three- to five-minute steady sound on a siren or horn. However, in many rural communities, the residents think that the *Wavering* siren signal means that the volunteer fire department has been summoned to respond to a fire or ambulance call. Most would never suspect that it is supposed to mean that the United States is under enemy attack and that they should take protective action immediately.

When the *Steady* siren signal is heard, the folks glance at their watches to see if the dispatcher is early or late with the noon siren test signal. Most have no idea that this is supposed to mean that local government wants them to turn on their radios or TV sets to get information concerning a peace time disaster. Although many people have heard the *Emergency Broadcast System* tests on their radios, most are not aware that this system may be activated to transmit instructions in day-to-day emergencies posing a threat to life and property.

Emergency Broadcast System

The Emergency Broadcast System is the accepted means of releasing vital civil defense information to the general public during periods of emergency. Facilities and personnel of the entire nongovernment communications industry are used on a voluntary basis to provide a nationwide functional system operated under appropriate government regulations and in a controlled manner consistent with security requirements during a national emergency.

DEVELOPMENTS IN COMPUTERS AND INFORMATION MANAGEMENT

Information Management

The fire chief of today is called upon to use computer and electronic data processing equipment throughout his organization so that the additional demands placed on the fire department with fixed resources can be met. In essence the fire chief of today can use contemporary computing technology to "work smarter not harder."(17) Most fire departments have introduced microcomputers throughout their office structure. Many new types of software allow the application of computer technology not only to administrative tasks but to emergency field operations. Operating a computer is easier than ever before, computer hardware is more portable and rugged than ever before, and the range of possibilities for effective computer utilization continues to grow at an astonishing rate.

The use of computer technology has caught on more rapidly in administrative applications in the fire service than in emergency or field operations. This is changing, however, as departments are beginning to adapt more and more computerized hardware for operational tasks. In a February 1989, article Hank Christen indicated that in a modern fire department or emergency management agency a tools/equipment chart might read: *Compartment One*—breathing apparatus and spare cylinders, *Compartment Two*—hydraulic extrication tool and accessories, *Compartment Three*—cellular telephone with an error-correcting modem, laptop computer, software, and printer paper.(18)

Because computers can search for and receive information in seconds, they can prove to be invaluable when hazardous materials data and guides for emergency action are needed rapidly. Databases are also available for varying types of equipment and personnel resources. "Recent hardware development has made it possible to bring computer databases directly to the incident scene.

"Even before a fire agency arrives on the scene, computers can prove helpful. They can be used to provide optimal response routes and directions to any preprogramed location in a given jurisdiction. "The screen output allows the user to simply step through a map, street segment by street segment, with all useful data being included in tiled display."(19)

Effective decision making during a major crisis is heavily dependent on information that is timely, accurate, and packaged in a useable form. Computers are having a major impact not only in increasing the amount of available information, but in providing methods for managing large amounts of data so that they can be analyzed and quickly retrieved.

Developments in Personal Computers

Developments in computers and information management continue to progress at a rapid pace. Computers continue to become smaller, more portable, and more powerful. A few years ago we saw portable, AC powered personal computers (PCs) the size of portable sewing machines. Now we are seeing battery-powered laptop portables the size of attaché cases. The most recent models, utilizing 80386 processors and 20 MHz clock speeds, can process information incredibly fast with remarkably powerful software.

Advances in data capacities of fixed disk, floppy diskette, and microdiskettes are permitting very large amounts of data to be stored and accessed in the field. CD-ROM and optical drives are now available that can store and access over 600 megabytes of data.

Data Communication

The expanded use of portable and laptop computers has led to increased communication of data between on- and off-site computers. For example, computers in the field can access detailed chemical inventory records stored in a central mainframe or computer-aided dispatch computer.

A variety of remote commercial databases, such as Hazardline, are available. They can be accessed from the field and provide extensive, up-to-date information on thousands of chemicals.

Data can also be communicated between any two personal computers using ordinary telephone lines and modems (modulation-demodulation). For

Personal computers have proliferated in the fire service, both in fire department offices and in the hands of individual fire service members. (Photo courtesy of the National Fire Academy.)

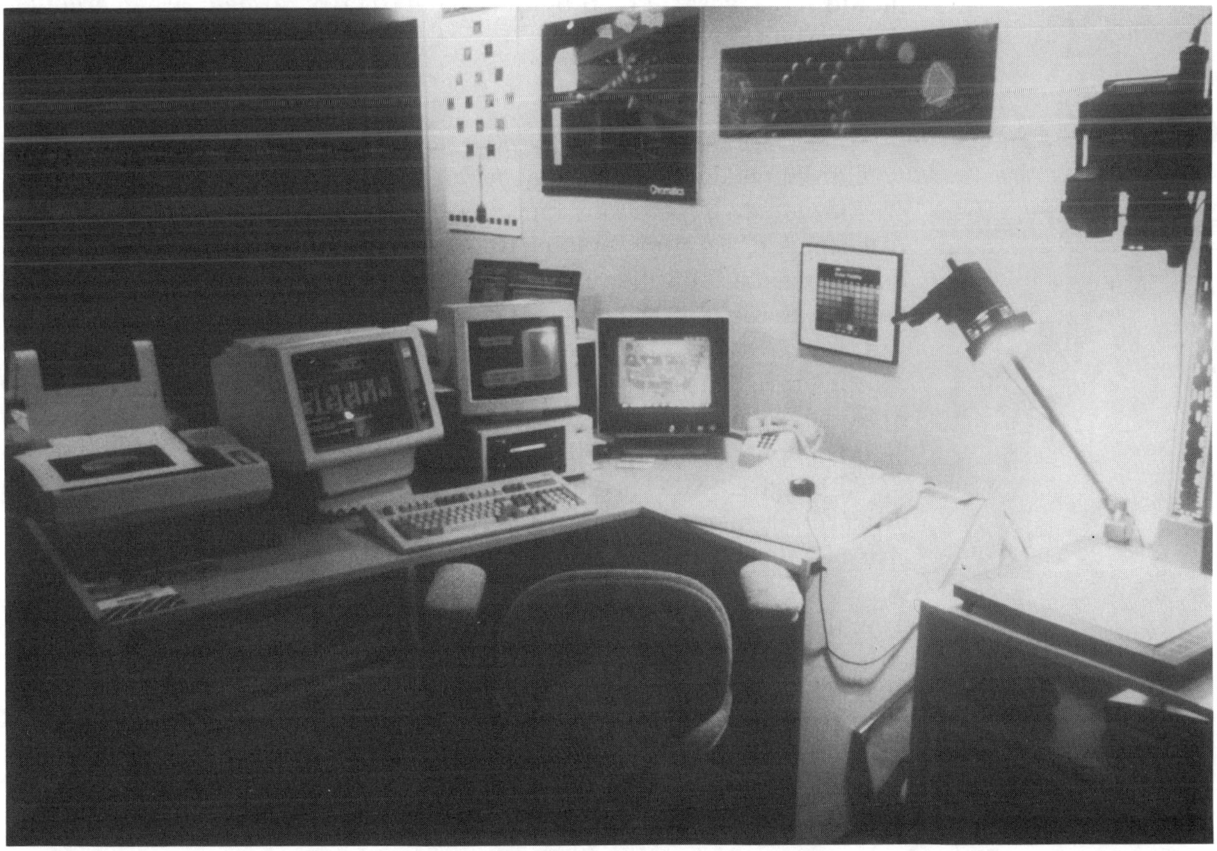

technical reasons, standard modems are not yet fully compatible with cellular mobile telephones, however, cellular modems are now available that operate at transmission rates up to 9,600 baud.

Another development originating in amateur radio circles is packet radio. This technology involves communication between two computers utilizing ordinary two-way radios and inexpensive electronic devices called Terminal Node Controllers (TNCs)—one at each end. These TNCs group data into "packets" and transmit them in a way that does not significantly interfere with voice communications. The technology employs a sophisticated scheme for eliminating errors that can arise from radio interference or variations in signals. Your local ham radio operators can provide this service during an emergency.

Packet radio modems are now available that can transmit computer data over dedicated frequencies in the VHF, UHF, and 800 MHz public service bands at transmission speeds of up to 4800 baud. These packet modems do not work with trunked radio systems. A separate frequency along with necessary relay equipment must be provided for the system. Many computers can, however, share the system as each packet of radio computer data carries the digital address of the specific computer with which you wish to communicate.

Developments in Computer Software

A number of software computer programs have been developed recently including several PC programs specifically aimed at emergency management.

CAMEO™

CAMEO is the trade mark name for the Computer-Aided Management of Emergency Operations developed by the National Oceanic and Atmospheric Administration (NOAA) in conjunction with the U.S. Environmental Protection Agency and is based on techniques developed in weather predicting. The program currently runs on both Apple MacIntosh and MS-DOS computers and costs a nominal amount to cover the costs of computer diskettes, manuals, and user support.(20) CAMEO is available from the National Safety Council, 444 No. Michigan Avenue, Chicago, IL 60611. Phone: (800) 621-7619 ext. 6900.

Although CAMEO is intended for use in a variety of emergencies, it was primarily developed for hazardous materials incidents. The program consists of a database of some 3300 chemicals cross-indexed with more than 60,000 synonyms and a set of local maps digitized by the user or available from the Census Bureau and stored on diskettes. In the field CAMEO can be accessed to obtain detailed technical information about each of the chemicals. This chemical database, in conjunction with the digitized maps and current weather information (entered at the time of the incident) can then be the basis for constructing an air dispersion model to project formation and movement of a toxic gas cloud.

CAMEO is configured to store all information required by the SARA

Title III and can be used in pre-incident planning for major facilities in which specific high-risk chemicals are known to be present. It's graphic capabilities also make it an excellent pre-incident planning tool for nonhazardous materials facilities.

CAMEO MacIntosh employs a user-friendly operating system that consists of a mouse that allows the user to point to an icon on the screen and then click a button to select the corresponding element of the program. This system is much simpler than having to remember precise codes that must be entered from a keyboard. Figure 6-3 depicts the initial Navigator screen in which each icon represents an element of the program that can be selected.

Figure 6-3. Initial Navigator Screen Showing Icons to Select Program Elements.

Figure 6-4 shows the screen that appears when the Codebreaker element has been selected. This element allows the user to search for detailed chemical information when only incomplete information is available (such as trade name, United Nations number, or DOT classification).

Figure 6-5 is an example of a digitized map with the projection of a chlorine plume (shaded area) extending from the incident site near Boeing Field to the southwest. The projection will change with time, changing weather conditions, and changing air concentrations of the chemical. CAMEO can also be used for preplanning for specific facilities.

EIS

The Emergency Information Systems (EIS) program actually consists of several software versions marketed by Research Alternatives, Inc., of Rockville, Maryland. EIS is an IBM-based program very similar to CAMEO, but considerably more elaborate. It, too, consists of a chemical database, digitized color maps, resource information, and facility and community planning information. It has various screen forms used for tracking multiple individual incidents such as fires, utility outages, injuries, gas leaks, and

Figure 6-4. Codebreaker Screen to Search for Chemicals.

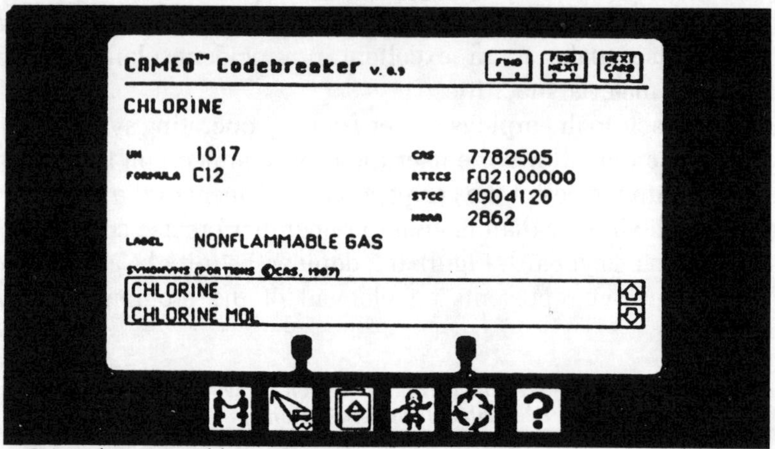

Original Version: Chlorine Example

Newer Version: Generic

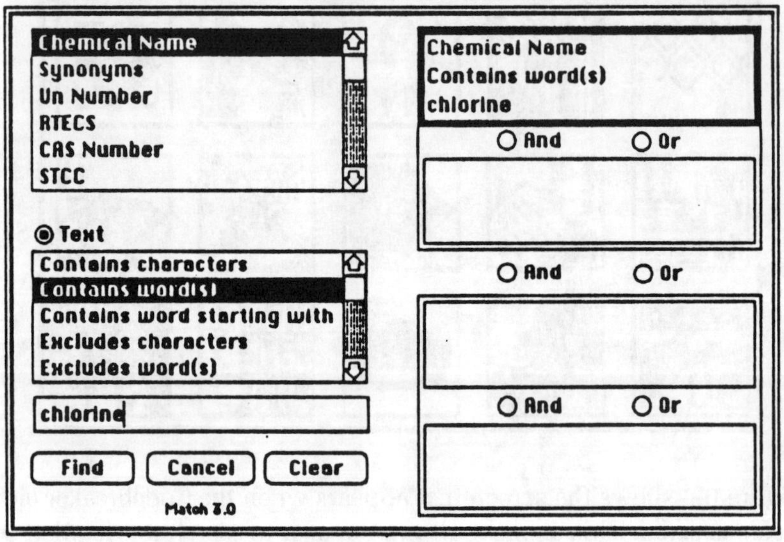

Figure 6-5. Digitalized Map with Projected Path of Chlorine Cloud.

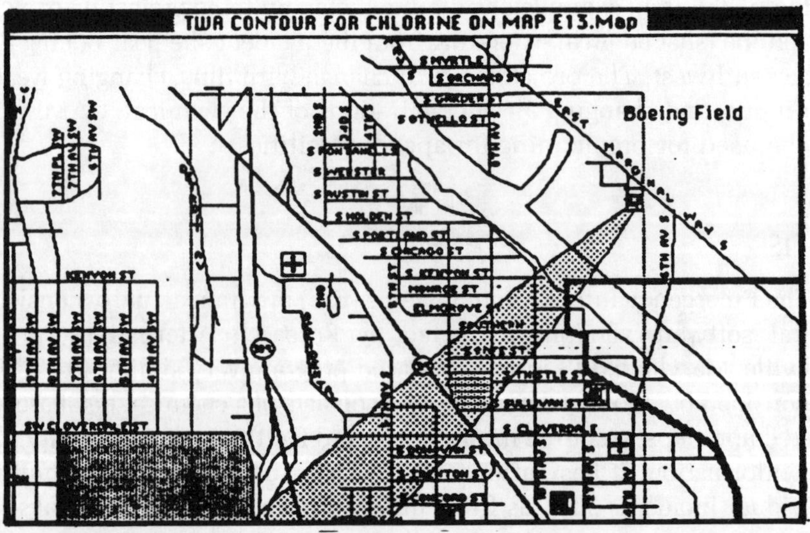

other episodes that develop during long-term, widespread disasters such as an earthquake or tornado touchdown.

EIS/C is a version developed specifically for SARA Title III. It contains additional forms and capabilities for entering data from fixed facilities. A special version is also available for use in EOCs. One drawback of EIS is that the program cost can run from $4,000 to more than $20,000, depending primarily on the number of digitized maps required.

Emergency Scene Reception of Computer Data

One of the most serious problems associated with hazardous materials accidents is the gathering of information. Very accurate information is necessary to formulate a plan to reduce or eliminate a hazard created by a hazardous materials accident. Manuals are available providing information on several hundred chemicals that are shipped or used in large quantities and are involved in many accidents. Unfortunately, accidents occur involving materials that are new, rare, or shipped under a trade name not included in most manuals.

Two cities (among others) that have documented success in the reception of computer data at an emergency scene are Dayton, Ohio, and Seattle, Washington. The Dayton, Seattle, and other fire departments are able to receive CAMEO on a microcomputer in hazardous materials response vans. "Now, instead of spreading books and prefireplans across the desk in the hazmat van, fire fighters push buttons on a computer keyboard and get critical information in seconds."(21)

Captain Dennis McCloskey, commander of the Dayton, Ohio, hazmat unit at the time we went to press, was an early pioneer in emergency scene reception of computer data and provides much of the information in this section.(22)

When information is needed on a material that may be hazardous at the scene of an emergency, past practice has been to contact Chemtrec and have it provide the information. Chemtrec has one of the largest files on chemicals in the United States. This is usually accomplished through a dispatcher, as telephones are usually not initially available at an emergency scene command post. Most of us have participated in command or hazmat communications training exercises where information is relayed through several people in different rooms over two-way radios. These exercises demonstrate that it is almost impossible to relay anything more than one simple sentence over a radio accurately. Other problems occur when labels or shipping papers are damaged and only incomplete information is available to identify the material.

Chemtrec has been aware for quite some time that mistakes are common when technical information is relayed verbally. These errors do not occur when data are transmitted between computers over telephone lines. To improve the accuracy of its service, Chemtrec initiated its HIT (Hazardous Materials Transmission) Service. Chemtrec planners figure if a complete and accurate printed copy of the information could be sent to the emergency dispatcher, the dispatcher could read it as many times as necessary to the fireground commander. This method would also enhance Chemtrec's end of the operation, as computer transmission of the data would speed up the process and free up its personnel for other tasks.

To receive a Chemtrec HIT, a personal computer equipped with a modem connected to a telephone line is needed. Chemtrec felt that because this equipment is so common in the business community, most fire departments already have the equipment or would have little trouble getting it. The computer could be used for other purposes if necessary, as long as the other uses could be suspended and it could be set up to receive a Chemtrec HIT in a few minutes. The computer does not need to be one of the expensive models, as a $700 portable system is capable of receiving and then printing the transmission.

As Chemtrec was developing this system, the computer industry was developing and marketing battery-operated portable laptop models that could be carried in a briefcase. These laptop portables can be equipped with a modem and are capable of receiving data via telephone lines. As laptop computers became popular, cellular telephones also became big sellers. Someone who owned a laptop computer and a cellular telephone no doubt tried to transmit data from his or her computer over a cellular telephone and found that the cellular telephone network caused errors in the data transmission. This problem became well known, and was a problem for both the portable computer and the cellular telephone industry. Soon, an error-checking modem was developed by Spectrum Cellular for Ameritech, the telephone company. This made laptop computers truly portable, capable of transmitting or receiving information anywhere a regular or cellular telephone can be used.

Ameritech marketing personnel studied the potential uses for its newly developed error-checking cellular modems and realized that the Chemtrec HIT program would be much more valuable to emergency responders if a printout could be provided at the scene of the emergency rather than at the dispatcher's position. This is accomplished with a cellular telephone and Bridge and Span error-checking modems so that Chemtrec HITs can be received at the emergency scene. It is a simple matter to add Bridge and Span compatibility to a cellular telephone specification.

It is suggested that the system utilize a MS-DOS compatible personal computer in the dispatcher's office connected to a dedicated telephone line. The computer would store the data and transmit it to a hazardous materials field unit at the convenience of the hazmat crew. See Figure 6-6 for a schematic diagram.

A data terminal device can be used in the field rather than a portable personal computer, because standard computers and printers are voltage, dirt, and shock sensitive and may not survive a fire vehicle environment. Needed is a fast, rugged terminal printer that is equipped with a keyboard, can be operated from a 12-volt dc electrical system, and can be connected to a cellular telephone modem. A personal computer operating in a clean, air conditioned, safe environment at a dispatcher's headquarters operating with the rugged 12 volt terminal printer in a field unit is a good combination.

One problem is that line fire fighters with no computer experience are unable to operate a system with reliability. This problem is alleviated somewhat by communication software capable of automatic operation if someone writes command and script files. These files are user created computer instructions that tell the communication software when and how to make the computer do something. It is possible to simplify the operation so that no commands need be entered by a hazmat crew in the van. The only task the crew is required to master is turning on the cellular telephone, terminal printer, and cellular modem. The staff in the dispatch center needs to learn how and when to depress escape, return, and F1 function keys.

Two of the new CAMEO DOS elements will work especially well with this setup. Codebreaker can search a data file of over 60,000 synonyms to provide the proper chemical name of over 3,300 chemicals if a synonym,

Figure 6-6. Chemtrec to Hazmat Links.

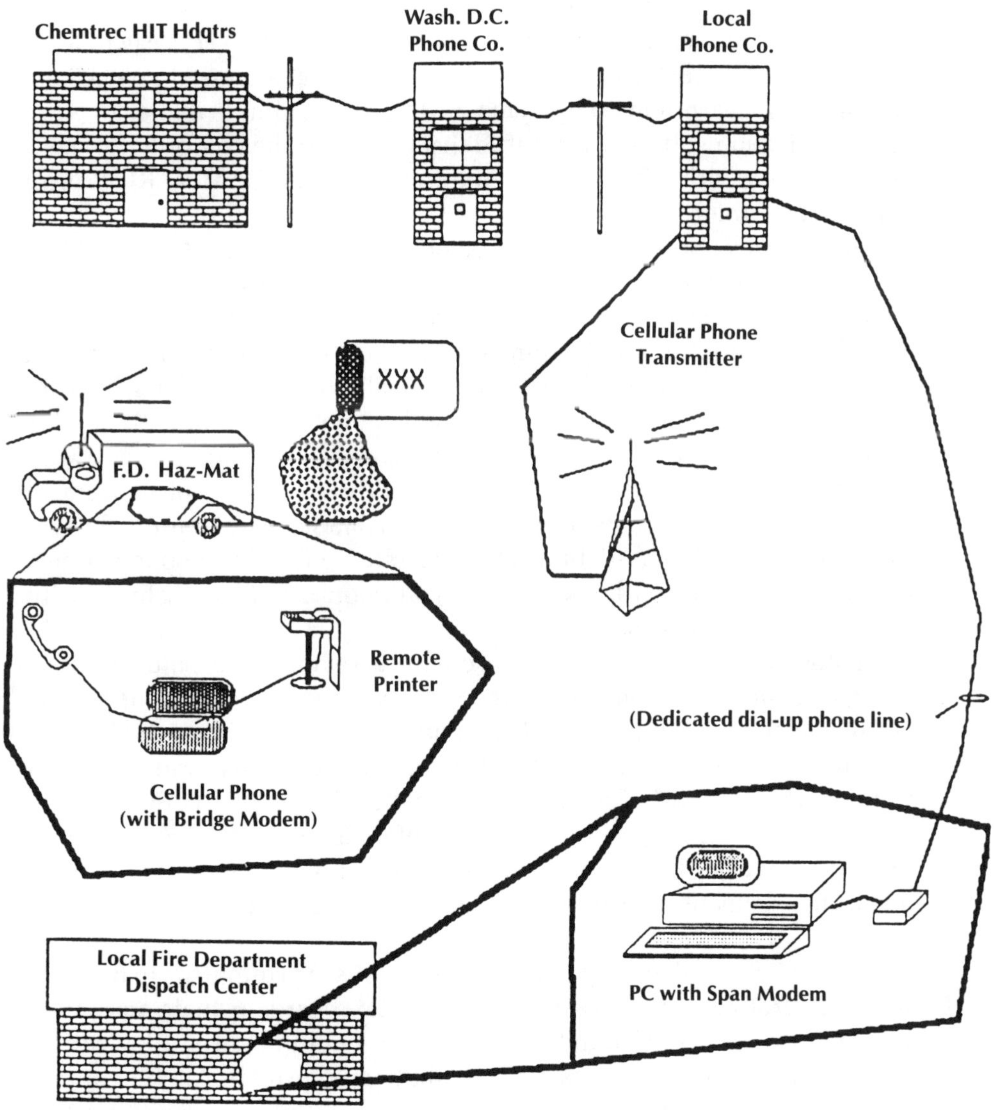

trade name, chemical formula, CAS Number (Chemical Abstracts System), RTECS Number (Registry of Toxic Effects of Chemical Substances), or UN (DOT) Number can be found. Chemtrec can provide a HIT only when the proper chemical name is known.

Once the proper chemical name is known, the CAMEO RIDS (Response Information Data Sheet) element can be used to provide a printout of emergency response information. A CAMEO RIDS print is similar to a Chemtrec HIT. You should try to obtain information from at least two sources before attempting a dangerous operation. Most hazmat units have a library in the vehicle, but some books may become obsolete. Chemtrec HIT files are updated daily and NOAA has promised to update the CAMEO chemical database frequently.

An MS-DOS computer with a 40 megabyte hard disk drive is more than sufficient for the program and data files necessary for the Chemtrec HIT, CAMEO Codebreaker and RIDS database.

We have already discussed CAMEO in some detail. We should note that the programs are user friendly and menu driven. They are very easy to learn as the next choice of steps are always listed. The staff at most dispatch centers can easily learn how to operate CAMEO. It is a simple matter to write a communication software script file that can order the computer to make the connection with a field printer and then print the CAMEO RIDS print on the field printer.

If cellular telephone and fax equipment are available, another important and valuable Chemtrec service can be economically utilized in the field. Chemtrec HITs are only available for pure chemicals, they are not available for trade name mixtures and commercial products. Chemtrec can, however, fax an MSDS of most trade name or commercial products. Chemtrec has over 420,000 MSDSs electronically stored and immediately available for fax transmission.

To receive a fax in the field, the only additional piece of equipment needed is an inexpensive item available from Ameritech or Motorola known as a Modular Telephone Adaptor. This accessory will allow you to connect any telephone device with a standard RJ-11 connector (fax machine) with most cellular telephones.

Another valuable information resource that should not be overlooked during a hazardous materials emergency is your local or regional poison control center. Many of these facilities and hospitals utilize comprehensive computer databases such as TOMES or TOMES Plus trademarks of Micromedex Inc. While Chemtrec HITs and MSDS faxs or CAMEO RIDS may contain up to four pages of chemical information, a TOMES Plus printout can be forty pages or more. Detailed information on medical treatment, hazard management, dangerous chemical reactions, and legal disposal is included.

The most convenient way to receive one of these printouts is to have the poison control center or hospital install a fax board in their computer. Installation of a SatisFAXtion board, trademark of Intel Corp., allows them to fax you the document just as easily as they could print it. Another option is to use modems and have the document transmitted as a computer text file similar to the way Chemtrec sends HITs. This can be slightly more difficult as the document must first be saved as a computer text file, and some databases are not set up to do this easily.

Fire departments that have used the CAMEO program at the scene of several emergencies to date have found CAMEO to be of more value that originally anticipated. During one incident involving Mercaptan, in Dayton, Ohio, CAMEO warned that there are eighteen different chemicals that may be shipped under the trade name Mercaptan or a synonym, several of them being very toxic pesticides. Additional investigative work was conducted by Captain Dennis McCloskey and his hazmat crew until they were sure that they had identified the proper Mercaptan. Benzene and acetone are other words that fire fighters have found to be very difficult to deal with, as many different chemicals may be shipped under these names. Additional investigation is needed to identify them by their proper chemical name. See Appendix F for complete software instructions for CAMEO and emergency scene reception of computer data.

Computer-Aided Dispatch

Computer aided dispatch (CAD) systems have been utilized by large city fire departments for a number of years. Advances and cost reduction in small business computers have recently brought many features of computer-aided dispatch within reach of the medium sized and smaller departments.

The original CAD systems utilized a computer to help dispatchers make decisions about what apparatus to send to particular alarms. Information about specific occupancies or various occupancy types such as industrial, mercantile, apartment, or residence were programmed into the computer, which also tracked the status of apparatus. When an alarm was received, the address and occupancy type were keyed into the computer which recommended an apparatus assignment. This assignment could be modified by the dispatcher.

Eventually CAD systems were integrated with radio systems allowing them to automatically transmit tones to activate station alerts, turn on lights, and raise bay doors. CAD systems were also integrated with silent status-keeping systems. These systems utilized apparatus radios and could send various routine messages (such as in- or out-of-service) by transmitting sub-audible tones rather than voice messages. This permitted the computer to track the status of vehicles with no input from the dispatcher.

The sophistication of CAD systems has continued to improve. The following are some key developments of the 1980s and 1990s.

- CAD computers can now store large amounts of data regarding specific occupancies including the presence of hazardous materials, disabled residents, or other special considerations.
- Mobile data terminals mounted in fire apparatus or mobile command centers can tap information in any of the databases maintained on the CAD computer. This can be critical if a large amount of information must be transmitted accurately.
- CAD computers can provide useful historical and administrative information. For example, it may be helpful for fire officers to know

that they are responding to an address that has had two arson fires in the last week or one where the police have responded twice in one day on domestic disturbances.

- CAD systems are being integrated with enhanced 911 systems that can automatically provide the telephone number and address of an emergency caller. Thus apparatus assignments can be computer generated automatically as soon as a call is received.

In Cincinnati, a computer-aided dispatch system improves a dispatcher's incident entry equipment, status monitoring, and general dispatch capabilities through use of video display and computers. This is accompanied by an apparatus status system that enables fast and reliable data transmission from apparatus to the control center by digital signalling techniques. This system helps the dispatcher keep accurate records and current status reports and records on department apparatus. A printer communications system enables a dispatcher to transmit radio messages to remote apparatus, vehicles, and facilities which are then transformed into permanent hard copy. Added directly to a department's two-way radio system, each of these "command and control" systems is designed to be incorporated into the dispatching operation individually, in any order, at any time. The only determining factor is department need. Systems not only have this flexibility of installation, but also flexibility of performance. The capabilities of each can be specifically tailored to meet individual department needs.

Many fire departments are having their dispatch centers upgraded for the second or third generation of CAD equipment. Common problems are lack of standardization, outdated equipment, insufficient space, and "make do" repairs.

Second-generation CAD software contains more than dispatching programs. Systems of the 1990s are reporting the inclusion of management information systems, database systems, and mapping/graphics systems. They not only provide linkage to other operating systems such as 911, mapping, fire prevention data, preplanned drawings/data, and resource/contact databases, but also can operate separately or in tandem. For example, a fire prevention database can operate in one office and the CAD system operate in another. The 1990s CAD systems include:

- dispatching
- hazardous materials
- vehicle status
- networking.

Jeff T. Chorba, systems integrator for a leading company involved in installing CAD systems, states, "Systems that I dreamed about three years ago are already here. Systems that were termed futuristic in 1987 are now routinely installed."(23)

It is recommended that fire departments considering upgrading communication and dispatching equipment maintain constant contact with agencies and other vendors who have made recent sales or purchases of technologically current equipment. All the most sophisticated equipment in the world is useless, however, if the human element—call taker or dispatcher—is not trained properly and has not had a job design integrated into the newer system.

With centralized 911 dispatching, all calls for assistance, fire, police, and medical usually arrive at the same point. A choice then for a system designer is to hand the call off immediately to the responsible agency so that a dispatcher can talk directly to the caller or to have the call taker collect all information and then transfer the information to the proper agency. While each of these systems has advantages, a given jurisdiction will have to decide which of the two it prefers and then design the system for the most efficient transfer of information from the caller to the responding units. In Cincinnati, a newly installed CAD/911 system uses the latter approach. While it has the advantage of only one conversation between caller and public complaint operator (PCO), it has the disadvantage that the dispatcher to whom the call is transferred from the PCO via video screen has not personally talked to the caller.

Computerized communications are not limited to large fire departments. They are also employed in volunteer fire departments. The CADRE (Computer Assisted Dispatching and Reporting) system developed for the volunteer fire departments in New Castle County, Delaware, provides the same services as the system in Cincinnati, which covers many times the same size area.

In Worcester, Massachusetts, a radio teleprinter dispatching system uses a scanner that reads eighty characters per card, converts the information to sound, and then sends it to receivers in fourteen stations and four vehicles. The printout received at the fire station can be recorded and paged if the message involves need for personnel or units housed there. In case of notice of alarm, the hard copy can be torn off and taken to the scene of the emergency. The entire procedure is completed in seconds.(24)

One of the greatest aids to effective communications in emergency operations is radio, personal and mobile types. The most important piece of equipment carried by an incident commander is his portable radio. Long ago, speaking trumpets were replaced by portable radios, and they have continued to improve in range, clarity, and capability ever since. Over the years varying types of noise-attenuating headsets and ear microphones have been tried with varying degrees of success. Currently, due to OSHA awareness of hearing damage, attenuating headsets are becoming increasingly popular. *Fire Engineering* reported that "high quality headset systems contain acoustical filtering, audio speakers, a microphone, and a radio interface module."

Mobile repeaters have increased the effectiveness of portable equipment in situations where the officer is in a "dead spot" inside a building and his portable is unable to contact the control center. With a mobile repeater, the portable equipment activates the mobile transmitter which in effect makes the portable unit equal to the power output of the mobile transmitter.(25) Currently, however, mobile repeaters are less popular as technological improvements have led to powerful transmission capabilities in handheld portable radios.

Mobile radios on pumpers with noise-cancelling microphones and all-weather speakers have greatly improved the ability to communicate with pump operators. Tape cartridges can be inserted into a player at the rear of the fire truck's cab, as described in an article about volunteer fire fighters in Little Ferry, New Jersey; when their quint responds to an alarm the tape gives them a two-minute briefing on the location of hydrants, hazards in factories and other buildings, locations of utilities shut-offs, the size of water mains, and other relevant information.(26)

COMMUNICATIONS EQUIPMENT

Communications equipment may be housed in a structure, a vehicle, or in your pocket. If in a structure, the building should be fire resistive, on high ground (if in a potential flood area), preferably in an isolated location, secured against attack by rioters or vandals, and otherwise meeting specifications for construction and emergency power supplies required by NFPA Standard 73, *Public Fire Service Communications.* To obtain protection against nuclear attack and radioactive fallout, some cities are locating control centers in underground facilities, designed to be self-contained for living requirements. The trend is to incorporate computers, radio teleprinters, data processing equipment, apparatus status systems, and other forms of automation to assist in dispatching operations.

Communications System Vulnerability

As communications hardware becomes more sophisticated, it also becomes more vulnerable to damage caused by the effects of disasters. Many of today's radio, telephone, and dispatch systems are partially or totally dependant on computers. Brief power interruptions or fluctuations can render them inoperable. Although most systems employ uninterruptable power sources (normally sophisticated battery systems) these backups can only last a certain period of time without normal or generator-produced AC power.

Some of the newest communications technology relies on microwave or fiber optic technology, which can be very sensitive to certain environmental conditions. For example, an earthquake can interrupt microwave communications by knocking dishes out of line.

Telephone systems are generally well prepared to deal with disasters, in part because of their importance to the national security system. Widespread telephone service outages can be caused by long-term power interruptions, fires, or other damage to key telephone switching sites.

System Overload

Probably the majority of communications problems encountered during disasters are due not to hardware damage but to system overload.

Radio systems can become overloaded for several reasons.(27)

- Lack of or ineffective incident command systems can result in no clear channels for information flow. Agencies and officials begin functions independently and attempt to communicate with practically everybody.

- Radio channels may be shared with other agencies. This may not cause a major problem until a large natural disaster occurs in which almost all such agencies are attempting to communicate at once.
- A generalized type of radio interference can occur when the majority of government and business radio communications systems become active over a large area. Peculiar weather conditions that frequently accompany disasters such as a tornado or hurricane can also cause interference.
- Additional communications needs that occur during major emergencies include
  - the need to communicate with agencies that one does not communicate with normally
  - the need to make a very large number of decisions by individuals at a lower organizational level
  - the increased need for information such as situation status, resource status, and technical advice
  - the need for higher levels of coordination with surrounding agencies brought on by extreme shortages of resources.

Telephone systems can easily become overloaded during wide spread disasters. This results in delays in obtaining a dial tone or failure to get one. During a recent earthquake in the Midwest the telephone system in some urban areas experienced dial tone delays of two to five minutes.

Most emergency dispatch centers, hospitals, and other essential agencies have high-priority status for accessing dial tone during emergency situations. However, this status is maintained by a telephone company computer, which can fail, causing serious problems in contacting outside agencies.

IMPLICATIONS FOR DISASTER PLANNING

Communications sections of disaster plans should identify the communications problems most likely to occur. Policies should be established to minimize communications overload and provide backup and auxiliary systems. Some specific steps include the following.

- Procedures should be developed for alerting the public to avoid use of the telephone unless necessary.
- Meetings with the telephone company should be held to assess its vulnerabilities and backup systems and procedures. Improvements should be agreed upon, if indicated.
- Sections of the plan addressing damage assessment should include reconnaissance by fire, police, or organized civilian groups rather than simply reacting to service demand received by phone. The most heavily damaged areas may be without phone service.

- ○ Backup communications systems should be in place that minimize reliance on the following:
  - computers
  - normal power service
  - extended use of battery backup systems
  - satellite receiver stations
  - a single transmitter or receiver site
  - the telephone system.
- ○ Totally separate auxiliary communications capabilities should be identified. Most communities have several active groups of FCC-licensed amateur radio operators with state-of-the-art communications systems, including portable radios, repeater networks, packet systems (for transmitting data), and amateur television.

These groups can be tremendously valuable for

- ○ amassing widespread damage assessment information
- ○ collecting requests for public safety services from the public
- ○ spotting severe weather
- ○ establishing communications links among groups not normally associated with emergencies
- ○ compiling casualty lists.

Citizens Band (CB) radio has been used successfully in many areas. CB hardware is generally less sophisticated and the groups frequently less well organized that those in amateur radio.

CONCLUSION

In this chapter we have seen how communications and information management have grown in complexity, as have the number and types of incidents the fire department is called upon to control. New technology exists, however, and is being harnessed by contemporary incident commanders to assist them in their role of handling disasters. We have considered advances in computer technology, portable radio equipment, transmission technology, and other forms of hardware and software, which are available and are rapidly being adopted by progressive fire departments everywhere. We have also touched upon the human element, the efficient utilization of which brings out the very best in technology. Leaders in the fire service arena are finding that in order to successfully control major incidents, including all types of disasters, one of their greatest allies is efficient telecommunications.

CHAPTER 6 REFERENCES

1. Kimball, Warren Y. *Fire Service Communications for Fire Attack*, Boston, MA: NFPA, 1972, p. 188.

2. Christen, Hank. "Computer Tech: Fiber for the Fire Service" in *Firehouse*, Vol. 16, No. 4. (April 1991), p. 70.
3. De Lawter, Diane. "The Art of Communication" in *Firehouse*, Vol. 16, No. 2. (February 1991), p. 50.
4. Collier, W. Steve. "Communications and Information Management for Fire Defense and Disaster Operations" in *Disaster and Fire Defense Planning*, Emmitsburg, MD: National Fire Academy, 1989.
5. Meyer, Steve. "Organizing Crews in Incident Management" in *Fire Engineering*, Vol. 144, No. 3. (March 1991), p. 113.
6. "Nine-One-One and Communications Centers," A.I.A. Special Interest Bulletin No. 322, July 1970.
7. Landers, Ann, in syndicated column, appearing various dates in daily newspapers, 1990.
8. "Logistics" in *Incident Command System*, 1st ed. Stillwater, OK: Oklahoma State University, 1983.
9. Knight, Jerry. "Cable TV—A New Resource for the Fire Service" in *Chief Fire Executive*, Vol. 2, No. 6 (December 1987), p. 14.
10. Bahme, Charles W. *Fire Officer's Guide to Disaster Control*, 1st ed. Boston, MA: NFPA, 1978, p. 70.
11. Watson, Max E. "You Get Gray Hairs Allotting Radio Frequencies" in *Fire Engineering*, May 1976, p. 82.
12. Mueller, Andrew J. "10-91—or—Is This Radio Transmission Necessary?" in *Fire Command!* October 1973, p. 26.
13. "Communicating—FAX Machines in the Firehouse and the Apparatus" in *Firehouse*, Vol. 15, No. 2. (1990), p. 23.
14. "Computer Vehicle to Keep Tabs on Men, Fire, Weather" in *Fire Engineering*, October 1974, p. 75.
15. "Radios Cut Forest Fire Losses" in *Fire Engineering*, October 1972, p. 57.
16. ABC News Report, April 13, 1991.
17. Johnson, Phil. "Work Smarter, Not Harder" in *Chief Fire Executive*, January/February, 1990.
18. Christen, Hank. "Emergency Computing" in *Firehouse*, Vol. 14, No. 2, (February 1989).
19. Rosenhan, A. K. "Computers" in *Chief Fire Executive*, January/February/March, 1989.
20. Harris, L., J. Richard, J. Galt, D. Kummerbowe, D. Payton, D. Chan, and F. Migaiolo. Chapter I: "Guided Tour of CAMEO" in *The CAMEO Manual*, Seattle, WA: National Oceanic and Atmospheric Administration (NOAA/HAZMAT), 1991.
21. Jacob, Dave. "Cameo Appearance" in *Firehouse*, Vol. 13, No. 2 (February 1988), p. 49.
22. McCloskey, Dennis. "Emergency Scene Reception of Computer Data," thesis prepared for the University of Cincinnati's Fire Science Program, May 1991.
23. Christen, Hank. "Computer-Aided Dispatch Comes of Age" in *Firehouse*, Vol. 16, No. 2. (February 1991), pp. 32–35.
24. Thompson, Ralph R. "Radio Teleprinter Dispatching in Worcester, Massachusetts" in *Fire Engineering*, April 1974, p. 68.
25. Mello, Alfred J. "Coming Changes in Communications" in *Fire Command!* June 1971, p. 19.
26. "Taped Information Played in Cab During Response" in *Fire Engineering*, October 1973, p. 52.
27. "Voice-Scrambling Two-Way Radio" in *Fire Engineering*, Vol. 144, No. 1 (January 1991), p. 85.

CHAPTER 7

CATASTROPHIC FIRES

THE TYPE OF DISASTER FIRE service personnel are most likely to encounter, and are most familiar with, originates with fire.

It is not within the purview of this study to set forth all the principles of fire fighting already covered in books on fire suppression, such as those obtainable from NFPA, *Fire Engineering*, and other sources. The emphasis here will be upon the unusual requirements necessitated by the magnitude of the operations and the number of lives that may be endangered. Most of our attention in this chapter will be toward the three types of fires most likely to take on disastrous dimensions: urban conflagrations, wildland/urban interface fires, and occupied highrise building fires. We term these "catastrophic fires." Common characteristics of all three include an extensive volume of fire, serious life hazard, and difficulty in containment. Several other categories of fires with potential for heavy loss of life are also covered in this chapter.

Certainly among the most disastrous of fires from the fire fighter's perspective are those involving the loss of several brothers or sisters in multiple fire fighter fatality incidents. In 1989, we had just finished mourning the loss of five fire fighters in Hackensack, New Jersey, killed as the result of truss-roof collapse in a controversial incident, when six fire fighters were killed in an ammonium nitrate fire and explosion in Kansas City, Missouri.

Not long thereafter, on February 23, 1991, the nation mourned the loss of three Philadelphia fire fighters trapped while fighting a spectacular highrise fire at One Meridian Plaza.(1) While fires involving multiple life loss among fire fighters may not be considered disastrous by the general public, they are indelible in the minds of us in the fire fighting profession, probably because they serve as a reminder of the dangers faced daily in our chosen profession.

FREQUENCY OF FIRES

About 2.2 million destructive fires occur every year in the United States, in which about 6,000 people lose their lives. Relatively few of these fires become "catastrophic." On a positive note, the numbers of reported fires and deaths are declining, according to a five-year analysis available in 1990. The number of multiple-death fires is decreasing also in direct proportion.(2) One category that is increasing, however, is fire loss, which was up to 7 billion dollars annually at the time of publication. When this figure is adjusted for inflation, however, it is also declining. Fire loss and fire deaths are expected to stabilize or decrease throughout the remainder of the 1990s.

A contributing factor to these situations is a corresponding increase in residential smoke detectors, which now protect close to 80 percent of all American households. Since detectors have the potential to discover fires in their incipient stages they are a powerful force in reducing the North American fire loss.

Of the 2.2 million or so fires each year less than ten develop into catastrophic fires or conflagrations. Those that do, however, cause hundreds of millions of dollars per year in fire loss. When a fire has reached the catastrophic magnitude of a "disaster," an incident commander must scramble to mobilize forces for defensive tactics, and face what is probably the most difficult tactical assignment of his or her career.

URBAN CONFLAGRATIONS

An "urban conflagration" is a large, destructive fire that spreads beyond natural or artificial barriers; it can be expected to result in a large monetary loss and may or may not include fatalities. Often called simply a "conflagration," it is usually larger than what is considered a "group fire," the latter referring to a fire confined to buildings in a single industrial plant, or within a single block. Urban conflagrations are often continuous fires involving buildings on many city blocks at once.

Unfortunately, conflagrations have been a part of American history since the destruction of Jamestown, Virginia, by fire on January 7, 1607.(3) During the nineteenth century, many cities, including New York, Boston, and Chicago, experienced conflagrations that reached catastrophic proportions in terms of property damage and human suffering. As a result of improvements in fire prevention codes, fire protection technology, fire-resistive construction, and suppression methods, the incidence of conflagrations has declined sharply. The fire protection community, however, should not allow the infrequency of such fires to lull us into complacency. There have been a number of urban conflagrations in the past few decades, notably, Oakland–Berkeley, in October 1991; Brooklyn, New York, in February 1991; Paterson, New Jersey, in February 1991; Providence, Rhode Island, in March 1990; Chelsea, Massachusetts, in October 1983; Anaheim, California, in April 1982; Lynn, Massachusetts, in November 1981; and Houston, Texas, in July

1979.(4) Most major cities contain some sectional area that under the most "ideal" circumstances could produce an urban conflagration. There have been and will be others for decades to come.

Causes of Urban Conflagrations

Fires caused by lightning, earthquake, volcanos, floods, and high winds have resulted in tremendous losses, and while the proximate cause of these fires cannot be said truly to originate with any human activity, it was often human failure to provide means of protection and control that allowed such fires to increase and become conflagrations. The usual cause of a conflagration is a fire, and fires are caused by acts that are deliberate, accidental, or negligent.

Among the reasons predominant in man-made fires is carelessness, such as occurs in the use of open flames and smoking materials. However, the principal reasons for ordinary fires to escalate into the conflagration class are as follows:

1. availability of a large amount of combustible materials over a wide area, and
2. lack of adequate means to interrupt the fuel supply, once the fire gets started, such as the lack of fire protection forces, fire breaks, water supply, and heavy stream appliances.

Types of Urban Conflagrations

Based upon a study of conflagrations that have occurred since the turn of the century in the United States, it is possible to make a general classification, though some might fall into more than one class. The five general types of conflagrations described by the National Fire Protection Association are as follows.(5)

1. Those originating in hazardous industries lacking adequate means for confining or extinguishing incipient fires; the structures usually contain large quantities of highly flammable or explosive materials and are located in a congested district among other combustible occupancies without exposure protection or adequate private or public water supplies or fire protection. In the Chelsea, Massachusetts, conflagration 300 buildings were involved, with a resulting loss estimated at 1.3 million dollars. The main contributing factors in this conflagration were poor water supply, dry weather, high winds, closely stored combustive materials, narrow streets, no fire breaks, and delayed reporting.
2. Those occurring in closely built-up residential sections containing combustible houses with wood shingle roofs. We should include all residential areas with combustible construction and multiple exposures in this classification. Even where the fire protection forces may be superior, the lack of adequate water supply, coupled with high winds, permits the fire to get out of control. Examples of this type

The Great Chelsea (Massachusetts) conflagration, which took place in October 1983, casts an eerie silhouette against the evening sky. (Photo courtesy of the National Fire Academy.)

include an apartment building conflagration in Dallas, Texas, on March 21, 1983.

Another is the large fire that occurred in the Westwood section of Los Angeles in late 1989 but which received little publicity because it occurred the same day the Berlin Wall was officially opened. Chief Manning stated this fire could have extended "all the way to the ocean" before being controlled. It almost did.

3. Wildland, forest, and brush fires that sweep into a town or towns over a wide front. The most extensive timber and brush fires in southern California's history—ten of them burning at one time—ravaged more than a half-million acres, killed eleven persons, and injured about 350 others. They occurred on the outskirts of Los Angeles, San Diego, and the southern part of the Sequoia National Forest in September 1970. More than 1,000 buildings valued at 170 million dollars were destroyed in a six day period.(6) In the late 1980s and into the 1990s wildland fires burned out of control in all parts of the United States, severely taxing firefighting resources. They captured the attention of the American public when a western fire burned a large section of Yellowstone National Park.(7) These are now referred to as urban/wildland interface fires and examples occurred in the southeastern, western, and northwestern states during the early 1990s, and one of the worst in the nation's history in late 1991. They will be covered in greater detail as a separate classification.
4. Conflagrations due to explosions causing fire to spread over a wide

area, for example, the explosion of butane railroad tank cars in Crescent City, Illinois, on June 21, 1970, resulting in fires that destroyed most of the business district and injured more than sixty persons.(8) On Thanksgiving Day, 1987, in McDonald, Ohio, high pressure induced into material gas lines caused 156 simultaneous fires. A conflagration was narrowly avoided and serious damage occurred to only six houses and one business, due to quick action by the fire department and swift arriving mutual aid.

5. Multiple fires in one area or city, caused by an earthquake or a riot, resulting in hampered fire fighting; examples include the Watts riot in Los Angeles of 1965 and the 1989 San Francisco earthquake. In the latter, a large fire in the Marina district was controlled by remote pumping from the San Francisco Bay using fireboats and extended relays of 5″ hose.(9)

Contributing Factors in Urban Conflagrations

Inferior and combustible construction is a significant factor in many conflagrations in municipalities. The next factor in order of frequency in recent years was high winds (velocity in excess of 30 mph). For many years following the turn of the century, the factor of wood shingle roofs was more frequent than any other secondary factor, but this now is tied in fourth place with inadequate public protection. We will discuss wood shingle roofs in greater detail in the separate category of wildland/urban interface fires. The next important contributing factor after high winds is inadequate water distribution systems, followed by lack of exposure protection. After all of the above, but almost equal in frequency during these periods, is the factor of unusually hot, dry, weather conditions.(10)

Duration of Urban Conflagrations

Mass fires in the form of conflagrations move downwind as long as favorable conditions exist. This is generally true despite concerted efforts of trained fire fighters. If the winds stop or reverse direction, or if rain falls, or the fire comes to an area where there is insufficient fuel to sustain it, its spread is likely to end, but this may be after several days.

Fire Storms

Under certain conditions a stationary mass fire called a "fire storm" might develop. A fire storm is not strictly a conflagration, since its behavior is quite different.

CAUSE OF FIRE STORMS

Although the easing of tensions between East and West makes the prospect of global warfare less likely then in past decades, lessons should nonetheless be learned from past experience. Also, continuing conflict in the

Middle East, involving the United States, directly or indirectly, shows that war and conflict continue to be a reality in today's world. Wars invariably lead to massive fires. During World War II fire storms were created by mass incendiary bombing, though one was caused by a single atom bomb. This phenomenon could be described as a mass of fresh air that rushes into the burning area to replace the superheated column of rising air. In a fire storm, therefore, many small fires converge over a large area to form one great fire of severe intensity, creating its own winds of terrific velocity, rushing toward the center of the fire area from all directions. Around the edges of the burning area such winds may exceed a velocity of 60 mph and increase as they near the center. A fire storm causes a complete burnout of all combustible material within the area involved, unlike any ordinary conflagration where some buildings may escape damage owing to open spaces, change of wind, or efforts of the fire department.

EFFECTS OF A FIRE STORM

One peculiar effect of a fire storm is the death of many persons within the fire area not from burns but of suffocation from lack of oxygen (being consumed by the fire) and from carbon monoxide poisoning.

Fire storms also cause severe property damage over a wide area. In Hamburg, Germany, in World War II, within twenty minutes after one of the extensive incendiary air raids, two-thirds of the buildings in a 4.5 square mile area were completely involved in fire. Two days later, when the area had cooled down sufficiently to permit the fire department to enter, it was estimated that more than 55,000 persons had perished. Before the series of attacks were over, some 30 square miles had been damaged, with a complete burnout of 12.5 square miles; 750,000 people were made homeless as the result of the total loss of about 300,000 dwelling units.(11)

DURATION OF FIRE STORMS

Although the movement of air in a fire storm can reach hurricane forces, uprooting trees and destroying buildings in the periphery of the fire, they are likely to burn out within three to four hours without spreading much beyond the initial fire areas.(12)

EXAMPLES OF FIRE STORMS

Fire storms were reported in Germany in World War II in Hamburg, Darmstadt, Kassel, Leipzig, Stuttgart, and a few more cities, with a great loss of life. This happened, even though construction in most German cities is such that conflagrations in the technical sense are not considered possible, owing to subdivision of structures by good fire walls. Incendiary air attack started thousands of fires, but these burned in individual buildings with little spread from one building to another, except where fire storms developed.

In Japan allied military air forces had already destroyed sixty-five Japanese cities before the atomic bombs were dropped, and because of the highly combustible construction many fires resulting from air attacks were true conflagrations. Some of these attacks in which incendiaries were used alone caused fire storms. Mass fires started by the atomic bomb at Hiroshima also resulted in a fire storm. Destruction in German and Japanese cities in World War II varied from 12 to 96 percent of the area, with a median in individual cities, wholly due to fires, about 50 percent.(13)

On a smaller scale than the wartime conflagrations, the author Charles Bahme witnessed a fire storm created as a result of thousands of flying brands falling upon wood shingle roofs of many homes located in one of the Bel Air canyons. The simultaneous ignition of dozens of roofs created an in-rushing of air toward the center of the mass fires sufficiently strong to reverse the general wind direction, making the total destruction of involved frame houses almost inevitable despite the efforts of fire fighters.

WILDLAND/URBAN INTERFACE FIRES

On Sunday, October 20th, 1991, the most costly fire in American history as measured by property losses, roared through the hills of Oakland and Berkeley, California, killing 24 persons including one firefighter, destroying 3,354 homes, and causing 1.5 *billion* dollars in damage. This fire had all of the elements of a wildland/urban interface fire, a conflagration and a fire storm rolled into one devastating inferno. If ever there was an example of the close overlap between serious fires and disasters, this was the textbook case. Investigators speculated that sparks from a crude barbarque pit may have ignited a grass fire that was contained one day but rekindled into the deadly Oakland blaze. Due to the picturesque nature of the real estate, much of which overlooks the beautiful San Francisco Bay, the area eventually will be totally rebuilt, but the scars of this deadly fire will remain indefinitely into the future. (13a)

Since 1984 attention has been given to the term "wildland/urban interface" fires. As greater numbers of our population have moved into forest and wildland areas and as subdivisions have been constructed on the fringe of wildlands in all parts of the country, attention is now turning toward the real threat to life and property that occurs in the homes adjacent to the wildland areas. While wildland fires can themselves burn huge tracks of acreage and be quite spectacular, unless they infringe upon populated areas, they pose little threat to society. In certain parts of the country, including the national parks, a policy of controlled burn is permitted, recognizing that forest and brush fires are a regular phenomenon of nature designed to clear the land for new growth.

Some of the largest fires on record have involved forested and brush areas, and these were at one time confined primarily to rural regions. During the thirty-year period of 1940 through 1970 in Los Angeles, only seven of hundreds of brush fires developed into wildland/urban interface fires in the 108 square miles of mountainous areas within city limits. These fires destroyed more than 1,300 homes, with losses running into millions of dollars, not counting the value of the watershed destroyed. Since 1970 into the

1990s, however, wildland fires that interface with residential areas have become all too commonplace, and because of extensive drought in the late 1980s and early 1990s have occurred in many different parts of the country.(14)

Destructive wildland urban interface fires have occurred in the Carolinas, Florida, in the Dakotas, Nevada, Oregon, California, New England, New Mexico, Arizona, and in many other states. Hardest hit seems to be the state of California which, year after year, due to the Santa Ana winds and dry conditions, is susceptible to devastation from wildland fire.(15) Scanning the pages of history, we find that 200 homes burned in 1980, several hundred burned again in 1985, and several thousand were destroyed in the Oakland and Berkeley hills in 1991 in one of the most costly wildland/urban interface fires in American history. It took a score of lives and was viewed by millions on television over a 24-hour period. *U.S. News and World Report* (November 4, 1991) described the conditions that led to the fire as follows. "It was a suburbanite's dream and a firefighter's nightmare. The homes were nestled in the hills of Oakland and Berkeley, many with views of San Francisco Bay—but their seclusion put them out of reach of firefighting equipment. Many were topped with wood-shake roofs, lovely to gaze at but quick to burn. They have the Western, bushy look Californians crave—but much of the bone-dry brush, instead of being cut back from the houses, grew unmolested, as did the combustible pines and eucalyptus trees."

The year 1987 was also bad year for wildfire in California. Nine wildland fire fighters lost their lives, the fire fighting force was augmented to tens of thousand, and $240 million in damage was done. In Pebble Beach, beginning on May 31, and fanned by winds up to 30 miles per hour, a fire burned for five days before control was achieved. Thirty-one homes were destroyed despite the efforts of 885 fire fighters manning 38 structural engines, 54 brush engines, and 7 helicopters.(16)

Brush fires can result in loss of life among residents who may be trapped in a dead-end canyon, and fire fighters and others who may be caught unaware. For example, in 1933, in Los Angeles' Griffith Park, twenty-eight workmen, whose knowledge of fire behavior was so limited that they tried to run uphill through dense dry grass to escape the rushing flames, perished in the attempt. Even experienced fire fighters have lost their lives in sudden "blow-ups" of dense, oily, dry brush, unexpectedly ignited by flying brands from a fire thought to be at a safe distance. The lives of many fire fighters have been saved by the use of forest fire shelters, or fire tents. These light weight aluminum enclosures can be deployed in 30 seconds or less providing a shield against an overtaking fire.(17)

Cause of Wildland Fires

Most wildland or forest fires in the eastern United States are the result of carelessness of people using the forests, while most fires in western states are caused by lightning, logging, and sawmill operations. Use of machinery and railroad engines in forested areas also cause fires.

Wildland Fire Behavior

A number of factors affect the behavior of wildland fires, including terrain, weather, wind, fuels, and seasons. It is well known that fire travels faster uphill than down and is more difficult to fight on steep slopes than on level ground.

When the weather is damp or wet, fuels soak up the moisture, and brush or grass is difficult to burn; when the weather is hot and the humidity is low, wildland fires can explode with the intensity of rapid combustion, making crown fires in timber almost impossible to stop.

Even in the absence of strong winds, such as a "Santa Ana," a fast-moving fire can generate its own updrafts, particularly in canyons, causing burning brands to be carried high in the air and drop a long distance ahead. This results in spot fires over a wide radius as the wind changes its heading. When strong, desiccating winds are prevalent, the fire moves with such rapidity that nothing short of a well-coordinated ground and air attack, coupled with wide natural barriers and clearings, can control its spread.

"'Santa Ana' winds in southern California create one of the most hazardous, difficult, wildland fire control problems in the world. Strong air flow, with very low humidities, quickly desiccates both living and dead fuels."(18)

When high velocity winds blow over a steep, rough, broken topography, a highly turbulent air flow is created in canyons; wind directions may switch 180 degrees and back again in only a few seconds. Dense smoke clings to the ground and burning embers are driven into unburned, dried-out fuel ahead of the fire. If such fuels happen to be finely divided and have a high oil content, plus a large surface-to-volume ratio as is the case with most chaparral fuels, it becomes understandable why a dozen members of a highly skilled U.S. Forest Service fire crew could be burned to death in a chimney-like canyon while fighting a brush fire in the Angeles National Forest.(19) This unfortunate incident, which also seriously injured twelve others (two of whom later died of burns), emphasizes the need for personnel trained in evaluating properly fire behavior in critical weather, fuel, and topographical conditions.

Natural fuels vary greatly in the rate and intensity of burning; fires in heavy brush and timber burn with great intensity but more slowly than in dry grass and leaves. Dense fuels propagate fire better than sparse fuels.

The forest fire season in the southwestern U.S., such as Arizona and New Mexico, is in May and June; whereas, in California the season extends from June to December. The season generally ends in September in the western United States. On the other hand, the eastern states have two fire seasons—the first in the spring after the ground cover dries out and grass becomes a good fuel, and the second season after leaves fall in October.

Forest Fire Protection Responsibility

Responsibility for fire protection on federal lands is centered primarily in the Department of Agriculture, Department of Interior, and to a lesser extent in such agencies as the Department of Defense and the Tennessee

Valley Authority.(20) Numerous laws have been passed by the states recognizing their responsibilities for state and private forest lands, and state agencies have been established to carry them out. In addition to local fire departments, which handle wildland fires within their own territory, private interests have their own fire control organizations. By participating in state and federal fire defense plans, fire chiefs are able to obtain the important information needed to carry out their fire prevention and suppression responsibilities.

Community Preparation

Fire chiefs of small communities near forest areas should equip and train their fire department personnel for brush and forest fire fighting, and should engage in fire prevention and education activities that will result in the residents having a better knowledge of how to safeguard their own homes, before and during the threat of an approaching forest fire. Use of local radio and television fire prevention programs, junior fire department instruction in the schools, fire weather forecasting available from the National Weather Service, and mutual aid compacts with adjacent communities and private agencies are all helpful in preparing a community to cope with fire. In communities having only a few paid fire fighters the fire department often budgets for employing extra personnel during the dry summer season. Sometimes the older high school students who have been trained as volunteers are hired for the summer vacation period to augment the paid personnel.

An excellent publication that shows how the fire department can enlist the various segments of the community to come to its assistance in times of disaster and major emergency is a National Fire Academy Open Learning Course entitled "Community and the Fire Threat."(21)

Organizing and Equipping for Rural Fire Control

When properly organized and trained for combatting outdoor fires, small communities can do a remarkable job in achieving fire protection in rural and wildland areas. A pamphlet that describes the fundamentals to be considered in such an organization, the type of equipment that is necessary or useful, and some basic tactics that are essential for the safety of personnel and the successful control of forest, grass, and brush fires is NFPA Standard 295, *Wildfire Control by Volunteer Fire Departments.*

Action to Be Taken When a Wildland Fire Threatens the Community

The fire chief is often head of the disaster control organization in small communities. In any case, he is certain to be called upon for advice and consultation when a forest or wildland fire threatens a community. Actions that should be taken are listed below:

1. Notify the mayor and civil defense director, so that the emergency operations center can be activated.
2. Establish contact with the nearest National Weather Service to insure adequate forecasting support.
3. Keep in communication with the incident command center at the scene to obtain strategy and the tactics intended plus requirements for additional manpower, equipment, and supplies.
4. Assess the fire threat to your own community, based upon the report of the incident commander, the speed and direction of the wind, fire breaks and natural barriers intervening, and the measure of control being achieved at the present time.
5. Provide the public information officer with appropriate information for news releases.
6. Line up assistance under existing mutual aid agreements by alerting department commanders who may need to be called upon.
7. Designate areas that may have to be evacuated and get prepared to "pass the word" in the event this action is required.
8. If the fire spread requires that help be sent to the state and federal agencies combatting the fire, coordinate your own EOC functions with those of the other agencies that may be conducting direct and indirect fire attack, fire break construction, back firing, and use of air support under the coordinated direction of the operations commander of incident command.
9. Be sure that operations officers in each local, area, and state EOC maintain communications upwards and downwards at all times to keep appraised of what the fire is doing, where it is "spotting," where crews have been deployed, where residents are being evacuated, where backfires are being started, where air drops are being made, apparatus pools being established, command posts being set up, etc.
10. Secure law enforcement assistance to cordon off the area of operations and to limit access only to emergency forces and their support services. Have roads blocked off and traffic rerouted as necessary.
11. Maintain reports to the EOC for developing periodic releases to the public to reduce apprehension, describe escape routes for evacuation, etc. The Civil Defense Preparedness Agency has prepared suggested instructions which might be broadcast to the public over the radio and TV in the event that a wildland fire threatens an urban area.(22)

Warning to Be Given the Public

Prior to the commencement of the annual season of high fire danger, citizens who live in the mountainous areas should be reminded of the constant threat of conflagration that may become so serious as to warrant their getting out of all canyons and ridges having only one path of egress, even before a fire starts.

At the start of the fire season, fire fighters can go from door to door, handing out a brochure on the subject and explaining how residents can judge when conditions are especially bad, even in the absence of hearing a public broadcast of a fire hazard warning.

If people look out windows and see trees being whipped by the wind, and if they feel static shocks upon touching the car door or light switch, then they can surmise that weather conditions are conducive for fast fire propagation. They should also be told that if their homes are surrounded by explosively dry brush, on dead-end streets, on steep slopes where an up-rushing fire can ignite a whole block of houses at once, or in the bottom of canyons where a downpour of flaming embers can ignite fifty houses at a time, then the fire department may not be able to save them in case of a major outbreak of fire.

They should also be warned that high winds not only spread fire through brush at a rate of two or three blocks per minute and through the air for a distance of two or three miles, but can also blow down power lines and cause the water pumping stations to fail, with the possible result that tanks supplying fire hydrants can go dry.

Whenever humidity drops to a very low level and hot dry winds, such as the Santa Anas, reach a dangerous velocity, the public should be warned by radio and television alerts, and the residents of extremely hazardous areas should be advised to take every precaution to safeguard their homes and, if practicable, find temporary quarters elsewhere until the peril diminishes.

Among the steps that householders can be advised to take is removing reed matting from patios and porches, putting garden furniture and autos in the garage, closing windows and venetian blinds, removing combustibles from windows and from under eaves of the roof, getting leaves out of gutters, hooking up garden hoses, placing ladders to combustible roofs, wetting down shingles, and listening to the radio for directions from the authorities as to the route to take in event that evacuation becomes necessary.

As preparation for evacuation, they should be advised to assemble the items that they want to take with them, such as important documents, irreplaceable photos, manuscripts, cash, and jewelry, in a small handbag or case that they can carry by hand in the event it becomes necessary to walk away from their homes.

If they plan to escape by auto, they should be reminded to learn how to open electrically operated garage doors manually if power fails. They should also be asked to refrain from using telephones as much as possible and to be sure that batteries in portable radios and flashlights are in working order.

Able-bodied persons who may wish to stay and protect their homes, notwithstanding an order to evacuate, should be warned that water supply to garden hoses may be lost at the height of the fires, and even though their homes may have rock, tile, cement shingle, or composition roofs, flying embers can enter unscreened attic vents, and can ignite vegetation alongside the house, spreading to wood siding or eaves and the attic.

Homeowners with swimming pools can be advised to make this reserve water supply available in case the mains go dry and the power fails. Gasoline-engine-driven pumps can be used to supply garden hoses, or a gasoline-engine-driven generator can be used to operate an electric sump

A fallen tree will not stop the advance of this wildland/urban interface fire as it enters a populated area in southern California. (Photo courtesy of the National Fire Academy.)

pump or the swimming pool pump supplying a garden hose off of its filter cleanout opening.

Wildland Fire Control Methods

It is an old axiom that fires require fuel, heat, and oxygen, and an uninhibited chemical chain reaction to be self-sustaining, and if any one of these elements is removed, the fire will go out. Since smothering a wildland fire is not feasible, common methods of fire control involve interrupting the fuel or removing the heat. Sometimes attack is made directly on the fire; at other times indirect attack is made through construction of fire breaks, backfires, and dirt barriers.

Direct attack may employ water from handlines and chemical fire retardants from aircraft, or dirt thrown from shovels. Many forest fires are controlled by constructing fire breaks to cut off or surround the fire and, where appropriate, to burn out fuel between the line and the fire.

Air operations include dropping water or fire-retardant solutions from aircraft, reconnaissance and command from small airplanes or helicopters, transporting equipment and supplies to fire camps or isolated crews, transporting trained crews, smokejumpers, injured victims, and evacuees.

The capabilities of air tankers for delivering retardants to the fire are constantly improving. An air tanker system was used to attack a 5,000 acre timber fire, in which 3,000 gallons of fire-retardant were dropped from a C-130 aircraft in less than six seconds, as a follow-up on a B-17 drop, using a lead plane to guide the big tanker to the target.(23) New reconnaissance technology adapted from the stealth bomber and other developments in the field of aviation will provide more sophisticated airborne tools for wildland/urban interface fire fighting.

The airborne infrared fire mapper gives the fire chief a useful tool for obtaining fire intelligence, day or night, even through dense smoke. The infrared scanner, which includes a detector printer, is mounted in the aircraft, and imagery is produced as the aircraft flies over the fire area. The energy emanating from the hot burning fuels is converted to visible light displayed on a computer terminal screen and photographed automatically. These photographs become a thermal map that shows the fire perimeter, hot spots, and terrain features.

Structural fire fighters have learned from the wildland forces in adapting thermal image and equipment for use in detecting internal fires not otherwise visible. At the Syroco Plastics warehouse fire in Syracuse, New York, in 1990, this device was attached to a helicopter and showed from the air varying densities of fire inside the structure.(24)

Helicopters are useful in scouting a fire, deploying attack forces, making an initial attack in inaccessible areas, dropping retardant in areas where tankers cannot operate, and effecting rescues. They can also lay hose and telephone wire and carry cargo suspended in slings. Ten- and fourteen-passenger helicopters are common for flying crews to initial attack positions.

Airplanes and helicopters can drop fire fighters into remote areas to build a landing spot for air-ferrying of ground crews. They can also transport smokejumpers close to the fire to start initial attack.

Systems originally developed back in the 1970s are used now for improving the pilot's night vision, including the infrared scanner, electro-optic light amplification, and goggles making use of moonlight. The latter two amplify starlight or moonlight ten to fifteen thousand times and permit pilots to operate aircraft without artificial illumination.(25)

Wildland fire fighting requires carefully organized command posts, preplanned services and logistical support, aerial reconnaissance, aerial and ground attack, frequent issuance of fire line maps, and communications channels to handle tactical and supply functions.

Wildfire History

The Peshtigo, Wisconsin, fire in 1871 started on the same day that Chicago, Illinois was devastated by a conflagration, and took more lives than any other forest fire in the United States. The loss of life has been reported as 800 to 1,500 with more than four million acres of forest destroyed.

A wildland fire swept into Berkeley, California, in 1923 and destroyed 640 buildings. Almost forty years later, the worst conflagration in North America since Berkeley occurred in the Bel Air and Brentwood residential sections of Los Angeles, California, with a loss of twenty-five million dollars. It was one of the most costly wildland fires of the twentieth century until the 1991 Oakland–Berkeley fire with a greater than one billion dollar loss. Pending an analysis of this fire, the Bel Air fire serves as a good study model representing the type of fire that has been and will be repeated many times. The fire started at 8:10 A.M. on November 6, 1961, near the top of the north slope of the brush-covered Santa Monica mountains. The weather had been extremely dry, with humidity as low as 4 percent, and the fire spread at 13 acres per minute, pushed by Santa Ana winds gusting up to fifty miles per hour. By the time 3,000 fire fighters were able to bring it under control, it had destroyed 450 homes and 180 other buildings in an area exceeding 5,000 acres.(26)

Initial response of five engine companies, immediately followed by the dispatch of fifteen more, was insufficient to halt spread of fire over the top of the mountain. A command post, established on top of the mountain, was later moved to the nearest mountain patrol fire station and became field headquarters.

Other command posts were established at intermediary points where major canyon roads intersected, and from these locations the next thirty companies were directed into side roads by assistant chiefs who were coordinating the efforts of the battalion chiefs assigned as sector commanders, with several companies each. These companies were then supplemented with other companies dispatched from a command post at the bottom of the mountain, to assist in suppression of fires breaking out in homes and brush along ridges and canyons.

Burning brands rose to 2,000 feet, crossing a wide freeway that normally was a good break, and fell into canyons over one and a half miles away.

Unfortunately, helicopters of police and fire departments were out of service, so initial reconnaissance was made on the ground. Five helicopters and twelve airplanes were used in later borate drops and command control. The borate drops proved effective within limitations; air tankers are not especially suitable for use on fires in inhabited areas, owing to the risk of collision with other aircraft in the same area. Their usefulness was hampered whenever the target area was obscured by the smoke cloud, when the fire was too hot or too fast-moving for the retardant drop, when convection currents scattered the drop, or when proper timing could not be achieved.

But some drops were effective in reinforcing bulldozers on the ground in stopping the advance of the fire. Drops were made close to dwellings, and occasionally on them, when necessary. Helicopters, with slower speed and ability to hover, were effective in making water drops on spot fires in canyons where fixed wing aircraft could not make an effective attack. They were also used for refilling portable tanks flown into inaccessible locations, for use by fire fighting teams dropped with portable pumps and small hose lines.

Backfiring was employed on a limited scale: mainly firing-out to protect exposed property or to improve a defensive line.

Overall strategy was to stop the fire advance in the north and east flanks, warn the 4,000 residents in the homes downwind (southwest) to leave, and save as many structures as possible.

Tactics in support involved use of police to assist in notification and evacuation of residents; conducting of a coordinated ground air attack along with fire break structure on the north and east perimeters; and "hit and run" attack on individual building fires spreading downwind.

Air tankers, bulldozers, and engine companies were able to establish defensive lines north and east of the fire; patrol crews and engine companies saved many homes. This "leap-frog" tactic worked fairly well until water mains went dry, making it necessary to establish tanker relays to refill engine and patrol company tanks.

As the fire spread westward, additional command posts with apparatus pools were established at several mile intervals along the southerly perimeter. Men and apparatus from twenty-four cities and Los Angeles County, along with several hundred military personnel, assisted in saving homes and preventing loss of life.

Many factors made control of this conflagration extremely difficult. These were the primary obstacles:

- Adverse weather conditions
- Adverse structural conditions (wood shingle roofs and large window openings)
- Insufficient clearance between buildings and highly flammable brush, tree, and dry vegetation
- Water supply failure in critical areas (owing to power loss to pumps that supplied storage tanks)
- Inability of communications facilities to permit a greater exchange of information between command personnel, field headquarters, command posts, companies, aircraft, and the many operating agencies.

To further complicate the problem, arsonist actions began in other canyons, one of which, designated as the Santa Ynez fire, involved 6,000 acres of sparsely inhabited, brush-covered mountains west of the Bel Air–Brentwood fire. This fire was also three dimensional in nature, spreading in all directions along the ground, vertically up the canyons, and through the air, with flying brands on convection currents high above the surface.

These fires, along with forty-three structural fires in other parts of the city (twelve required greater alarm response) had to be suppressed simultaneously while coping with the Bel Air–Brentwood conflagration. The Santa Ynez fire was not fully controlled until two days after the Bel Air–Brentwood fire had been contained.

When the study of the Oakland–Berkeley conflagration of October 20, 1991 has been completed, undoubtedly many of the contributing causes will be identical to those in the Bel Air fire.

The January 1990 issue of *Fire Command* reported the "Black Tiger Gulch" fire which caused more than $10 million in damages and included the destruction of 65 homes and buildings. This Boulder County, Colorado, fire took 500 fire fighters from as far away as North Carolina a week to control, and cost $1 million.(27) It can be seen that wildland/urban interface fires are deadly, occur every year, and should be one of the high-priority incident types for disaster planning.

Predictability

Conditions conducive to wildland/urban interface fires and conflagrations are known, such as wind velocity of 13 mph or more, humidity of 40 percent or less, and rainfall of 0.01 inches (or less) on the day of the fire and for three days preceding it. A fire protection engineer may forecast a major fire, for example, for a dense brush area in a hot dry climate, where high winds occur regularly, and where hundreds of homes with wood shingle roofs are nestled in highly flammable vegetation.(28)

Limiting Loss from Wildland/Urban Fires and Conflagrations

Having knowledge of the leading factors that contribute to wildland/urban interface fires and conflagrations, it is not difficult to suggest actions that can prevent them. These might include elimination of inferior construction, such as non-fire-resistive roofs; installation of automatic fire detection and extinguishing systems in buildings of combustible construction; and enforcement of recognized building and zoning laws. The elimination of overgrowth, brush, and high grass for a minimal safe distance from all structures can minimize their involvement in fire.

Other methods of keeping little fires from growing into large ones can

Fire departments must concentrate on placing water on uninvolved buildings where savings can be made, rather than totally involved structures. (Photo courtesy of James Lanagan.)

be used, such as subdividing large area buildings with fire walls, and subdividing blocks of buildings with unpierced, parapeted, division walls and open spaces. Fire breaks can be created within a community by means of parks and wide streets. More automatic alarms to central stations or directly to fire dispatch headquarters can provide quicker warning, faster response, and extinguishment while fires are still relatively small.

An adequate and reliable water supply and distribution system, having alternative sources for obtaining the water, is very important.

The provision of adequate public and private fire fighting forces with necessary equipment, including heavy stream appliances, is a *sine qua non* (absolutely essential) for keeping little fires from growing into big ones.

Where local fire forces are inadequate, a good working agreement with mutual aid companies can help offset limitations when a major fire occurs. Department-wide training in large-scale operations and efficient use of heavy stream appliances are other requirements.

OCCUPIED HIGHRISE BUILDINGS

We have discussed at length urban conflagrations and wildland/urban interface fires. A third category of fire which can also become disastrous or catastrophic is a fire in an occupied highrise building. On May 4, 1988, fire enveloped the upper floors of the First Interstate Bank Building in Los Angeles. The fire enveloped the 12th through the 16th floors in the 22 story building and was stopped only through the heroic dedication and tactical decision making of the Los Angeles Fire Department. This fire required 400 fire fighters from 58 companies—40 percent of the fire department's on-duty strength. Mutual aid backup protection was provided by 20 engine companies from the Los Angeles County Fire Department.(29) One person died and 37 were injured. Most highrise building fires, however, have been far more deadly in terms of fatalities and destruction. As you will recall, we opened this chapter with a discussion of the March 1991 Philadelphia highrise fire that resulted in massive destruction of the building and claimed the lives of three fire fighters.

A highrise building is one in which emergency evacuation is not practical and in which fire must be fought internally.(30)

Since 1970, fire has been responsible for loss of many lives in tall buildings, even though these buildings were modern in design and complied with building codes at time of construction. Why did this happen? There were several reasons why lives were lost in fire-resistive buildings with enclosed stairways and without large areas or heavy fire loads. In the absence of automatic fire detection and alarm there can be delay in notifying occupants of a fire and of the need to escape or seek refuge in a safe place. In the absence of automatic sprinkler protection, smoke and heat can spread through air conditioning ducts despite automatic dampers that are supposed to prevent this but sometimes fail to do so. Moreover, fire travels from floor to floor via exterior walls in buildings that are not tightly cut off at each level. For example, at One New York Plaza on August 5, 1970, a fire starting on the 33rd floor caused heat and smoke damage on the 35th floor, resulting in the death of two building guards.(31) Pipe shafts and "poke through" openings for conduits that are not tightly sealed provide vertical openings for upward travel of fire.

In the absence of subdividing each floor into separate fire-resistive compartments, which would permit occupants of unsprinklered buildings a place of refuge in case of fire, or in the absence of smoke barriers or smoke-proof enclosures in all stairways, which would permit them to move to a floor not involved in fire, the occupants' chances for survival are not very good. They are not likely to be able to open windows, for in air conditioned buildings windows are not designed to be opened. Nor can they use elevators safely; these may stop at the fire floor because of the effect of heat on the call button. Once the door opens, smoke may obscure the photoelectric cell in the door-closer and hold the door open. This would not be the case where a detector has been installed that deactivates the elevator controls and permits it to pass the fire floor.

The use of emergency exits in highrise buildings is not always practical. Assuming 240 persons per floor and a two-unit exit stairway system, about 90 persons per minute can enter a stairway. Under optimum escape conditions it has been estimated that it would take about 66 minutes to evacuate a 40-story building.(32) If electric power fails and there are no standby generators to restore lights in the halls and stairways, evacuation would be greatly impeded. The same would be true if fleeing occupants were to encounter fire fighters ascending the stairs on the way up to the fire.

Fighting a fire in a tall building presents some problems not encountered in controlling wildland/urban interface fires and conflagrations and can put as much demand on the fire department. Manpower equivalent to an entire battalion can easily become committed.

Building fires usually have to be fought from the inside, since the upper floors are beyond reach of aerial ladders and elevating platforms. Windows are not always openable, but there is a hesitancy to shatter glass because of hazard to pedestrians below. Even fires on lower floors may need to be attacked from the inside. In any case, but especially where strong winds are blowing, breaking out windows can accelerate the fire by increased air flow to the interior of the building. In the absence of flame barriers required by some building codes, flames issuing from broken windows may extend upward, cause further glass breakage, and enter the next floor above.

Where fire fighters must go inside a building having unopenable windows to fight a fire that is giving off much heat and smoke, causing automatic shut-down of the air conditioning system, ventilation becomes a major problem. It is like walking into an oven. If one of the elevators has not been designed for safe operation by the fire department, it becomes necessary to walk up to the fire floor. It can be exhausting to carry heavy equipment up flights of stairs. It is not always possible to attack fire from the floor above by landing fire fighters on the roof with helicopters, owing to obstructions, convection currents, and dense smoke above the building.

A building that might be considered low compared to modern highrise structures, since it was only ten stories, was the scene of a disastrous fire on February 7, 1967. It was located in Montgomery, Alabama, and 25 persons died in its penthouse restaurant when the rapidly spreading flames in the combustible furnishings and decorations cut off their escape to the single stairway.

Four fire fighters and two officers who had entered one of the two elevators to ascend to the restaurant were also trapped when it stalled on the top of its climb. The other elevator, with the chef inside, was stalled alongside. Upon prying open the doors slightly the fire fighters were faced with a blast of heat and flames, so they stuffed coats and salvage covers in the crack to keep out the fire while they climbed out the escape hatch on top of the elevator car. After rescuing the chef through the car's hatch, they escaped by sliding down elevator cables to the basement. The fire eventually was extinguished using a large stream from an aerial ladder pipe directed from the street and hose connected to inside standpipes. Using fog streams to form a water curtain, they rescued survivors who had escaped cremation by breaking out windows of the restaurant and fleeing to an open patio on the roof of the building.(33)

In São Paulo, Brazil, two tall building fires occurred about two years apart, on February 24, 1972, and February 1, 1974. The first started in the 31-story Andraus Building as a result of ignition of combustibles in a light well that opened to the fourth and fifth floors of this combination department store and office building. The fire spread so quickly in the combustible fiber ceiling tile that many persons were barely able to get out of the building, some with their clothing on fire. Open stairwells permitted rapid vertical spread to the seventh floor, and heat buildup was so great that it continued to spread up the outside of the building through the 31st floor, entering every floor on its way, until the whole structure became quickly and fully involved. There was little that the fire department could do to extinguish the fire, but about an hour and a half after it had burned out, the first helicopter landed and the evacuation of some 350 people from the roof commenced. A heliport parapet that extended five feet beyond the face of the building helped (along with a stiff breeze) to deflect heat and smoke from the roof. It was remarkable that only sixteen persons lost their lives in this disaster.

The second São Paulo fire in a highrise building about two years later started at about 9 a.m. on the 12th floor of the concrete Joelma Building. The lower 10 of its 25 stories were occupied by a garage that was not involved in the fire, but the upper floors, which had no partitions or fire doors separating the wings from the elevator corridor, were in use as offices for various business firms. The only means of egress from the building was a single unenclosed stairwell, accessible from the elevator corridor, and when this became filled with smoke and heat, the occupants fled to the roof or out on exterior ledges to await rescue. The fire did not spread as rapidly through the Joelma Building as it did through the Andraus, where there was almost simultaneous ignition of four floors in the department store; however it did proceed up the open interior stairway to the floors above, increasing in intensity until the entire office section of the structure became involved.

Unlike the Andraus fire, there was no wing or parapet to deflect heat from people trapped on the roof, and because of strong convection currents, updraft, heat, oxygen reduction (that affected the operation of aircraft

A fully involved highrise hotel in Sao Paulo, Brazil, which claimed numerous lives in 1974. (Photo courtesy of the American Consulate in Brazil.)

engines) helicopters were unable to land on the Joelma to make rescues. Of the 179 victims who died in this fire 80 were found on the roof.(34)

On June 24, 1980, a fire in a 42-story Manhattan, New York, skyscraper raged out of control for more than two hours, injuring 127 fire fighters and trapping 16 office workers who eventually were escorted to safety through choking smoke. In San Juan, Puerto Rico, a highrise building fire at the Dupont Plaza resulted in death for 95 patrons who were preparing to celebrate the coming of the New Year, 1987. Dupont Plaza had no sprinkler system because it is not required under Puerto Rican law. In the United States, guidelines vary greatly from one city to another. Among the first states to make installation mandatory in hotels were Nevada, Florida, and Massachusetts.(35)

Former Fire Commissioner John O'Hagen of New York stated "the main problem associated with the protection of life from fires in highrise buildings is the limitation of the size of the fire. Public fire protection services can usually contain a fire in an elevated portion of a building if the area is limited to 5,000 square feet or less. If the horizontal spread of the fire exceeds that limit, the heat development will increase the risk of vertical spread. If the building contains open vertical arteries, then a tragedy can be expected."(36)

Fire Fighting in Highrise Buildings

Some of the problems confronting the fire department in coping with three-dimensional fires in tall buildings have been mentioned. Suggested procedures have been outlined in one book on emergency action(37) and the NFPA has published a book by a deputy chief of the New York Fire Department that is devoted to this subject exclusively.(38) Several articles have appeared in the professional journals of the fire service dealing with tall building fire tactics. About the finest summary that can be found on this subject is contained in a paragraph from an early editorial in *Fire Command*.

"For years, the quick summation of fireground action has been expressed as 'Surround and Drown' tactics, 'Over and Under' attack, or 'Task Force Assault' and other different terms. But essentially the fireground commander knows that he is dealing with a fire in a relatively confined area, as long as he can cut off and protect exposures and other possible means of extending the fire's spread. He must move men and equipment under his command through different positions in this cube, front to back, bottom to top, sometimes diagonally up, down, sideways. At the same time he must assure that logistical supply is maintained to each and every individual and group within the area of fire control. Most important is the continuation of communications through each level of movement, to and from different levels of command. Training and preplanning help smooth these operations, but even for the most professional department a sudden change in the fire problem can interrupt, perhaps dangerously, the command and control of the fire fighting forces."(39)

Several new considerations have further complicated the complex job of fighting fires in highrise buildings. For example asbestos was discovered during the investigation of the Dupont Plaza Hotel, indicating its presence during the fire fighting operations. Although progressive fire departments have long advocated the use of complete respiratory equipment, this provides yet another reason why it is so vitally necessary where synthetics, plastics, and long-term dangers such as asbestos are present.(40)

Obviously, the better the design of the building, the easier will be the task of the fire department. A building fully equipped with an automatic fire protection system will provide the least problem providing that the system is kept in working order. Automatic fire detection systems that can activate fire dampers in duct systems and fire doors in fire wall openings can do much to help the fire fighter check the spread of the fire. Automatic fire alarm systems that can alert the occupants in the danger zone and simultaneously report the fire to a central control station within the building and to a supervisory service on constant duty in a nearby location outside the building can help avert a loss of life by the early warning and notification of the fire department.(41)

Fire departments might be well advised to conduct highrise mock disaster drills such as that conducted in Toledo, Ohio. A simulated fire on the 14th and 15th floors of the 30-story Owens Corning Fiberglass Tower proved to be exhausting for many drill participants, but provided valuable lessons.(42)

A control station within the building from which an alarm can be sounded manually anywhere within the building and a public address system permitting the fire commander to talk directly to any area will enable the commander to instruct occupants on the proper action in an emergency. The provision of special controls on one or more elevators that can be used exclusively by the fire department in case of fire in the building will help speed up operations, lighten the fire fighters' load, and increase their safety in the performance of arduous duties. Most modern elevators have the recall provision which brings them all to a lobby or other predesignated level in the event of a fire.

A system to move smoke directly to the outside of the structure will help achieve the ventilation needed in smoky fires, not only for the protection of the occupants, but to assist in fire suppression operations.

Enclosed vertical openings and fire barriers that prevent the upward propagation of fire, either inside or outside the building, are essential to fire control. The limitation of fire areas by fire walls with protected openings is a good way to limit the fire loading, fire loss, and fire spread.

The provision of standby electrical power for emergency lighting will assist the occupants in reaching a place of safety, as well as the fire fighters in performing their tasks.

The adoption of emergency prefire plans and methods of evacuation, with required periodic drills, can be of help in avoiding confusion and loss of life in a major highrise fire.(43)

Hotel Disasters

One special type of occupancy that deserves attention regardless of size is hotels. While most of them are highrise, and most disastrous fires have occurred in highrise hotels, they present a serious life hazard regardless of construction or configuration. Two Las Vegas hotel fires in the 1980s focused national attention on this problem. On November 21, 1980, 84 people died in the MGM Grand. Then on Tuesday, February 10, 1981, 8 more people died in the Las Vegas Hilton fire only a few blocks from the MGM Grand. Helicopters were successful in rescuing numerous victims from the MGM Grand, while more conventional rescue techniques were used at the Hilton.(44)

Helicopters were also used to rescue 6 guests of the Tae Yon Kak Hotel in Seoul, Korea, on December 25, 1971, along with 100 others. Elevating platforms and aerial ladders were used on this 21-story building which the hotel shared by a vertical division with several business firms. However, the fire department's valiant efforts could not prevent 163 fatalities, of which 40 resulted from falls or jumping from the building. Because the hotel's stairway was not enclosed, smoke and gases spread everywhere as the fire progressed for about six hours through most of the unsprinklered building from its point of origin in the second floor coffee shop. Most fire streams were directed from aerial ladders and platforms and were supplied from tank trunks because of the limited hydrant supply.(45)

Hotel disasters have dotted the history of the United States. A major hotel fire in 1946 took at least 122 lives on December 7, when guests of the 15-story Winecoff Hotel in Atlanta, Georgia, died in the fire or jumped or fell from the building. An open narrow stairway acted like an incinerator stack to carry smoke and heat from the ground floor to all the upper floors; open transoms in guest rooms undoubtedly contributed to the death toll, as toxic products of combustion penetrated openings. Absence of an automatic sprinkler system, and the twenty- to thirty-minute delay in reporting the mattress fire on the third floor permitted the fire to gain considerable headway, making the job of the fire department more difficult. There was no automatic alarm system that might have given the guests more time to escape. Some chose to jump rather than remain and hope to be saved. Fire fighters were injured and ladders dented or broken by falling bodies. By using hoselines from standpipes in buildings across a 10-foot alley and a 58-foot wide street, fire fighters were able to get above the fire and stop its spread before it got to the top stories.(46)

Places of Amusement

On March 25, 1990, the nation was jolted by the deaths of 87 patrons of the Happy Land Social Club, operating after hours in the Bronx, New York. Although the fire department of New York had ordered the club closed it had not yet been able to enforce the closing order.(47)

Other tragic fires have occurred where people are assembled for amusement purposes. The fire in Chicago's Iroquois Theater on December 20, 1903, resulted in the deaths of 602 patrons and 250 injuries. Although an asbestos curtain had been provided to prevent the spread of flames from the stage to the auditorium, an obstruction prevented it from falling at the crucial time, and the two automatic ventilators above the stage were nailed shut. The exits were not marked properly, and the gallery did not have a main exit of its own. There were no sprinklers, alarm boxes, or drilled attendants. Of the more than 1,800 persons in attendance, more than 200 were standing in the aisles.

The fire started from a spark of hot carbon from a floodlight on stage falling on flammable scenery. It was quickly drawn to the open vent over the auditorium and to the open exit door in the top of the gallery. When someone yelled "Fire" the panic started. Many of the patrons, of whom a large number were children, were trampled to death or died of suffocation from the smoke or direct action of the swirling flames above their heads. Most of the dead were found in the gallery and the balcony where the smoke and flames reached first. Fire safety precautions, such as keeping exits unlocked and lighted, were probably neglected because the building's fire-resistive construction created a false sense of security on the part of the management. The fact that its contents—highly flammable scenery and people—would burn seems to have been overlooked.(48)

The Cocoanut Grove nightclub fire, on November 28, 1942, provided another horrible example of a nightmare disaster. The fire started in the basement cocktail lounge's highly combustible decorations (artificial cocoa palms and cloth covered ceiling and walls). As the flames blasted through the air under the ceiling, there was a frantic rush for the exits, but the main doorway was blocked by a jammed revolving door, and 200 victims piled up behind it. Another 100 piled up behind an exit door that opened inward, and others behind doors that were locked. A few lucky ones were able to crawl out of a basement window, and some managed to escape through small windows in the rest rooms. Of 1,000 or more in this nightclub, which had an official capacity of 600, the death toll was 499.(49)

The fire department brought the fire under control in quick order, once the flash-over was ended, but the rescue operations were the most important and difficult task it had to accomplish. Criminal indictments were brought against the building owners, the Boston Building Commissioner, members of the police and fire departments, a building designer, a contractor, and the foreman who participated in building the new bar and lounge.(50) The lessons learned from this disaster resulted in improved enforcement of fire safety laws throughout the nation, and important knowledge was gained in the use of blood plasma and the treatment of burns.

History repeated itself in 1977 when a fire at the Beverly Hills Nightclub in Southgate, Kentucky, just south of Cincinnati, killed 164 patrons. Faulty wiring was blamed for the fire and state investigators turned up numerous fire code violations and safety hazards. There was no alarm system or sprinkler system, and the wall paneling burned freely. Like other major fires, this led to strengthening of fire codes throughout the United States, occupancy limits on places of assembly, and other improvements. The nonmeasurable silver lining around this cloud may well be the numerous lives that have since been saved but have gone unnoticed.(51)

The Hartford Circus disaster occurred less than two years after the Cocoanut Grove fire, and was the worst fire tragedy ever experienced by the "Greatest Show on Earth." On the afternoon of July 6, 1944, 163 persons died within minutes after the huge circus tent was suddenly enveloped by a flash fire. Severe burns were suffered by more than 200 others, many of them children and circus employees. There were about 7,000 spectators in the stands at the time of the fire, covered by a tent approximately 180 by 425 feet in dimensions. The Ringling Brothers and Barnum and Bailey combined shows had about 1,300 employees, many of whom were probably inside when the fire started on the ground outside the tent and spread up the side. The exact cause is not known, but it may have been a discarded cigarette or match.

Unlike the Cocoanut Grove and Beverly Hills fires, where many of the deaths were caused by suffocation, most of the fatalities in this tragedy were caused by severe burns; when the burning canvas fell on the fleeing crowd, clothing was ignited. Additional deaths were due to crushing by the panic-stricken crowd. The nine exits from the tent proved to be inadequate, and some of them were blocked by animal chutes. Instead of being flameproofed, the new tent had been waterproofed by the use of paraffin applied with gasoline as a solvent.(52)

Nursing Homes

About 6,000 fires occur in nursing homes in the United States each year, and hardly a year goes by that a fire disaster does not strike one of these occupancies that are largely populated by elderly people. On January 9, 1970, a Marietta, Ohio, convalescent home fire killed 31 of its occupants. In the late 1980s a rash of nursing home fires claimed numerous lives, as these occupancies gained residents and increased in populations. Fire records indicate that multiple death fires in these institutions caused 338 fatalities within 12 years,(53) pointing up the need for automatic fire protection systems. At an annual NFPA meeting in the early 1970s in Toronto, Canada, members in attendance voted overwhelmingly to amend the NFPA standards to require *both* fire alarm and automatic sprinkler systems in nursing homes, in an attempt to set an example for those states that had not already adopted comparable requirements.

In the intervening years, some states have adopted laws or have permit-

The Indianapolis Fire Department moves to a defensive operation as fire consumes a large warehouse. (Photo courtesy of Jim Lanagan.)

ted their state fire marshals to adopt regulations for the purpose of achieving greater life safety in occupancies housing non-ambulatory persons. Until every one of these institutions is fully equipped with automatic fire detection and fire extinguishing systems, the potential for fire disaster in homes for invalids will continue to exist. Thorough training of the attendants and extensive prefire planning on the part of fire departments may help to avert, but cannot always prevent, a large loss of life where no automatic protection is provided.

Department Stores

During the early morning hours of Saturday, December 19, 1964, fire of undetermined origin started in display materials near the open escalators in the second story of the seven-story Neiman-Marcus Company Department Store in Dallas. Because of a delayed alarm, rapid spread of the fire up the open escalator opening, and a closed valve controlling the sprinklers in the escalators, 368 fire fighters with 24 engines, three ladders and two elevating platforms were unable to prevent property damage estimated at $7,240,000 and a business interruption loss of $2,800,000. Following this fire, which was the fourth major one that this store had experienced, and the largest loss in the history of Dallas, Texas, the store officials decided to install a complete system of central-station-supervised automatic sprinklers throughout the building.(54)

Department store disasters have not been confined to the United States. In Mexicali, Mexico, such a fire resulted in the death of seven persons. Serious fires have also occurred in Colombia, New Zealand, Belgium, and Japan.

Because of open stairways, the six-story "'1' Innovation" department store in Brussels, Belgium, presented a disaster threat for many years. On May 22, 1967, at least 325 persons died, and countless others were injured as a result of a fire that started in the storage area on the second floor and spread so rapidly that almost all of the people above that floor were trapped, as were many on the second floor. After a delayed alarm and a futile attempt on the part of the store's fire brigade to extinguish the fire with portable extinguishers, the fire fighters arrived to find that 12 persons had already jumped from upper floors in a desperate attempt to flee the heat and thick black smoke that filled the building. Other people were sliding down ropes, hanging out windows, or preparing to jump, while some waited on ledges or in windows for fire department ladders and life nets. Civilians helped to hold nets and raise ladders, and though fire fighters were not able to enter the building, even with breathing apparatus, because of the intense heat and smoke conditions, about 200 shoppers were rescued. The $20 million property damage, as well as the great loss of life, might have been prevented had the building been equipped with automatic sprinkler protection.(55)

During Japan's Fire Prevention Week in 1973, a fire broke out on the third floor of the eight-story Taiyo Department Store in Kumamoto on November 29, killing 103 Christmas shoppers and injuring 119. There was no fire evacuation alarm system in this concrete building, and the sprinkler system that was being installed was not yet completed when the fire started in some cardboard boxes stored in a stairwell between the second and third floor. Employees tried to put it out with fire extinguishers, but it spread rapidly to combustible merchandise and up the open airways, trapping hundreds of customers on the upper floors. The building had enclosed stairways but most of the rolling fire doors failed to close or were blocked open, so the stairs were unusable due to the fire conditions on the lower floors. The loss of life would have been much greater had the fire department not been able to rescue many persons from the roof with helicopters and from the upper floors by means of their only 100-foot aerial ladder.(56)

Factors Responsible for Death or Injury

Ten principal factors contribute to injuries and fatalities in fire disasters:

- attempts to extinguish incipient fire without notifying the fire department, resulting in a delayed alarm;
- failure of occupancy management to alert employees, customers, patrons, guests, and residents at the first sign of smoke or fire, so that planned evacuation can take place;
- failure of customers, patrons, guests, and residents, upon seeing smoke or fire, to sound an alarm and alert others of the danger;
- failure of employees, customers, patrons, guests, or residents to use the designated means of egress;
- failure of these persons to close doors behind them to help confine the fire;
- action taken by some persons to break windows, permitting air to enter and intensify fire;
- lack of confidence of building occupants that the fire department will rescue them, so they jump out of windows;
- combustible interiors (walls, ceiling, carpeting, furniture);
- inadequate exits (inward swinging doors, open stairs, revolving doors, lack of panic bars, unlighted, locked and insufficient number);
- absence of automatic fire detection and fire extinguishing systems (which would overcome most deficiencies stated above).

Structural Fire Disasters

Contributing factors to large loss structural fires include: values too large in one area; vertical openings unprotected; alarms delayed; areas too large; exposures unprotected; and inadequate fire protection.

PLANNING FOR CATASTROPHIC FIRES

Despite the devastation caused by major fires it is still possible to find fire service personnel who will think, if not say: "Why must we plan for major fire emergencies? It will never occur here! We don't know where to start. Planning is too complex and too much work. Besides, wildland fires, conflagrations, and highrise fires occur so infrequently, it is not worth the effort. I am so hung up on the day-to-day administration for the fire department that I don't have the time to plan for future events."

Certainly motivating fire managers to plan is, in many instances, the most challenging component of the planning process. Fire departments should undertake pre-incident planning for large-scale fires for four reasons. A pre-incident plan will do the following:

- reduce the command burden to manageable portions,
- help to improve the quality of on-scene mitigation decisions,
- communicate to all involved agencies their role within the control effort,
- provide a lasting framework from which the mitigation effort can be tailored to a specific incident.

1. Reducing the Command Burden. Fire managers know that wildland/urban interface fires, highrise and urban conflagration fires are demanding fire emergencies that necessitate the commitment of immense amounts of resources from all levels of government. Simply stated, the task of managing such efforts is so complex, that the incident commander requires help. Resolution of such issues as resource identification, task assignment, communication procedures, command organization, inter-agency cooperation, and logistics presents a myriad of complex and technical decisions that must include delegation by the incident commander. A pre-incident plan, by addressing such issues before the event occurs, lessens the burden on the incident commander, thus freeing him or her to deal with the specific demands of the emergency.

2. Improving the Quality of On-Scene Decisions. Planning reduces the need for the incident commander to make on-the-spot, split-second decisions in incident management. Such a decision making system may be dramatic but deadly, producing decisions that are based on inaccurate or incomplete information. During the planning process, the fire manager has the time to consult with the appropriate parties, deliberate, and then reach a well-considered decision.

3. Communicating to All Agencies Their Role Within the Command System. Agency leaders, whether the chief of a neighboring fire department or the head of the regional disaster preparedness agency, need a clear and unambiguous understanding of their responsibilities in any joint effort. A pre-incident plan allows the fire manager to inform agency heads about what will be expected of their organizations.

4. Providing a Mitigation Framework. Pre-incident planning gives the fire manager the resources and command framework from which an effective mitigation effort can be constructed. Once such basic issues as resource identification and command and control procedures have been established, the fire manager is free to tailor the existing plan to the demands of the specific incident. Fire incidents are seldom alike; however, a pre-incident plan, by addressing issues common to all emergencies, provides the fire manager with the opportunity to address unique issues for each incident.

Those cities that have developed a good disaster plan should have an appendix designed to handle all of the planning functions necessary for a catastrophic fire. This appendix is most properly subdivided into the three major categories addressed here—urban conflagration, wildland/urban interface fires, and occupied highrise fires.

Critical Factors in Planning

The following five issues must be considered when developing an effective pre-incident plan or appendix for catastrophic fire emergencies.

1. Never underestimate the speed of the fire. Learning to estimate fire spread potential is probably one of the most important skills that the fire manager must acquire. It does not require a million-dollar computer to warn you that a small fire in dense brush or closely spaced combustible buildings on a 100°F day with low humidity and high winds is likely to achieve rapid fire extension. Brush fires under such conditions may advance at the rate of 25 acres of brush per minute in canyons and up to 150 acres per minute on uphill slopes.

In urban conflagrations, buildings can ignite, be totally involved, and be threatening the adjacent structures in as little as 20 minutes. In highrise buildings fire can spread vertically through such openings as pipe chases and unprotected stairwells at a rate of approximately 10 minutes per floor. One rule of thumb might be to consider that a fire unchecked will need double the resources needed for effective control every half hour it goes unchecked.

2. Don't assume that there will be only one fire at a time. Expect the unexpected and plan for all reasonable contingencies. Remember that when the conditions favor the outbreak of disastrous fires in your community, they also favor the occurrence of such fires in neighboring communities. This is especially true for wildland fires and conflagrations whose extension is influenced by wind, humidity, and temperature. Furthermore, such fires are often of incendiary origin, and the arsonist may ignite several fires in one or several jurisdictions. In this instance, the response of outside agencies may be reduced or delayed.

As your fire department responds to an emergency the fire department's normal work load must also be managed. When resources are limited who will answer additional alarms? Fire officials must expect and plan for the occurrence of multiple fires and the corresponding need for additional resources.

3. Don't underestimate the time it takes to muster a massive attack. Usually, the response to a major emergency will seriously deplete the suppression capabilities of a fire department. Additional capabilities will necessitate summoning companies further away from the fire and such units may incur substantial dispatch, turnout, travel, and set-up times. Mutual aid departments must still provide for services within their own jurisdictions. Therefore you may not receive all requested equipment and manpower. A strong argument for pre-incident planning is the establishment of procedures that will facilitate rapid response from needed resources.

An additional aspect of planning is a consideration of how to commit, for the best tactical advantage, the initial response assignment. A plan, which represents a consensus of fire department decision makers, should delineate specific policies and criteria that will guide the tactical commitment of the initial response while the massive attack is organized and the necessary resources marshalled.

In anticipation of a massive earthquake along the New Madrid fault, the State of Missouri produced a comprehensive statewide mutual aid plan in late 1990. Although the quake had not yet materialized, the statewide mutual aid plan was used in March 1991 to successfully control a railroad emergency involving a massive derailment and burning tank car of yellow phosphorus, not far from St. Louis.(57)

4. If you are chief, don't assume that you will be at the scene of the fire. In-depth training is necessary at the command positions, preferably at least three deep. Since the command officers may be incapacitated or unavailable, the chief officer has a primary obligation to teach assistants how to perform the tasks normally discharged by the command positions. This is most applicable to small career and volunteer departments where there may not be sufficient depth in command positions.

5. Don't assume that your prefire plans will work as expected. Each time a pre-incident plan is practiced new problems with it or difficulties with participating agencies are discovered. These can also occur as the plan is invoked for the present emergency. Perhaps the first due units, or one of the mutual aid companies, are operating at another incident. An effective plan must be flexible. The pre-incident plan is only a framework from which the response can be tailored to the incident at hand.

Improving Operations

IS MORE EVER ENOUGH?

Has the acquisition of more fire stations, more fire apparatus, and more sophisticated command and communications systems put an end to large-loss fires? No, primarily for two reasons. First, the rapid extension of many fires is greatly influenced by weather conditions that are beyond the control of the fire protective system. Second, other uncontrollable pre-incident conditions, such as closely spaced combustible construction, inadequate water supplies, and poor housekeeping practices, may exert a considerable influence on the fire growth process.

Fire service decision makers who desire to improve their response capabilities should adopt a pro-active approach to the mitigation of such emergencies. It is important to develop a pre-incident plan so that when the emergency occurs a department will be prepared.

INCREASED EFFICIENCY OF OPERATIONS

The planning document will identify resources, assign tasks, and develop a command structure, all of which will increase the efficiency of fire control efforts and ultimately reduce fire loss.

At this Sherwin-Williams warehouse fire, Chief Glen Alexander wisely permitted the fire to burn rather than poison the nearby water table. (Photo courtesy of the Dayton, Ohio, Fire Department.)

EXTENDED OPERATIONAL CAPABILITY

Since most fire department operations are of relatively short duration, many fire managers fail to realize that operations at such immense fires as highrise and conflagrations, may last for several days. However, through its assessment of the logistical needs of suppression resources and its provision for the fulfillment of such needs, a pre-incident plan will effectively extend the operational capabilities of the fire department.

IMPROVED ALLOCATION OF RESOURCES

By identifying the goals and priorities of the mitigation effort, a pre-incident plan will establish the policy for resources assigned to various divisions or sectors within the overall operation. Each arriving unit will be assigned to a specific task or sector, governed by a clearly defined objective.

LEARNING FROM THE PAST

It is impossible to promulgate principles and practices to cover every possible urban conflagration, wildland/urban interface fire and highrise fire. Nevertheless, we can look at the many past incidents documented in this chapter to determine the conditions that contributed to the inception and growth of the fire, and then actively seek state and local legislation to help eliminate the impact of such conditions. For example, the fire service may seek more stringent local fire prevention and zoning laws requiring property owners to cut the brush around their homes, increase the distance between combustible structures, or eliminate combustible wood shingle roofs, all of which have been major contributors to wildland and conflagration fires.

IMPROVING FIRE CONTROL

We must seek to improve our fire defenses, fire suppression land rescue capabilities, fire fighting organizations, communications, command systems, and agency coordination systems. The following tactical concepts can be incorporated into a pre-incident plan for catastrophic fires.

BARRIERS

Employ natural and man-made fire barriers that deny the fire new fuel supplies. Natural barriers consist of open spaces such as a vacant lot or a street in a congested urban area. Man-made fire barriers for wildland fires may consist of newly constructed fire breaks. For conflagrations and highrise buildings they consist of unpierced fire walls. These locations should be considered when deciding where to cut off fire spread.

WATER SOURCES

The size of conflagrations and wildland and highrise fires mandates the application of large volumes of water. Whether dropped from aerial tankers, sprayed from handlines, or deluged from master stream devices, a cooling punch must be applied. If municipal sources are insufficient, then natural sources must be tapped. The planning effort must assess water sources in terms of their capacity and duration. Recent technological improvements in the areas of large-diameter hose and tanker operations have the capability of dramatically increasing a fire department's ability to move large volumes of water considerable distances. If possible, these should be incorporated into a plan. If one department does not have the financial resources to purchase large-diameter hose, it can add to its plan the hose of a neighboring department.

COMMITMENT OF FORCES

When resources are limited, especially in the initial stages of a major fire, the incident commander must determine where the on-scene forces can exert the most influence on fire growth. Is it better to concentrate them or spread them out? Can any side of the fire be left unattended? The operations officer, through the establishment of the appropriate policies, can provide the initial incident commander with a yardstick to make such tactical decisions.

An apparatus is in proximate danger in this major block long fire in Baltimore. (Photo courtesy of the Baltimore City Fire Department.)

SECTORIZATION

In Chapter 3 we discussed the incident command system (ICS). This concept implies that an incident scene should be divided into manageable command divisions or sectors. Although the sectors are often geographical they may also be functional (e.g., emergency medical services, water supply, logistics).(58)

Sectorization reduces the responsibilities for the incident commander as sector-level command decisions can be made by the sector officer. The communications volume is also reduced since all units need not receive every order from the incident commander. In this manner, the flow of information up and down the chain of command is facilitated. Logistics and communications are of the utmost importance to meet great demands for fuel, oil, tools, and other consumables.

Keeping accurate records is a responsibility that should be assigned at the incident command level to a competent officer. In addition to advising the incident command of who is doing what and where, such records will be vital to any post incident review or investigation.

FLEXIBILITY

While command and coordination are important, fire officers should also be cognizant of the need to be able to shift personnel and equipment rapidly as the situation changes. The elements of span of control, delegation of authority, and chain of command should be established for all divisions or sectors. Staging areas and command post sites should be carefully selected so that they will not be vulnerable to the advancing fire. Apparatus should be placed where it is not likely to be trapped and demolished. Of course, the safety of personnel should be paramount. In urban conflagrations, principal dangers are falling walls, heat exhaustion, and stress. Fatigue may become an important factor, so resting and feeding personnel must be considered early in the operation.

PROTECTING TARGET HAZARDS

Finally, target hazards such as hospitals, nursing homes, hazardous materials sites, and civilian refuge areas must be protected. One effective tactic to protect exposures or check fire extension is narrowing the front of the fire. This concept advocates heavily attacking the flanks of the advancing fire front either by constructing fire breaks or applying master stream devices in order to channel fire extension into a "V" shape as shown in Figure 7-1. The intent of this tactic is to reduce the length of the front on which the fire is advancing to more manageable proportions.

While Figure 7-1 represents a theoretical urban conflagration, the same graphic can be applied to a high rise building fire as illustrated in Figure 7-2 if we envision the graphic to represent a vertical rather than horizontal configuration. It was just such a tactic that helped the Los Angeles Fire Department successfully control the First Interstate Bank fire on May 4, 1988.(59) Figures 7-1 and 7-2 illustrate the overlap and similarity between varying types of catastrophic fires.

CONCLUSION

This chapter has considered a variety of types of major fires, all disastrous or catastrophic in nature. Nonetheless, common characteristics were identified and elements necessary for effective disaster planning relative to each type were explained. Since the original mission of the fire service was to extinguish fires, this type of disaster should be the most natural for fire departments to successfully control. Indeed many success stories have been recorded where fire departments have successfully contained wildland, conflagration, and highrise fires. One such example is the Peachtree Plaza fire in Atlanta, Georgia where swift rescue and fire control measures were successful in limiting a highrise building fire to one floor.(60)

The more disastrous the fire, the more likely it is to become national news. And the more successful the mitigation effort, the less likely these fires are to be newsworthy. In a sense, it is the job of the fire officer to successfully stay out of the news.

Figure 7-1. Handling a Major Fire.

Narrowing the fire front. Chief forms his companies in the form of a blunted "V." Companies on the wings with heavy streams move in on the flanks of the fire until the fire front has been reduced to one or two blocks.

Figure 7-2.

Imaginary "Blunted V" for stopping the vertical advance of fire in high-rise building. At lower levels, side streams can be applied with aerial devices. At high levels, stairwells and protected perimeter voids can be used.

CHAPTER 7 REFERENCES

1. Duncome, Ted. "Philly Tower Inferno Kills Three" in *The Press Democrat*, February 25, 1991, page A-8.
2. "The 1989 Fire Picture at a Glance" in *Fire Journal*, Quincy, MA: NFPA, September/ October 1990 p. 48.
3. Coleman, Ronny J. (Ed) *Managing Fire Services*, Washington, D.C.: ICMA, 1988 p. 8.
4. "Year Book for 1991" of *World Book Encyclopedia*, Chicago, IL: Field Enterprises Educational Corporation, 1992.
5. *NFPA Handbook*, 17th Edition 1991.
6. *NFPA Handbook of Fire Protection*, 13th Edition 1962, p. 1067.
7. "Year Book for 1991" of *World Book Encyclopedia*, Chicago, IL: Field Enterprises Educational Corporation, 1992.
8. Isman, Warren E., and Gene P. Carlson, *Hazardous Materials*, New York: Macmillan Publishing Co., 1980.
9. "First Report—San Francisco Earthquake" in *American Heat* video journal, St. Louis, MO, November 1989.
10. Dunn, Vincent. "Fire Studies—The Danger of Wildfire" in *Firehouse*, Vol. 12, No. 10 (October 1978), p. 18.
11. Bond, Horatio. *Fire and the Air War*, NFPA, 1946 (out of print).
12. "Fire Aspects of Civil Defense," OCD TR-25, July 1968.
13. See note 11.

13a. Wilson, James D., "No More Home Sweet Home" in *Newsweek*, Vol. CXVIII, 19, November 4th, 1991, pp. 34–35.

14. Few, Edward W. "Tracking Arsonists in America's National Forests" in *Firehouse*, Vol. 15, No. 8 (August 1990), p. 34.
15. *Los Angeles Times*, November 25, 1980.
16. Cullom, Keith. "Timber Devastation" in *Firehouse*, Vol. 12, No. 12 (December 1987). p. 31
17. *Fire Command*, March 1986.
18. Countryman, C. M., M. A. Fosberg, and R. C. Rothermel. "Fire Weather and Fire Behavior in the 1966 Loop Fire" in *Fire Technology*, May 1968, p. 137.
19. Klaene, Bernard. "*Fire Behavior*," p. 126.
20. Disaster Civil Preparedness Agency, *Disaster Operations*, CPG July 1972, p. 49, U.S. Government Printing Office.
21. Kramer, William M. "Meeting the Community Fire Threat" in *Community and the Fire Threat* (Open Learning Course), Emmitsburg, MD: National Fire Academy, 1991.
22. "Civil Defense" in *1988: The Year in Review*, Federal Emergency Management Agency, U.S. Government Printing Office, 1989.
23. Pope, Kerry. "New Air Tanker System Passes Tests by Fires" in *Fire Engineering*, April 1974, p. 64.
24. "Syroco Plastics Fire," *American Heat Video Productions*, St. Louis, MO, 1990.
25. Dektar, Cliff. "Night Vision Systems May Extend Use of Helicopters on Wildfires" in *Fire Engineering*, August 1974, p. 178.
26. Lyons, Paul R. *Fire In America!* NFPA, 1976, p. 218.
27. "Black Tiger Gulch Fire" in *Fire Command*, January 1990.
28. Dunn, Vincent. See Note 10.
29. *Firehouse*, October 1988.
30. Le Blanc, Edward. "Fire Fighter and Equipment Limitations in High-Rise Building Fires" in *Fire Command*, December 1973, p. 16.
31. Casey, James. "High-Risers—Big Problem" in *Fire Engineering*, April 1971, p. 27.

32. "The Tall Ones—Why Are They Special?" in *News Bulletin*, National Automatic Sprinkler and Fire Control Association, January–March 1973, p. 8.
33. Gaudet, Robert E., and Ernest E. Juillerat. "Fire at Dale's Penthouse Restaurant" in *Fire Journal*, May 1967, p. 5.
34. O'Hagen, John T. "Sao Paulo, Brazil, Adds to High-Rise Fire History" in *Fire Engineering*, July 1974, p. 18.
35. *Time*, January 12, 1987.
36. O'Hagen, John T. See Note 34.
37. Bahme, Charles W. *Fire Officer's Guide To Emergency Actions*, 2nd ed., p. 76, NFPA, 1976.
38. Mendes, Robert F. *Fighting High-Rise Building Fires*, Boston, MA: NFPA 1975.
39. "Fire Fighting in Three Dimensions" in *Fire Command!* July 1973, p. 17.
40. *Fire Journal*, March–April 1990.
41. *Fire Journal*, January–February 1990.
42. *Fire Engineering*, April 1990.
43. Kline, Fred W. "California Develops Fire Safety Rules For New High-Rises," *Capitol News Service*, October 1974, p. 4. Mr. Kline reported in this article that California's State Fire Marshal had adopted regulations that would require that the safeguards discussed in this section be made applicable to all existing buildings more than seventy-five feet in height by January 1, 1975.
44. Notes from on-scene reports by author William Kramer, April 1991.
45. "That Hotel Fire in Seoul" in *Fire Command!* March 1972, p. 32.
46. Bahme, Charles W. *Fire Officer's Guide to Disaster Control*, 1st ed. Boston, MA: NFPA, p. 137.
47. "New York Social Club Fire Kills 87" in *American Heat* video journal, St. Louis, MO, April 1990.
48. Bahme, Charles W. See Note. 46, p. 145.
49. *World Book Encyclopedia*, Vol. 5, p. 181, Chicago, IL: Field Enterprises Educational Corporation, 1969.
50. Bahme, Charles W. See Note 46, p. 145.
51. *Newsweek*, June 13, 1977.
52. Bahme, Charles W. See Note 46, p. 146.
53. "Nursing, Retirement Home Fires" in *American Heat* video journal, St. Louis, MO, January 1990.
54. "The Neiman-Marcus Fire" in *Fire Journal*, May 1965, p. 5.
55. "Fatal Store Fire, Brussels, Belgium" in *Fire Journal*, November 1967, p. 5.
56. "Department Store Fire, Kumamoto, Japan" in *Fire Command!* January 1974, p. 18.
57. Sachen, John. Commentary in *American Heat* video journal, St. Louis, MO, April 1991.
58. See Chapter 3 for a complete explanation of the incident command system.
59. *Firehouse*, October 1988.
60. Kramer, William M. Commentary in *American Heat* video journal, St. Louis, MO, August 1989.

CHAPTER 8

CIVIL DISORDERS AND RIOTS

THOSE OF US living in the 1990s and anticipating the turn of the century are well aware of the many forms of social and political discontent that exist in our country. Persons of all socioeconomic classes have always been publicly involved in demonstrating their collective feelings on various issues. Their protestations can be peaceful such as the simple planting of a grain flag in a field by Pennsylvania farmers as a show of support to preserve their farmland, and prevent the installation of a rock quarry in April 1990.(1) When protestations are taken to extremes, however, they may give rise to civil disturbances and acts of terrorism.

For example, on May 5 and 6, 1991, just blocks from the White House, stinging tear gas lingered over a twelve block area as restless youths were involved in rock throwing, car burning, and looting in a largely Hispanic neighborhood. This was triggered by the arrest of one resident for public intoxication. Fortunately, police were able to calm this riot before anyone died.(2) The residents of Port-Au-Prince, Haiti, were not as lucky. When they rampaged against a business with ties to the Duvalier regime, in early 1991, fifty people died.(3)

Louis Masotti and Don Bowen in the foreword to their joint research effort entitled *Riots and Rebellion: Civil Violence in the Urban Community,* advance the position that violence and conflict appear relentlessly in human societies.(4) The authors believe that it is an extremely rare society that is devoid of violence.

Many members of the fire service may associate the term civil disturbance with riot. The basis for that association may be the social upheaval of 1960s, as well as the antiwar and antimilitary demonstrations of the 1960s and 1970s. However, the events of the 1980s and early 1990s, particularly acts of international terrorism and sprees of incendiary activities, have altered and broadened the fire service's experience within the realm of civil disturbances.

Throughout the occurrence of a civil disturbance, the local fire department often expands its level of service delivery to handle the outbreak of numerous incendiary incidents. Accordingly, fire departments must be sensitive to the changes in the composition of civil disturbances. This chapter will examine fire department emergency planning within the civil disturbance context. It begins with a brief review of the causes and effects of civil disturbances, then discusses specific planning considerations, including fire control operations and personnel protection.

CIVIL DISTURBANCES: A SOCIAL PHENOMENON

Contemporary theory holds that the local fire service organization is a creation of and subordinate to the local social system. Nowhere is this theory more relevant than in the field of civil disturbance. For generations, fire fighters were perceived as heroes, a service that could do no wrong. Certainly within the last twenty years this concept has changed and many fire departments have experienced the wrath of the people they are sworn to protect.

Although many fire departments enjoy the respect and support of the communities they serve, some members of the community may feel that the fire department is allied with the police against their struggle for social change, and they may vent their anger against the fire department. This would be the case when rioters pelt the fire fighters with rocks. Also, as fire fighters in some cities conducted strikes and work stoppages, civilians saw buildings burn while the fire fighters either did nothing or actually thwarted their extinguishment. This caused some civilians to look with some disfavor on the fire service.

Within the scope of this chapter, a civil disturbance may be defined as a demonstration of popular unrest that may manifest itself in acts of violence and destruction against property and human life. For fire departments, civil disturbances may be characterized by violence against operating personnel in the performance of their duty and numerous acts of incendiarism that seek to destroy public and private property. The riot, the most serious variant of the civil disturbance, is a form of crowd behavior that is characterized by the destruction or threat of destruction of lives and property.

Although riots have received enormous media attention and did produce widespread devastation in a number of American cities during the late 1960s and early 1970s, they are just one type of civil disturbance with which fire department planners must be concerned.

Incendiary Activity

In recent years, several large urban fire departments have experienced periods of pronounced incendiary activity, which is a form of civil disturbance even though it poses less of a safety threat than a riot. Perhaps the most vivid example of this variant of civil disturbance is in Detroit, Michigan, where on "Devil's Night," the night before Halloween, hundreds of alarms are transmitted and dozens of serious structure fires are set each year.(5)

International Terrorism

International terrorism is another form of civil disturbance. Performed by perpetrators that are well organized and dedicated to their cause, acts of terrorism may produce events that are highly destructive in human and monetary terms. This fact, coupled with the need for the services of a number of agencies on all levels of government, makes planning for terrorist acts a necessity.

A locality does not have to experience a riot, a terrorist incident, or a violent protest demonstration for the fire department to be involved in a civil disturbance. Your department may be summoned to aid another department that is experiencing a period of civil unrest, or an event may occur within your community (e.g., the construction of a nuclear power station) that engenders anger and fear a fire department finds itself operating within a potentially explosive situation. Certainly, planning for civil disturbances no longer can be considered largely the exclusive concern of urban fire departments.

Riots

Our planning discussion will focus on the more serious instances of civil disturbance, specifically riots. Riots "appear to be more spontaneous, unplanned, and disorganized"(6) than other forms of civil disturbances. Because of their extreme threat to the safety of operating forces and the high incidence of incendiary activity, they represent one of the most demanding situations for fire fighters.

From a legal standpoint, the word "riot" may include a number of offenses, from "disorderly behavior" to "insurrection," depending upon the laws and decisions of the various states. In most jurisdictions, regardless of the legal definition, it refers to a noisy, violent outbreak of disorder by a group of people. Some of these disturbances have resulted in major disasters, involving death and injuries to persons and widespread property damage.

Riots may be planned or spontaneously triggered by some event that leads to mass violence. A riot may involve only a small number of agitators, as in the famous Boston Massacre, or thousands of rampaging men, women, and children, plus thousands of soldiers, police officers, and fire fighters, as in the racial riots of the 1960s.

Riots have occurred throughout the world from the beginning of time. "In ancient Rome, the poor rioted several times to press their demands for lower rents and lower food prices. In England, during the early 1880s, workers called Luddites staged a series of riots in which they destroyed labor-saving machines, which they feared would replace them."(7)

Ironically over a century later the British again rioted for a similar reason. On March 31, 1990, an antitax demonstration by 40,000 people turned into a riot when militants pelted police with rocks and bottles, set cars ablaze, and smashed window shops in a rampage through central London.

They were protesting a "poll tax" advocated by Prime Minister Margaret Thatcher. So successful was this protest, that the widely respected Mrs. Thatcher saw her worldwide intense popularity fall dramatically, almost overnight. She resigned gracefully but under pressure.(8)

Rioting spectators at many foreign soccer games have started fires and panic behavior that has resulted in mass casualties, including many killed in rival clashes and the crush of fleeing patrons.

SMALLER CITIES AFFECTED

Not all riots have taken place in large cities or in impoverished areas. Cambridge, Maryland, with its 100 percent volunteer fire department, had to cope with a rash of fires accompanying riot conditions on July 24, 1967. Protected by city and state police officers, as well as the National Guard, and assisted by volunteers on the hoselines, the Cambridge fire fighters were able to make a holding action on eighteen fire-involved buildings until mutual aid arrived from six neighboring departments.

RIOTS AT AND NEAR COLLEGES

Students in all parts of the free world tend to exercise a creative right to dissent, and occasionally it gets out of hand. On April 27, 1991, for example, police used water cannons to spray students outside Yonsei University in Seoul, South Korea. A student, Kang Kyung-Bae, 20, had died during a previous police confrontation occasioning the student protest.(9) Campus demonstrations and riots certainly are not new, and were commonplace in the United States especially during the years 1969 and 1970. Cambridge, Massachusetts; Berkeley, California; and Kent State, Ohio; are all examples of well-known campus riots that escalated out of control. Campus rioting was in large part responsible for a dramatic increase in the number of fire bombings reported nationally in the late 1960s and early 1970s. One statistic, for example, showed that between January 1, 1969, and April 15, 1970, there were 43,000 bombings in the U.S. resulting in 40 deaths, 383 injuries, and $22 million in losses.(10) Offensive weapons used by the campus rioters, in addition to dynamite bombs and incendiary devices, included railroad flares, fuses, guns, grenades, bricks, clubs, bicycle chains, and sling shots.

In the 1980s, the students turned to more peaceful means of political activism and, although their actions may not have been as flamboyant and noticeable, their more peaceful approach did permit genuine dialogue and student involvement in shaping world affairs. Unfortunately, the 1991 war in the Persian Gulf rekindled some student demonstrations but protest actions were not as violent in 1991 as those that had prevailed in the 1960s and 1970s. Some sociologists feel that the young generation always has a latent propensity for violence, regardless of socioeconomic class, that must be guarded against. One study showed that in the 1990s, on the streets, "disputes once settled with fists were now settled with guns. Every 100 hours, more youths die on the streets than were killed in the Persian Gulf."(11)

THE CAUSES OF RIOTS

Riots involving arson, bombing, and criminal destruction have been caused by a number of circumstances. Groups who have felt that they were being unfairly taxed, drafted into military service, or receiving an inequitable return on the fruits of their labor, have started riots. Others, who believed they were being discriminated against in education, housing, employment opportunities, or other matters of economic, social, and political concern, have participated in riots. When people believe that they are being mistreated by government or social agencies, when they feel that they are not permitted to participate in the decision-making processes that rule their daily lives, when they feel that they are being "ripped-off" by a judicial system that seems to mete out to them harsher treatment than is rendered to others in the community, they may begin to harbor smoldering grievances that might suddenly erupt into flames of anger, resentment, and violence. A seemingly unimportant act, such as the arrest of an individual for a traffic infraction, may be the event that causes a violent explosive social reaction.

Masotti and Bowen have identified two causes of civil violence: first, the general conditions of dissatisfaction and unrest; and second, the precipitating event. They claim that "both are necessary; as neither by itself is sufficient; together they are both necessary and sufficient."(12)

We know in advance when and where conditions of unrest exist. But we cannot know what kind of event will serve as the immediate explosive cause, although the National Advisory Commission on Civil Disorder (The Kerner Commission) noted the tendency for most urban riots in the summer of 1967 to be touched off by confrontations between police officers and civilians. The precipitating event must be viewed by a fairly large audience so that new participants can be drawn into the disturbance. Accordingly, fire departments that are on alert during periods of civil disturbances should strive to present a low profile to the community and to avoid the occurrence of a confrontational precipitous event.

EFFECTS OF RIOTS

Riot disasters have resulted in a great loss of life, untold injuries, and tremendous economic losses. For example, in the nation's worst period of civil unrest—between the summer of 1965 and the summer of 1967—riots in 82 U. S. cities resulted in 130 persons being killed, 3,623 injured, 28,932 arrested and $714 million loss.(13)

Riots have the obvious effects of fatalities, injuries, and property losses. However, riots have also scarred our social system by generating increased fear and distrust among blacks, Hispanics, and whites; between labor and management; rich and poor; the police and the community; voters and politicians; and students and administrators.

Not all effects have been negative. In some instances riots have caused people to more closely analyze the inequities of our society, resulting in laws and social programs aimed at eliminating or mitigating the conditions that precipitated the riots. Examples are the adoption of the Voting Rights Act of 1965 after the freedom marches in Alabama and the Civil Rights Act of 1968 after the murder of Dr. Martin Luther King and the ensuing riots.

Although riots may initially be police problems, the occurrence of incendiary activity soon invokes the response of fire department units. Under normal conditions, a fire response is aimed at suppression of a fire in an occupancy with the occupants eager to assist the fire department in its suppression efforts. Under riot conditions, the fire department may encounter a number of large simultaneous incendiary fires that require the commitment of the entire fire department to the area of unrest. Additionally, operating forces may be obstructed in their suppression efforts by actions ranging from the blocking and damaging of fire hydrants to sniper activity.

EXAMPLES OF MAJOR RIOTS

In the 1760s and 1770s, the American colonists rioted against tax collectors and other British appointed officials, in such notable events as the Boston Tea Party, Shay's Rebellion, and the Whiskey Rebellion.

During the 1800s, many native-born Americans had a strong resentment against blacks, Irish Catholic immigrants, and Orientals, and riots against these people were common. In the 1850s, the Know-Nothing, or American Party, staged uprisings against the Roman Catholic Church by attacking Irish Catholics in several cities, including Baltimore, Louisville, New Orleans, and St. Louis, in which several lives were lost. These riots were instigated by the majority against a minority because of fear that the Irish Catholics were getting too much political power.

In 1863 the United States had some of the most destructive riots in its history. It was during the war between the states, and many men were strongly opposed to being drafted into the Union Army. Armed mobs swarmed through downtown New York, setting buildings on fire, looting shops and homes, and shooting blacks, policemen, and federal troops. More than 1,000 persons were killed or wounded.(14)

During the 1800s and 1900s labor riots and race riots caused much destruction. Blacks and abolitionists were the targets in several northern cities and in the South, while Chinese immigrants were victims of mob violence in the economic depression of the 1870s in several western states. The great railroad strikes of 1877 precipitated riots in several cities, resulting in numerous deaths; the Haymarket riot in Chicago in 1886 started when someone threw a bomb during an anarchist meeting in protest of police brutality against strikers at an industrial plant. In 1919, violent riots erupted in the steel mills of Indiana, Ohio, and Pennsylvania, when labor leaders attempted to organize the steel industry. The cotton-textile industry unionization attempts were the cause of riots in Georgia, South Carolina, Alabama, Rhode Island, and other states, in 1934, resulting in the loss of about twenty lives.

The "Bonus Army," composed of about 15,000 unemployed World War I veterans, marched and demonstrated in several cities in the early 1930s, culminating in the June 1932 march on Washington, D.C., for the purpose of getting Congress to pass a bill that would immediately give them the bonus that was not due until 1945. President Hoover was opposed to the bill for fiscal reasons; after it was defeated in the Senate, General Douglas MacArthur and his troops swept down upon these veterans and drove them out of Washington.(15)

Clashes between blacks and whites have led to riots. Such a riot took the lives of 39 blacks and 9 whites in 1917 in East St. Louis. In 1919, a Chicago riot caused 38 deaths and 500 injuries. Racial riots broke out during World War II, the worst in 1943 in Detroit, Michigan, where 34 persons died. Riots erupted in 128 cities in the 1960s (75 in 1967 alone), the Watts riot in Los Angeles in 1965, those in Newark and Detroit in 1967, Cleveland and Washington, D.C., in 1968. The Detroit riot was probably the most destructive in terms of lives lost (43 lives) and property damage ($45 million), although Watts was a close second.

Before we look at these riots more closely, we should point out that the 1968 Chicago riots occurred during the Democratic National Convention. Thousands of young people gathered in the downtown area to protest the country's participation in the Vietnam War. Though several bloody clashes took place between the Chicago police and the demonstrators, no one was killed.

In the early 1970s, student riots occurred in many cities throughout the world, including London, Paris, Tokyo, Mexico City, Berlin, New York, and Montreal.

Looting and burning were commonplace during riots which rocked the country in the late 1960s and have sporadically occurred every year since. (Photo courtesy of Eugene A. Jackson.)

In 1971, one of the worst prison riots in American history occurred in Attica, New York, at the state prison. The rioters contended that the guards had mistreated them, and they seized the prison and held it for four days until state troopers finally charged in and obtained control. In doing so, 11 guards and 32 prisoners were killed.(16)

In the 1980s, riots erupted in various cities throughout the United States, among them Miami where riots broke out in 1980, 1982, and 1989. During the Super Bowl weekend in January 1989, the shooting of an unarmed black by a Miami police officer created a riot that lasted several days and drew national attention and network cameras away from the Super Bowl and into the heart of Miami.(17) In April 1990, another race-related riot broke out in Teaneck, New Jersey. Because of all the perceived inequities and counterpositions in the world, riots are expected in various quarters and at unforeseen times and places well beyond the year 2000.

During the Persian Gulf crisis, violent protests erupted in major cities of Asia, Europe, and the United States by thousands of persons objecting to the military build-up. Often started by flag-waving pro-war demonstrators, the anti-war activists burned flags, set fire to overturned police cars, blocked entrances to government buildings, bridges, highways, and military installations. Although fire fighters did not encounter serious resistance, large forces of law enforcement personnel were extremely busy attempting to preserve the peace and keeping access to roads and public buildings open.(18)

In March of 1991, the movie entitled *New Jack City* based on the operations of the drug world in a largely black community inspired numerous black youths into riot in the Los Angeles area.(19)

WATTS: A CASE STUDY OF RIOTS

Of all the periods of civil unrest in our country throughout the 1960s, 1970s, 1980s, and 1990s, none has equaled the racial riots of the 1960s in terms of death and destruction, magnitude, or ensuing social change. Of these "famous 1960s riots" the premier example was the Watts Riot in 1965. We will examine this riot from the perspective of the Los Angeles Fire Department, which was called upon to exercise its primary mission of life-saving and property damage control under the most difficult of circumstances.

An explosive condition existed in Los Angeles in August 1965, as in other cities, but it took a traffic arrest to touch off five days of destruction, looting, and arson. As a result of the riot, 34 persons were killed, more than 1,000 injured; about 4,000 of the 10,000 rioters were arrested and relieved of more than 2,000 guns. About 47 square miles were involved. Within the city limits, 261 buildings housing 318 separate occupancies were partially or totally destroyed (157 were totally destroyed). Throughout the county, more than 600 buildings were damaged by looting and burning; 200 or more of these were destroyed by fire.(20)

The interim fire chief who had been appointed to fill the vacancy left by the retired chief pending an examination was out of the country when the riot broke out, leaving deputy chiefs to fill billets of department commander and operations officer. The mayor, who has the official responsibility for declaring a state of local peril to activate the civil defense organization, was in San Francisco, and the governor, who is the official to normally call out the National Guard, was in Greece. Although they all returned to their posts of duty before the disaster was over, there was a delay on the part of the lieutenant governor in activating the military support needed to restore order to the city. This state of affairs demonstrated the need for having a plan of action that provides for depth of personnel with unmistakable and clearly understood authority to act in the absence of regular heads of departments and governments.

Violence Against Operating Forces

One problem not foreseen was assault on fire fighting forces with bricks, bottles, clubs, guns, and even fire bombs. Time and again fire fighters were attacked while fighting the rash of fires started by looters, whose initial targets were liquor stores, pawn shops, and any occupancy where guns might be found.

Of 104 fire trucks damaged, 45 had windshields broken (a chunk of concrete went through one windshield and broke the steering wheel), tires were shot flat, and Molotov cocktails were tossed into a hose bed; four of the rigs had bullet holes in them. Though four fire fighters were shot, only two had to be hospitalized with bullet wounds; of the 183 men who sustained injuries, only 38 had to be placed off duty. One fire fighter was killed when a structure collapsed upon him.

Why would anyone want to shoot a fire fighter? There are some psychologists who believe that hostility toward police officers rubs off onto fire fighters during a riot. This transference phenomenon causes the indigenous population to view fire fighters not as friends and providers of a vital municipal service, but rather as representatives of an unresponsive or oppressive government.

Following the riots of the 1960s, affirmative action programs were launched in both the private sector and governmental organizations throughout the United States. It is hoped that the affirmative action movement will minimize the threat of future race riots, and should they occur, the hostility toward the fire service should be greatly diminished due to the large number of black fire fighters now serving this country. It is possible, however, that another wave of riots could start by Hispanics or other minority groups seeking social justice. History could repeat itself.

In any case, the civil disturbance fire defense problem is one that requires careful and deliberate pre-incident planning, with special attention to methods of deployment, tactical control, and the safety of personnel.

Communications Overload

During the Watts riot the staff at the Operations Control Center was so swamped with calls that alarms from street-corner alarm boxes then in use were ignored, and telephone calls were screened so that dispatching was made only to serious fires that could present a possible life hazard; autos, fences, sheds were disregarded. Alarm board operators estimated that more than 3,000 alarms were received in the three most active days of the riot, during which approximately 2,000 fires occurred in grass, fences, sheds, autos, and structures. In the rest of the city the citizens were exercising such care that very few fires were reported. On the evening of the busiest day, Friday the 13th, at least 200 fires were in progress at the same time. As operations officer, co-author Charles Bahme had conferred with the deputy chief who had assumed the duties of department commander in the absence of the interim chief, and it was agreed that when 40 of the 112 fire stations had been depleted a recall of one of the off-duty platoons would be initiated. Within one and a half hours sufficient personnel had returned to duty to activate 20 reserve companies and 10 others made up of two-piece engine companies. Before the emergency was over an extra 75 companies were in service.

Resource Management

Many surrounding communities including a 15-square-mile area in the County Fire Department's jurisdiction were faced with similar problems, so calling for mutual aid would have netted little more in response than already volunteered by the County Fire Department and three small border cities.

To assist in dispatching apparatus, and to provide security in numbers, fire fighting forces were organized into 30 task forces (TFs) each having a chief officer with two engine companies and a truck company. Command posts were established in fire stations at each end of the riot area and at stations in other areas of the city. About 100 companies in TF formation were assigned to these CPs.

Fire fighting consisted mostly of hit-and-run tactics, using large streams for a quick knockdown, and with no attempt at overhaul or salvage work. If looters happened to get trapped in a fire, wagon batteries and ladder pipes would be shut down long enough to get them out, and then opened again.

On midnight of the third day of unrest, National Guard forces arrived to supplement the 900 to 1,000 police officers who had been assigned to riot control duties. Until that time the police department had not been able to provide sufficient officers to protect the fire fighters, and it became necessary to announce on public television that a decision had been made to withdraw all fire forces from a two-square-mile area; this area was later increased to four square miles. Sniper attack on fire fighters brought a police response, but after their departure, the sniping would resume. When the National Guard assigned troops to convoy the task forces and ride on the apparatus, the harassment was greatly diminished.

The beginning of control was noticeable following the imposition of a curfew on the fourth day; it covered a 46.5 square mile area. The lieutenant governor appeared on TV and told the public it would be a crime for anyone to be on the streets after 8:00 p.m. This curfew was not lifted by the governor until the sixth day when the situation had calmed.

On viewing the disaster area from a helicopter, it could be seen that numerous "charcoal alleys" had been created in various business districts. Many of these burned-out businesses would never be rebuilt. However, when reconstruction got underway, half the unemployed in Watts found jobs, bus service was expanded, and a hospital was built there.(21)

The National Advisory Commission issued a report on civil disorders in March 1968, giving results of its study of twenty-three cities in which rioting occurred. It indicated that most disorders were accompanied with incendiary fires, which ranged from burning of police barricades in Jackson, Mississippi, to 2,000 fires set in the Watts riot.(22)

PREPARATION FOR RIOTS

Precautions a fire department may take for riot-caused fires include the following: evaluating existing resources; developing an organization for effective action; equipping the department to do the job; agreeing with police officials on the limitations on your law enforcement activities and the amount of protection you may expect for your forces when attacked; reviewing mutual aid agreements with neighboring fire departments; reviewing the civil defense plan with the coordinator and director to be certain that provisions relating to the declaration of local peril, closing of gasoline service stations, liquor stores and bars, summoning of the National Guard, imposing a curfew, and the authority of those in the chain of command are thoroughly understood.

The following guidelines provide an outline for preparation.

1. Community Survey. Identify areas where riots are apt to occur. Estimate the number of units or task forces you may have to deploy in such a contingency and select command post sites. Some fire departments have designated certain fire stations as command posts and have assigned task forces for any type of response, including riot-caused fires. A battalion headquarters may meet the requirements of a command post, especially if it is large enough to house two engines and a ladder company, along with a battalion chief. Otherwise, the facility should provide for a command room, easy access, egress and parking space for apparatus, communication facilities, and space to take care of personnel for extended time.

2. Logistical Study. Review the measures that can be taken to offset lack of water supply, damage to fire hydrants and sprinkler control valves from sabotage, damage to tires and windshields on fire trucks, injuries to fire fighters from missiles and bullets. Tank trucks can be mobilized to supplement weak water systems; special adapters can be used to connect hoses to damaged hydrant threads; hose lines can bypass closed and damaged sprinkler valves; a tire from one of the rear duals can replace a flat front tire; tape can be placed on windows and windshields of fire trucks before they are broken to mislead rock-tossers; plastic faceshields can be put on helmets, and protective cages on apparatus, to protect fire fighters from missiles. Bullet-resistant vests could be provided for fire fighters operating streams from aerial ladders or elevating platforms.(23)

Estimate in advance the amount of spare hose, batteries, fittings, food, blankets, cots, extra radios, telephones, and other logistical support you may need for your command posts.

3. Annex to Plan. Prepare a "Riot Bill Annex" for your major disaster plan, outlining the arrangements and understandings for this contingency. Identify conditions under which you will activate your emergency operations center and put your plan into effect.(24)

4. Safety Study. In addition to determining the legal and operational courses of action available to the fire manager, pre-incident planning for civil disturbances must develop operational procedures that combine effectiveness with personnel safety. Review the general disaster planning concepts of previous chapters. Then tailor your operations plan to your fire department's needs during the type of civil disturbance that you are most likely to face, incorporating safety factors outlined in NFPA Standard # 1500.(25)

5. Authority Determination. If the laws of your state, city, county, or fire district already impose on your fire department the duty to prevent and extinguish dangerous fires, and authorize you to protect life and property endangered as a result of fire, explosion, man-made or natural disaster, then you have the legal authority to formulate a basic plan for fire department operations in any major emergency, including a civil disturbance. This authority may even extend to enforcing all laws relating to fire safety, but not to suppressing insurrection and civil disturbances.

6. Objective Focus. During periods of pronounced incendiary activity, the fire department's objective remains the protection of life and property from destruction by fire. However, operational realities as well as the number of incidents that may occur make the attainment of this objective exceedingly difficult, especially in the absence of increased fire control and law enforcement resources.

In such emergencies, fire planners must realize that they may not be able to suppress or even operate at all fires that occur or are reported to emergency operations centers. We have seen that during the 1965 riots within the Watts section of the city of Los Angeles, over 200 fires were reported to be underway at the same time. Your plan should have a mechanism that governs the assignment of suppression resources to enable operation at the more serious incidents or those where the operating forces have the highest probability of achieving fire control. Further, some fires may indeed be too large for the assignment that is dispatched, and the plan may then be to protect surrounding exposures. Nevertheless, since the fire department will not be able achieve control of all reported fire incidents, the pre-incident plan must have a mechanism through which suppression resources are allocated to fire incidents on a priority basis.

7. Changes in the Basic Disaster Plan. Basic disaster plans that recognize that the fire department must operate at multiple incidents may be adapted for civil disturbances. Conflagrations and wildland/urban interface fires are, with regard to the simultaneous occurrence of multiple fires, similar to civil disturbances, except that in civil disturbances, the populace may openly interfere and prevent fire suppression activities.

If your community survey discloses likely areas where incendiary fires might develop during a civil disturbance, your plan could designate locations for setting up command posts at the perimeter of the area. These should also include sufficient room to allow for a police command post as well as available rest rooms, apparatus, augmented telephone service, and similar considerations.

8. Study of Other Civil Disturbance Plans. Before you draft your own plan, be sure to consult and carefully study the plans of other local, county, and state emergency service organizations. Your review of their plans may produce a number of good ideas. Second, such a consultation will afford the leaders of several emergency organizations the opportunity to sort out such important matters as command, liaison, and communications. Now is the time to determine which agency will be in overall command of the operation, the extent of the fire department's responsibility, and the provisions for the protection of fire department personnel.

9. Conferring with Other Agencies. When composing your plan, you should confer with an appropriate law enforcement official within your community (or whomever would be in overall command of an incident of civil disturbance or terrorism) to determine the actual amount of resources that will be dedicated to the support of fire department operations. Since your operational plans and the tactics employed during suppression operations are both contingent on the degree of protection your forces receive, this level of protection should be defined before an event. In the 1965 Watts riots, the police officers did their best to provide protection when called, but since the law enforcement resources were stretched to the limit, they were only capable of temporarily suppressing sniper fire.

This is also a good time to establish the law enforcement/fire department command liaison procedures, including how fire calls will be forwarded from the emergency operations center to the fire command post and how fire department calls for police protection will be relayed to the law enforcement command posts.

In planning, be careful not to overemphasize the role of the National Guard. Its primary responsibility is law enforcement, not fire suppression, and the Guard is not always readily available. Forty-eight to seventy-two hours is a realistic estimate of the time required for the Guard to arrive on the scene. So when the law enforcement commander tells you that resources will be pretty slim until the National Guard arrives, expect to be on your own for a considerable period of time.

10. Conferring with Neighboring Fire Chiefs. The fire chief and ranking fire officials should confer with the chief officers of neighboring fire departments. Now is the time to decide on such issues as

- what resources will be needed?
- where should departments report?
- what will be the activation procedure?
- what are the command and communications procedures?

If you do not intend to commit outside resources to fire suppression within the troubled areas, they may be employed to handle the normal work load of your department in quiet areas, thus freeing other companies.

11. Preparing the Civil Disturbance Plan. Your plan should provide guidelines for department action during the civil disturbance and for the preceding period of alert. The alert stage of your plan will seek to prepare all resources so that when the incident occurs the fire department is able to function immediately. During the alert stage, the chief fire officer should review plan provisions with the various unit commanders, check on command and communication liaison with other agencies, review personnel protection, review tactical operation procedures, and, above all, avoid any confrontational incidents that might precipitate the occurrence of the civil disturbance.

To minimize the risk of injury to personnel and damage to apparatus during the response to incidents, all units should respond as a group, avoiding the separation of one piece of apparatus from the others. To keep response routes relatively traffic free, responding units should refrain from the use of sirens, headlights, or other warning devices that would draw undue attention to the responding apparatus and should observe normal traffic laws. During the response, since "one Miami fire fighter was nearly pulled from the tailboard as the apparatus slowed down,"(26) all crew members should be in the cab, in the jump seat area, or within other enclosed vehicles such as police or chief cars. This is suggested by NFPA Standard 1500 safety guidelines for even routine daily responses. No one should ever ride in a standing position anywhere on an apparatus and certainly not on the tailboard. Additionally, response routes should be altered to avoid establishing a response pattern in which apparatus could be lured to underpasses or other sites that maximize ambush and sniper opportunities.

POTENTIAL PROBLEMS FOR FIRE DEPARTMENT OPERATIONS

A review of past civil disturbances shows a number of potential problems in civil disturbances that should be addressed in the pre-incident planning activities.

Incendiarism

Even under normal conditions fire departments respond to numerous arson fires, but it is not often that the arsonist comes back to the scene of a fire that you have just extinguished and touches it off again while you are up the street working on another fire that he set earlier. This happened often in the major urban riots experienced in the last several decades.

In future riots there is apt to be less gasoline purchased at the going high prices for use in jug bombs, but a practice of breaking into gas stations and stealing the incendiary fuel will be difficult to stop unless the installation of secret electric switches for the gas pumps turns out to be a deterrent. A simultaneous order to lock up all service stations at the time the mayor declares a state of extreme emergency may discourage forcible attempts to obtain gasoline for use in incendiary bombs.

The most familiar incendiary device used by rioters is the traditional

Automobiles are a common target for incendiarism during riot conditions. The problem is compounded when fires spread to adjacent structures. (Photo courtesy of the Federal Emergency Management Agency.)

"Molotov cocktail" utilizing any glass container filled with gasoline and a homemade improvised wick. In the early 1990s, however, subversive groups within the population have resorted to "booby trap Molotov cocktails" designed to look like the traditional gasoline-filled bottles but engineered to explode when lifted by an unsuspecting person. Filled with a mixture of gasoline and sulfuric acid, and containing a tea bag with potassium, chlorate, and sugar, this device is activated when picked up or tipped over.(27) See Figure 8-1 for a diagram and description of this newer, more legal incendiary device.

During civil disturbances and certainly riots, many fires will be of incendiary origin. Given the prospect of harassment and the large number of fires, it will be virtually impossible to initiate any investigative efforts. Also, incendiary fires during riots are subject to reignition, oftentimes while the fire department is operating at another fire. Preventative actions to curb incendiary activity must focus on the securing of gasoline stations so as to prevent the large-scale production of fire bombs.

Obstructing Operations

During civil disturbances, fire department operations may be impeded by the local populace. Such activities may include obstructing fire hydrants with parked cars, damaging fire hydrants, sniper fire at fire fighters and apparatus, and throwing fire bombs and other missiles at the apparatus. Fire department planners should equip their apparatus with noncombustible hose bed covers and enclosed crew cabs.

Personnel Protection

Although law enforcement agencies have some responsibility for providing operating forces with adequate protection from harassment, obstruction, and injury from bystanders, they are often constrained by inadequate personnel and competing demands for their services. Therefore, your planning effort should assume that law enforcement resources will be insufficient for your protection and should address the issue of personnel protection through the policies of appropriate response and operational procedures, adequate personal safety equipment, and the discretion of on-scene commanders.

Second, departmental personal safety policies and equipment can also play a large role in personnel protection during civil disturbances. Apparatus innovations such as the sedan or crew cab will reduce responding and returning injuries during both normal operations and civil disturbances.

Third, discretionary withdrawal should be part of an effective personnel protection policy. Task force commanders must have the authority to evaluate the degree of fire severity, the amount of available protection, and the level of risk to which operating personnel will be exposed.

Last, because the law enforcement authorities are responsible for the restoration of law and order, your plan should establish a liaison with the

---

Figure 8-1. Booby-Trap Molotov Cocktail.

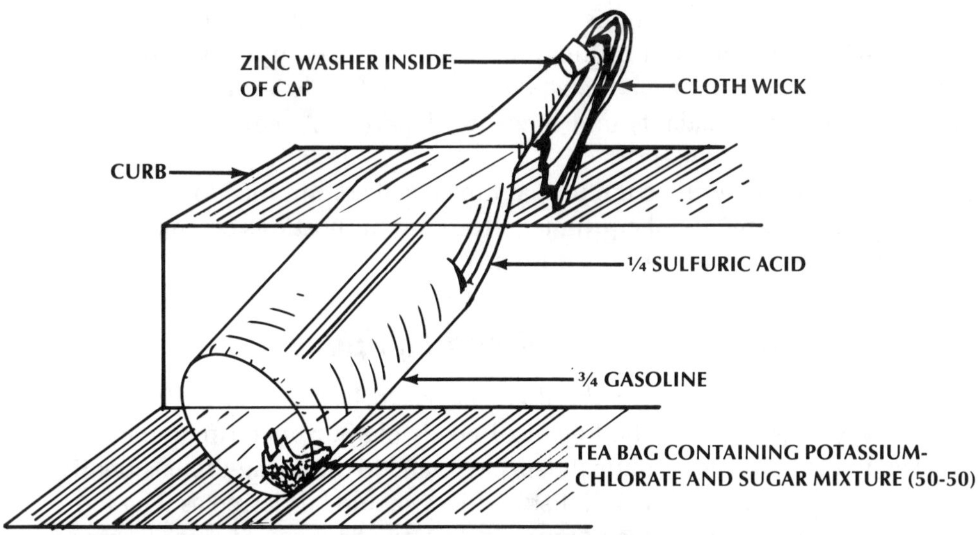

This incendiary device is activated upon being picked up or tipped over. The gasoline-sulfuric acid mixture fills the bottle within an inch of so from the bottle cap when lying against a street curb, or another such angle. When the mixture makes contact with the zinc washer, hydrogen is produced which casuses the bottle to burst. Upon bursting, the gas comes into contact with the potassium-chlorate/sugar mixture which in turn creates a fiery explosion.

appropriate law enforcement authorities. Send a liaison officer to the law enforcement command post; this officer can represent the interests of the fire department in command decisions, and communicate your needs to the law enforcement commander. Further, request that the law enforcement authorities also assign a liaison officer to your command post. This will improve interagency communication and allow you to indicate your concerns directly to persons from the appropriate law enforcement agencies.

Accurate Records

Civil disturbances are demanding and expensive operations. Consequently, accurate records of equipment lost or damaged, as well as expenditures for overtime, fuel, consumables, and food should be maintained. Additionally, any injuries that occur to department members should be documented because the information may be a factor in awarding compensation, hospitalization, and retirement plan payments.

Accurate records of the incidents can also be useful for future operations. Lists of the incidents responded to, the tactics employed, and the equipment used can help future planners refine their planning efforts, which may result in safer and more effective fire suppression operations.

Communications

Interagency, intraagency, and interunit communications, as described in detail in Chapter 6, are of paramount importance during a civil disturbance. The planning effort should attempt to equip all operating forces with two-way radios and, if possible, all operating personnel should have portable radios. In a civil disturbance, a portable radio will be the means through which help can be summoned.

Interunit communication is primarily an issue of equipment, but inter- and intra-agency communications are both issues of equipment and procedures. Realistically, your plan must describe how fire department units will be dispatched to incidents and how fire department requests for police assistance will be processed.

Radio probably will be the primary means of communication. As such, the appropriate amount of equipment, the communications center, and the correct frequency numbers must be provided.

Replenishment of Personnel and Equipment

Your plan should expect the disturbance to last several days, during which you must address the logistical needs of your operating personnel and equipment. Incident commanders often tend to think one-dimensionally about finishing the job, but civil disturbances should be viewed as long-term campaigns. Accordingly, the planning effort must recognize a fire fighter's needs for rest, drink, and food as well as the need for motor fuel, apparatus maintenance, replenishment of consumable supplies, and replacement of damaged and lost equipment.

Hours of nonstop fire fighting will take their toll on operating personnel.

As fatigue sets in, the chances of personal injury increase. Also, personnel may express fear of sustaining personal injuries resulting from potential violence directed toward them, and this will further aggravate the fatigue problem. Operational and incident commanders must plan rest periods for operating units. The incident commander should resist the temptation to recall all off-duty personnel for the emergency. A far better course of action is to stagger the solicitation of help so that operating units can be replaced by fresh units as the incident progresses.

Specialized Training

Training in multiple company, and even in multiple department, operations is not uncommon, but the special problem of task force training for civil disturbances is mainly one of simulating a fire attack on a multiplicity of large and small fires over a large area, while arson, bombing, looting, and criminal destruction is taking place, and fire fighters are being fired upon.

At least the mechanics of deployment to command posts on the perimeter of an assumed trouble area, with aerial reconnaissance and dispatching of fire apparatus from the resource pool to simulated fire locations can be practiced. The use of a taped scenario to simulate the frantic calls from frightened store owners who describe their blazing businesses, broken windows, flaming jug-bombs, marauding "shoppers," with the sound of gunfire in the background, might add a touch of realism to the drill, but nothing can create in fire fighters the feeling of being shot at until they actually become targets.

The training effort must be tailored to the demands of your operational plan. There will be two training tasks: first, units and individuals must be trained in their roles within the plan; second, department members must be trained in the skills and procedures they need to carry out their responsibilities within the plan. Accordingly, the training effort should focus on quick attack procedures, hit and run tactics, master stream operations, and multicompany operations. For some fire departments that are organized on an independent company basis, the training problem in multicompany operations may be significant. However, the advantages of the task force concept in these situations far outweigh any training problems.

If your fire department can realistically expect to summon outside aid during a civil disturbance, then you must incorporate the mutual aid departments into your training effort.

An additional training problem will be preparing the command staff. Although these officers are not operating in line units, they are nevertheless discharging significant responsibilities within the effort to control the situation and must be prepared for their task.

Sabotage

Your planning efforts must include an assumption that equipment is likely to be damaged or rendered inoperable. Not only are fire trucks damaged by rioters with gunfire, sling shots, clubs, bricks, and paving stones, but in one instance a Molotov cocktail was hurled into a hose bed during the Watts riot resulting is a mass of flames that required prompt extinguishment to save the hose and the fire truck.(28)

During the student riot in Isla Vista, California, hundreds of prefabricated tire wreckers were thrown in the path of fire apparatus; these resembled twisted tenpenny nails that were fashioned so as to always land with a sharp point upward.

During the Watts riots, in addition to damaging threads and valve stems of fire hydrants, sprinkler control valves in some of the large occupancies were closed by the arsonist before he went to work on kindling the fires. Hydrant adapters may be of some value where threads have been damaged, and a pre-riot educational program in security measures for building occupants may result in the institution of safeguards that will keep sprinklers in service and the building more resistant to sabotage and arson attempts.

Mutual Aid

Where riots are breaking out in one town they are apt to spread to nearby communities, so do not place too much emphasis in your plan on mutual aid. When the chips are down it may not be forthcoming.

In one instance during the Watts riots, the city attorney of a municipality bordering on the City of Los Angeles told the fire chief to keep all of his companies in his own back yard, because of his fear that a segment of his inhabitants might grow more restless and start a riot there. Three other neighboring fire chiefs were not prevented from sending companies to cover for Los Angeles in the harbor area where L.A. had pulled all of its own companies out to operate out of the command post at the south end of the Watts riot area.(29)

Personnel Protection

The problem of obtaining adequate police protection for fire fighters during riot response and fire defense is touched upon in several sections of this chapter, but no solution for the problem is easy. With the trend toward tighter municipal budgets and reduction of personnel in public services, there is little prospect that protection forces will be augmented in the foreseeable future. It has been suggested by some persons that fire fighters be deputized as sheriffs, be given police training, weapons, and provide their own protection. Law enforcement and fire fighting are separate and distinct professions that are not easily combined even in normal periods of routine calm.

Logistics

A problem that is not peculiar to riot fire defense operations is one of feeding large numbers of personnel at command post locations, often including police and military personnel, until such time as they are able to get their own field kitchens in operation. A plan should address the inability or unwillingness of some food caterers to enter an area where they may be danger of being attacked or hijacked enroute to the command post. Escorts may be necessary.

Even if you have your own mobile field kitchen, or your command post is located in a facility having a kitchen, you may still have a problem of obtaining sufficient food from nearby stores, for most store owners lock them up when a riot threatens their security. Some fire departments stockpile large containers of beef stew and similar easily prepared food for use in major emergencies in their mobile kitchens.

In any case, you may need a police escort for service personnel, such as telephone, gas, and electrical utility repairmen, as well as for drivers of gasoline tank trucks enroute to refuel the apparatus in your resource pool at the command post.

Public Information

The need for giving periodic assessments of the situation to the director of civil defense, mayor, or to whomever is the head of your political entity is common to all major emergencies. Some departments have a public information officer to keep the news media informed.

Include one in your plan; it will present you with the opportunity to demonstrate the preparedness and effectiveness of your organization.

You may have to go on television to advise the public when you are about to pull all of your forces out of a given area and also advise them that they

Public service announcements can be helpful if issued by a city or community official who has the support of the rioting elements within the community. (Photo courtesy of American Heat *Video Productions.)*

may stay and protect their own property if they wish; but you may also advise the public that your response to calls for help will be limited to rescue-ambulance emergencies if you are able to obtain a police escort to accompany the responding ambulance personnel.

Estimating Losses

Another problem that is nearly always present in a major disaster is the one of estimating losses. For example it was not too difficult to multiply 600 homes by $250,000 in the Santa Barbara, California, fire in July of 1990, to come up with a "ball park" figure that was sufficient until a more thorough study of the losses could be made.(30) But it was virtually impossible to estimate the loss to the hundreds of buildings and vehicles in the Detroit riot until a detailed survey of all the damaged and destroyed structures was completed sometime later. It was difficult enough to compile fire department losses due to lost, stolen, and damaged fire equipment. Losses are not limited to material items. In the Watts riot losses included 183 fire fighters shot or injured, of which 38 had to be hospitalized. Fortunately only one fire fighter was killed and that was caused by a collapsed structure.

In those fire departments where overtime is awarded for extended service in a major emergency the cost to the city could be astronomical, especially since the Fair Labor Standards Act, in the wake of the Garcia case, now mandates time and one-half. In addition to this, if the city is billed for the costs incurred by assisting fire departments in accordance with their mutual aid agreement, the amount of money needed for reimbursement purposes could be tremendous.

FIRE DEPARTMENT OPERATIONS AT CIVIL DISTURBANCES

Fire department operations during civil disturbances must be predicated on two factors: (1) the fire fighter is no longer viewed as a friend of the community, but rather as a manifestation of a hostile authority; (2) the incidence of fire alarms and fires may exceed the suppression capabilities of the fire department. Consequently, the planning document must recognize that units will be operating at a number of serious fires, under conditions of harassment that may range from rock throwing to sniper fire. As the events unfold, the fire manager will be responsible for the suppression of fires and the safety of operating personnel.

Captain Robert Obermayer of the Albany, New York, Fire Department has prepared an excellent set of guidelines that a fire department can use to prepare for operations at civil disturbances.(31) These have been combined with a previous listing prepared by author Charles Bahme, and the merger of the two provides a comprehensive set of 13 guidelines which provide the modern-day fire chief with a workable set of principles to ensure his readiness for handling civil disturbances and riots. These are listed on the pages that follow.

1. Evaluate the Situation. Because the two factors that will govern the commitment of fire suppressive resources are the severity of the fires and the degree of risk to operating forces, the fire commander must evaluate the fire situation as well as the level of protection that can be given to operating forces. This is in stark contrast to the normal fireground command process where the incident commander is concerned solely with the specifics of the fire at hand.

An aerial reconnaissance is an excellent way to determine the number, location, and severity of existing fires, as well as the location and direction of travel of groups who may be participating in the disturbance. Helicopters are standard equipment in large fire departments; for example, Los Angeles City Fire Department operates seven Bell helicopters of various models. Helicopters may also be available from police agencies or the National Guard. Though subject to sniper fire, the helicopter's ability to survey large areas makes it an excellent reconnaissance tool.

When vandalism, breaking of windows, overturning of autos, setting fires, and harassment of fire fighters commence, the number of false alarms (both automatic alarms and telephone) also increases. Aerial surveillance can also help sort genuine from false alarms.

If aerial reconnaissance confirms reports of police, and the situation appears to be more serious than a localized outburst, notify your staff of the information you have obtained and your intention to put the Emergency Operations Procedure into effect. Send a message to all stations, or otherwise inform all members on duty. Volunteer fire fighters should be notified to report for duty immediately.

Without a helicopter, the fire commander must assess the fire situation by monitoring requests made to the incident command post for fire assistance and must use information provided by law enforcement authorities to assess crowd conditions.

2. Confer with Law Enforcement Authorities. With an understanding that the primary objective of the law enforcement authorities is the restoration of order—not necessarily the protection of fire fighters—you may want to confer with the law enforcement commander. Obtain an estimate of how many officers can be assigned to the protection of fire forces. This estimate will be the cornerstone of your operating strategy. If the law enforcement commander decides that your force cannot be protected and there are reports of violence, then you should not allow your forces to operate in the area of the disturbance except for the rescue of fire-trapped persons; after that they should withdraw to a safer location.

3. Activate Your Plan. Since most civil disturbances, especially riots, do not start all at once, but escalate gradually, the decision to activate your civil disturbance plan may be a difficult one. Many fire commanders are understandably hesitant to activate their plan at the first sign of trouble. Perhaps the most appropriate solution to this dilemma is to escalate your level of readiness as the civil disturbance escalates. At the initial indications of a disturbance, activate your emergency operations center so you can monitor the situation while remaining in contact with your subordinate commanders. As the disturbance progresses and on-duty resources are deployed, you may go to a higher level of readiness. Lastly, when major escalation of the disturbance is imminent, you can activate your plan.

4. Activate Your Emergency Operations Center. In large fire departments this center will be separate, but a part of the main communications center for your dispatching operations. In smaller communities, especially if the fire chief is also the civil defense coordinator, the disaster operations center will be activated in accordance with your defense civil preparedness plan, to which representatives of fire, police, civil defense, and military departments report.

5. Deploy Your Forces. The task force, since it is a self-contained fire fighting force, is the preferred means of deploying forces during civil disturbances. A task force consisting of two engines, a truck, and a task force commander will provide considerable fire suppressive power while maximizing personnel safety by consolidating units into one fire suppression team and avoiding fire fighter isolation. The commander would usually be a battalion or district chief in a metropolitan area.

In many suburban and rural areas, the fire department may not have enough companies to form a task force, so companies from several departments may have to be combined. The pre-incident planning process should have addressed such command issues as the assignment of units to each task force, the individual in charge, and the location of the task force.

Where resources are in short supply, a task force should consist of at least two engine companies and a commander, with the optional addition of a truck or a squad company. With one engine located at the source of the water supply and the second at the scene of the fire, a task force can rapidly apply the large volumes of water that will be needed to quickly control advanced incendiary fires. If available, a truck company can provide unmanned aerial stream application.

If the stations you have designated as command posts (CPs) do not already have permanently assigned companies that comprise a task force under your plan, (e.g., two engines, truck, and a chief), or if your plan calls for using a facility such as a playground recreational building for a CP, then required companies should be dispatched to these locations to form task forces. Your law enforcement agency should be requested to dispatch equipment and personnel (at least two squad cars) to each CP, in accordance with your plan.

6. Relocate Fire Companies. Evacuate fire stations in the threatened areas of violence and relocate companies to preplanned locations.

7. Activate Command Posts. CPs designated on the periphery of the riot area should be activated. Preferably, these should be near a school, which can be used in feeding personnel and has a gymnasium for use as a barracks. These CPs will be used for all four elements of incident command.

A. Operations: Dispatching companies to emergencies.
B. Logistics: Maintaining an apparatus pool.
Feeding and resting members.
Meeting operational needs of companies.
C. Finance: Recordkeeping expenditures.

Recording buildings and property damaged or lost to fire.

D. Planning: Carrying out the operational phases of the riot annex to the disaster plan.

8. Recall Off-Duty Personnel. To augment on-duty forces, your dispatchers should follow the plan for recalling personnel and activating reserve apparatus. Recalled members can report to the manpower pool nearest to their homes for transportation by department vehicles (with police escort) to CPs in the riot zone. Where the hazard of being attacked enroute is minimal, members can be instructed to report directly to the station or facility serving as a CP.

9. Prepare Personnel, Apparatus, and Equipment. Fire chiefs may wish to swap white helmets and coats for regular ones. Plastic faceshields may be attached to helmets.

Remind fire officers and fire fighters that guns are not to be carried except by those who qualify under the laws of the state as peace officer, e.g., arson investigators.

Have axes, crowbars, and other tools removed from the side of apparatus and placed in locked compartments.

Have hose laid out in fire stations from standpipe outlets or hose bibs to use in case of Molotov cocktail attack; also keep a large dry chemical extinguisher ready for use.

If a station is near a riot area or is to be used as a CP, have underground fuel tanks topped off now, in case local suppliers are reluctant to enter the area later; also obtain sufficient food to last until other food logistics can be established.

Lock stations vacated by move-up companies; assign guard personnel to stations when security forces, such as the military, become available.

10. Use Appropriate Fire Fighting Strategy. The overall strategy for coping with riot-caused fires is simple but its implementation may be complicated since the strategy that you employ will be commensurate with the available level of police protection. When police protection is furnished, operating forces should confine and extinguish accessible fires and protect exposures; without it, the only strategic alternative is withdrawal. During riots that include numerous incendiary fires over a wide area, fire attack and exposure protection can best be accomplished through the application of large volumes of water via apparatus-mounted master streams and preconnected handlines.

As the civil disturbance escalates, the number and severity of fire incidents will grow, while the level of protection available to the operating forces may decrease. In this instance, the fire plan must delineate the criteria through which the decision to attack and operate at a given fire will be reached. Here are suggested considerations in the decision to operate at a given fire, in order of decreasing priority.

1. The degree of protection available to operating forces.
2. The degree of danger to occupants.
3. Is the structure occupied?
4. Does the fire building have occupied exposures?

5. Is the fire building vacant?
6. Is this a nonstructure fire?

11. Employing Hit-and-Run Tactics. Under the conditions just described, fireground tactical decisions will depend on hostility encountered, level of protection available to operating forces, availability of suppression resources, and water supply. In most instances, task forces will employ hit-and-run tactics, using preconnected short lines and master streams. The tactics identified within the body of your plan represent the tactics that your forces will employ throughout the duration of the incident. Just because your forces are now operating without harassment is no reason to assume that the situation will not change dramatically.

Your plan should have communications provisions that keep the task force commanders updated with information about both the fire and law enforcement situations. Each task force commander should have the authority to terminate fire suppression operations and withdraw. If the level of personnel protection deteriorates, the task force commander should sound a predetermined signal, and all units should withdraw together.

12. Follow Your EOP (Emergency Operating Plan) in Dispatcher's Office and Emergency Operations Center (EOC). Dispatchers will notify personnel designated in the EOP to report to the Emergency Operations Center (EOC); these staff members should be sufficient in number to provide reliefs at twelve-hour intervals (at least); unless your department has limited personnel, do not call back at once *all* the staff who are trained in this work; save some for reliefs, if possible.

Try to arrange your EOC layout so that your operations officer and his staff are in a room apart from the dispatchers, with a third room provided for reporters, visiting fire chiefs, and public relations people.

If you have sufficient radio frequencies to do so, assign one or two frequencies solely for riot area communications.

Try to have mobile communications units of the fire and police departments set up near each other at their respective CPs, to facilitate exchange of information in the field.

In the EOC, assign members to monitor the police and CD (civil defense) receivers; if additional receivers are available, put them at each battalion headquarters office so that these chiefs can be informed of the spread of the riot and may anticipate being called.

Because of attacks on fire department personnel and their vehicles, it may be necessary to discontinue dispatching of squad and ambulance companies (lone vehicles) into the riot zone except with a police escort.

When designated engine companies have been dispatched to CPs to join task forces, initiate your plan for recall of off-duty members for reserve apparatus; truck company personnel can activate reserve pumpers pending arrival of off-duty enginemen.

Notify caterers or other suppliers of emergency food of the number of meals needed at what time and places.

Dispatch gasoline and diesel refueler trucks to CPs.

Keep information updated on the situation map and status board; when officers are relieved who have been in command of CPs, task forces, and pools, post names of successors.

Use telephones or separate radio frequency to confer with CP commanders and to relay fire reports.

Relay requests from CPs to police for protection of task forces under attack.

Advise the public by radio and television if it becomes necessary to withdraw forces from the riot area.

If the National Guard is called out to quell the riot, request its commander to assign fire department personnel who have been recalled to Guard duty to specified CPs to direct deployment of task force (TF) guard convoys.

Request the mayor or other appropriate authority to order all bars, liquor stores, and gasoline stations to be closed and locked for the duration of the emergency; request the appropriate official to order a curfew until order has been restored.

If the situation demands, request mutual aid from jurisdictions not already (or about to be) engaged in similar problems.

Demobilize forces as conditions permit.

13. Follow Riot Plan Annex. In responding to alarms, obey traffic laws and do not use red light and siren during the last few blocks of response (to avoid attracting attention).

Stay out of alleys as much as possible to reduce your vulnerability to sneak attack.

Be prepared at all times to extinguish fire bombs that may land on apparatus.

In attacking fires, use large streams to get deep penetration and a quick knockdown.

Do no salvage or overhaul work; when the fire is out, have TF commander report his availability to take another call.

Keep TF units together; TF commanders should not return part of the force, as would be customary under normal circumstances.

If TF members are severely harassed or fired upon, notify CP, shut down handlines, keep monitors in operation, and take cover until police or other help arrives. If no help is available immediately, shut down completely, disconnect hose and withdraw to a safer location with apparatus. Later, when police escort becomes available, return to the fire scene and resume operations or pick up hose (if fire is already burned out).

Except when a separate frequency has been designated or when replying to messages directed to them, or in case of extreme emergency, TF companies should maintain radio silence and let the TF commander make all radio communications to the CP. (CP will talk to the EOC by land line or separate radio frequency.)

Do not use hose streams and monitor nozzles to subdue rioters; but if under attack and in fear of serious bodily harm, use whatever force is necessary (including heavy streams) to effect a safe retreat.

TERRORISM: A NEW CHALLENGE

Increasingly during the 1980s, and continuing into the 1990s the news has featured reports of terrorist events that caused immense human suffering and substantial monetary loss; indeed, causing enough loss to make the headlines seems to be the goal of the terrorist. Winning a war, such as the one in the Persian Gulf in 1991, will not give the allied victors any assurance that worldwide terrorist activities will cease; on the contrary, they are almost certain to increase in number and violence. As long as terrorist events occur, and the fire department is entrusted with the task of the preservation of life and property, the fire department can expect to operate at the scene of such incidents. The fire manager must recognize the need for effective pre-incident planning.

Definition

Terrorism is not a new phenomenon although the worldwide incidence of terrorism has increased markedly in recent years. Many societies have, over the ages, practiced or suffered from some form of terrorism. For our purposes, **terrorism may be defined as the use of nonmilitary violence by individuals and organizations for the attainment of political ends**. In reality, "terrorism is a form of warfare in which unpredictability and surprise are major weapons."(32)

There are four reasons for the contemporary increase in the incidence of terrorism.

- The ease of international travel enables individuals from unfamiliar cultures to journey, without attracting undue notice, halfway around the world to commit an act of terrorism.
- Despite the high tech security measures now in effect, current travel sophistication allows terrorists to reach their target, perform the act of terrorism, and escape relatively rapidly.
- Improved worldwide communications present terrorists with the opportunity to accomplish their primary objective, namely the widespread communication of a threat to someone other than the victim.
- Technological advances have reduced the size and weight of weaponry available to the terrorist, while increasing its effectiveness.

Distinction from Civil Disturbances

What then are the differences between acts of terrorism and civil disturbances? Briefly, the terrorist possesses a definite political motivation; there is a political goal toward which the act of terrorism is directed. The civil disturbance, although it may be a reaction to a specific political event, lacks a definitive political focus. Second, terrorism is well organized and has a specific target; this is in sharp contrast to civil disturbance, which is usually a spontaneous expression of social discontent. Finally, whereas the civil disturbance is characterized by the disorganized participation of large numbers of individuals exhibiting varying degrees of motivation, the act of terrorism is carried out by a dedicated few who are highly motivated toward the attainment of their political purposes.

Degree of Danger in Terrorist Incidents

The dedication of terrorists to the accomplishment of their objectives represents the primary danger for fire department forces. Since the terrorist combines ideological fanaticism with cold rational professionalism in the planning and execution of the act, no one, not even a child, and certainly not the fire service, is immune to violence or regarded as innocent.

The fire department may be seen as a mitigating agency that seeks to reduce or eliminate the destructive effects of the terrorist's act. Accordingly, operations may be disrupted as in the civil disturbance, but given the dedication of the terrorist, such disruptions will take the form of well planned acts of violence. Rocks and homemade missiles are not as likely to be thrown at fire fighters and fire apparatus; instead, the fire department is more likely to be involved with explosive devices.

Interagency Cooperation

At incidents of terrorism, law enforcement agencies will be the primary mitigating forces, with the fire department serving in a ancillary role. Although its role may be secondary, it is of paramount importance because the fire department may represent the only agency that is entirely concerned with the effects and mitigation of the present event.

Realistically, the fire department can expect to be called upon to provide the following:

- extensive fire suppression services
- emergency medical services for a single victim or mass casualties
- assistance in evacuating area residents
- support functions such as lighting and electrical power generation
- assistance in the investigation of incendiary devices.

Additionally, fire department units may be requested to stand by while law enforcement units search for and dispose of explosive devices. Most certainly, fire department units will participate in the extrication of victims from any rubble or wreckage.

The Planning Process

The disaster planning process as it relates to acts of terrorism requires a dual focus. First, emergency planners should perform a threat analysis in which they identify the most likely sites within their jurisdictions at which terrorist incidents might occur. If you protect a large city, certainly the consul offices or embassies of foreign governments are potential targets. Additional potential targets include all government offices, headquarters of activist political groups, and certain private sector installations such as nuclear power plants or laboratories.

Once you have identified potential sites of terrorist acts, obtain an assessment from appropriate law enforcement agencies of the potential for danger and what expectations they may have for your organization. Most law enforcement officers are uninformed about fire department capabilities and might have unrealistic expectations of your service and equipment inventory. Ask the law enforcement people what they might want you to do; then tell them just what your capabilities are.

The emergency planner should also consult with the owners and occupants of the selected target sites to identify potential risks to personnel, assess the hazard that the facility poses, and learn about the in-house system that may be used to suppress and retard fire growth.

Following the threat analysis, the planning effort should focus on establishing the foundation for effective interagency cooperation. Remembering your subordinate role in overall incident mitigation and your obligation not to expose fire department personnel to undue risk, the fire department must help establish the system through which all agencies effectively cooperate to mitigate the event at hand. Furthermore, the fire manager must identify the potential level of risk and the means that will nullify or reduce the identified risk factors.

The Associated Press reported that 13 Marion City and county officials participated in Operation Onset (Ohio National Security Exercise on Terrorism) concentrating on potential disasters. Don Caprino, Marion County Director of Disaster Services pointed out that the disaster exercise, designed around a terrorist group setting off bombs downtown, showed numerous weaknesses that could be evaluated and alleviated prior to the real thing. Factors considered were deaths, crowded hospitals, hostage taking, panicky crowds, and a demand of the terrorists for a plane to leave town.(33)

CONCLUSION

This chapter has dealt with the elements of society that "boil over" during times of social or political unrest. Even though riots and civil disturbances dominated the landscape in the 1960s and 1970s, this chapter has shown how the 1980s and 1990s have not been immune to civil disturbances of their own, even if less dramatic, less extensive, and less newsworthy.

While civil disturbances have decreased in magnitude as we entered the 1990s we saw an increase in individual acts of terrorism, tied to a corresponding increase in the political unrest worldwide. These can prove to be even more deadly. Looking ahead, for example, in Chapter 11 we will examine Pan Am Flight Number 103 which exploded over Lockerbie, Scotland, in 1988 killing a total of 258 persons in the airliner and 17 on the ground;(34) and a similar crash of an Australian airliner in northern

Thailand on May 27, 1991, which claimed 223 lives and which was also traced to a terrorist bomb.(35)

It is the fire department that will be called upon to protect all elements of the population when civil unrest, riots, or acts of terrorism interrupt the peaceful routine of daily existence. Once again, planning and preparation are a sure way to lessen the negative impact of a sudden violent event in the community.

CHAPTER 8 REFERENCES

1. *Wall Street Journal*, Vol. LXXI, No. 128 (April 17, 1990).
2. Associated Press Report in *Cincinnati Post*, May 7, 1991, p. 2A.
3. ABC News broadcast January 9, 1991.
4. Masotti, Louis, and Don Bowen. *Riots and Rebellion: Civil Violence in the Urban Community*, Beverly Hills, CA: Sage Publications, 1968, p. 13.
5. "Devil's Night" featured in CNN broadcast, November 1, 1991.
6. Masotti, Louis, See Note 4, p. 29.
7. *World Book Encyclopedia*, Vol. 16, p. 324, Chicago, IL: Field Enterprises Educational Corporation, 1976.
8. *Atlantic City Press*, April 1, 1990.
9. Associated Press Report in *Cincinnati Enquirer*, April 28, 1991, p. A-2.
10. *Newsweek*, August 13, 1970, data were also obtained from FBI agent Tom Howard who addressed the Northern California Fire Chiefs' meeting in Monterey, California, on February 19, 1971.
11. "Kids Who Kill" in *U.S. News & World Report*, April 8, 1991, p. 26.
12. Masotti, Louis, See Note 4, p. 33.
13. Bahme, Charles W. *Fire Officer's Guide to Disaster Control*. 1st ed., Boston, MA: NFPA, p. 309.
14. Ibid. p. 309.
15. Ibid. p. 310.
16. Ibid. p. 311.
17. Because the national network news crew was already in Miami to cover the Superbowl, Miami city officials contend that this incident drew an unfair amount of negative publicity.
18. Alter, Jonathan. "The Necessity of Dissent" in *Newsweek*, September 3, 1990, p. 40.
19. CNN broadcast, March 15, 1991.
20. Bahme, Charles W. "The Watts Fires and Their Lessons" in *Fire Journal*, Volume 60, No. 2 (March 1966), Reprint No. FJ 66-27.
21. *World Book Encyclopedia*, Volume 14, p. 112c, Chicago, IL: Field Enterprises Educational Corporation, 1976.
22. "Report of Civil Disorders," National Advisory Commission, Government Printing Office, 1968.
23. *Fire Fighters Fight Fires, NOT People*, International Association of Fire Fighters, AFL-CIO, Washington, D.C., January 1969.
24. Bahme, Charles W. *Fire Officer's Guide to Emergency Action*, Boston, MA: NFPA, 1976.
25. NFPA Standard #1500 is the recognized safety standard throughout the United States and Canada and is especially relevant for the expected dangers of riots in the 1990s.
26. *Civil Disturbance—Strategic Analysis of Fire Department Operations*, Emmitsburg, MD: National Fire Academy, 1984, p. 6.

27. "Booby-trap Molotov Cocktail" described in internal memo, Cincinnati Fire Division, 430 Central Avenue, Cincinnati, Ohio 45202, May 1, 1991.
28. Bahme. See Note 20.
29. Ibid.
30. "Wildland Blazes: Santa Barbara, California" in *American Heat* video journal, St. Louis, MO, August 1990.
31. Obermayer, Robert, "Civil Disturbances" in *Disaster and Fire Defense Planning*, Emmitsburg, MD: National Fire Academy, 1988.
32. Anzouin, Steven, ed. *Terrorism*, New York: H.W. Wilson Company, 1986, p. 72.
33. "Mock Disaster Reveals Flaws in Defense," *The Cincinnati Enquirer*, August 30, 1990, p. C4.
34. "Bomb Destroys Pan Am 747 In Blast Over Scotland," *Aviation Week & Space Technology*," January 2, 1989, pp 28–32.
35. "Officials Link Bomb to Thai Crash," *The Cincinnati Enquirer*, May 28, 1991, p. A-10.

CHAPTER 9

WEATHER-RELATED NATURAL DISASTERS

EVERY EARTHQUAKE, tornado, hurricane, typhoon, cyclone, or flood is not necessarily a natural disaster. The disaster aspects of these acts of nature occur only when they result in loss of life, injury, property damage, or impact on the social structure.

This observation sets the tone for considering the natural disasters that are presented in this chapter. Even though all these spectacular displays of nature will not be catastrophic, the fire service, to fulfill its role of responsibility, must plan and be prepared for response to deadly natural disasters.

This chapter deals with the description and impact of tornadoes, hurricanes, typhoons, cyclones, floods, and other weather-related natural disasters, as well as the role of the fire service in response to problems that may result. Earthquakes, volcanoes, landslides, and other "terrain-based" natural disasters will be considered in Chapter 10.

"Severe lightning, heavy rains, destructive winds, and heavy hail are often 'advance men' for one of the most violent storms on earth—the tornado."(1) Let us start our discussion with lightning and thunderstorms, proceed to tornadoes, hurricanes, and winter storms, and conclude with floods.

LIGHTNING AND THUNDERSTORMS

Severe local storms often begin as thunderstorms. The sound of thunder is caused by violent expansion of the air after it has been heated by lightning. By counting the time between the flash of lightning and the sound it is possible to tell how far away the storm is; since sound travels about 1,100 feet per second, if it is five seconds between the flash and the sound of thunder, the storm is about a mile away.

Lightning is an electrical discharge in the sky of tremendous current—2,000 times greater than ordinary house current which is generally less than 100 amperes. A potential difference of 100 million volts may exist between a cloud and the ground and under certain conditions may cause a bolt of lightning. Every year about nine million strikes to ground objects occur in the United States.(2)

A lightning bolt is classed "hot" or "cold" according to whether its flow of current lasts long enough to start a fire in combustible materials. About one third of them are "hot." In most years, lightning accounts for somewhat less than ten percent of the forests burned in California. As this book was being readied for revision in 1991, however, lightning-caused fires destroyed about 20,000 acres of Yosemite National Park and another 70,000 acres elsewhere in California.(3)

A fifty-mile-wide swath of land that cuts across Florida from the Gulf Coast near Tampa to Daytona Beach on the Atlantic Ocean is called "Lightning Alley." From May through September, thunderstorms make Florida the worst state in the nation for lightning casualties. In one three-year period, more than 112 people were killed or injured in this "alley."(4)

Thunderstorms include heat energy converted into wind, electrical discharge, and violent upward motion of the air. Severe local storms may develop from heating of the ground or colder air above warm, moist, less dense air.

Storms are generally most severe when they occur along squall lines generated when cold, dense air and warm, moist air clash along borders called

Lightning signals the start of thunderstorms, tornadoes, and numerous similar forms of devastation. (Photo courtesy of C. Clark.)

"fronts." The greater the contrast in temperature and moisture across the front, the greater the energy content, and the more violent the thunderstorm. You can judge something of the severity of the storm by the amount of lightning, thunder, hail, and heavy rains. The more lightning, the more intense the storm, and the closer it is, the louder the thunder will be. Hail is precipitation in the form of lumps of ice, called hailstones; the larger they are, the more intense the thunderstorm is likely to be. Heavy rains accompany thunderstorms at their maturity.

Effects of Thunderstorms

Lightning in many of its forms—forked, chain, or zigzag—is capable of great destruction; not only are forest fires kindled, power generating stations "knocked out," buildings destroyed, but in the United States alone, lightning kills one person and injures four others every day. Sheet lightning is really light from a flash of chain lightning that takes place beyond the horizon. Heat lightning is the same, but fainter. Ball lightning has been witnessed by many observers, but does not appear to be dangerous.(5)

Hail can range from the size of peas to baseballs; the largest hailstone on record in the United States measured seventeen inches around and weighed one and one-half pounds. Hailstorms occur most often near Cheyenne, Wyoming, and Denver, Colorado, and generally in the spring. They occur elsewhere and have caused many injuries to people and animals. Severe hailstorms damage crops and greenhouses and have also produced extensive damage to vehicles and aircraft.(6)

Some transportation disasters have been attributed to severe storms. Lightning has struck aircraft not equipped with fuel tank automatic explosion prevention systems, with disastrous results. A Boeing 727 jet with 123 persons on board crashed during a thunderstorm on June 24, 1975, into sandy marshland, just short of the runway at John F. Kennedy International Airport in New York. At least 109 persons were killed when the aircraft hit a tower while descending onto a runway approach and then burst into flames. It was one of the worst domestic airplane disasters in U.S. aviation history.(7)

On August 2, 1985, Delta Flight #191 crashed at the Dallas/Ft. Worth airport as a result of severe "wind shear" in a storm.(8)

Lightning Protection

Lightning protection systems can be installed, not to prevent lightning from striking, but to provide a means whereby its discharge may enter or leave the earth without passing through a nonconducting part of a building or structure.(9)

To safeguard aircraft against lightning and to help solve the problem of thunderstorm interference with aircraft communications, protection systems have been developed for antenna systems and radomes, as well as fuel systems and structural elements. Special high-energy artificial lightning discharge facilities have been set up to reproduce typical recorded flight damage and to test various methods of damage control. Little can be done to prevent lightning from striking an aircraft, but a graded resistance lightning diverter rod can be used for controlling somewhat the specific localized point at which lightning might strike.(10)

Prior to the first U.S./Soviet space mission, scientists dropped millions of metal-coated fibers over the Cape Canaveral launch site in the hopes of taming lightning that could threaten their mission. The objective of dropping these four-inch long aluminized fibers into cumulus clouds was to dissipate the buildup of electrical charges that often develop into thunderstorms. Weather records showed a twenty-three percent statistical chance of thunderstorms in the launch area during the scheduled launch hour, and it was feared that lightning strikes near the site might jolt the sensitive equipment of the *Saturn* rocket and *Apollo* spacecraft.(11)

Lightning—How to Avoid Getting Killed

If it's necessary for you to go outside in a thunderstorm, keep the following precautions in mind. Inside an automobile or fire truck with a metal top is a fairly safe location from lightning. Since the NFPA 1500 Standard requires all fire trucks to be equipped with enclosed crew cabs, fire fighters aboard such apparatus will be relatively safe.(12)

If there is a choice of shelter, choose in the following order: large metal or metal-framed buildings; buildings that are protected against lightning; large unprotected buildings; small unprotected buildings.

If outdoors, avoid the following: small sheds and shelters in an exposed location; isolated trees; wire fences; hilltops and wide open spaces. If on a lake in a small boat, get to shore as soon as possible; if on an ocean or lake in a ship, get inside and amidship as much as possible.

If indoors, avoid doorways and open windows. Do not touch plumbing fixtures, electric wiring, telephones, radios (except when battery operated), and television sets. (If lightning strikes the power lines leading to these appliances, they could become highly energized.)

Through your emergency medical or paramedic training, you may have received instructions on how to take care of a person who has been rendered unconscious by lightning. If your fire department has not received such instruction, ask your fire department training officer to help in getting the information.

Generally, here are the essentials.

If someone in your presence is injured and made unconscious by lightning, first, make certain that the victim is not contacting any electrical wiring or fixtures that may be retaining a residual charge. If such contact is being made, use a dry pole or other nonconductive object to free the victim from the potentially energized object or material.

As soon as this has been accomplished, examine the victim to see if respiration is evident; if not apply cardiopulmonary resuscitation (CPR), including mouth to mouth breathing and chest compression. Loosen the victim's clothing and examine the skin for indication of burns. Try to get medical assistance immediately, and use radio communication to contact a doctor in those vital first five minutes when the victim's breathing may be stopped.

WIND STORMS

"Winds are by nature capricious. For example, even those of low velocity, the simple squalls and gales, have been responsible for eighty percent of all wind damage incidents and thirty-eight percent of the dollar loss reported at properties insured by Factory Mutual. But destruction by the more forceful, less frequent hurricanes and tornadoes has averaged about sixty-two percent of the total dollar loss."(13)

Buildings are destroyed and seriously damaged by high winds because it is not economically feasible to design an average structure to withstand all possible wind pressures to which it may be subjected. Winds striking buildings are deflected upward and across the roof. The uplift forces are greatest at the eaves on the windward side. If several windows are broken on the windward side, or if a large door is open, additional uplift force is added inside the building to that already being exerted above the roof and alongside the building.

This hapless resident cannot decide whether to work on his house or work on his truck since the parts for both have become interwoven during a tornado. (Photo courtesy of Ron Buckingham.)

Wind Damage Prevention

Much wind damage, even near the path of tornadoes, can be prevented by good design, preventative maintenance, and emergency planning. Main roof structures should be designed to withstand wind uplift forces; new roof coverings should be securely anchored to roof decks. Cranes and moving bridges should have breaks, clamps, and special devices to resist wind forces. Preventative maintenance should include the repair of loose window panes and frames; repair of broken or loose flashing, shingles and roof coverings; tightening of slack guy-wires on stacks, antennas, and ventilators; and strengthening of supports for exterior wiring and piping.

Emergency planning should cover the actions to take when windstorm warnings are received, procedures to follow during the danger period, and measures to clean up and restore any damaged equipment or property after the storm passes.(14)

Closely akin to a tornado is a cyclone, which is a storm or series of winds that rotate about a center of low atmospheric pressure, clockwise in the southern hemisphere and counterclockwise in the northern. It advances at a speed of about 20 to 30 miles per hour and often brings abundant rain. When a cyclone occurs in the region of the Philippines or the China Sea, it is called a typhoon.

The term "typhoon" refers more to the origin of a storm than to its composition. Essentially it refers to storms that originate at sea. It has been used to refer to hurricanes in some parts of the world such as in Bangladesh and India and to tornadoes that originate at sea. It is used almost universally in this latter context. Suffice it to say that we will have covered all of the destructive forces of a typhoon in our treatment of hurricanes, tornadoes, and floods.

SANDSTORMS

High winds sometimes blow sand that can wreak havoc on structures and vehicles and can bury highways, making them impassable.

A sandstorm may result from heat of the sun on deserts or broad plains, causing heated air to rise and become quickly replaced by cooler air. The resulting winds may be strong and constant or intermittent and gusty. In the southwestern United States, sandstorms may result from the strong winds and turbulence that sometimes accompany a passing cold front.

On June 18, 1975, winds up to 90 miles per hour swirled enormous clouds of dust and sand across southern California desert areas, forcing thousands of motorists to abandon their vehicles and seek shelter wherever they could. Many of the cars were buried up to their headlights in sand. One motorist and his wife had to be rescued by California Highway Patrol Officers when the rear window of their car was shattered and sand began filling up their car like water. In the communities of Desert Hot Springs and Palm Desert, ambulances and fire trucks were kept busy throughout the night carrying to hospitals scores of elderly people who complained that the dust caused them great difficulty in breathing. Despite closed doors and windows the sand and dust seeped into homes and apartments. "The demand for respiratory help was so great that fire fighters in some cases had to administer oxygen to persons in their homes for an hour and a half before an ambulance or fire truck became available."(15)

On the highways, the sand reduced the visibility to zero at times, and some disoriented motorists wound up driving off the road. Wind-warning signs had been posted along the highways, following advisories issued by the National Weather Service, but they were largely disregarded by vacationers with campers, motor homes, and cars pulling boats and travel trailers; they apparently thought that the storm would soon blow over. Highways got so clogged with sand that they had to be closed until bulldozers could get them cleared. The winds also broke windows, uprooted trees, and downed power lines throughout the Coachella Valley.

CYCLONES

A cyclone is a low-pressure area in the atmosphere in which the winds spiral inward. Cyclones may be just a few hundred miles across or they may cover areas larger than the entire United States. In the northern hemisphere cyclonic winds blow counterclockwise, and in the southern hemisphere they blow clockwise. Depending upon atmospheric pressure, small cyclones may move rapidly or slowly, but large ones usually move slowly, perhaps a few hundred miles per day.

A *tropical cyclone* occurs over the warm ocean waters in the tropics. If they form in the West Indies they are called *hurricanes*; if they form in the Pacific Ocean they are called *typhoons*. These storms, which may bring winds up to 150 to 180 miles per hour, are called *baguios* in the Philippine Islands area.

They are said to be formed by the rotational effect of the earth and the differential heating of the sun on large masses of air.(16)

TORNADOES

On April 26, 1991, at least 48 tornadoes swept Kansas and Oklahoma killing 23 people in all. Characteristically, mobile home parks were hit hardest and suffered the most casualties. About 1500 Andover, Kansas, residents were left homeless by one tornado, the first to strike the town since 1958. This is a story that has repeated itself, one place or another, every spring in the United States, and will likely strike somewhere every spring in the future. The places and death tolls may change, but tornadoes are a somber reality.(17)

No region in the world suffers so many tornadoes as the continental United States, which averages 800 per year. The areas of highest vulnerability are in those states that lie in the eastern plains and the Mississippi

Table 9-1. Major United States Tornadoes 1925 to 1991.

Date	*Location*	*Deaths*
1925 Mar. 18	Missouri, Illinois, Indiana	689
1927 Apr. 12	Rock Springs, Texas	74
1927 May 9	Arkansas; Poplar Bluff, Missouri	92
1930 May 6	Hill, Ellis County, Texas	41
1932 Mar. 21	Alabama (outbreak)	268
1936 Apr. 5	Tupelo, Mississippi	216
1936 Apr. 6	Gainesville, Georgia	203
1938 Sep. 29	Charleston, South Carolina	32
1942 Mar. 16	Central to northeast Mississippi	75
1942 Apr. 27	Rogers, Mayes County, Oklahoma	52
1944 June 23	Ohio, Pennsylvania, West Virginia, Maryland	150
1945 Apr. 12	Oklahoma, Arkansas	102
1947 Apr. 9	Texas, Oklahoma, Kansas	169
1948 Mar. 19	Bunker Hill, Gillespie, Illinois	33
1949 Jan. 3	Louisiana, Arkansas	58
1952 Mar. 21	Arkansas, Missouri, Tennessee (outbreak)	208
1953 May 11	Waco, Texas	114
1953 June 8	Flint to Lakeport, Michigan	116
1953 June 9	Worcester, Massachusetts	90
1953 Dec. 5	Vicksburg, Mississippi	38
1955 May 25	Udall, Kansas	80
1957 May 20	Kansas, Missouri	48
1958 June 4	Northwestern Wisconsin	301
1959 Feb. 10	St. Louis, Missouri	21
1960 May 5–6	Southeastern Oklahoma, Arkansas	30
1965 Apr. 11	Indiana, Illinois, Ohio, Michigan, Wiconsin	271
1966 Mar. 3	Jackson, Mississippi	57
1966 Mar. 3	Mississippi, Alabama	61
1967 Apr. 21	Illinois	33
1968 May 15	Arkansas	34
1969 Jan. 23	Mississippi	32
1971 Feb. 21	Mississippi Delta	110
1973 May 26–7	South, Midwest (outbreak)	350
1974 Apr. 3–4	Alabama, Georgia, Tennessee, Kentucky, Ohio	350
1977 Apr. 4	Alabama, Georgia, Mississippi	22
1979 Apr. 10	Texas, Oklahoma	60
1980 June 3	Grand Island, Nebraska (outbreak)	4
1982 Mar. 2–4	South, Midwest (outbreak)	17
1982 May 29	Southern Illinois	10
1983 May 18–22	Texas	12
1984 Mar. 28	North Carolina, South Carolina	67
1984 Apr. 21–22	Mississippi	15
1984 Apr. 26	Oklahoma to Minnesota (outbreak)	17
1985 May 31	New York, Pennsylvania, Ohio, Ontario (outbreak)	90
1987 May 22	Saragosa, Texas	29
1989 April & May	14 states, Arkansas to New York	30
1990 June 3	Ohio, Kentucky, Indiana	12
1990 Aug. 29	Illinois	24
1991 Jan. 15	Midwest U. S. of America	14
1991 Apr. 26	Kansas, Oklahoma	23

and Ohio River valleys. Even though some areas of the country may be relatively free from the tornado threat, tornadoes do occur in all fifty states.

Tornado season begins in February and continues into summer. About half occur during the months of April, May, and June. Most often they happen in the afternoon and the direction of travel is from the southwest to the northeast.

As shown in Table 9-1, the United States experiences dramatic loss of life from tornadoes almost every year.

A tornado is often called the most violent of all windstorms. It is a rotating column of air pendant from a thunderstorm cloud and touching the ground. If it occurs over an ocean or lake it is called a *waterspout*; a waterspout is not a column of water rising out of the body of water, but a whirling column of air and watery mist. Some sea or lake water may be drawn up at the base of the column, but even in ocean waterspouts, most of the water is fresh water formed by condensation.(18)

A tornado, sometimes called a *twister*, nearly always starts as a funnel-shaped attachment from the cumulonimbus clouds overhead and then builds downward to the ground. When it picks up earth and debris it gets dark in color. Sometimes two or more funnels may form from the same thunderstorm clouds at the same time; as these clouds move along, tornadoes may form, travel along in touch with the ground for a few miles, dissipate or lift, and then touch down again—several times.

Nature of Tornadoes

Tornadoes almost always form a conjunction with a thunderstorm. Heavy hail may also accompany these storms. A single thunderstorm may generate numerous tornadoes or twisters as they are sometimes called. On April 3–4, 1974, 148 tornadoes developed in 11 midwestern states and Canada. This was known as the "Super Outbreak."(19)

Tornadoes are categorized from Class 1 to Class 5, with Class 5 being the most severe. A Class 1 tornado has wind speeds less than 100 miles per hour, a path width of less than 100 yards and may stay on the ground for only 100 yards at a time.

A Class 5 tornado may have a single large vortex or multiple vortices rotating about one another. Wind speeds may reach 300 miles per hour, cutting a path up to a mile wide. These systems may stay on the ground up to an hour.

The width of the touchdown path averages about 300 to 400 yards, and the length about four miles, though it may go as far as 300 miles. In 1917 a tornado lasted over seven hours and traveled across Illinois and Indiana for a distance of 293 miles. The average speed is from 25 to 40 miles per hour, and its sound has been described as a roaring, rushing noise, like a flight of jet planes or train speeding through a tunnel. The horizontal wind speed in the center of a tornado may be more than 300 miles per hour, and its destructive power is due to the combined action of its strong rotary winds and the partial vacuum in the center of its vortex. As the "eye" of the tornado passes over a building, pressure builds up inside, the walls topple outward, windows explode, and debris of this destruction is driven through

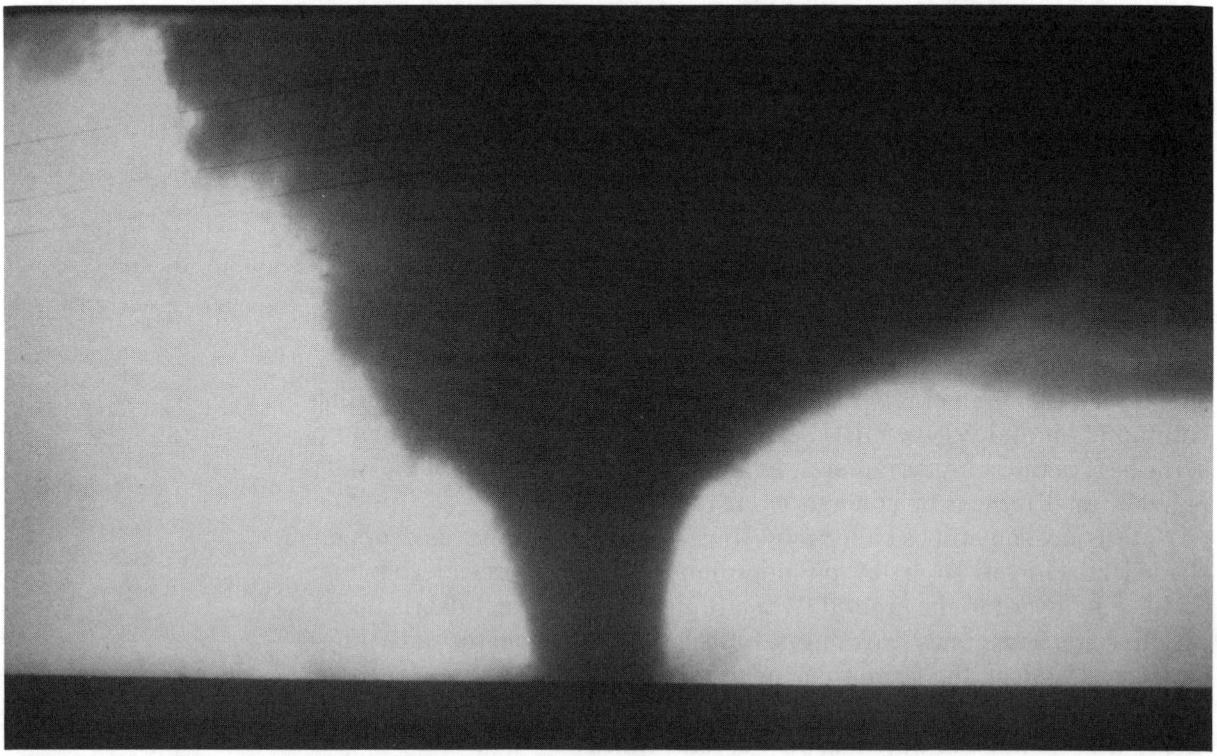

The unmistakable funnel-shaped cloud of a tornado roars across Seymour, Texas, in April 1979. (Photo courtesy of Ron Buckingham.)

the air in a dangerous barrage. Heavy objects like railroad cars and machinery have been lifted and carried for considerable distances. Winds of destructive force extend for several hundred feet outward on both sides of the vortex.

What Are the Odds of Getting Hit By a Tornado?

The fact that your hometown has never had a tornado does not mean that you can disregard the possibility. Every year, places that have never been struck before join the list of tornado disaster locations. The disaster potential is greatest where the most tornadoes hit densely populated areas. The greatest tornado threat is in Massachusetts, though the greatest number of tornadoes per 10,000 miles occurs in Oklahoma. Texas has had more tornado-caused deaths than any other state.(20)

The mathematical chance that a specific location will be struck by a tornado in any one year is quite small, but they have occurred with exceptional frequency at particular locations: Oklahoma City has been struck 26 times since 1892; Baldwyn, Mississippi, was struck twice in 25 minutes on March 16, 1942; some cities, such as Irving, Kansas, and Austin, Texas, were struck twice in rapid succession; and Codell, Kansas, was struck on May 20, 1916, again on May 20, 1917, and a third time on May 20, 1918!

Examples of Tornado Disasters

As this chapter was being prepared, on April 26, 1991, a tornado wiped out an entire neighborhood, formerly called Hayesville, near Wichita, Kansas. At least 100 people were expected dead in Kansas and Oklahoma. Andover, Kansas, a town of 3,000 was also completely leveled. Twenty-three of its inhabitants died and the injuries were too numerous to assess.(21)

On August 29, 1990, 24 people died and more than 300 were injured by tornadoes that rolled across "tornado alley," about 35 miles southwest of Chicago. Devastation was so widespread that President Bush immediately declared the region a disaster area. Typical of the problems confronting the fire service was a seven building apartment complex near Plainfield, Illinois, where six people were missing. At least nine of these storms' deaths occurred near apartments, where some victims were blown 400 feet out of the building.(22)

On June 3, 1990, midwestern storms spawned 101 tornadoes. Hardest hit was the tristate area of Ohio, Kentucky, and Indiana. Typical of the devastation was the small town of Harrison, Ohio, which lost 426 homes and businesses, cutting out the heart of this small community.(23)

In November 1989, several tornadoes hit the United States when frigid air howled out of the Arctic to collide with record balmy weather pushing northward from the Atlantic and Gulf of Mexico. This unseasonable clash generated a hopscotching barrage of twisters through 14 states from Arkansas to New York that killed at least 30 people. Some of the winds were up to 250 miles per hour.

The worst single tragedy in that catastrophe hit the small community (population about 1,200) of East Coldenham, New York, 40 miles northwest of New York City. More than 120 children were eating lunch when a blast of wind estimated at 100 miles per hour struck the yellow brick and glass building. A massive section of the south wall crashed into the children in a hail of shattered glass, concrete and falling bricks. When the frantic rescue ended, seven children were dead and 18 hospitalized, three with critical injuries.(24)

In Huntsville, Alabama, a school was also struck by a tornado. Yet, although it was leveled, the timing was fortunate as it hit after the school day was over, around 4:30 P.M. The twister roared through the town killing 18, ranging in age from 2 to 67, and demolished 119 houses.(25)

The outbreak of 37 tornadoes on Palm Sunday, April 11, 1965, killed 257 people and injured more than 5,000 persons in the Midwest.

In March, 1932, a series of tornados occurred in Alabama killing 268 persons and injuring 1,874, with property damage of about 5 million dollars.

The series of tornadoes that caused the greatest recorded loss of life occurred in portions of Missouri, Indiana, Illinois, Kentucky, and Tennessee on the afternoon of March 18, 1925. Eight separate tornadoes were observed, one of which killed 689 persons, injured 1,890, and caused more than 16 million dollars in property damage. An additional 41 lives were lost in the seven other tornadoes, along with added property damage.

The Xenia Tornado—A Case Study

Occasionally an entire city is devastated by a tornado, taxing the resources of the fire department to the fullest and causing it to be integrated into its community to an extent never before realized. This happened in Xenia, Ohio, at 4:41 P.M. on April 3, 1974, and will serve as a case study showing how a fire department's operations are conducted during the time of an unforeseen tornado of disastrous proportions.

The tornado that struck Xenia illustrates well the problems that can confront a fire department. Xenia is a city of 27,000 people covering seven square miles. Its 36-man paid department operates out of two stations with an average of ten men on duty. The southwest area of Ohio had been under a tornado watch since about 1 P.M. and a tornado warning was put into effect around 3:15 P.M. At 4:40 P.M. a patrolling police officer reported that a tornado was approaching the city from the southwest and it hit a minute later. It cut a path of total destruction about six miles long and three-fourths of a mile wide, leaving 33 persons dead and about 1,000 injured. In the ten minutes that it took to sweep through the city it hit the City Hall, took the roof covering off one of the fire stations, seriously damaged the second, destroyed a high school and wiped out (or damaged beyond repair) 1,347 buildings).(26)

Within a half-hour all off-duty men were back on duty including five who had lost their homes; but the work of fire fighters was not without handicaps. Streets were filled with rubble, and a train that had been derailed by the tornado blocked all the major east-west streets. It was impossible to travel very far by car, so the off-duty fire fighters started working in areas nearest their homes, joining any emergency unit operating in the area.

Telephone service had been knocked out, and the fire department's main and auxiliary antennas were destroyed, seriously cutting the range for use of its radio. Fire apparatus could not respond from the second fire station because cars, trees, pieces of buildings and other debris blocked the ramp.

Fire apparatus and ambulances responded from nearby communities, without being called—since without telephone service or a radio link, there was no way to request help. Without compatible radio frequencies there was no way to communicate with them after they arrived, making search efforts difficult to organize. According to Xenia's fire chief at the time, John Troeger, his biggest problem early in the disaster was getting incoming units to report to the disaster control center. In retrospect, he thought it might have been well to set up a second command post at the western end of the city during the early search and rescue operations, for that would have overcome the problem of the physical separation of the city caused by the derailed train.

Other problems included the fact that all power to the city was disrupted, water pressure was low or absent due to ten missing fire hydrants and broken pipes in about 800 buildings, including those supplying sprinkler systems. During the 32 hours of getting the water system restored, neighboring communities supplied water tankers to help fill the need. To control natural gas leaking from the many broken pipes, it was necessary to close valves that cut off gas to the whole community.

Ironically, the only deaths from fire during these disaster control oper-

Shattered homes and shattered lives come in the wake of the 1974 tornado in Xenia, Ohio. (Photo courtesy of the Dayton, Ohio, Fire Department.)

ations were suffered by two National Guard soldiers who were trapped in their improvised basement shelter—a three-story furniture store that had become fully involved before they could escape.

The lessons learned from this tornado disaster led to some improvements. To improve the dependability of the communications system, an additional auxiliary radio antenna was added and three vehicles were equipped with 100-watt radios which could be used for a mobile base station. All fire department vehicles were equipped with four-channel radios to permit communications with all fire departments in the county and some in neighboring counties. A good disaster plan was adopted that incorporated flexibility and simplicity; each operating division of the city developed its own operational details subject to the approval of the city manager; these plans were coordinated at countywide meetings of the emergency services chiefs, so that large-scale disaster drills could be conducted.

Many other jurisdictions have learned valuable lessons from the Xenia tornado. A situation where a good mutual aid plan and a countywide radio network paid-off was described by the Berkshire County Fire Coordinator following a tornado strike in West Stockbridge, Massachusetts. The county's radio base station for the fire network had been knocked out by a thunderstorm a few minutes before the tornado hit, but portable radios were adequate to direct incoming mutual aid companies in search and rescue operations. Leaking propane gas from damaged cylinders infiltrated the piled-up cars, trucks, and structural debris. Also a gusher of fuel oil issued from a ruptured pipe from an underground tank. A "no-smoking" warning had to be issued repeatedly to people in the area. Search parties had to be on the lookout for live electrical wires that might arc and ignite the leaking gas or give someone a deadly jolt.(27)

In the wake of the "Super Outbreak" of tornadoes, including the Xenia tornado, municipalities and jurisdictions all across the United States installed or improved early warning systems, such as building-mounted sirens, and began testing them regularly. Ironically, even though the sirens had been tested monthly, they failed to work properly when needed most in Andover, Kansas, in 1991, when devastating tornadoes wreaked havoc on this small town.(28)

Tornado Predictability

It is possible to predict the most probable times and places for tornadoes. The areas where tornadoes are likely to occur most often are well known (Massachusetts, Kansas, Iowa, Texas, Oklahoma, Arkansas, Missouri, Mississippi, Ohio, and Alabama). The months of the year when at least half of them occur is a matter of record (April, May, and June); also predictable is the time of day (afternoon) and the usual direction of movement (southwest to northeast). Even the type of thunderstorm clouds (cumulus and

Figure 9-1. Average Number of Tornadoes and Tornado Days Each Month in the United States.

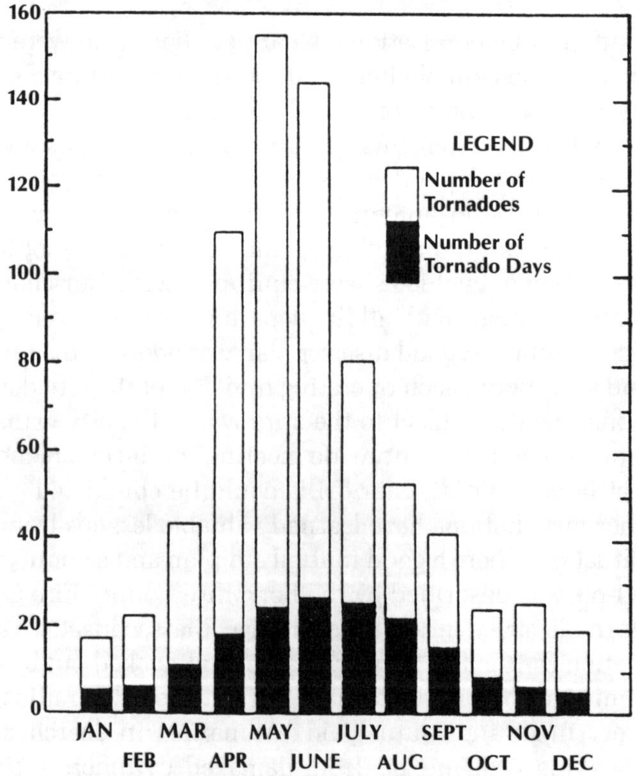

cumulonimbus), the weather conditions (heavy rain and hail), and the ominous sound (roar of planes) are sufficiently established that the National Weather Service has been able to provide its severe local storm forecast and warning service since 1953. Since then, with the help of the emergency services and individuals who watch for all the indications and make quick reports to the National Weather Service, plus the cooperation of the news media that make the information available to the public, the average number of deaths and injuries from tornadoes has decreased considerably. Figure 9-1 illustrates how the probability of a tornado increases in the spring and early summer.

Tornado Preparedness Planning

A tornado preparedness plan can be developed to fit any size community or industrial plant, and such planning should not be limited to an area that has been struck before. Because tornadoes strike so suddenly and move on, it is important to have cooperative arrangements for detecting them as soon as possible and for warning those in the path; only in this manner will people have time to get to shelter quickly. As soon as the tornado hits, planned rescue and damage control operations should be ready to start. The National Weather Service has regional offices that can be contacted, and the representatives will be glad to assist your local authorities in developing your community's tornado preparedness plan. Its publication *Tornado Preparedness Planning* gives the addresses and telephone numbers of all its regional headquarters.(29)

In Appendix G can be found a list of tornado facts provided by State Farm Insurance and the National Coordinating Council on Emergency Management. This is an excellent example of public information literature available.

A major factor in preparedness is public education for the citizens of a community. They must be made aware of how advisory information is transmitted over radio and television. Many communities also utilize a special signal on horns, sirens, or whistles to alert the public.

Impact on the Community

Damage caused by tornadoes as they cut a path through a community can vary dramatically. A Class 1 tornado may overturn mobile homes and cause erratic and freakish damage to sign boards, store fronts, and power lines. These storms often form before warnings can be issued, but the resulting property damage and injury rate is generally low.

A Class 4 or 5 tornado, on the other end of the spectrum, can cause near total destruction to everything in its path with the exception of only the most sturdy structures. Major damage to above ground power lines and antenna towers typically occurs. Severe injuries and loss of life can be anticipated even with an early warning.

Roadways and access routes within the touchdown area will be impassable to vehicular traffic due to fallen trees, power poles, and structural debris. Power and telephone communication will be disrupted. Water pressure may be extremely low or absent because of broken pipes throughout the area. Hazardous chemicals or flammable liquids may be released posing special problems for both the public and emergency services.

The fire potential may be extreme following the devastating effects of a tornado. Possible fire causes include inoperative built-in protective systems (sprinklers and alarms), the drop in water pressure from broken pipes and mains, the possible delay in reporting, the delayed response, and lack of access. They all add fuel to the potential for multiple large-scale fires.

An in-depth Hazard Identification of a community should determine its potential for a tornado emergency and identify problems to consider.

The Capability Assessment should determine what resources are available and what additional resources are needed. Finally, a Multi-Year Development Plan should be implemented to overcome capability shortfalls.

A fire department's tornado emergency planning should consider those special operational concepts (over and above standard operating procedures) that will provide effective command and control during a large-scale tornado emergency.

A tornado can be a devastating event, obliterating almost everything in its path. Loss of life, injuries, and property loss can be beyond comprehension within a community. When the fire department responds appropriately to the myriad of problems, the planning, the training, and the preparation will, to a large extent, influence the performance. At times like this, the real worth of a fire department is measured.

Emergency Action—Tornadoes

When a tornado approaches, your immediate action may mean the difference between life or death. Fire fighters and civilians should be listening to radio and television stations for the latest advisory information. Remember that a *Tornado Watch* means that tornadoes are expected to develop, and *Tornado Warning* means that a tornado has actually been sighted.

Fire department personnel will be familiar with the advice radio and television networks broadcast to the public when a tornado watch has been set, and whether they are at home or at the fire station will know what to do. The following paragraphs are typical of the kind of suggestions that might be given the public.

If you see a funnel cloud heading your way, or hear something that sounds like a mill whistle or a tank coming toward you, seek shelter, preferably in a tornado cellar, underground excavation, or a steel-framed or reinforced concrete building of substantial construction. Stay away from windows!

If inside an office building, stand in an interior hallway on a lower floor, preferably in the basement. If in an industrial plant, join other workers in moving quickly to the predesignated location that offers the greatest protection. Stay there until the "All-Clear Bulletin" is received.

In your home the basement usually offers the greatest safety. Get under a heavy table or sturdy workbench if possible. Where there is no basement, take cover under heavy furniture in the center part of the house. Keep some windows open, but stay away from them. Mobile homes are particularly vulnerable to overturning during strong winds; keep the tie-down cables' turnbuckles tightened. Get to the community shelter if one has been provided.

Students should go to an interior hallway on the lowest floor of their school. They should get out of auditoriums and gymnasiums or other structures with wide, free-span roofs.

In open country, move away from the tornado's path at a right angle. If there is no time to escape, lie flat in the nearest depression, such as a ditch or ravine.

This advice, though broadcast before a tornado may be formed, is intended to be acted upon after a tornado is formed and headed your way. If the situation is such that only a Tornado Watch has been set for your area, you may have time to activate your emergency operations center and make preparations in accordance with the tornado annex in your disaster plan.

As a fire chief, you will want to make sure that your condition of readiness is such as to be able to operate in the absence of public electrical power, telephones, or normal water pressures. Your fire stations will be alerted to make everything secure against the force of strong winds, and your off-duty personnel will know what your plan calls for in the matter of report-

In the aftermath of a weather-related natural disaster, sorting of debris, autos, and private property of all sorts becomes a major problem. (Photo courtesy of Xenia, Ohio, Fire Department.)

ing to the nearest fire station or emergency service location for assistance in search and rescue operations, shutting off broken gas and water mains, reporting fallen power lines, and carrying out fire control and suppression activities.

Fire departments within your mutual aid areas, if not already alerted, will be notified of the possible need for assistance if a destructive tornado touches down in your community.

If a tornado hits, the officer in charge of each fire station and department facility should provide information to the emergency operations center by whatever means are available, stating damage sustained,and any impairment of response capability. If station doors are obstructed by heavy debris, assistance should be obtained quickly to clear the ramp with chain saw, bulldozers, cranes, or other debris-removal equipment. If streets are obstructed by fallen walls, heaps of autos, or derailed trains, an aerial reconnaissance, if feasible, can help determine a by-pass route for emergency equipment pending clearing of normal routes of travel.

If fires break out, or leaks and spills of dangerous chemicals, flammable or toxic liquids, and gases occur where access by fire apparatus is not readily possible, it may be necessary to evacuate occupants of buildings in the zone of danger. If winds die down, it may be possible to fly in crews and lay lines with helicopters to handle emergencies in inaccessible locations. Measures suggested for chemical emergencies discussed in Chapter 12 may be useful following tornado damage to heavy industrial occupancies, refineries, and chemical processing plants.

As soon as practical at the conclusion of emergency operations, a post-disaster inventory should be made of fire equipment lost or damaged, and steps taken to restore response readiness to a condition of normalcy.

HURRICANES AND STORM SURGES

In 1988 on September 10 to 17 Hurricane *Gilbert* began in the Caribbean and hit Jamaica and the Cayman Islands before it finally unleashed its fury on the Yucatan Peninsula in Mexico. This storm killed 400 and left 750,000 homeless. It was a classic Class 5 hurricane with 175 miles per hour winds.(30)

And then on September 21–22, 1989, what began as a tropical depression off the west coast of Africa and whirled across the Atlantic gaining strength from the moist tropical air, grew into a 150 miles per hour hurricane. Hurricane *Hugo* carved an awesome arc of destruction in a 2,300 mile sweep that first hit the Virgin Islands, Montserrat, Guadalupe, Puerto Rico, and finally hit the South Carolina coast, doing the most damage to Charleston and the outer islands. Hurricane *Hugo* killed at least 48 people and caused more than $2 billion in property damage.(31)

Hurricanes recur every several years on the average and have become a fearsome fact of life for coastal dwellers. A hurricane is a tropical cyclone originating in the West Indies. The name is from the Spanish *huracan*, probably derived from the Mayan storm god, Hunraken; the Quiche god of thunder and lightning, Hurakan; and numerous other Caribbean Indian terms for evil spirit, big wind, and the like. "They are tropical children, the offspring of ocean and atmosphere, powered by heat from the sea, driven by the easterly trades and temperate westerlies, the high planetary winds and their own fierce energy. In their cloudy arms and around their tranquil core, winds blow with lethal velocity, the ocean develops an inundating surge, and as they move toward land, tornadoes now and then flutter down from the advancing wall of thunderclouds."(32)

The outer extremities of a hurricane can vary from 60 to 1,000 miles in diameter, and its center may move as much as forty miles in a day, or it may remain fixed in one place for hours. On the average, the great spiral of a hurricane covers an area some 100 miles in diameter with winds greater than 74 miles per hour; it spreads gale-force winds (winds over 40 miles per hour) over a 400-mile-diameter area. Their broad spiral base may dominate weather over thousands of square miles. While the winds on the outer band of the storm may only be 30 miles per hour, within 30 miles of the center winds may have velocity exceeding 150 miles per hour. Near the very center where ascending air releases heat to drive the storm, winds may reach 200 miles per hour.

Hurricanes begin as tropical depressions and when the large counterclockwise circulating wind pattern increases in intensity to 74 miles per hour, a hurricane has developed. Whereas tornadoes have a normal duration of minutes or hours, the devastating forces of a hurricane may persist for days or weeks. "No other atmospheric disturbance combines durations, size, and violence more destructively."(33)

The powerful winds of a hurricane may cause extensive collapse of structures and downing of trees and power poles; however, most damage from a hurricane results from flooding and the storm surge. Six to twelve inches of rain is typical and causes flooding in the low lying costal areas as well as inland. The storm surge, however, is the most lethal aspect, as the sea water level may rise fifteen feet or more along with crushing storm waves reaching heights of fifty feet or more.

In a hurricane the winds flow toward the low pressure in the warm, comparatively calm core; convection currents swirl the air upward around the "eye" of the storm, which on the average is around 14 miles in diameter, though 25 miles is not unusual. Winds in the eye may be as low as 15 miles per hour, but as the storm moves over a given location, the winds increase again at the opposite wall; however, they will be moving in the opposite direction because of the cyclonic circulation of the storm.

Effects of Hurricanes and Storm Surges

In an average year there will be fewer than ten tropical cyclones, of which about six will develop into hurricanes. These will usually kill about fifty to one hundred persons between Texas and Maine and cause property damage of more than $100 million. If it is a worse than average year, there will be several hundred deaths and the property damage will run into billions of dollars.

The hurricane brings devastation by the impact of high wind pressure

In a matter of minutes, this beach house is flooded, leaving residents little time to evacuate. (Photos courtesy of Ron Buckingham.)

causing collapse of structures, breaking of windows, and knocking down trees and poles. It creates flood-producing rain, and most lethal of all, the storm surge (the increase in water level that, along with abnormal waves and currents, produces extensive "wave-wash" damage to coastal structures).

Floods produced by hurricane rainfall are more destructive than the winds. Hurricane *Diane* of 1955 caused little damage as it moved onto the continent; but long after its hurricane winds subsided, it brought floods to Pennsylvania, New York, and New England—floods that killed 200 persons and cost an estimated $700 million in damage. Hurricane *Camille*, in 1969, produced winds of 190 miles per hour, 10 inches of precipitation on the Mississippi Gulf coast, and waves up to 25 feet above sea level. Then sweeping inland and curving to the east, it also caused heavy flooding in Virginia.(34)

The hurricane's worst killer comes from the sea. Ocean waves generated by hurricane-force winds may reach heights of fifty feet or more (See Figure 9-2). The water surged in a few seconds as a result of the Galveston hurricane in 1900 and wiped out 6,000 persons—almost half of all the lives lost in the United States from hurricanes since that time. Because much of the United States' densely populated coastline along the Atlantic and Gulf coasts lies less than ten feet above mean sea level, the danger from storm surge is great.

Another effect of hurricanes is sudden low atmospheric pressure that produces an internal pressure inside tightly closed buildings, causing their roofs to come off or structural collapse. This happened to the Navy hangers at Richmond, Florida, in 1945. Other effects include flying missiles from broken glass and dislodged roofs and fires started by overturned stoves and electric shorts.

Figure 9-2. Seashore Profile of a Hurricane.

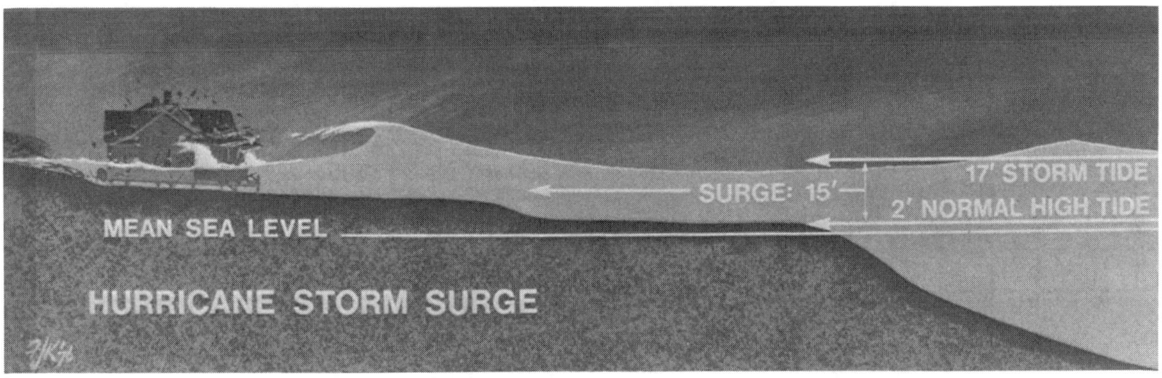

The hurricane is moving ashore close to the beach area. It is high tide time again. This time, however, there is a 15-foot surge added to the normal 2-foot astronomical tide creating a 17-foot *storm tide*. This great mound of water, topped by battering waves, is moving slowly ashore along an area of coastline 50 to 100 miles wide. Winds are now over 130 miles an hour. Much ocean-front property will be unable to withstand this combined assault of wind and water.

The combination of storm surge, battering waves, and high tide is the hurricane's most deadly killer.

Impact on the Community

In addition to direct destruction already discussed, secondary problems mount quickly. Roadways and access routes are blocked with debris. Power and telephone lines are down causing a failure of telephone communications. All but battery powered radios and television sets are out. There may be a tremendous impact on the fire department communication system with no incoming calls to report emergencies and no capability of the dispatch office to transmit radio messages. Threat from fire is intensified as the result of the hurricane's destructive forces. Many fire causes are generated, reporting is delayed, access is restricted or impossible, and response is delayed. All of the above add to the possibility of large-scale fires.

Low-lying population centers can be inundated by waves causing near total destruction in their path. (Floods are discussed later in this chapter.) The greatest loss of life in hurricanes is from drowning by a ratio of about 9 to 1. Electrocution caused by fallen power lines is another extreme danger.(35) Evacuation and sheltering of thousands of people affect not only the involved community but also the inland communities that act as hosts. Evacuations generally involve some holdouts who stay on and eventually present additional concerns to the emergency services.

Hurricane Planning and Preparation

Planning procedures for hurricanes are similar to those for tornadoes. An in-depth Hazard Identification of a community should determine its potential for a hurricane emergency and identify problems to consider. A Capability Assessment should determine what resources are available and what additional resources are needed. A Multi-Year Development Plan should be implemented to overcome capability shortfalls.

A fire department's hurricane emergency planning should consider those special operation concepts (over and above standard operating procedures) that will provide command and control during a large-scale hurricane emergency.

Hurricanes—Where and When

The areas of highest vulnerability to hurricanes in the United States are those states that border the Gulf of Mexico and the entire eastern seaboard. Areas subject to flooding from the excessive rainfall include the eastern, midwestern, and New England states. Even southern California is susceptible to severe squalls resulting from hurricanes off the Pacific Coast of Mexico.

Hurricane season for the Caribbean and Gulf of Mexico is early summer through fall. The hurricanes that form southwest of Mexico occur in the spring and summer.

Major United States hurricanes are shown in Table 9-2.(36)

Table 9-2. Major United States Hurricanes: 1960 to 1991(36)

Date		*Name*	*Location*	*Deaths*
1960	Sept. 4–12	*Donna*	Caribbean, Eastern U.S.	148
1964	Oct. 4–7	*Hilda*	Louisiana, Mississippi, Georgia	38
1965	Sept. 7–10	*Betsy*	Florida, Mississippi, Louisiana	74
1966	June 4–10	*Alma*	Honduras, SE U.S.	51
1966	Sept. 24–30	*Inez*	Caribbean, Florida, Mexico	293
1967	Sept. 5–23	*Beulah*	Caribbean, Mexico, Texas	58
1969	Aug. 17–18	*Camille*	Mississippi, Louisiana	256
1970	Jul. 30–Aug. 5	*Celia*	Cuba, Florida, Texas	31
1972	June 19–29	*Agnes*	Florida through New York	118
1975	Sept. 13–27	*Eloise*	Caribbean, NE U.S.	71
1979	Aug.29–Sept. 7	*David*	Caribbean, Eastern U.S.	2,000
1980	Aug. 4–11	*Allen*	Caribbean, Texas	272
1983	Aug. 18	*Alicia*	SE Texas	21
1985	Oct. 26–Nov. 6	*Juan*	SE U.S.	97
1988	Sept. 10–17	*Gilbert*	Jamaica, Mexico, Gulf Coast	400
1989	Sept. 21–22	*Hugo*	Virgin Islands, Puerto Rico and SE U.S., S. Carolina	36

Predictability of Hurricanes

The National Hurricane Center (NHC) at Miami, Florida, has overall responsibility for operations and meteorological analyses related to the position, intensity, and movement of hurricanes in the Atlantic, the Caribbean, and the Gulf of Mexico. The National Hurricane Center coordinates the Hurricane Warning Offices (HWOs), which are responsible for maintaining a close watch on potential hurricane development in their respective areas and for issuing bulletins and advisories to the general public and to disaster and rescue agencies. Based upon information from other government agencies, supplemented by weather data from ships, aircraft, satellites, and special reconnaissance flights, a detailed hurricane forecast is prepared by the NHC or by an HWO after consultation with the NHC.

Once a hurricane nears landfall, a coastal radar network extending from Maine through Texas keeps constant surveillance on the hurricane's position and movement. Forecasts and warnings are given immediate widespread distribution by teletype, radio, television, and newspapers.

Hurricane Prediction/Warning

Four kinds of notices are issued through the National Hurricane Center (NHC) and the Hurricane Warning Offices (HWOs).

- Advisories—giving information on location, intensity, and movement of the storm, along with precautions to take
- Bulletins—similar to advisories, but more general or newsworthy
- Watches—announcing specific areas threatened by the hurricane
- Warnings—announcing that one or more of the following dangers are expected in a specified area:

  winds in excess of 74 miles per hour

high water
exceptionally high waves in a specified coastal area within twenty-four hours or less.

The primary methods of warning the threatened areas are through radio and television. "Last appeal" types of notification are frequently made by loud speakers from helicopters or ground patrol units.

Two major gaps have existed in prediction: the distance to be covered by the hurricane in twenty-four hours and the landfall location. This is further complicated by the difficulty of pinpointing the exact center of the hurricane. With respect to distance traveled, the average error is about 100 nautical miles; the same is true for the average landfall error, which could easily result from a 10-degree error in direction of motion (e.g., 310 instead of 320 degrees).(37)

Present-day hurricane warnings are issued to cover about 200 miles of coastline, while the segment identified as subject to storm surge is usually about 100 miles long. A warning is issued whenever possible to allow up to a least twelve hours of daylight for emergency protection of property and for population evacuation before the hurricane strikes. Officials of the NHC have reiterated their concern for safety of the increasing populations along the Atlantic and Gulf coasts, where many areas lie less than ten feet above sea level. The sledgehammer effect of pounding, surging waves over twenty-three feet in height, whipped by winds up to 230 miles per hour near the hurricane's eye, as in Hurricane *Camille*, can destroy any structure not specifically designed to resist such forces. Officials have predicted that in a hurricane emergency ten years from now, unless proper planning is done to prevent population concentration and to provide for evacuation, the limited capacity of escape routes could cost 20,000 to 50,000 lives.

Historic Hurricanes

Certainly one of the most memorable hurricanes in recent history was Hurricane *Hugo* which devastated the east coast in late 1989. In addition to the devastation previously described in this chapter, the hurricane was noteworthy in that it was half of a two-part calamity to strike our country when it was followed shortly by the Bay Area earthquake affecting San Francisco, Oakland, and the surrounding areas. These two calamities served as "book ends for disaster" for our country in late 1989.

Two decades previously, in 1969, Robert H. Simpson, then Chief of the National Hurricane Center, said of Hurricane *Camille*, also mentioned previously, "by any yardstick, *Camille* was the greatest storm of any kind that has ever affected this nation." Structures that withstood the winds risked being burned down as fires raged out of control because fire fighting equipment could not reach them. Highways were impassable as the 200,000 persons who had fled inland from the storm returned to see what was left of their homes. The 600-square-mile area was deprived of means of communication and power and its inhabitants confronted with water and food shortages, along with the threat of a typhoid epidemic.(38)

In 1972 by the time *Agnes* disintegrated over western Pennsylvania, she had dumped an estimated 28.1 trillion gallons of water over the eastern seaboard, the equivalent of 25.5 cubic miles of water. Most of it was poured over the Middle Atlantic states. Streams and rivers were flooded—13,500 miles of them—and more than half a million people suffered losses. About 116,000 homes, 2,400 farm buildings, and 5,800 businesses were damaged or destroyed. At least $2.5 billion was lost in private and public property. Damage to highways, bridges, and public buildings amounted to more than $700 million. Considering the violence of the winds and the extent of the floods, it is amazing that only 118 lives were lost. The Army Engineers credit this to the fact that most flood control measures were adequate. Where they were insufficient, the results were devastating.(39)

Fifi is an example of a hurricane that took a large toll in lives; on September 18 and 19, 1974, she killed between 7,500 and 8,000 persons, and left more than 300,000 Hondurans homeless.(40)

In 1935, the barometer fell to 26.35 inches during a hurricane that struck Tampa, Florida. This was the lowest reading ever recorded in the western hemisphere. More than 400 persons were killed in the storm, and winds reached from 150 to 200 miles per hour in the Florida Keys.

In 1938 a hurricane killed about 600 persons in southern New England and on Long Island, New York. New England was badly hit again in 1955 when Hurricane *Diane* caused billions of dollars of damage along the Atlantic coast, commencing with North Carolina and spreading northward. Only a year later, Hurricane *Audrey* struck from Texas to Alabama and killed 390 persons.

Hurricane *Donna*, in 1960, became the first recorded storm to strike with hurricane force winds in Florida, the Middle Atlantic states, and the New England states at the same time.

In 1963 Hurricane *Flora* killed about 5,000 persons in Haiti, over 1,000 in Cuba, and left about 750,000 persons homeless.

Earlier in this chapter we reported on Hurricane *Hugo* which slammed across the South Carolina shore wreaking havoc in Charleston and numerous other southern cities on September 23, 1989. By April of 1990 the city had reported not only the obvious physical damage but numerous other emotional scars left in the wake of the storm.(41)

1. During the first three months of 1990, 70 rapes were reported in the Charleston area, an increase of 112% from the prior year.
2. The number of domestic abuse and divorce cases filed in Charleston County was up 17%, an unusually large increase.
3. An estimated 20,000 South Carolinians suffered "Post-Traumatic Stress Syndrome."
4. The number of people who sought help at the Charleston Office of the State Department of Alcohol and Drug Abuse was up 25%.
5. Young children from the Sea Islands and Mount Pleasant, which bore the brunt of the storm's 135 miles-per-hour winds and 15-foot tidal surge, suffered a variety of emotional symptoms.

It can be seen that a hurricane—and for that matter any primary storm—causes physical and emotional damage on the community.

Hurricane Control

For the permanent inhabitants of the Gulf and east coasts of the United States, there is so far no practical way to avoid a hurricane. Public education and research seem to offer the most useful way to counter this threat.

During the last several decades, experiments have been conducted to find ways to modify or neutralize hurricanes to reduce their destructive forces. Since 1962, experimental work under Project Stormfury has been carried out every year by the National Oceanic and Atmospheric Administration. Seven flights into three Atlantic hurricanes were made and confirmed that giant storms have developing cloud towers in the right places to make cloud seeding feasible. Storm seeding is based upon the premise that supercooled water (water cooled below the freezing point but still in liquid form) in a cloud can be induced to freeze by seeding it with silver iodide crystals. As the water freezes, it releases energy, causing the cloud to grow. If seeding clouds away from the storm's center would cause them to grow at the expense of the clouds around the eye, then an intense storm would tend to convert into a moderate one.(42)

Research to improve techniques of forecasting has been conducted in an attempt to improve their accuracy. Other programs were conducted to improve the accuracy and efficiency of the observing and communications network.

Hazard Reduction—Hurricanes

The Office of Emergency Preparedness, in its report to Congress, pointed out that certain safety rules can be followed to reduce peril to human life. Anyone living in areas frequently affected by hurricanes should stock up with emergency supplies at the beginning of each hurricane season. If a hurricane is expected, windows should be boarded or protected with storm shutters, outdoor paraphernalia should be secured or placed inside, automobiles should be fueled in readiness for swift evacuation, and the progress and position of the storm should be monitored. People should stay inside until notified that the hurricane has passed and it is safe to go outdoors. They should not rush outside when the wind stops, for it may only be the eye of the storm passing, and they might get caught out in the open when the other side of the storm strikes.(43)

Fire Department Preparation

When the warning comes, review the control measures that have been adopted in your "Hurricane Bill" annex to your disaster plan. The general precautions which fire departments should make do not differ materially from those outlined under the subjects of "Floods," "Windstorms," and "Tornadoes."

Emergency Action–Hurricanes

The handicaps under which fire departments will have to operate following a hurricane are about the same as those encountered in a flood, windstorm, and tornado–all at once! The added danger of a storm surge in coastal communities could make shore highways impassable even before they become littered with fallen trees, demolished structures, and other debris. But on the whole, the search and rescue activities, fire and dangerous chemicals control procedures, as well as other emergency actions will not differ substantially from those discussed for other storms.

WINTER STORMS

Winter storms include blizzards, ice storms, and snow storms. These vary in intensity and may affect only part of one state or many states. These storms are generated from disturbances along the boundary between cold polar and warm tropical air masses; fronts are formed where air masses of different temperatures and densities conflict. The disturbances may become intense low pressure systems, churning over tens of thousands of square miles in a great counterclockwise sweep.

Effects of Winter Storms

Winter storms can disrupt water, power, communications, transportation of people and the necessities of life, and cause the deaths of many persons and livestock. Buildings can collapse under heavy snow loads, and entire cities can become paralyzed under blizzard conditions, causing the cessation of all commerce. Everything may reach a standstill, including the emergency services called upon to suppress fires, provide ambulance service, apprehend criminals, release victims from stalled elevators, deliver babies, and handle all the other routine emergencies, plus new ones created by the storm.

Examples of Winter Storm Disasters

Blizzards are possibly the most dramatic and perilous of all winter storms, characterized by low temperatures and by strong winds bearing large amounts of snow. Most of the snow in a blizzard is so fine and powdery, and in such great quantities, that it can greatly restrict visibility as it is whipped around by the wind.

In January of 1991, a series of devastating winter storms dipped south into the heart of California, paralyzed New England, and affected states in between from coast to coast. When the citrus crop in California was destroyed, the ensuing high prices for produce affected all Americans, and demonstrated the far-reaching secondary, or indirect, costs of a weather-related disaster.(44) In 1977, blizzards crippled Buffalo, New York, and other northeastern cities, and brought record snowfall, and record-low temperatures to most of the northeast and Midwest.(45)

When a blizzard raked the West and Midwest, commencing on Good Friday, 1975, Chicago's O'Hare International Airport was closed for twenty-four hours, all its schools were closed, persons ate and slept in parked planes, restaurants, banquet halls, and department stores. In the Loop, hundreds of persons stayed overnight in their offices and many stranded travelers sought refuge in churches. On a 13-mile stretch of the Kennedy expressway there were 350 abandoned cars and trucks, stuck in drifts, out of gasoline, or otherwise disabled. Most of the 31 storm fatalities were heart attack victims stricken while trudging through deep snow, trying to free stuck vehicles, or shoveling walks. As the storm moved eastward it snapped power lines and tree limbs in parts of western Pennsylvania and spread freezing rain and snow across a wide area of New York. At the same time, heavy snow and high winds caused blizzard conditions in much of northern New England.(46)

On February 9, 1969, a severe snowstorm, covering the entire Eastern Seaboard from New England to Virginia, virtually paralyzed the City of New York. The storm dropped 15 inches of snow on the city and 20 inches on the suburbs and was responsible for at least 43 deaths. More than 50 fatalities were attributed to the snowstorm elsewhere in New York State, at least 64 in New England, and 13 in New Jersey, making it one of the deadliest storms of the century.(47)

New England had barely recovered from this storm, when it was hit by another on February 24 to 27 that lasted 90 hours and left behind a record 25 inches of snow in Boston and 60 inches in many sections of Maine. At least 40 deaths in the area were attributed to this second storm.(48)

A third northeaster followed the preceding two, in March of that year (1969), producing a record snowfall of several feet; drifts of 15 to 20 feet were not uncommon on many roofs, and millions of dollars of collapse damage occurred. "Much of the damage could have been avoided had plant management organized squads of men operating in shifts to manually remove the snow from critical roof areas."(49)

Winter Storms—Predictability

Weather forecasters can usually tell when a blizzard is coming and warn people what to expect. But occasionally, blizzards may shift direction unexpectedly and strike a city that is unprepared. "Blizzard Warnings" are issued when winds with speeds of at least 35 miles per hour are accompanied by considerable falling or blowing snow and temperatures of 20 degrees Fahrenheit or lower are expected to prevail for an extended period of time. "Severe Blizzard" warnings are issued when blizzards of extreme proportions are expected and indicate wind with speeds of at least 45 miles per hour plus a great density of falling or blowing snow and a temperature of 10 degrees Fahrenheit or lower.

Other warnings include a "cold wave warning" (for a rapid fall in temperature within twenty-four hours), "Hazardous Driving (Travelers) Warnings" (indicating falling, blowing, or drifting snow, freezing rain or drizzle, sleet or strong winds), and "Stockmen's Warnings" (to alert ranchers to protect livestock from snow, ice, wind, etc.). The National Weather Service also provides frost warnings to farmers in areas where grapes, citrus fruits, and other crops might be damaged by an unexpected cold wave. In certain areas the Agricultural Weather Service and Fruit-Frost Weather Service effectively meet the needs of agricultural interests to cope with the annual threat of frosts and freezes.

Emergency Preparation—Winter Storms

A heavy snow storm or blizzard, with the accompanying cold temperatures made worse by the wind chill factor, can give fire fighters problems, causing delay in response and adversely affecting fire fighting operations.(50)

Even rather routine single-occupancy fires can often become so taxing because of cold temperatures that numerous additional resources are needed. For example, on December 4, 1990, the Ostego Lake, Michigan, Fire Department needed mutual aid from five neighboring departments to help combat a blaze in a tire store during an early winter freeze.(51)

The amount of preparation that may be advisable depends upon where you are located and the kind of foul weather conditions that can be anticipated. When the fire department of London, Ontario, Canada, was confronted with a blizzard on a January morning, and visibility was reduced to 50 feet, the fire chief asked his communications division to advise all fire stations to accommodate any people who might be stranded. That night it had to fight a $100,000 fire under these severe storm conditions, and the next day the visibility was zero and continued that way for 36 hours. The snowfall was 20 to 25 inches, with heavy winds piling drifts up to 5 feet. As the blizzard got worse, the chill factor dropped to 30 below zero, and off-duty fire fighters reported for duty in four-wheel-drive vehicles, some with plows, and others in snowmobiles and sleighs.

The storm did not abate, and it was necessary to solicit assistance over three radio stations to obtain volunteers with snowmobiles and four-wheel-drive vehicles to man each fire station. The response was excellent. The next morning the mayor declared a limited state of emergency which brought in military personnel with tracked personnel carrier vehicles that could be equipped to fight fires and traverse difficult areas at speeds up to 40 miles per hour. The police department obtained radio-equipped four-wheel-drive jeeps from the Canadian Forces, to be manned with police and Canadian Forces personnel, and put one at each fire station. After the third day, the storm seemed to let up, but it returned again in a few hours, and the police returned with their jeeps to the fire stations. Four days after the storm started, it finally left for good, but during that interval response had to be made to 77 alarms, and the emergencies were successfully handled.(52)

Use of Weather Advisories

It might be useful to make special arrangements between the fire department and the local weather bureau office to determine as quickly and accurately as possible in advance what the storm might bring, how much snow is expected to fall and how low a temperature may be expected, the duration, and whether it will be accompanied by sleet or ice. An ice storm can put heavy loads on telephone and signal wires, coat antennas and roads, and cause tree limbs to break off and block the streets.

Activate the Emergency Operations Center

Where the forecast is for heavy snow, blizzard, ice storm, or such severe cold weather that the declaration of a state of emergency is warranted, the community disaster control center will be activated. The fire department will want to activate its emergency operations center and put its winter storm bill annex of its disaster plan into effect. Actions might include the following.

- ○ Hold over or recall fire fighters to assist in carrying out precautionary measures, such as attaching stakes to fire hydrants that have not already been designated by markers high on poles or trees, to be able to find them when covered by snow.
- ○ Test hydrant-thawing devices and place them on fire apparatus, ready for use.
- ○ Mount snow tires or chains on fire department vehicles.
- ○ Patrol schools, public buildings, and business and residential districts, for possible outbreaks of fire, or roofs dangerously overloaded with snow.
- ○ Uncover hydrants deeply buried by snow plows.
- ○ Study the emergency snow routes established under the city's disaster plan and augment response assignments where necessary in anticipation of blocked streets and other unusual conditions that may cause delay.
- ○ Have fire fighters, paid or volunteer, who are likely to have difficulty reporting to duty in the storm, come in early and stand by until the emergency is over. Extra cots and sleeping bags can be kept for extra personnel in fire stations. Extra manpower can augment strength of in-service companies and man reserve apparatus.
- ○ Get an extra supply of provisions for feeding members during the emergency.
- ○ Check fire apparatus for peak conditions of readiness; batteries lose substantial power at low temperatures, so keep them fully charged. Condensation can affect air brake systems, so have condensation drained at the start of cold weather and check periodically. Tire pressures decrease in cold temperatures, so check tire pressures in extreme cold weather.
- ○ Obtain extra four-wheel-drive vehicles and snowmobiles to supplement regular fire apparatus and to carry fire equipment to emergencies not accessible by standard apparatus.
- ○ Ask the news media—television, radio, and the press—to request cooperation of the public, such as not to obstruct fire lanes on the emergency snow routes; not to abandon cars in main thoroughfares

if they can be removed to other locations; to remove snowdrifts that block emergency exits from places of public assembly; to uncover fire hydrants in front of their places of business to speed an attack on any fire that may break out in their buildings; to be especially careful of fire in their homes, offices, and places of business.(53)

FLOODS

Dating to the earliest settlers of the North American Continent, people in the United States and Canada have relied on natural water supplied by rivers. Logically populations grew in close proximity to rivers. These rivers which are the lifeblood of developed population centers can, however, turn deadly in time of flood.

In early 1991 flood waters threatened to spill over an area stretching from northeast Texas into the Great Lakes regions of southern Michigan and western New York. In the center of this was Indiana, Ohio, and Kentucky, which bore the brunt of flood waters reaching the highest levels in more than a decade. In Indiana alone, more than 3,000 residents were forced from their homes. In Cincinnati, the north-south runway at Lunken Airport was closed because of high water, forcing air traffic controllers to work with only two

A flood in Greenport, New York, in 1982 threatened to overtake this operating pumper. (Photo courtesy of Ron Buckingham.)

runways. The highest water in more than a decade left a devastating mark on the Ohio River valley that will long be remembered.(54)

In May 1990, floods ravaged Texas and parts of Oklahoma when torrential rains hit the area. Three people were killed in Brownwood, Texas, as a result of the flooding. The waters rose over 13 feet overnight and completely destroyed the main business district. Eight people were injured when a tornado ripped through Paris, Texas, in northeast Texas, demolishing a nightclub, flipping vehicles, uprooting trees, and tearing roofs. Flooding was also widespread in Oklahoma and Arkansas. In the wettest spring in Texas since 1922, flood waters closed highways and sent lakes surging over dams in the Dallas-Ft. Worth area. The west fork of the Trinity River, which flows through Ft. Worth and Dallas was six feet over flood stage and many people were rescued by boat or helicopter.(55)

Floods may originate with hurricanes, dam breaks, lengthy rainfall, or other persistent precipitation causing overflow of rivers, lakes, and seas onto the land, but this is not the case with flash floods; to have a flash flood all that is required is a sudden heavy rain and a gully to channel it somewhere. If that gully leads to a parking lot as it did in Las Vegas on July 3, 1975, the result can be destruction of 300 cars, dozens of homes inundated, and two people killed in less than an hour.

For action in flood waters, fire department personnel must be prepared for a variety of tasks, but one of the first acts of the fire officer should be to verify which members of the fire company can swim, which ones can row or operate a power boat, and what equipment will be needed to carry out rescues, fire control, and other operations while flood waters move dangerously and inexorably through the area. In these situations, handportable two-way radios are essential; so are rope, protective clothing, ground ladders, pike poles, and lights.

The fire officers must learn quickly where the flood currents are most dangerous and how to prepare for work in these areas. Rescue personnel must be ready to work as teams in case people are swept into the flood waters and are in danger of drowning. It will be very important for the fire department to work efficiently with all other agencies assigned to the flood area.

Finally, in the aftermath, the fire department probably will be using portable pumps to move water from submerged homes, or portable generators to provide standby electrical power to important buildings like hospitals and fire stations or halls where flood victims may be housed temporarily.

Floods—Where and When

Virtually every state in the country has areas that are subject to disastrous flooding conditions. Flooding occurs when the soil and vegetation are unable to absorb the falling rain or melting snow, or when the runoff exceeds the capacity of the normally adequate control channels.

Communities that are particularly vulnerable to flooding emergencies are those located in the following places:

- ○ in low areas in proximity to rivers or flood control channels
- ○ in the watershed of barren mountains that lack vegetation due to environment or fire
- ○ in the path of water flow from a dam or reservoir failure
- ○ in low coastal areas subject to storm surges

In the past several years flooding has occurred around Lake Michigan and at the Great Salt Lake, primarily because of years of uncontrollable rise in the water level of the lakes.

The season for flood threat varies according to locale throughout the country. The peak for flood threat on the West Coast and Pacific northwest occurs during winter and early spring. The Rocky Mountain area peaks in late spring. The Mississippi and Ohio River valleys and the southern states peak in the winter and early spring. The Great Lakes region and New England states peak in early springtime, while Florida is subjected to flooding in the fall.

Earthquake-related flood threats are not associated with weather or season. The potential collapse of dams or reservoirs may pose a severe flash flood threat that needs to be addressed. Earthquakes on the ocean floor may generate gigantic ocean waves called tsunamis. The major threat from tsunamis in the United States lies in the Hawaiian Islands and on the west coast, including Alaska.

The major floods and the resulting life loss of the past century in the United States are depicted in Table 9-3.

Table 9-3. Major United States Floods: 1889 to 1991.

Date		Location	Deaths
1889	May 31	Johnstown, PA	2,200
1900	Sept. 8	Galveston, TX	5,000
1903	June 15	Heppner, OR	325
1913	March 25–27	Ohio, IN	732
1915	August 17	Galveston, TX	275
1928	March 13	collapse of St. Francis dam, Santa Paula, CA	450
1928	Sept. 13	Lake Okeechobee, FL	2,000
1937	Jan. 22	Ohio, Mississippi river valleys	250
1972	Feb. 26	Buffalo Creek, WV	118
1972	June 9	Rapid City, SD	236
1976	June 5	Teton dam collapse, ID	11
1976	July 31	Big Thompson Canyon, CO	130
1977	July 19–20	Johnstown, PA	68
1980	Feb. 13–22	Southern CA, AZ	26
1982	June 6	Southern CT	12
1982	Dec. 2–9	IL., MO, AK	22
1983	Feb.–March	Calif. coast	13
1983	April 6–12	AL, LA, MS, TN	15
1984	May 27	Tulsa, OK	13
1990	May 5–10	TX, AK, & OK	Est. 20
1991	Jan 1–20	OH., ID, IL, LA	Est. 16

Nature of Floods

There are generally two types of flooding, the slow-rise flood and the flash flood. The onset of a slow-rise flood, as the name implies, is a relatively slow process with several days of buildup. Long periods of heavy rainfall or melting snow combined with rain is normally the cause for rivers to overflow their banks. Flood warnings in such cases can be issued many hours or even days in advance of the flood crest.

Flash floods normally occur after torrential rains of short duration, causing violent flooding of a river, stream bed or even a normally dry wash or gully. Heavy rainfall running off a mountain range can be the perpetrator as in the Las Vegas situation previously mentioned. The breakup of an ice jam also presents the threat of flash flooding.

Causes of Floods

Heavy rainfall causes most floods, but sometimes other factors are involved. Rain that normally soaks into a mountainside can cause a flood if timber and grass have been destroyed by fire. Rain can loosen rocks to cause a landslide that obstructs a river and causes a flood. A cloudburst can cause flash floods and fill a dam so fast that it breaks and floods the countryside. Tidal waves and high winds can cause seacoast floods. Even an earthquake can cause a flood. An earthquake caused the Mississippi River to flow upstream in 1811 and 1812 and created Reelfoot Lake in Tennessee.

As an example of a landslide that obstructed a stream, on July 22, 1970, falling rocks diverted the course of the Alaknanda River in India, causing a sudden flood 200 miles northeast of New Delhi; the rushing waters carried away buses and other vehicles taking pilgrims home from Badrinath, a holy city, killing about 600 persons.

As an example of a flood caused by too much rain, about two months after the Alaknanda River flood, another 600 persons died and 20,000 were made homeless in India when heavy rains caused flooding from Bombay to Calcutta.(56)

Ten inches of rainfall in two days sent South Dakota's Rapid Creek over its banks and collapsed a dam. The flood that swept through the city killed 237 persons and caused an estimated $120 million in property damage.(57)

A dam can break to produce a flood even if there has been no rainfall for months. In the Baldwin Hills area of Los Angeles on the afternoon of December 14, 1963, a small crack in the dam developed into a large break, allowing 300 million gallons of water to gush forth and wipe out a large residential area.

A dam need not be of the earth-filled type nor as large as the one in Baldwin Hills to cause havoc upon release of its water. The coal-slag dam that gave way in February 1972 held only about 130 million gallons of water, but by the time the Buffalo Creek flood subsided 125 West Virginians were dead and 4,000 others were homeless.(58)

In the most devastating disaster in more than a decade, 200,000 people died in Bangladesh in early May 1991 as a result of a typhoon which carried 20-foot waves and 145 mile-per-hour winds ashore leaving behind acres of flooded countryside. This was just one of a series of disasters that have continued to plague the 110 million people in this densely populated country.(59)

Another cause of floods is the denuding of a water shed by fire. The western states were especially affected in 1991 after the extensive series of wildland fires previously reported left large tracks of land without forestation. A series of floods and mud slides killed over 100 persons and made about 10,000 homeless.

Causes of Seacoast Floods

Floods from the sea usually are caused by high winds that pile water up against the harbor and drive the waves onto the land. In some countries, such as the Netherlands, dikes used to wall off the sea generally have been successful in preventing this type of flood. However, after the great flood in Galveston, Texas, the city built a sea wall, which but the storms of 1913 went over.(60, 61)

Tidal Wave Floods

A tidal wave is an ocean wave or series of waves of great height. They have no connection with true tides, but are responsible for seacoast floods. These gigantic waves are referred to as "tidal waves" when they are caused by submarine earthquakes (seaquakes), submarine volcanic action, underwater atomic bomb burst, or by a hurricane.

The scientific name for such a wave caused by a seaquake is *tsunamis*, the Japanese word for storm wave. When at sea, the wave is about 100 to 400 miles from crest to crest, and it does not pile up until it hits shallow water. Tsunamis occur mainly in the Pacific Ocean along the west coast of Central America, northwest coast of South America, the Hawaiian Islands, and the east coast of China and Japan.

As a result of the Good Friday, 1964, earthquake in Alaska with the epicenter in Prince William Sound, seismic ocean waves were created. In Valdez (50 miles distant), the harbor was emptied of water three times with ensuing waves up to 50 feet high refilling it. Similar activity occurred in Anchorage, Seward, and Kodiac with 40-foot waves. The impact of the wave action caused damage as far away as Crescent City, California and the Hawaiian Islands.

Tidal waves are somewhat predictable. In the areas where the contour of the ocean floor is known to be favorable, and within the rather narrow belts of the ocean where they have previously occurred, a large underseas earthquake, recorded on a seismograph ashore, may give rise to the expectancy of a tidal wave in five to thirty minutes, especially if the tremor is followed by an abnormally low tide. However, such waves are not distinguishable from others when viewed from afar, and since they move at a speed of between 300 to 500 miles per hour, they do not give much warning.

An example of a tidal wave flood caused by a volcano occurred following the eruption of Krakatoa, an Indonesian volcano that "blew its stack" in

August 1883, causing one of the world's worst disasters. Much of the island was blown to bits, and the resulting tidal wave that washed over the nearby islands killed about 36,000 persons. One ship, carried a mile and a half inland, was left 39 feet above the sea.

A tidal wave caused by a hurricane that battered the island of Galveston, Texas, for 18 hours in September 1900, produced the worst natural disaster ever recorded in the North American continent. Over 6,000 of the city's 37,000 inhabitants were killed, and nearly half the residences—3,000 homes—were demolished, leaving the site looking like a lumberyard strewn with furniture.(62)

Tidal waves caused by seaquakes create great destruction. One occurred in the Sanriku area of Japan in 1896; 27,122 people lost their lives in about 30 seconds when a wall of water, 75 to 100 feet high, rolled over their festival celebration.

Flash Floods

A sudden, violent flooding of a river, stream bed, gully, or wash, called a "flash flood," is touched off by torrential rains filling natural and man-made drainage systems to overflow with raging waters, uprooted trees, smashed structures, boulders, mud, and debris.

In the summer of 1990, the small town of Shady Side, Ohio, witnessed utter devastation when the small normally peaceful creek that flows alongside this city suddenly became a raging river sweeping away most of the homes along its river bank. Dozens were drowned or missing, and the small Shady Side volunteer fire department could do little to help, even though it was well trained in handling water emergencies.(63)

On July 31, 1976, a wall of water surged through a Rocky Mountain river canyon, killing more than 100 campers, hikers, picnickers, trout fishermen, and residents living along the Big Thompson River, following a weekend storm that dumped 10 inches of rain on the area in a matter of hours. Around 8 P.M. the sheriff, his deputies, and Colorado highway patrolmen jumped into their cars and raced through the area to warn of highwaters and order an evacuation, but the people would not listen. Three times the sheriff's men toured a crowded trailer camp and it was only after the third time that people began to hook up their trailers, but it was too late. Helicopters rescued 925 persons from the canyon in two days, but more than 800 were missing. President Ford declared the 2,700-square-mile Larimer County a major disaster area, making residents eligible for federal funds to rebuild.(64)

Flash flood warnings are transmitted via radio and television. The National Weather Service Hydrologic Forecasts and Warnings uses the following terminology.

- Flash flood: the occurrence of a dangerous rise in water level of a stream or over a land area in a few hours or less caused by heavy rain, ice jam breakup, earthquake, or dam failure.
- Flash flood watch: heavy rains occurring or expected to occur may soon cause flash flooding in certain areas. Citizens should be alert

to the possibility of a flood emergency that will require immediate action.

- Flash flood warning: flash flooding is occurring or imminent on certain streams or designated areas and immediate precautions should be taken by those threatened.

Tsunami warnings can be issued after an earthquake has occurred if sufficient lead time exists. Speed of the surge can be calculated as well as its direction of travel, thus giving preparation time to those coastal areas that may be threatened.

Effects of Floods on the Community

Annual flooding of marshes, deltas, and tidelands are sometimes beneficial to farmers and wildlife, but flash flooding situations can create utter devastation. A wall of water 10 to 20 feet high carrying massive boulders, trees, and large structural debris is virtually unstoppable in its destructive attack on a community. In addition to destroying houses and other structures, the raging torrents usually damage or demolish roadways, bridges, and any utilities within the path. The possibility of loss of life from drowning is always present with the swift currents.

Damage resulting from slow rise flooding is generally caused, not only by the inundation of water, but also by silt and debris passing through the affected community. In the United States, in an average year, flooding will force 75,000 people from their homes, about 80 people will be killed, and property loss will be about $1 billion.

Unexpected floods can cause drownings, render people homeless, wash away buildings, erode land from one place and deposit it elsewhere, pollute drinking water, spread disease, heave underground flammable liquid storage tanks and cause them to spill their contents and float away, and overturn above-ground flammable liquid and flammable gas storage tanks, causing fires. Flood waters may inundate storehouses containing water-reactive chemicals, causing violent reactions, explosions, and release of toxic gases.

"Flooding creates an atmosphere favorable to fire, either by causing it or by raising problems in efforts to fight it. The flood itself damages goods, equipment, buildings, fire protection equipment and machinery either through wetting or by foreign material plugging and damaging moving parts and by floating debris slamming into objects. Damage brought about by water, mud, silt, and debris is serious enough, but devastation that could be brought about by fire, when the property is unapproachable or normal protection has otherwise been interfered with, can be far worse."(65)

Even though we think of flooding as primarily a water problem, there are often compounding problems with which the fire department must deal. In early March 1991, the two balmy states, California and Florida, each experienced heavy thunderstorms and flooding. In Florida, heavy thunderstorms raged across the southern and central portions spawning tornadoes and causing structural damage and power outages. In northern Florida, high waters from two days of rain flooded homes and roads and forced evacu-

Fire fighters in Montgomery County, Ohio, work to save the Sinclair Community College Library during flood conditions. (Photo courtesy of the Dayton, Ohio, Fire Department.)

ations. One tornado there caused one million dollars in damage to houses in Paradise Shores.(66)

Industrial fires were caused by flash floods resulting from Hurricane *Diane* when it moved northward into the New England states about two decades ago.(67) One fire broke out in a flooded magnesium powder plant in Putnam, Connecticut, and burning drums floated from the building, exploded, and set fires in buildings downstream.

Another serious effect of floods is the restriction of activities of fire fighters through inaccessibility. This was demonstrated in a warehouse fire in Dubuque, Iowa, where the fire department had to move ladders, hose, and personnel by Coast Guard boats to the fire scene, hook up to submerged hydrants, effect rescues, and after removing sand bags and dikes to get inside, extinguished the fire.(68)

The effects upon fire department response were also demonstrated when a tropical storm put much of Boundbrook, New Jersey, under flood waters; when volunteer fire fighters were summoned to help fellow residents they found that they were among the victims. Many were trapped in their homes. Two of their four pumpers were under water and one stalled in water while leaving quarters. Fire apparatus could get only within one block of a residential fire because of the flooded streets. Flood waters reached almost six feet in depth on one street before they started to recede. Police used power boats to patrol the town to prevent looting. When the water receded a large military wrecker was used to retrieve the submerged pumpers. Mutual aid companies assisted in pumping out hundreds of cellars, unclogging storm drains, washing down streets and warning the public of the contaminated water supply.(69)

In addition to the loss of heat and power during flood, the ripping away of bridges to which telephone cables are attached can also disrupt communications. This happened at Corning, New York, during the 1972 flood; dikes 25 feet above normal water level of the Chemung River were not high enough to prevent water from flooding the computer room of the Corning Glass Works and within an hour it was only 12 inches from the ceiling. Tapes, disc packs, and everything were watersoaked with a slick covering of mud and silt residue. Water silt infiltrated safes in another building where back-up tapes were stores for remote security. The communications center where all computer input was received was inundated and all building services, including power, were out. The entire Corning area was without power, heat, water, and telephone and was under National Guard security, with surface travel limited to priority needs.

Flood Prevention and Precautions

A Capability Assessment should determine what resources are available to deal with the flooding problems and what additional resources are needed. A Multi-Year Development Plan should be implemented to overcome capability shortfalls.

A fire department's flood emergency planning should consider those special operational concepts (over and above standard operating procedures), that will provide effective command and control during a large-scale flood emergency.(70)

Measures that help prevent floods include dams, levees, dikes, storm drains, soil erosion programs, and forest fire prevention. The U.S. Department of Agriculture has found that barren land lost 30 percent of its rainfall and 65 tons of soil per acre each year. But nearby forest land lost only about 1 percent of its rainfall and four pounds of soil per acre each year. Tree planting is important in modern flood control. The Soil Conservation Service warns that size of floods is affected by the way in which farmers plow their land, the kinds of crops they plant, and how they control drainage of their land.(71)

Other precautions include firmly securing or anchoring buried tanks, above ground tanks, and other objects that might float away; placing valuable property on upper floors of buildings; and checking to make sure that building foundations and retaining walls will withstand flood conditions.

Prepare a flood control and evacuation plan. Select routes of response to areas in your community by roads that will avoid low spots and paths of torrents. Your plan could provide that when off-duty members have made their own homes secure, they could report for duty to standby while the on-duty members make their homes secure.

Flood Predictability and Warnings

Except for flash floods, the onset of most floods is a relatively slow process, and often with adequate warning. Buildup usually takes several days, and progressive reports are available from NOAA (National Oceanic and Atmospheric Administration), through its Weather Service River Forecast Centers and River District Offices. Warnings are given by NOAA when rainfall is enough to cause rivers to overflow banks and when melting snow may combine with rainfall to produce similar effects.

Flash flood warnings are the most urgent type. They are transmitted to the public over radio and television, and by other signals, such as sirens, horns, whistles, telephone alerts, or by means of emergency service personnel using loudspeakers on their vehicles or helicopters.

Forecasts of impending floods reach people in the form of warning messages about the severity of flooding (minor, moderate, or major), the affected river, and when and where flooding is expected to begin. Area radio and television stations usually broadcast the latest flood information and warnings. It is advisable that local government give more specific advice and instructions over local stations, in accordance with your disaster plan. The Defense Civil Preparedness Agency has issued instructions that might be included in your community plan. These advise the actions to take before the flood, upon receiving the warning, during the flood, and after the flood.(72)

If your community has a history of recurring floods, continuing communication should be established with the National Weather Service, so that flood forecasts and warnings can be transmitted to local emergency services headquarters at stipulated periods. "During the flood-control planning phase, communities should coordinate closely with the nearest office of the Army Corps of Engineers, in accordance with established state procedures. Also, the Corps of Engineers can provide preventive assistance for flood control as authorized under Public Law 99."(73)

A flood that occurs because of a dam break is not as predictable as one that results from overflowing rivers. In the absence of heavy rains that might throw an added burden on a dam or an earthquake which might shake its foundation, it is sometimes possible to observe trouble signs, such as leaks around an earth-filled dam on volcanic terrain riddled with caverns and crevices. These signs might cause safety officials to anticipate possible weakening and break. This was believed to be the situation in the case of the Teton Dam collapse in southeastern Idaho on June 5, 1976, where eleven people were killed, 4,000 homes wrecked or damaged, and 100,000 acres of rich farmland stripped of topsoil by the torrent of unchecked water that swept the dam away.(74)

Emergency Action—Floods

When a flood threatens, the fire department should be prepared to assume its designated role under your community's flood annex to its disaster plan. When the civil defense control center is activated, and the situation calls for a twenty-four hour flood watch, the fire department should activate its emergency operations center and carry out planned measures for maintaining operational readiness under flood conditions. Check emergency power supplies, supplemental communications systems, including the use of "ham" operators; obtain emergency fuel and food supplies sufficient to cover your need for the probable duration of inaccessibility of roads serving your fire stations. Stations that are apt to be flooded to a depth that would make use impractical should be rendered as impervious to damage from water as possible, and evacuated. Routes of response should be planned from temporary field headquarters established on high ground.

The depth of water in which fire trucks can safely operate should be determined in advance. Tractors and bulldozers capable of operating in deep water should be stationed where they can be readily available to tow disabled apparatus. Although winches, cables, and towing eyes on fire trucks can drag stalled light vehicles out of flooded thoroughfares, most fire trucks are not geared sufficiently low to provide towing or pushing service to other heavy vehicles without mechanical damage to themselves.

Before water rises in the business district, colored flags on poles can be attached to fire hydrants to mark their locations, as was done in Dubuque, Iowa, and in Cincinnati, Ohio.(75) All boats, portable generators, flood lights, emergency food, fuel, and water supplies for drinking that may be needed during search and rescue operations should be brought to your command post or field headquarters on high ground.

Establish emergency communications procedures to ensure that National Weather Service reports are obtained by your dispatcher and operations officer. Dike patrols and helicopter reconnaissance are useful for observing conditions.

Following a flash flood, the use of hand-held and airborne infrared devices to detect body heat of victims buried under silt and rubble have been found useful. In the Big Thompson River flash flood, described earlier, twelve hand-held and two airborne infrared devices were used to quicken the tediously slow process of recovering bodies buried under tons of debris scattered down the 35-mile length of the canyon.

Portable generators can supply power to gasoline pumps to get fuel out of underground tanks when power supply has been cut off, and they can operate freezer equipment in food establishments, meat markets, and warehouses, to prevent spoilage pending restoration of the normal power.

Using specially equipped vehicles, boats, and helicopters, fire fighters can relay evacuation messages, rescue persons stranded on upper floors and roofs, notify homeless persons where refuge centers are located, and do other tasks for saving life and property.

The fire department can assist in placing salvage covers and vinyl sheeting over earth-filled dikes to prevent erosion and aid in stabilizing spongy dikes. Fire stations are often used as a supply depot for sandbags. In areas likely to suffer flood and mud slides following a large brush fire, the public works department can dump loads of sand in the street near the threatened homes for use in filling sandbags for home protection.

Following recession of the flood waters, fire fighters can restore services and salvage damaged property. The use of siphon ejectors to drain flooded basements is preferable to using pumps on first line fire apparatus, for mud and debris will not have to go through the pump itself, thereby reducing the chance of damaging it. With chain saws, power winches, and other tools, fire fighters can assist public works and street departments in clearing the streets of debris, fallen trees, and other obstructions.

CONCLUSION

In this chapter we have covered the major disasters caused by weather conditions—storms, tornadoes, hurricanes, and floods. While it is weather that provides us with fresh breezes, adequate drinking water, and other necessities of life, it is also weather unrestrained that causes some of our most serious disasters. In this chapter we have noted the devastating effects of the weather-related types of disasters and the almost impossible task facing fire fighters who are the first line of defense against these disasters. We will continue our discussion in the following chapter where we turn our attention away from the weather-related types of natural disasters to these that are "terrain-based."

CHAPTER 9 REFERENCES

1. *Spotter's Guide for Identifying and Reporting Severe Local Storms*, U.S. Department of Commerce, National Oceanic and Atmospheric Administration, 1973, U.S. Government Printing Office, Washington, D.C. 20402, p. 5.
2. Frydenlund, M. "Modern Lightning Protection" in *Fire Journal*, July 1966, pp. 10–12 at p. 11.
3. CNN broadcast, May 10, 1991.
4. Panati, C., and W. Schmidt. "Lightning Alley" in *Newsweek*, October 27, 1975, p. 65.
5. *World Book Encyclopedia*, Chicago, IL: Field Enterprises Educational Corporation, 1969, Vol. 12, pp. 263–265.
6. *World Book Encyclopedia*, Chicago, IL: Field Enterprises Educational Corporation, Vol. 9, p. 8.
7. Goldman, J., and R. Dallas. "Jetliner Crashes in New York Storm; 109 Die" in *Los Angeles Times*, June 25, 1975, Part I, p. 1.
8. *Chief Fire Executive*, September/October, 1988.
9. *Lightning Protection Code*, NFPA No. 78, cited in George H. Tryon, "The Philosophy of Lightning Protection" in *Fire Journal*, November 1967, p. 59.
10. Newman, M., J. Robb, and E. Yonkers. "Aircraft Protection from Thunderstorm Electromagnetic Effects," Conference Paper No. CP 62-1133, American Institute of Electrical Engineers, 1962, p. 4.
11. "Lightning Tamers Used Over *Apollo* Launch Site," *Los Angeles Times*, July 9, 1975, Part I, p. 11.
12. NFPA 1500 was the most prominently recognized safety standard as we went to press in 1991.
13. "Wind and Buildings" in *Factory Mutual Property Conservation Engineering Management Record*, Volume 52, No. 6, November–December 1975, p. 14.
14. The address of Factory Mutual's office where the Loss Prevention Data Sheet 1–7 can be obtained is Factory Mutual Engineering, P. O. Box 688, Norwood, MA 02062.
15. West, R. "Thousands Stranded by Sand Blizzard in Desert," *Los Angeles Times*, June 19, 1975, Part I, p. 1.
16. See note 5, Vol. 4, p. 960.
17. Associated Press Report in *The Cincinnati Enquirer*, April 29, 1991, p. A-3.
18. See note 5, Vol. 20, p. 116.
19. Author William Kramer was a fire fighter on Cincinnati's Engine Co. 3 the day of the "Super Outbreak." With strong wind in his face he remembers an eerie ride in

the aisle of an open-bed pumper 20 miles to the western edge of Cincinnati. A five-alarm response was enroute to the suburb of Saylor Park, where a tornado had just touched down, causing massive destruction. He and 70 other Cincinnati fire fighters worked through the night helping victims and securing damaged homes.

20. *Tornado Preparedness Planning, Skywarn*, U.S. Department of Commerce, National Oceanic and Atmospheric Administration, National Weather Service, 1973, p. 24. U.S. Government Printing Office, Washington, D.C. 20402.
21. Associated Press Report in *The Cincinnati Enquirer*, April 29, 1991, p. A-3.
22. "Death Toll Reaches 24" in *The Cincinnati Enquirer*, August 30, 1990, p. A5.
23. "Scores of Homes Destroyed" in *The Cincinnati Enquirer*, June 4, 1990, p. A1.
24. CNN broadcast November, 1989.
25. *Time*, November 27, 1989.
26. Troeger, J. "Anatomy of a Major Disaster" in *Fire Command!* May 1975, pp. 18–20. "Tornado Hits Xenia, Ohio, Leaves 33 Dead, 1,000 Hurt" in *Fire Engineering*, November 1974, pp. 30–32.
27. Villanova, R. "Radio-Controlled Mutual Aid Speeds Tornado Rescue Work" in *Fire Engineering*, November 1974, pp. 26–27.
28. CNN broadcast April 27, 1991.
29. *Tornado Preparedness Planning, Skywarn* (see No. 20).
30. Hurricane *Gilbert* ravaged the Caribbean, September 10–17, 1988.
31. Hurricane *Hugo* struck the southeast coast of the United States wreaking havoc, especially in the Carolinas, September 21–22, 1989.
32. *Hurricane, The Greatest Storm on Earth*, U.S. Department of Commerce, National Oceanic and Atmospheric Administration, 1971, p. 3., U.S Government Printing Office, Washington, D.C. 20402.
33. Ibid.
34. "The Hurricane—Its Effects" in *The Sentinel*, Factory Insurance Association, July–August 1971, p. 7.
35. *Disaster Preparedness*, Report to Congress, Office of Emergency Preparedness, January 1972, Vol. I, p. 47, U.S Government Printing Office, Washington, D.C. 20402.
36. *World Almanac and Book of Facts*, 1988, "Hurricanes."
37. See note 31.
38. See Note 5, *Year Book*, 1990, p. 314.
39. This booklet, *Agnes Was No Lady*, which gives no publisher's name or date, was obtained by author Chuck Bahme from a rack outside the office of the U.S. Army Corps of Engineers in Chicago's Federal Building, where copies were available to the public without charge.
40. See Note 5, *Year Book*, 1975, p. 300.
41. Associated Press Report in *The Cincinnati Enquirer*, April 17, 1990.
42. "Pacific May Get Storm-Taming Project" in *Los Angeles Times*, July 6, 1976, Part IA, p. 7.
43. Bahme, Charles W. *Fire Officer's Guide to Emergency Action*, NFPA, 1976.
44. *Winter Storms*, U.S. Department of Commerce, National Oceanic and Atmospheric Administration, NOHA/PL 70018, 1990.
45. Ibid.
46. "31 Die as Storm Rakes Chicago Area," *Los Angeles Times*, April 4, 1975, Part I. p. 17.
47. See Note 5, *Year Book*, 1990, p. 316.
48. See Note 5, *Year Book*, 1990, p. 315.
49. "Let it Snow, Let it Snow" in *Record*, Property Conservation Engineering and Management, Factory Mutual System, Vol. 48, No. 6, November–December 1971, p. 5.
50. "Wind Chill" is an estimate of how cold the wind makes a person feel in cold weather, e.g., if the temperature is 10 degrees F. and the wind is blowing at a speed

of 10 mph, the wind chill temperature is equal to minus 9 degrees F.; with a 20 mph wind it would be minus 25.

51. "Hot Shots" in *Firehouse*, Vol. 16, No. 2 (February 1991) p. 71.
52. Wills, O. "Snow Emergency in Canada" in *Fire Engineering*, p. 33.
53. Bahme, Charles W. *Fire Officer's Guide to Emergency Action*, Boston, MA: NFPA, 1976.
54. *Cincinnati Post*, Wednesday, January 2, 1991, p. 1.
55. CNN broadcast May 15, 1991.
56. *The World Book Year Book*, 1971, Chicago, IL: Field Enterprises Educational Corporation, p. 296.
57. *The World Book Year Book*, 1973, Chicago, IL: Field Enterprises Educational Corporation, p. 307.
58. Keerdoja, E., and E. Shammon. "After the Flood," in *Newsweek*, July 12, 1976, p. 7.
59. *Washington Post*, May 4, 1991.
60. *The World Book Year Book*, 1970, Chicago, IL: Field Enterprises Educational Corporation, p. 317.
61. *World Book Encyclopedia*, Volume 7, 1969, p. 210.
62. Rhodes, L. M. "The Great Galveston Hurricane of 1900" in *American Legion Magazine*, September 1968, pp. 12–16, at p. 15, 46–49.
63. "Shadyside Floods" in *American Heat* video journal, St. Louis, MO, September 1990.
64. Chu, D., and J. Huck. "River on a Rampage," in *Newsweek*, August 16, 1976, p. 30.
65. "Flood Damage Prevention," in A.I.A. Special Interest Bulletin No. 323, October 1970.
66. *USA Today*, March 4, 1991, p. 6A.
67. Babcock, C. I. "Flood and Fire," in *NFPA Quarterly*, October 1955, pp. 153–157.
68. Kemps, J. "Flood and Fire," in *Firemen*, August 1965, p. 6.
69. Newhall, R. "N.J. Volunteers Win Fight Against Flood Waters" in *Fire Engineering*, July 1972, p. 37.
70. "Disaster and Fire Defense Planning," open learning course, National Fire Academy, Emmitsburg, MD, 1988.
71. See note 61, p. 22.
72. Disaster Operations, Defense Civil Preparedness Agency, CPG 1-6, July 1972, pp. 41–47. U.S. Government Printing Office, 1973 0-521-819, Washington, D.C.
73. Disaster Planning Guide for Business and Industry, Defense Civil Preparedness Agency, CPG 2-5, May 1974, p. 2. U.S. Government Printing Office, 1974, pp. 525–539.
74. "A Bureau Covers Its Eyes," editorial, in *Los Angeles Times*, July 20, 1976, Part II, p. 1. Based upon an analysis by Times Staff Writers Gaylord Shaw and John Kendall.
75. Kuhn, Dan, William Brewster, and Bernard Klaene. "Water Emergencies" in *Procedures Manual*, Section 203.17, Cincinnati Fire Division, 1992.

CHAPTER 10

TERRAIN-BASED NATURAL DISASTERS

NATURAL DISASTERS TAKE many forms. While we examined weather-based natural disasters in the previous chapter, we turn now to natural disasters that occur in or on land directly—"terrain-based" natural disasters. Specifically, we will identify disaster planning factors in earthquakes, volcanoes, and landslides.

EARTHQUAKES

The prevalence of earthquakes was illustrated in early 1991 when on January 3 at least two dozen quakes shook Carson City, the state capital of Nevada. These earthquakes pointed out how quakes can be of long duration and yet cause minimal damage. The series of quakes began just after 4 A.M. and continued for hours but the largest quake registered only 4.5 on the Richter scale. Most are far more deadly. On June 28, 1991, two people died and more than 100 were injured in an earthquake that registered 6.0 on the Richter scale in Pasadena, California. Property losses were estimated at $18 million. In a geological rarity, no aftershocks were reported following the quake. More than 300 buildings were damaged in Pasadena, Monrovia, Sierra Madre, Arcadia, and other communities.(1)

In April of 1990, aftershocks were still occurring from the earthquake that shook the San Francisco Bay area in October 1989, killing 60 people. An earthquake is the trembling or shaking movement of the earth's surface. Of all natural disasters, earthquakes can inflict the greatest loss of life and property. They have a long history of devastation of people and their works. As long ago as 1556, Shensi, China, suffered the loss of 830,000 lives; as recently as 1990, the loss of life in an earthquake in Iran was estimated at 25,000. In 1737 in Calcutta, 300,000 were killed in 1920, in Kansu, China,

180,000, and in 1933 143,000 in Tokyo and Yokohama.(2) If the seers and scientists are correct in their forecasts, more devastating quakes are likely to occur in the future.

Causes of Earthquakes

Most earthquakes are caused by accumulation of strain in the earth's interior resulting from convective movement of massive tectonic plates covering its surface. Rocks that make up outer layers of the earth are continually being compressed and stretched by forces within the earth. The forces are believed by some geologists to be the result of enormous heat in the earth's core; one theory is that radioactivity of the earth over a period of millions of years produces heat inside, which cannot readily escape through the crust, especially through continents, so it breaks through near the sea. This could account for the fact that earthquake belts and volcano belts almost coincide around the perimeter of the Pacific Ocean.

When the strain becomes too great, a massive cracking of interior rocks takes place and the earth shakes. These fractures in the rocks are called faults. When slippage subsequently occurs along these faults, additional earthquakes are produced, the intensity depending upon the length of the fault and amount of the slip. Most earthquake faults lie beneath the surface, but some, such as the San Andreas Fault in California, can be seen from the air for a distance of 600 miles.

According to some geophysicists, the combined gravitational pull of the sun and the moon may be strong enough to trigger an earthquake. The Tehachapi earthquake on July 21, 1952, may have been caused this way. The sun and moon were in almost perfect line when this 7.7 magnitude shock occurred, killing fourteen people and causing about $60 million damage. About twenty aftershocks were magnitude five on the Richter scale or greater.

Shrinkage of the earth's crust, the collapse of cave roofs, volcanic action, the pumping of fluids into the crevices of subterranean rocks, and the detonation of nuclear devices beneath the earth's surface are also said to cause earthquakes. The subterranean point of origin is called the *focus*, and the point on the surface directly above the focus is called the *epicenter*. The release of energy at the time of the quake produces earth waves which travel along the surface of the earth; others pass through the earth at a depth of about 1,800 miles.

Power of Earthquakes

On April 18, 1990, a strong earthquake battered central China killing 115 people and leaving thousands homeless. The quake and more than 600 aftershocks caused collapse of over a thousand homes. This type of destruction, which measured a relatively moderate 6.9 on the Richter scale, is representative of the tremendous power that earthquakes can unleash.(3)

The power of earthquakes is estimated in various ways. The Richter scale is used to measure energy released as recorded on seismographs, and the magnitude of a quake is assigned a number from one to ten. No shock has been assigned a magnitude over 8.6; the intensity increases in geometric ratio—an increase of one number means the energy released is ten times greater. Thus, an earthquake having a magnitude of 8.5 is ten times greater than one of 7.5, and some experts say damage would be thirty times greater.

In populated areas, an earthquake of 3.5 on the Richter scale can cause slight damage in the local area; 4 can cause moderate damage; 5 considerable damage; 6 can be severe. A reading of 7 is a major earthquake, capable of widespread, heavy damage. A reading of 8 is a "great" earthquake, such as the one in San Francisco in 1906 which was 8.3 on the Richter scale; and in Alaska in 1964 which registered between 8.3 and 8.7. The Alaska earthquake released energy equivalent to 100 underground 100-megaton nuclear explosions on one line.(4) A shock of the magnitude of 9.5 would release about forty times the energy of the largest recorded shock, and a shock of 10 would be felt over the whole earth.(5)

The Modified Mercalli Intensity scale has twelve divisions; moderate damage is produced at VII and people will run out of buildings; VIII will cause major damage.

The Rossi-Forel scale has ten divisions, and people leave their dwellings on a class six earthquake. (See Appendix I for a description of these two earthquake scales.)

Location of Earthquakes

Violent earthquakes occur roughly in two zones of the earth's surface. One follows a line drawn through the Mediterranean, Asia Minor, the Himalayas, and the East Indies; the other follows the western, northern, and eastern shores of the Pacific Ocean. In the United States, 80 to 90 percent of all earthquakes occur in the California-Arizona-Nevada region, where hundreds of them, usually of minor intensity, occur daily. Table 10-1 shows the record of major earthquakes of the world.

Frequency of Earthquakes

In the past two decades the United States has experienced several severe earthquakes (Alaska, San Fernando Valley, and the San Francisco Bay area in California), though thousands of small earthquakes occur in the nation every year. Seismic risk maps of the United States depict the relative damage expectancy from earthquakes and show that several population centers are located in high-risk areas.

Earthquake Predictability

Anyone can predict that an earthquake will occur within an hour, someplace, and would be right, because about a million occur every year—about 800 in the range of 5 to 5.9 on the Richter scale, about 6,200 in the range of 4 to 4.9, about 49,000 in the 3 to 3.9 range, and about 100,000 having a magnitude of 2.5 to 2.9. About 150,000 perceptible shocks are felt each year.(6) However, the ability to predict an earthquake of specific magnitude at a specific location is open to question. The areas where earthquakes are

Table 10-1. Major Earthquakes of the World

Year	*Locality*	*Deaths*
856	Greece, Corinth	45,000
1038	China, Shensi	23,000
1057	China, Chihli	25,000
1268	Asia Minor, Silicia	60,000
1290	China, Chihli	100,000
1293	Japan, Kamakura	30,000
1531	Portugal, Lisbon	30,000
1556	China, Shensi	830,000
1667	Caucasia, Shemaka	80,000
1693	Italy, Catania	60,000
1737	India, Calcutta	300,000
1755	Northern Persia	40,000
1755	Portugal, Lisbon	60,000
1759	Lebanon, Baalbek	30,000
1783	Italy, Calabria	50,000
1797	Ecuador, Quito	41,000
1811	U.S., New Madrid, Mo. US	Several
1812	San Juan Capistrano, Calif. US	40
1819	India, Cutch	1,543
1822	Asia Minor, Aleppo	22,000
1828	Japan, Echigo (Honshu)	30,000
1868	Peru and Ecuador	25,000
1868	Hayward, Calif. US	30
1872	Owens Valley, Calif. US	27
1886	Charleston, S.C. US	60
1875	Venezuela and Colombia	16,000
1886	U.S., Charleston, S.C. US	60
1896	Japan, Sea Wave, Sanriku Coast	22,000
1897	India, Assam	1,542
1899	San Jacinto, Calif. US	6
1905	India, Kangra	20,000
1906	U.S., San Francisco, Calif. US	700
1906	Chile, Valparaiso	1,500
1908	Italy, Messina	75,000
1915	Italy, Avezzano	29,970
1915	Imperial Valley, Calif. US	6
1918	Puerto Rico (tsunami from earthquake in Mona Passage) US	116
1920	China, Kansu	180,000
1923	Japan, Tokyo; Yokohama	143,000
1925	Santa Barbara, Calif. US	13
1926	Santa Barbara, Calif. US	1
1932	Humboldt County, Calif. US	1
1932	China, Kansu	70,000
1933	Long Beach, Calif. US	115
1934	Kosmo, Utah US	2
1935	Helena, Mont. US	4
1935	Pakistan, Quetta	60,000
1939	Chile, Chilian	30,000
1939	Turkey, Erzincan	23,000
1940	Imperial Valley, Calif. US	9
1946	Hawaii (tsunami from earthquake in Aleutians) US	173
1946	Eastern Turkey	1,300
1946	Japan, Honshu	2,000
1948	Japan, Fukui	5,131
1949	Ecuador, Pelileo	6,000

1949	Puget Sound, Wash. US	8
1950	India, Assam	1,500
1952	Kern County, Calif. US	14
1953	Northwestern Turkey	1,200
1954	Algeria, Orleansville	1,657
1954	Eureka-Arcata, Calif. US	1
1955	Oakland, Calif. US	1
1956	Northern Afghanistan	2,000
1957	Northern Iran	2,500
1957	Outer Mongolia	1,200
1957	Western Iran	2,000
1958	Khantaak Island and Lituya Bay, Alaska US	5
1959	Hebgen Lake, Mont. US	28
1960	Hilo, Hawaii (tsunami from Chile earthquake) US	61
1960	Morocco, Agadir	12,000
1960	Southern Chile	5,700
1962	Northwestern Iran	10,000
1963	Yugoslavia, Skopje	1,100
1964	Prince William Sound, Alaska (tsunami) US	131
1965	Puget Sound, Wash. US	7
1965	Chile, El Cobre	400
1966	Eastern Turkey	2,529
1967	Venezuela, Caracas	236
1968	Northeastern Iran	11,588
1970	Western Turkey	1,086
1970	Northern Peru	66,794
1971	San Fernando, Calif. US	65
1971	Los Angeles US	64
1972	Nicaragua	10,000
1972	Southern Iran	5,000
1973	Mexico	539
1974	Pakistan	5,200
1975	China	2,000
1975	Turkey	2,000
1976	Guatemala	23,000
1976	China	750,000*
1976	Italy	1,000*
1976	Indonesia	9,000*
1976	Philippines	4,000*
1976	Guatamala	23,000
1976	Western Iran, Indonesia	8,000
1976	Northern China	250,000
1976	Philippines	8,000
1976	Turkey	4,000
1977	Eastern Europe	1,500
1978	Eastern Iran	25,000
1980	Algeria	4,500
1980	Southern Italy	4,800
1981	Southeastern Iran	8,000
1985	Central & SW Mexico	7,000
1986	El Salvador	1,500
1987	Los Angeles, Calif. US	25*
1988	Soviet Armenia	25,000
1989	San Francisco Bay area	62
1990	Central China	115
1991	Georgia, Soviet Union	84
1991	Central America	59

Data thru 1970 by National Oceanic and Atmospheric Administration; data from 1971 to 1991 from World Book Encyclopedia and News Reports.

*Estimated

likely to occur are known, but this does not preclude them from occurring elsewhere. As for the specific time when a strong earthquake will occur, there are better answers available today than the advice given to author Charles Bahme in his college geology class, i.e., "the longer it has been since the last, the closer you are to the next!"

There are scientists who have achieved moderate success in predicting earthquakes. "In localized areas small and moderate quakes are being predicted accurately today by United States and Soviet scientists. In quake-prone mainland China, a major earthquake was predicted, and measures taken, averting a catastrophe even worse than that in Guatemala."(7) The earthquake with a 7.3 magnitude occurred on February 4, 1975, in the town of Haich'eng, and nearly every house was flattened. Since people had been warned at 6:00 P.M. to leave their homes with their animals, only those who ignored the warning were killed when the quake struck one and one-half hours later. Two or three days prior to the shock the warning signs began to multiply: water rose in wells suddenly, farm animals refused to enter their folds, and school children skating on frozen lakes noticed that frogs were jumping through the holes in the ice.

In 1987, Tom Tugend reported on research being conducted by UCLA on the west coast. One of the country's leading geophysicists, Leon Knopoff, discovered, for example, that earthquakes do not happen at random. Over certain periods of time and geographical space, they actually occur in clusters and by analyzing these patterns of clustering earthquakes can be predicted.(8)

The Chinese have accumulated knowledge over the past 3,000 years about events that usually precede earthquakes. These include minute earth movements across fault lines which are sensed by lasers; slight changes recorded in tiltmeters; changes in electrical resistance in the earth's crust recorded on gages (signaling changes in its density and the amount of water in the rock); decreases in amount of radioactive gas; radon in well water; and variations in speed of seismic waves. Seventeen fully equipped seismic centers and 250 auxiliary stations have been installed; land earthquake data are being gathered at 5,000 separate stations by a trained corps of earthquake professionals. Three of their successful predictions were based in part on abnormal animal behavior; rats leaving buildings, snakes crawling from their holes in large numbers, and fowl refusing to go to roost.(9)

Their system is not always successful, for there apparently was a failure to predict a strong quake in Yunan Province in May 1974, and the two quakes that struck a three-city area of 15 million population in July 1976; more than 700,000 were estimated to have been killed in these quakes, the first of which, measuring 8.2 in magnitude, was the strongest felt anywhere in the world since the 8.5 quake devastated Anchorage, Alaska, in 1964.(10)

The first valid U.S. earthquake prediction was in 1973 by Dr. Yash P. Aggarwal, who discovered changes in speed of seismic waves from seismometers placed around Blue Mountain Lake, a region of frequent but small tremors in the Adirondack Mountains. His calculation that a 2.5- to 3-magnitude tremor would occur in two to four days in that area proved correct when one of 2.6 hit within that time interval.(11)

Other predictions, specifying place, time, and magnitude, have been made in the United States. In 1974, Dr. James H. Whitcomb, a California Institute of Technology geophysicist, came close to meeting the criteria—he predicted the time and place, but missed the magnitude. And in northern California, Menlo Park scientists monitoring the network of instruments south of Hollister informally predicted a magnitude of 5.2 shock that occurred on schedule but slightly off in location. Dr. Whitcomb made another prediction. He stated in April 1976 that there is evidence of stress building up in subsurface rock layers to the point where a 5.5 to 6.5 magnitude earthquake could be expected to occur within a roughly circular area of southern California, about eighty miles in diameter, centering in the same region as the site of the 1971 Sylmar tremor. There are at least twenty different active faults in that zone, and the tremblor could occur on any one of them, he explained, but the evidence, in the form of variations in velocities of sound waves traveling those stressed layers, does not permit a more precise location. His prediction may have been realized in the 1986 quake in Los Angeles.(12)

In assessing Dr. Whitcomb's forecast, the California Earthquake Prediction Council could not determine that probability of an earthquake in that area was significantly higher than for similar geologic areas in California.(13)

Various prediction systems are in use, some based on earthquake surveillance, provided by seismometers and strong-motion seismographs. Comparison of surface motion on various soils and rocks in the vicinity of strong-motion stations during an earthquake can be made with seismoscopes. Several hundred seismograph stations are operating continuously throughout the world, under control of government and private agencies.

When doomsayer Iben Browning predicted that a major earthquake would occur along the New Madrid Fault on December 2, 1990, city officials, school officials, and residents from north to south took his prediction seriously. There was a run on home insurance, a concerted effort to conduct school quake drills—and even to call off school in some southern counties—and people generally prepared for a genuine earthquake threat. The New Madrid Fault runs about 120 miles from Marked Tree, Arkansas, to Cairo, Illinois, and includes Tennessee, Kentucky, and Missouri. One of the reasons that persons took this prediction seriously is that one of the three of the nation's worst earthquakes happened in New Madrid, Missouri—named after the fault. In western Kentucky, school officials in eight districts cancelled classes December 2 and 3. Residents went on buying sprees and stocked their basements with water, food supplies, and survival gear.

Although the quake failed to materialize the prediction certainly had value in arousing public awareness to the real dangers that earthquakes possess and the preparations that can and should be made in anticipation of them.(14)

Earthquake Surveillance Equipment

The three types of earthquake surveillance generally provided are worldwide seismography, local seismic recording, and strong-motion seismography. In addition to a worldwide network of standard seismographs, there are micro-earthquake seismic networks, one of which consists of about fifty stations in the area from San Francisco to Hollister, California. This network locates about 2,000 earthquakes per year. Silent earthquakes which occur when the land along a fault slips so slowly and smoothly that it cannot be felt may be precursors to strongly felt tremors, and the recording of data on this kind of "creep" may enable scientists to calculate the repeat time of large quakes. The "repeat time" is the approximate rate of reoccurrence of large earthquakes, which evidence suggests is about every 100 years off Japan. Although the technique might not be applicable to Southern California because of its particular geology, that is also the time interval thought by some scientists to be applicable to large tremors along the San Andreas Fault.(15)

A phenomenon believed to take place in deep underground rock layers subjected to great strain, *dilatancy*, has been proposed as a very promising way of predicting the time, place, and magnitude of an earthquake; dilatancy is the action of rock under extreme pressure suddenly opening with tiny cracks and expanding. This process is believed to occur when two large land masses try to move past each other. Initially the total volume of these cracks is greater than the volume of underground water available to fill them, but when the cracks fill up, the water lubricates the interface between the land masses, and an earthquake occurs. Dilatancy explains other precursors—variations in electrical sensitivity of the rock, increase of radon in the well water, and uplifting of land under strain. In the Palmdale area northeast of Los Angeles, the Seismic Safety Commission has been asked to consider the possibility that a large swelling that extends for about 100 miles along the San Andreas Fault, varying from 2 to 10 inches in height over the 4,500 square miles involved, is a precursor of a coming major earthquake. Although this uplifting of the land has taken place since the early 1960s, its significance is difficult to judge, since there have been instances in which a sudden uplift of land preceded a large and destructive earthquake and other cases where no tremors followed.(16)

Soviet and American scientists continue to work to achieve a reliable earthquake-forecasting system. Dr. Robert Hamilton, director of the Office of Earthquake Studies in the U.S. Geological Survey, was reported as saying that lack of funds has retarded research aimed at developing techniques that would give 75 percent accuracy in quake predictions.

Legislation was adopted unanimously by the U.S. Senate in the late 1970s to authorize $150 million over a three-year period to improve earthquake prediction techniques. The 1976 devastating earthquakes in Guatemala, Italy, the Soviet Union, and China, along with the discovery of the Palmdale bulge, apparently spurred the action. In addition to furthering prediction research, the bill aimed to reduce loss of life and damage caused by earthquakes by identifying communities near fault lines, or those likely to suffer from landslides or tsunamis; development of building codes and construction methods to prevent such future occurrences as the collapse of the Sylmar Veterans Hospital in the 1971 southern California quake or the collapse of the double-deck freeway in Oakland in 1989 (both of which had been designed to existing earthquake-resistant standards); and development of warning capabilities and emergency services.

The article states, "Under the worst combination of an inaccurate prediction and ill-conceived public response, the prediction and quake together might be even more costly than an unpredicted quake would have been."(17)

It also asserts that while scientists should develop, assess, and issue *predictions* of earthquakes, *warnings* should come from elected officials; if mass evacuation is deemed necessary, it would be the decision of governmental authorities to issue such orders to residents of vulnerable areas. It was believed that all steps in the development of a prediction should be made public, to prevent possibility of "inside" information being used by some people to the disadvantage of the general public.

How will people react to quake forecasts? There has been speculation that, following prediction of a major quake and issuance of official warnings, the panic that would follow could be more disastrous than the physical damage. "The body of scholars headed by UCLA's Dr. (Ralph H.) Turner doesn't agree with that alarming idea. But their report does see some very serious problems involved in earthquake forecasting, especially if government doesn't begin to formulate strategies for action. And thus far, there's not much evidence of public policy planning for the inevitable eventuality which lies down the road."(18) The report pointed out that fewer than 10 percent of Americans can safely ignore the earthquake hazard, but most use or occupy buildings that are substandard for seismic safety.

Our failure to correct unsafe conditions, which must be done no matter how accurate the science of earthquake prediction becomes, gives a possible clue to one response that the public might make to such warnings; apathy. Perhaps the editor of *Nature* magazine was not too far off base when he suggested in the January 1973 issue that earthquake prediction was useless if the evacuated inhabitants of a destroyed city had no homes to return to. Better they should perish with the act was the clear implication.(19)

Often earthquakes strike in rapid succession and in distant places simultaneously. On May 30, 1990, 14 people died in an earthquake that rocked eastern Europe. On the following day, approximately 100 aftershocks were felt from Moscow to Istanbul and caused serious structural damage to buildings in Romania and Bulgaria. On the very next day, June 1, 1990, a moderately strong earthquake rattled Mexico City, while aftershocks were still reverberating around eastern Europe. Although the Mexico City earthquake measured 6.1 on the Richter scale, it caused no serious damage or injuries.(20)

Effects of Earthquakes

One important result of great earthquakes is the production of subsequent great earthquakes, known as aftershocks, sometimes in the same region, sometimes thousands of miles away and months later. It is believed now that there may be a link between the huge earthquakes that periodically erupt all around the rim of the Pacific Ocean basin. Scientists now are beginning to doubt the "local" nature of large quakes—those with a magnitude of five or more. "A huge tremor that shakes Japan, the scientists suspect, may well be the trigger that sets off another earthquake months later in Peru, Mexico, California, or some place on the edge of the Pacific basin."(21)

Recent evidence indicates that material from the upper mantle is welling up along the mid-Atlantic ridge and causing the movement of large areas of the earth's surface called "plates." These plates are thought to interact in one of three ways: spreading where new crust is formed, subduction where one plate plunges under another, or fault action where two plates rub together.

Whether the earthquakes cause the movement of these massive land masses, or vice-versa, a large earthquake is one of nature's most devastating phenomena. "The energy released by a magnitude of 8.5 earthquake on the Richter scale is equivalent to 12,000 times the energy released by the Hiroshima nuclear bomb. While these cataclysms have their foci well beneath the earth's surface, cities have been destroyed and thousands of lives lost in a few seconds as the result of great earthquakes in the past."(22)

Among the primary effects of an earthquake's surface waves is the violent shaking of the ground. *Shaking* causes the greatest amount of damage from earthquakes in highly populated areas. Probably the majority of structural failures in earthquakes result from shaking that directly damages the structure, or causes soil failure beneath its foundation (as by slope failure or liquefaction of water-saturated soils), or which causes the soil beneath the foundation to densify and settle, allowing the structure to fall. Sometimes shaking will open up fissures and cause great displacement; displacement in the 1906 San Francisco quake was 21 feet horizontally, while the one in Yakutat Bay in 1899 was 47 feet vertically.

Shaking can cause vibrations to be set up severe enough to cause collapse or shearing of dams, bridges, tunnels, or buildings. Chimneys, highrise buildings, water tanks, and bridges are especially vulnerable to resonant vibration motion. Walls of highrise buildings without adequate lateral bracing can fall outward, allowing floors to cascade one on top of the other, crushing occupants. In countries where construction is largely of unreinforced mud and adobe brick, collapse of buildings is apt to be total.

Breaking of the earth can be caused by four different processes: faulting (displacement of two earth masses along the fault plane), lurching (surface cracking of soils at places other than directly along faults), liquefaction (causing soils to lose their strength), and slope failure (causing landslides).

The movement during an earthquake, lateral and vertical, severs pipes as well as underground power and communications cables, releasing flammable and toxic gases and liquids, and knocking out fuel, power, sewer, and water systems. Elevators can be stalled, jammed, or their suspension systems may fail.

Water damage can be caused by an oscillation known as "seiching" that occurs in lakes and bays, causing water to surge from one end to another, reaching great heights and overflowing the banks. As a result of the 1864 earthquake in Alaska, water rose six feet in Memphis, Tennessee, 5,000 miles from the center, due to this type of action. Collapsed water tanks and broken dams can produce floods, the effects of which we discussed in the previous chapter.

Another secondary effect is a tsunami, the seismic sea wave generated

An earthquake often causes a redesign of building structures against the wishes of the property owner. (Photo courtesy of Ron Buckingham.)

rather infrequently by seaquakes or submarine disturbances. Since 1900, 181 have been recorded in the Pacific area. It is estimated that about 600,000 people would be endangered by a tsunami in Pacific Coast states if the wave were to be 50 feet high; if the wave were 100 feet high, about 1,141,000 would be endangered.(23)

Tsunamis, major and local, took most of the 122 lives in the Alaskan earthquake of 1964, including those lost in Hawaii and on the west coast. In 1755, a seismic wave 50 feet high wrecked Lisbon, Portugal.

Local waves in confined areas generated by earthquake action can be very destructive. When a landslide caused by an earthquake fell into Lituga Bay, Alaska, a wave 1,700 feet high was created. Even without creating a water wave, landslides caused by earthquakes can result in a great loss of life. At least 70,000 persons were killed—most of them buried under a fast-moving wall of water, rock, and mud—when an earthquake struck western Peru in 1970. About 800,000 were left homeless.(24)

Earthquakes can also cause subsidence. About 30,000 square miles of Missouri sank 5 to 15 feet in 1812 during seismic activity that centered near New Madrid, and the Mississippi River flowed upstream to form Reelfoot Lake in Tennessee. Some areas in that region were raised at the same time. Part of Japan's Sagami Bay dropped 689 feet and part raised 850 feet, creating a seismic wave 35 feet in height in the 1923 earthquake that completely destroyed the city of Yokohama, 71 percent of Tokyo, and 51 percent of Yokosuka.

Subsidence was a major cause of loss in the 1964 Alaska quake, when almost the entire residential subdivision of Anchorage's Turnagain Heights slid from its original site toward Turnagain Arm. Another example in the same quake was the slide on 4th Avenue in Anchorage.

Fire was the secondary effect that caused 80 percent of destruction in the great San Francisco earthquake of 1906. Of 143,000 who died in the 1923 Kwanto, Japan, earthquake, in which Tokyo lost all its records and art collections, 40,000 were residents of Tokyo who took refuge in an open space across which flames were driven and none escaped. The monetary loss from fire and shock was $2.5 billion.(25)

Fires caused by overturned stoves, broken electric wiring, and spilled molten materials are fed on escaping gas and flammable liquids from ruptured fuel tanks and pipe lines, as well as debris from damaged buildings. Control of fire becomes especially difficult when water mains are broken, streets blocked by fallen buildings, fire stations and fire apparatus damaged in the quake, and fire fighters are killed, injured, or otherwise unable to answer emergency calls. Where highways and rail systems are put out of service, not much help can be rendered by mutual aid companies responding from areas outside the zone of damage.

Property damage from quakes has been considerable in the United States—billions of dollars since 1865, as indicated in Table 10-2. Earthquakes have been responsible for loss of hundreds of thousands of lives, as shown in Table 10-1.

Earthquakes affect the economy of the nation, destroying means of production, distribution, and transportation of commercial products, and disruption of communications and public utility services. Many businesses destroyed never resume, while most suffer delay until factories can be rebuilt and disruptions in the labor force and public utilities have ended.

No one knows how much damage is done to the ecology by violent earthquakes, such as in Peru (1970), Alaska (1964), Montana (1959), and San Francisco (1989), but it is believed that transient and permanent inundation and large landslides strongly affect flora and fauna. Large forests are destroyed by violent shaking, uprooting, or inundation with sea water, with adverse effects on wildlife. Crustal changes of the ocean shore can exterminate marine organisms.

Examples of Major Earthquakes

One of the most deadly earthquakes struck Iran on June 21, 1990. The quake, which killed over 25,000 and injured tens of thousands, registered 7.7 on the Richter scale shaking the densely populated Caspian Sea region of northwest Iran. Heaviest damage occurred in a wide swath of Zanjan and Gilan provinces. Electricity, telephone lines, and water supplies in most of the region were cut off by the quake which lasted only a minute but was followed by a dozen aftershocks, one of them 6.5 on the Richter scale. It is interesting to note that the United States set aside its political differences with Iran and spearheaded an international relief effort that included

Table 10-2. Earthquake Monetary Losses in U.S.*

Year	Locality	Damage
1865	San Francisco, Calif.	.5
1868	San Francisco, Calif.	.4
1872	Owens Valley, Calif.	.3
1886	Charleston, S.C.	23.0
1892	Vacaville, Calif.	.2
1898	Mare Island, Calif.	1.4
1906	San Francisco, Calif.	24.0
	Fire loss	500.0
1915	Imperial Valley, Calif.	.9
1918	Puerto Rico (tsunami damage from earthquake in Mona Passage)	4.0
1918	San Jacinto and Hemet, Calif.	.2
1925	Santa Barbara, Calif.	8.0
1933	Long Beach, Calif.	40.0
1935	Helena, Mont.	4.0
1940	Imperial Valley, Calif.	6.0
1941	Santa Barbara, Calif.	.1
1941	Torrance-Gardena, Calif.	1.0
1944	Cornwall, Canada-Messena, N.Y.	2.0
1946	Hawaii (tsunami damage from earthquake in Aleutians)	25.0
1949	Puget Sound, Wash.	25.0
1949	Terminal Island, Calif. (oil wells only)	9.0
1951	Terminal Island, Calif. (oil wells only)	3.0
1952	Kern County, Calif.	60.0
1954	Eureka-Arcata, Calif.	2.1
1954	Wilkes-Barre, Pa.	1.0
1955	Terminal Island, Calif. (oil wells only)	3.0
1955	Oakland-Walnut Creek, Calif.	1.0
1957	Hawaii (tsunami damage from earthquake in Aleutians)	3.0
1957	San Francisco, Calif.	1.0
1959	Hebgen Lake, Mont. (damage to timber and roads)	11.0
1960	Hawaii and U.S. West Coast (tsunami damage from earthquake off Chile)	25.5
1961	Terminal Island, Calif. (oil wells only)	4.5
1964	Alaska and U.S. West Coast (tsunami damage from earthquake near Anchorage—includes earthquake damage in Alaska)	500.0
1965	Puget Sound, Wash.	12.5
1966	Dulce, New Mexico	.2
1969	Santa Rosa, Calif.	6.3
1971	San Fernando, Calif.	553.0
1987	Los Angeles, Calif.	10.0
1989	San Francisco Bay area	5,200.0
1991	Carson City, Nevada	0.5

* Property Damage in Major U.S. Earthquakes, in millions of dollars (actual)—from *Earthquake Investigation in the United States*, U.S. Department of Commerce; OEP data used.

Britain, Japan, France, Switzerland, and other members of the European community.(26)

Nearly two decades earlier, on April 10, 1972, in southern Iran an estimated 5,000 persons were killed and about 60 villages devastated by a major earthquake and about 1,000 aftershocks.

We will now trace the history between these two great quakes, noting that the 1970s and 1980s were deadly periods in earthquake history.

On December 23, 1972, a series of quakes hit Managua, Nicaragua, killing at least 10,000 persons. Two light shocks were felt on the 22nd, causing the fire department to move apparatus out of fire stations; following the second shock, all rigs were put back inside and then the big one hit thirty minutes after midnight on the 23rd, collapsing the main fire station that had been built only eight years before (using earthquake-resistive standards). Four of the latest fire trucks and four rescue vehicles were destroyed with collapse of the fire station's second floor onto the first, and two fire fighters were killed at the same time. The second fire station withstood the shock, leaving about 35 percent of the fire equipment in the department in operable condition.

All on-duty fire department members remained on duty (unlike many government employees who left immediately to attend to their families). Within thirty minutes, 70 of the department's volunteer fire fighters reported for duty, and 230 of them were on duty by seven o'clock in the morning, the total force for this capital city of 1.5 million population. In the absence of communications, other than vehicle radios, the fire chief had to drive about 18 miles to request help from the nearest neighboring community; as soon as word got out about their situation, additional assistance was forthcoming without further requests.

To handle fires that developed into a mass fire in an area of 9 by 18 blocks in the center of the city, water had to be relayed from 3,000 feet away; only 1½ inch lines were used to provide perimeter control, since there was insufficient water for use of master streams. With the help of the U.S. Army's Corps of Engineers, heavy equipment and explosives were used to clear a fire break around the involved area, and the 13 fire trucks with 135 personnel from other towns in Nicaragua, plus 9 fire trucks from 6 neighboring countries, were deployed around the major fire area to prevent spread.

The final result of the quake showed that 295 city blocks were severely damaged, with the mass fire confined to 162 blocks. Of the city's 70,000 homes, 53,000 were destroyed. Of light shops and commercial buildings, 95 percent were destroyed, 250,000 people were displaced, and the damage caused by the quake was over one billion dollars.(27)

A major earthquake in 1973 killed more than 500 persons in a wide area south and east of Mexico City. In 1974 there were at least two quakes of devastating proportions: the first at Lima, Peru, on October 3, killed 73 persons and injured 2,000 others; the second near Rawalpindi, Pakistan, destroyed 11 villages and killed about 5,300 persons.

In 1975 two massive earthquakes shook parts of China and Turkey. A tremor struck a heavily populated industrial area in Manchuria on February 4, and the force of the tremor shook Vladivostok, Russia; Communist officials did not release their casualty figures. The quake that hit Lice, Turkey, on September 6 leveled parts of the city and surrounding towns and villages, killing more than 2,000 persons.

The Soviet Asian republic of Uzbek was hit by powerful earthquakes in April and May 1976 and, while there was said to be property damage in this area which contained the Soviet Union's largest gas fields, no report

A super crane is quite helpful when major large structures suffer significant structural damage in an earthquake. (Photos courtesy of the Federal Emergency Management Agency.)

of casualties was given. Both of these quakes were stronger by several times than the one that devastated northeast Italy in early May 1976, killing more than 900 people. A Moscow seismologist reported by television that the Soviet quakes measured 7.3 and 7.0 on the Richter scale. The Italian quake measured 6.5 to 6.9. The Moscow report said the reason that the Soviet's largest of their two quakes caused no injuries was because seismologists gave early warning of the tremor and the inhabitants were evacuated.(28) The Italian quake left death and destruction in the towns near the border of Yugoslavia—Buia, Maiano, Forgaria, and Gemona. A few chimneys were knocked down in Venice, but in western Italy, the famous Leaning Tower of Pisa, though shaken, had suffered no damage and its tilt of 17 feet off center remained unchanged.(29)

On July 9, 1976, more than 9,000 people died in an earthquake that caused huge landslides and wiped out 17 communities in the remote central area of Indonesian New Guinea. The quake measured 7 on the Richter scale, the worst ever recorded in Indonesia, and a series of further shocks were felt over a 1,000 square mile area.(30)

On August 16, 1976, two major earthquakes struck Asia in a two-hour period—one in central China and another in the Philippines. Seismic reporting stations said that the Chinese quake was centered near the Kansu-Szechwan province boundary in central China, about 650 miles west-southwest of Peking and far from the northeast China area devastated by a quake in the preceding month, killing an estimated 750,000 people in and around Tangshan. The August Chinese quake was measured at 7.3 on the Richter scale, the one in July was 8.2.

The National Earthquake Information Center in Golden, Colorado, reported a tremor measuring 8.0 on the Richter scale in the Moro Gulf of the southern Philippines, near Mindanao, on August 16, 1976. The quake struck more than a dozen cities, from Jolo, 550 miles south of Manila, to Legaspi City, 300 miles to the north. The Philippines National Geophysical Observatory reported the quake as 6 on the Rossi-Forel scale (see Appendix I). More than 4,000 were killed and 90,000 made homeless in this worst quake ever to hit the Philippines Islands.(31)

One of the worst disasters to hit the Western Hemisphere was the 30-second earthquake of 7.5 magnitude that struck Guatemala on February 4, 1976 at 3:00 A.M. The major shock covered an area of 3,500 square miles, and in the days that followed some 23,000 people were found dead, about 77,000 injured and over a million homeless. Of 1,000 to 1,500 aftershocks that followed the main shocks, more than 100 were felt in Guatemala City, during the first few hours after the initial quake. In that city, more than 58,000 homes were destroyed; many cascaded down a hill during the violent shift of the earth.(32)

As we have described, the numerous earthquakes that struck during the 1970s made it a rather lethal decade. Moving into the 1980s, history repeated itself with only the names of the places changing.(33) In 1980, 4,500 people died in Nigeria, and 4,800 in southern Italy. In 1981, 8,000 people died in the Iranian earthquake earlier described. In 1985, 7,000 people died in central and southwestern Mexico, followed in 1986, by 1,500 deaths in El Salvador. Then came 1988 and the devastating earthquake in

Soviet Armenia that claimed 25,000 lives and taxed the relief services of the world. Both the Mexico City and Armenia incidents highlighted enormous search and rescue (SAR) problems following a major earthquake and the need for well-equipped and trained personnel for locating victims and extricating them from heavy debris.(34)

The decade closed with the 1989 San Francisco Bay area earthquake, not as deadly in terms of lost lives, but certainly the most devastating to the American landscape, and taking the greatest toll in property damage. Sixty deaths were attributed to this earthquake, but property damage ran into the billions of dollars.(35) During the first year of the decade of the 1990s we experienced an earthquake in central China claiming 115 victims; in Peru, 200; in Iran, 40,000; and in the Philippines, 1,650. (Figures come from *1991 World Book Yearbook*, p. 274.)

Can the Damaging Effects of Earthquakes Be Avoided?

There appears to be no likelihood that methods will be devised to prevent earthquakes. Research is being conducted on the "lubrication" of rock interfaces, by pumping fluid down a bore hole under pressure, but the possibility of generating sufficient microtremors to avoid a macrotremor is purely speculative at this time. Nevertheless, the hazards of earthquakes can be greatly reduced and their disastrous effects minimized by application of seismology principles in building design and construction.

In regions known to be earthquake-prone, places of public assembly should be of quake-resistant construction and should not be placed over faults or in areas of unstable slopes.

Building codes should be revised to require that existing structures be strengthened to resist earthquake damage. The need for such legislation to avoid the inevitable result of natural catastrophe in densely populated areas is illustrated in the estimate by the Office of Emergency Preparedness, when it calculated that five hundred million people are in imminent danger of death, serious injury, or substantial destruction of property due to earthquake.

In 1991, a handbook, *Rapid Visual Screening of Buildings for Potential Seismic Hazards*, was developed by the Federal Emergency Management Agency (FEMA) dividing the country into three areas of earthquake risk, low, moderate, and high. These figures are based on geological and seismological investigations into the 1990s as well as locations of the historical earthquakes we have just studied.(36) This is found in Figure 10-1.

In Los Angeles, engineers and scientists estimated that 200,000 people could die in the metropolitan area in the event of a major earthquake that would collapse the 28,000 potential death traps—pre-1933 unreinforced masonry structures.(37)

Since the 1933 Long Beach earthquake, in which many schools collapsed—fortunately about 6 P.M. when they were unoccupied—the State of California required that new construction conform to minimum earthquake-resistant standards, and existing schools conformed or by law were abandoned by 1975.

Figure 10-1.

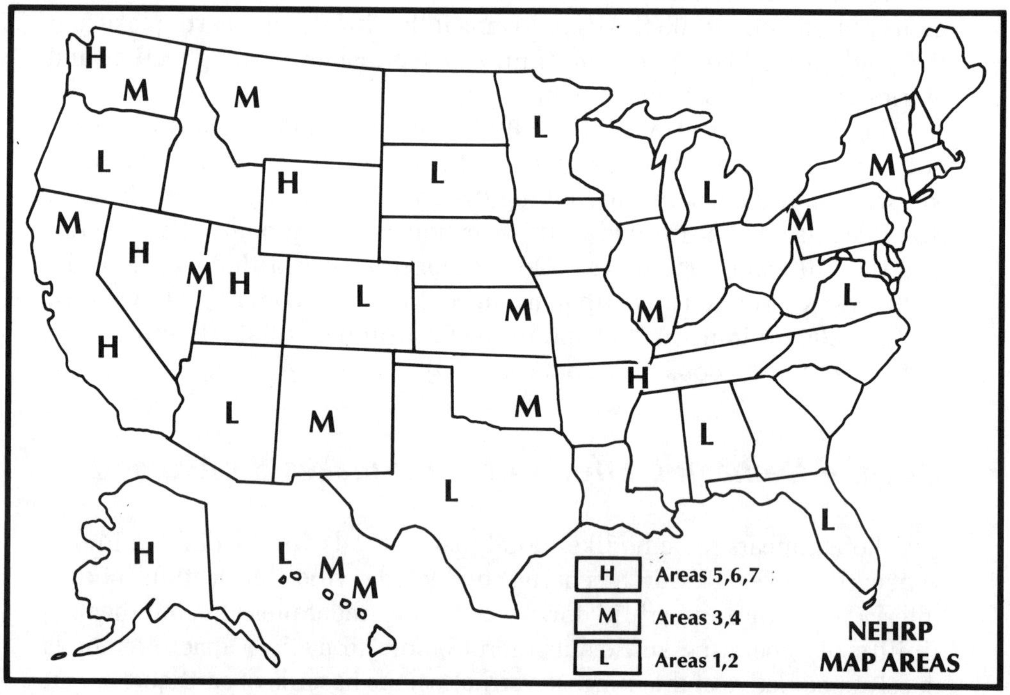

The Building and Safety Committee of the Los Angeles City Council estimated that 14,000 commercial and residential structures did not conform to seismic safety standards in the city's building code. Of these, about 300 were churches, theaters, libraries, nightclubs, restaurants, meeting halls, and assembly buildings. The Council Committee proposed that, in lieu of ordering these 14,000 buildings to be brought up to code or closed down, a requirement be adopted for posting conspicuous signs on them to warn the occupants that they were potential death traps in a severe quake. The building owners would have to notify occupants of the risk; to have the signs removed, they had to make improvements to bring the buildings up to code, the cost of which averaged about 80 percent of the cost of new building.(38) As expected most building owners had to live with the signs and most of these buildings survived into the 1990s.

Hazard Reduction

Hazard reduction measures that would increase fire department usefulness in a major quake include the following:

1. Increasing the ability of water systems to withstand severe earth vibrations. In the Sylmar earthquake of February 9, 1971, the Los Angeles water system lost the use of about 2,000 fire hydrants, 25,000 individual services, and had about 1,500 distribution main breaks. During 14 days required to restore the system, emergency water was supplied from 53 tanks from various agencies, and by 17 fire department pumpers pumping water into the uphill systems as breaks were repaired. (This method reduces the possibility of pollution from water from sewer breaks near the water system breaks.)

Speedy restoration of the water system is important. It was found that purchase of additional skid-mounted diesel pumps, capable of 350 head, that could be moved to various locations as needed, plus one-cubic-foot booster pumps for local areas, would remedy deficiencies. Increasing the amount of repair materials on hand, such as clamps for repairing small mains, is a good precaution.

Following the original San Francisco earthquake fire, approximately 150 underground tanks or cisterns were constructed at key intersections. Holding approximately 75,000 gallons or about an hour's supply for a pumper, these are maintenance free and have been so successful that in 1991, a decision was made to add 94 cisterns to additional portions of San Francisco.(39)

San Francisco has also pioneered the development of a Portable Water Supply System (PWSS) incorporating 16,000 feet of five-inch-diameter hose carried on special hose wagons, portable hydrants, and reducing valves. This system is capable of complex gridding and can create a water curtain several city blocks in length. Cities in earthquake-prone areas might consider the development of a PWSS.(40)

2. Increasing the earthquake resistance of public utility systems to lessen the chance of power outages from violent earth movement is another measure that would help the fire department. Broken power lines are a life and fire hazard, and the absence of regular power supplies makes rescue work, especially at nighttime, more difficult, owing to the necessity for reliance on portable generators to operate some of the rescue tools and provide needed light.

The experience obtained in the Sylmar quake, which had a magnitude of 6.6 and caused an area of approximately 200 square miles in the San Gabriel Mountains to move southward and rise permanently several feet, is illustrative of the time required to restore public utilities after a moderate quake. (It released only one one-hundredth as much energy as the 1964 Alaska earthquake.) Power was restored in two days to all areas with the exception of the evacuated area below San Fernando Dam, which regained power in three days. Gas service was restored to most customers in five days and to the remainder in eleven days. The telephone system was back in service for 8,000 customers in six days, with complete restoration of service in twenty-one days. Repair of the sewer system took longer because of the difficulty in locating the many breaks, but emergency repairs were virtually completed within two weeks.(41)

3. Increasing earthquake resistance of streets, tunnels, bridges, freeway overpasses will improve the fire department's response capability. Even if highways are not damaged, the fire service will have difficulty trying to respond on streets jammed with vehicles carrying people evacuating the area after a major quake. With only three or four ways for residents of Los Angeles County to escape from the metropolitan basin, the highways following an eight-magnitude quake would become a seven-million-car parking lot. In the Sylmar quake five bridges collapsed, fortunately early in the morning before the normal traffic load was present; streets were also ruptured by gas line breaks and washed out by major water main ruptures.

4. Increasing fire protection measures for hazardous occupancies will lighten the work of the fire department in case of a major tremor.

5. Increasing earthquake resistance of structures whose use is critically needed after a disaster, such as fire and police stations, hospitals, disaster control, and emergency communications centers. "Communications centers are often surprisingly vulnerable, due to nonsecured computers, inadequately raised floors, unsecured battery racks, etc."(42)

6. To reduce the amount and difficulty of rescuing the occupants of large assemblage buildings, and to provide safe places of refuge for displaced persons following a disaster, increase the earthquake resistance of schools, churches, theaters, large hotels, and other highrise buildings housing large numbers of people.

Emergency Action in Earthquakes

An earthquake is like an enemy attack in many respects—you don't know when it will be made, so there is no warning. You don't know where the main strike will hit, or with how much force, so you can't tell how much damage has been caused until it is over and you have had a chance to look around—you may be only in the periphery of the heaviest area of damage. You don't know if or when you will receive another strike, perhaps greater than the first.

If you have not already adopted a plan of action that has been widely disseminated and understood by your emergency forces, your problems will be multiplied. Any plan you adopt after the quake—perhaps sitting on the curb in front of your collapsed headquarters building—will be difficult to promulgate if your personal computers, telephone, and radio base station are knocked out. If your fire stations were not earthquake resistive before the strike, it will be too late to do anything about that afterward; those that have not collapsed may do so in aftershocks that follow.

Massive destruction confronts first arriving fire fighters following an earthquake. (Photo courtesy of the San Francisco Fire Department.)

Assuming that you are lucky and the quake has not deprived you of all your response capability and you still have some uninjured fire fighters and apparatus to cope with problems, what should you do first?

There will be people trapped under fallen buildings; fires at broken gas mains; overturned molten materials; fires started by building owners who have no earthquake insurance, but do have fire insurance; fires set by looters, to cover up their thefts. You are also likely to have one or two large fires that could develop into a conflagration.

You could have a dam break, or a dam weakened enough to require removal of everyone, including your fire department personnel, from the zone of danger, before a second major shock completes the destruction. If yours is a coastal city, there could be a seismic wave on the way capable of wiping out your forces and people in the harbor area and along the seacoast.

You are not likely to be flooded with calls for help, for your communications will probably consist of whatever portable and vehicle radios you still have available. Unless you have a common radio frequency with other agencies and other communities, and enough signal strength to reach them, the only way you may be able to request help is to send someone on horse or helicopter, for the expressways will be blocked by fallen overpasses, bridges will be down, and roads will be blocked with debris of collapsed structures. For several hours you are apt to be on your own. It will take some time for the Civil Defense organization to get its disaster control center in operation; in any case, the mayor or whoever is the head of it probably is going to ask you what steps should be taken, in what order, to save lives and property.

To make a decision, you need facts. Your earthquake disaster annex should include provision for activating and staffing your Emergency Operations Center; there should be no need to telephone off-duty members to notify them to report; the violent shaking will serve as better notification than a dead telephone system.

You need facts on response capability and the extent of damage to the community. After you learn what personnel and equipment can respond and have assessed the total situation, you can evaluate the need for deploying forces to the most critical locations. In the Sylmar earthquake, three Los Angeles County fire stations were destroyed, and in the station at Olive View Hospital it took fifteen minutes to break down the station's door to free the apparatus.(43)

Use your best judgment in establishing priorities for emergency service. Some structures may be so badly damaged or so well involved in fire that you will have to write them off and direct efforts where you have a chance of stopping fire spread.

If you have a helicopter available, you should be able to survey the damage and plan for deployment of your forces. But whether survey is by air or vehicle or only based on reports received from station officers, you will still need to plan for concerted action. If you leave discretion to your subordinate officers, they may commit units to a local rescue or fire suppression action. But you may realize that such action will be futile if a sweeping fire headed their way is not stopped.

Stabilization of structures following an earthquake is of immediate concern to fire fighters and rescue personnel. (Photo courtesy of Ron Buckingham.)

Your biggest problem may involve rescue of victims trapped inside buildings that have partially or completely collapsed, though some fire chiefs feel that the primary goal of the fire department is to extinguish fires and prevent development of a conflagration.(44) The OEP Report to Congress seemed to concur with this view when it closed its findings as follows, "The capability to mount effective search and rescue (SAR) operations in an earthquake disaster is marginal." As evident from the San Fernando case in 1971 and the San Francisco case in 1989, there is need for readily available special equipment and special procedures to locate and extricate buried persons.

As with most of the major obligations facing the fire service, preparation for earthquake begins with properly sizing up the potential for earthquake in a given jurisdiction. Cities with buildings constructed before 1970 could easily face simultaneous building collapses, major fires, and compounding emergency events, all without warning. Size-up starts with estimating the impact on your own fire department. What will happen to the fire stations, apparatus, and personnel? It should be noted that fire stations are often vulnerable due to large spans and the openness of apparatus floors. When San Francisco conducted structural surveys of all of its fire stations in 1991, more than 40% were judged "high collapse hazards."(45) Another study pointed out that fire departments need to pay more attention to SAR (Search and Rescue Capabilities) involving heavy debris.(46) All fire fighters should be given minimal training and in high risk areas special teams are advised.

The advice was referring to the 1971 Sylmar quake rescue operations at the collapsed Veterans Administration Hospital where 49 persons lost their lives. Personnel of 18 engine companies, using 3 bulldozers, 1 motor grader, a demolition truck, ladder trucks, wreckers, and 4 helicopters accomplished the tasks. The fire departments also used power saws, acetylene torches, hydraulic jacks, spreaders, "jaws," and similar tools during the four days it took to recover the last body in the wreckage. In addition, the U.S. Army Corps of Engineers supplied dump trucks, loaders, backhoes, cranes, jackhammers and concrete-cutting equipment. A field kitchen and a morgue were set up, and medical supplies were brought in by four teams of fire fighter paramedics who took EKGs of patients removed from demolished buildings; two patients in cardiac arrest were defibrillated and their lives were saved. Doctors and nurses were flown in helicopters to and from Harbor Hospital forty miles away; helicopters also transported casualties.

The small town fire chief who has been training volunteers for years will be on familiar ground if it becomes necessary to put citizen volunteers to work on a moment's notice. Even fire fighters who are unable to report to their regular places of duty may be called upon to form volunteer companies to assist in search and rescue work and to extinguish fires.(47)

Three years prior to the Bay Area earthquake, San Francisco Fire Chief, Emmit Condon, stressed the need for community self reliance. "When it occurs, the damage, injuries, and death will overwhelm our capability to deal with them. That's why we teach self-help. People who can take care of themselves and families can do more to save lives than all the emergency equipment in the state."(48)

The Bay Area, due to its extensive population and the relatively recent

Figure 10-2.

Bay area earthquakes

Earthquakes with a magnitude of 5.0 or greater on the Richter scale which have occurred in the San Francisco Bay area since the great quake of 1906 that was estimated at 8.3

DATE	TOWN	RICHTER SCALE
March 11, 1990	Watsonville	5.5
July 1, 1911	Coyote	6.6
Nov. 9, 1914	Los Gatos	5.5
Oct. 24, 1926	Santa Cruz	5.5
April 25, 1954	Watsonville	5.3
Sept. 5, 1955	San Jose	5.5
Oct. 24, 1955	Antioch	5.4
March 22, 1957	San Francisco	5.3
March 2, 1959	Gilroy	5.3
Dec. 18, 1967	Watsonville	5.3
Oct. 2, 1969	Santa Rosa	5.7
Aug. 6, 1979	Gilroy	5.7
Jan. 24, 1980	Livermore	5.8
April 24, 1984	Morgan Hill	6.2
Jan. 26, 1986	Hollister	5.5
March 31, 1986	Fremont	5.3
Oct., 1989	San Francisco & Bay Area	7.1

Reprinted with permission from *The Press Democrat,* Santa Rosa, CA earthquake with aftershocks felt into 1990, serves as a focal point for study of earthquakes. Figure 10-2 gives an illustration of the earthquakes felt in this one region alone. Earlier in this chapter, Table 10-1 shows major historic and recent earthquake activity. Table 10-2 shows earthquake monetary losses in the United States Note in Table 10-1 how, in recent times, loss of life in the United States. is relatively low compared with other parts of the world.

VOLCANOES

After more than 600 years of dormancy, the Mt. Pinatubo volcano in the Philippines began erupting on June 9, 1991. After a series of massive explosions, tremendous ash storms, and ongoing tremors, 310 people had died and the nearby Clark Airbase and Subic Bay Naval Base used by the United States military were both heavily damaged.

Approximately one week after the eruptions started, they grew to such cataclysmic proportions that this was considered among the major volcanic events of recorded history. According to scientists at the National Oceanic and Atmospheric Administration, the volcano pumped enough ash and gas into the air to color sunsets and lower temperatures slightly throughout the Northern Hemisphere, perhaps for years.(49)

In May of 1990, a lava flow 2,000 feet wide crept toward the heart of the village of Kalapana, Hawaii, having destroyed 120 homes since 1983 including 44 after a new lava flow began in April of 1990. A small historical church was moved by the townspeople before their village was overrun with lava; they feared that their local history and geography would be changed forever. This is an example of the massive destruction and disastrous consequences that can result from the powerful release of molten energy from volcanoes.(50)

What Is a Volcano?

A volcano is an opening or number of openings in the surface of the earth at the top of, and sometimes on the sides of, a coneshaped hill or mountain connected to the interior of the earth by a funnel or crater through which steam, gases, ashes, rocks, and sometimes molten materials are expelled in periods of activity.

The word volcano comes from Volcanus, the name of the Roman god of fire. The name "Volcano" was first used to identify one of the Lipari Islands in the Mediterranean Sea where the god was supposed to have lived.

History of Volcanoes

From ancient times volcanoes have terrified people with the violence and roar of their eruptions and have been responsible for many disasters. The close correlation of volcanic lines with the epicenters of earthquakes makes their location and activity of significance. According to Dr. Charles Richter, it is possible that massive subsidence and volcanic activity of the sort experienced off Iceland is a manifestation of tectonic forces that may sooner or later generate earthquakes. It may be significant, in view of Guatemala's disastrous earthquake, that more than half of the eighty volcanoes in Central America rise in Guatemala.(51)

Moreover, the Office of Emergency Preparedness has declared "The people, property, economy, and ecology of the area surrounding the active volcanoes of Hawaii, the Pacific Northwest, and Alaska are endangered by the threat of future volcanic activity."(52)

The OEP Report states that debris, pyroclastic and mud flows, floods and lava flows could create danger for people and property in the immediate vicinity of volcanoes. Fallout of airborne ash and flood-transported debris affect people within a radius of fifty to seventy-five miles. Flying ash could cause people serious respiratory problems or even suffocate people; ash and debris ruin crops and reduce the productivity of the land for many years. If fallout is sufficiently great, buildings collapse and people and animals may be trapped. "In short, all life and property is endangered by volcanic eruptions, and the extent of the disaster is chiefly a function of the magnitude of the particular event."(53)

Some relatively recent adverse affects from volcanoes include the pelting of an airliner with fist-size rocks, cracking cockpit windows and spraying ash into engine intakes. This caused an Al Nippon Lockheed Tri-Star Airliner carrying 317 passengers over Japan to make a forced landing at Sapporo. Since volcanoes are often located close to the oceans, when their lava flow reaches the sea it creates an acid-laden mist known as "laze."(54)

Cause of Volcanoes

The cause of volcanoes is not clearly identified, but in the last 100 years scientists have gained better understanding of their origin. Since they are generally located near the sea, it was once thought that the entrance of sea water into the supposedly molten interior of the earth was responsible; this is not the accepted view today. A more recent theory asserts that weakness in the earth's crust combined with pressure exerted in the formation of folds and faults is responsible for starting volcanoes. It is believed that chambers of extremely hot magma (molten material) lie from tens to hundreds of miles below the volcano, and the passage called the *conduit* carries lava gases and rock to the earth's surface.

The extreme heat and magma contribute to the process of *orogenesis* (mountain building). Shifting rocks are forced to great depth, where increased temperature and pressure cause them to be dissolved and converted to magma (*anatexis*). Once formed, these magma exert pressure on the surrounding solid material. Where a fissure in the solid material extends to the surface, the molten material surges upward, and a volcano is born.

The explosive eruptions are due to gas pressures built up when the conduit becomes plugged and is finally blown out. As the lava (molten rock) flows out of the earth it cools and forms a cone which can eventually become a high mountain. Cotopaxi in Ecuador rises to a majestic 19,347 feet and Argentina's Aconcagua to 22,831 feet.

Occasionally when a volcano erupts its majestic height can be lessened considerably. For example, when Mount St. Helens erupted in the state of Washington during May 1980, twelve hundred feet—or one-seventh of its height—was blown away.(55)

Volcanic Emissions

The gaseous materials ejected from a volcano include steam, carbon dioxide, hydrogen chloride, hydrogen fluoride, hydrogen sulfide, sulfur dioxide, and hydrogen. In the presence of water vapor most of these gases form toxic, irritating, and, in some cases, foul smelling acid vapors. Other emissions include pyroclastic ejecta (flying ash), pyroclastic flows (rocks thrown into the air but too heavy to remain aloft, so they flow down hill like a hot liquid, at speeds up to 60 mph, like an avalanche, setting every-

A massive volcanic eruption spews volcanic ash and dust into the air contaminating the landscape for miles around. (Photo courtesy of Ron Buckingham.)

thing on fire in their path), and lava flows (streams of molten rock that rise from the vent, explosively or quiescently, and flow down the side of the mountain at speeds up to 30 mph).

Solfataras and *fumaroles* are volcanic vents that expel only gases and magmatic vapors; a volcano may have a number of such openings in addition to the main conduit.

Kinds of Volcanoes

Volcanoes have been classified into four groups: Hawaiian, Strombolian, Vulcanian, and Peleean. Hawaiian volcanoes have a fluid magma that usually erupts without violence. The Strombolian has a thicker lava from which gases are released with explosions. The Vulcanian has magma almost as thick as tar; it also has violent explosions that produce heavy clouds of volcanic dust. The Peleean volcano is the most violent of all, and the clouds above its eruptions glow as though on fire.(56)

Some volcanoes are constantly active, such as Stromboli, which erupt so regularly they are like lighthouses on their little islands. Mt. Etna is an intermittent volcano that erupts at fairly regular intervals; this type of volcano erupts in cycles, with the duration depending on the length of time it takes to build up enough heat and pressure to reactivate.

Dormant volcanoes are those not active at present, but no one knows when they will erupt again. They have not been quiet long enough to be designated *extinct*, as have those which have not erupted since the beginning of recorded history.

Another grouping of volcano types shows three classifications:

1. shield volcanoes
2. cinder-cones volcanoes
3. composite volcanoes.

The first of these, shield volcanoes, are formed by free flowing lava spilling from a vent and spreading widely. This produces a low, broad, dome-shaped mountain. The second type, cinder-cones, build up when *tephra* (cinders) erupts from a vent and returns to the earth near the event area. The third type, composite volcanoes, are formed by a combination of lava and tephra from a single vent. The materials accumulate in alternate layers around the vent and form a towering cone-shaped mountain.(56)

Location of Volcanoes

Most volcanoes are distributed in belts of weakness in the earth's crust near the great mountain chains. About 79 percent of known volcanoes are contained in the Circum-Pacific System, though their distribution is irregular, with 45 percent in the western Pacific islands and only 17 percent in the Pacific coastal region of North and South America. (See Figure 10-3) The United States has 27 active volcanoes in the Aleutian chain of Alaska and 2 in Hawaii. Along the Pacific Coast of the United States is a chain of 12 volcanoes, in Washington, Oregon, and California. They are all poten-

Figure 10-3.

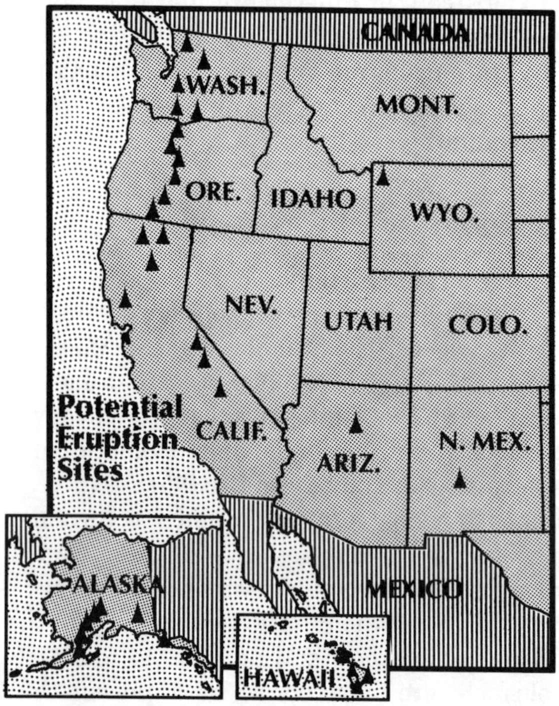

Figure 10-4.

Where volcanoes occur

This map shows the location of many volcanoes. It also shows the earth's large, rigid plates. Volcanoes usually occur along the edges of the plates.

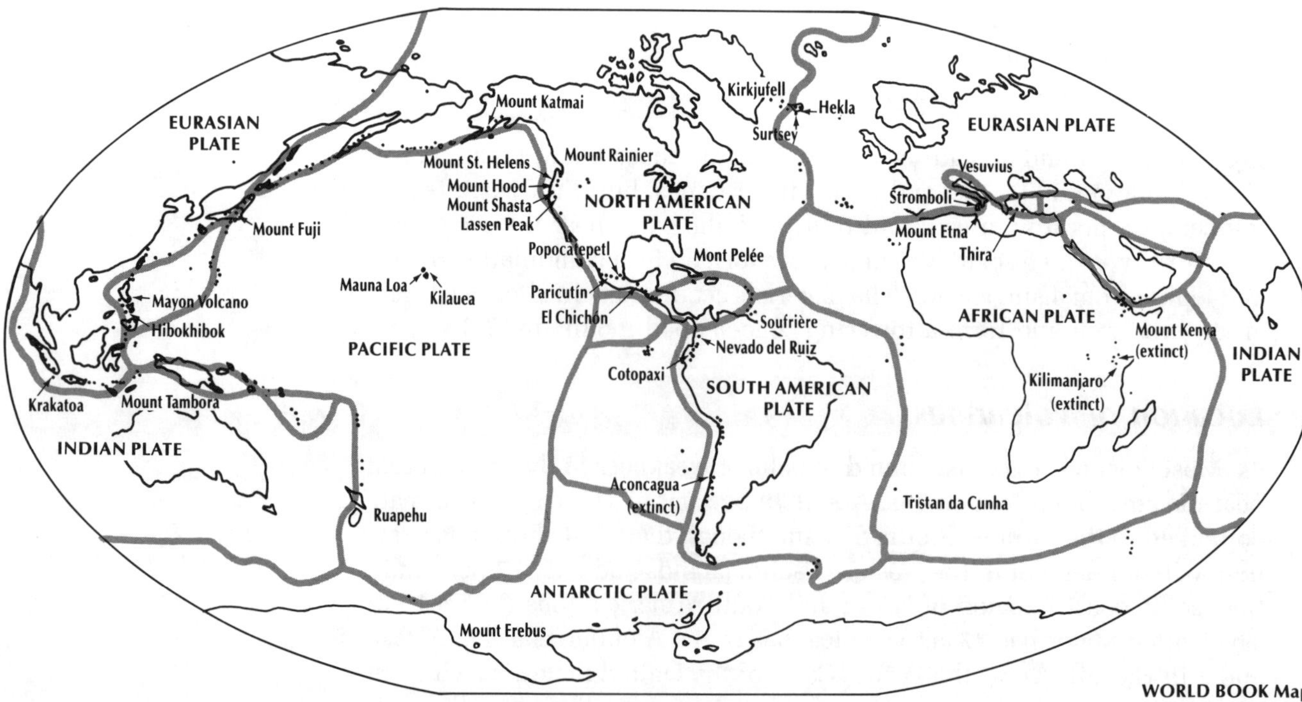

WORLD BOOK Map

tially dangerous to the urban and agricultural centers of the Pacific Northwest. See Table 10-3 for the most prominent U.S. locations and Figure 10-4 for a world map showing the relationship between the earth's large rigid plates and volcanoes.

Table 10-3. Volcanoes in the U.S. Pacific Northwest

Washington	*Oregon*	*California*
Mt. Baker	Mt. Hood	Mt. Shasta
Glacier Peak	Mt. Jefferson	Cinder Cone
Mt. Rainier	Three Sisters	Mt. Lassen
Mt. St. Helens	Crater Lake	Mt. Konocti
Mt. Adams		

Effects of Volcanoes

A violent eruption can produce a lava fountain a half-mile high. Airborne ash, containing cinders and pyroclastic rocks, is blown into the air and remains suspended for long periods before eventually falling to earth. About one million cubic feet of these materials were produced every second in the first half hour of the 1947 eruption of Hekla in Iceland; before it subsided it ejected 225 trillion cubic feet, enough to cover a 100,000 square-mile area with a one-inch deposit. An eruption of Vesuvius in 1779 caused a river of lava 65 feet wide that flowed at a rate of 25 to 30 feet per second.

When solid particles are suspended in a gas, a pyroclastic flow occurs which acts like a liquid, capable of moving with great speed down the mountain. In the Katmai, Alaska, eruption of 1912, 300 million cubic feet of ignimbrite (rock layers formed from the deposit of these materials) were produced in 60 hours. These ignimbrite sheets may be as deep as 1,600 feet.

The choking fine ash thrown into the air from the Crater Lake volcano about 6,600 years ago covered central Oregon with a layer more than six inches deep. "If similar eruptions were to take place today and the wind happened to be blowing in an adverse direction, Mt. Ranier or Mt. Hood would shower a layer of debris more than six inches deep over downtown Seattle or Portland. In such an eventuality, the loss of life and property would, of course, be enormous."(57)

Most of the largest eruptions have occurred in uninhabited areas, such as the Valley of Ten Thousand Smokes (1912) and Bezymianny, Kamchatka (1956); but, as the OEP Report (cited in reference 4) points out, such an eruption in a densely populated area would produce a "catastrophe of unheard-of proportions."(58) Catastrophic enough was the Mount St. Helens eruption on May 18, 1980.

When the Mount St. Helens volcano erupted, a stratospheric cloud of grit, ten miles up, fanned out across the country and dumped ash as far east as Montana. Fifty-seven people died in the Mount St. Helens eruption. Shock waves leveled 44,000 acres of fir trees and boiling mud traveling 30 miles per hour overflowed local river banks, smashed bridges, swallowed homes, and killed millions of salmon and trout. Some rivers got so hot that fish actually jumped out. So much mud poured on to one stretch of the Columbia River, normally 40 feet deep and six hundred feet wide, that the passage shrank to 14 feet deep and two hundred feet wide—trapping 31 ocean going ships in Portland, Oregon, and Vancouver, Washington.(59)

Examples of Volcanic Activity

The most active volcanoes in the United States, located in populated areas, are Haleakala on Maui and the Hualalai, Mauna Loa, and Kilauea volcanoes on the big island of Hawaii. Since 1750, Kilauea has been constantly active, with 32 eruptions on the record, but Mauna Loa has erupted only about once every 3.6 years on the average since 1832; Haleakala last erupted in 1750.

Eruptions of many volcanoes have been sufficiently close to inhabited areas to cause disasters. Mount Katmai in Alaska has been made a national monument since its last eruption in 1912. In 1963, Mt. Agung on the Indonesian island of Bali erupted and killed about 1,500 persons. In the same year a lava flow started building the new island of Surtsey about twenty miles south of Iceland; by the time the lava stopped flowing in 1967 it had risen to a height of 567 feet and covered over a square mile.

After twenty-five years of inactivity, Mauna Loa in Hawaii started spouting yellow and red molten rock in 1975, creating a fountain over 40 feet high along the northeast flank of this 13,680-foot mountain. Two major lava flows issued from the northeast rift area, one moving north and the other in the direction of Hilo, the island's principal city of 35,000 persons; it lies a distance of thirty-five miles from Mauna Loa's summit. In a 1942 eruption, lava flows stopped twelve miles from Hilo. Prior to the 1975 eruption, its previous eruption was in 1950, when it put on a twenty-three-day display of fountaining, sending 600 million cubic yards of lava into the sea on the southwest side of the island.(60) See Table 10-4 for a listing of some famous volcanoes.

Predictability of Volcanic Eruptions

Despite regular monitoring of volcanoes in the Cascade Range, geologists admit that they cannot predict when one might erupt. Volcanic eruptions usually are preceded by earthquake activity. One U.S. Geological Survey report indicated that increased volcanic activity was detected on Mount Baker in Washington. The report predicts that if this mountain erupts it could send millions of cubic yards of rock, mud, and ice hurtling down the mountainside. "The mud flow could move at speeds of more than thirty miles per hour, engulfing hundreds of acres of surrounding forest. If it reached Baker Lake the mud flow would create a twenty-foot-high wave that might completely wash out the surrounding beaches, campgrounds, and resorts."(61)

In regions where volcanic activity poses a threat to safety of residents, the U.S. Geological Survey is carrying out programs to warn of imminent danger. In Hawaii, the prediction and warning of volcanic hazards is performed by the Hawaiian Volcano Observatory (HVO) on the rim of the

Table 10-4. Some Famous Volcanoes

Name	*Location*	*Height above sea level*		*Interesting facts*
		In feet	*In meters*	
Aconcagua	Argentina	22,831	6,959	Highest mountain in Western Hemisphere; volcano extinct.
Cotopaxi	Ecuador	19,347	5,897	Eruption in 1877 produced mudflow that traveled about 150 miles (241 kilometers) and killed about 1,000 people.
El Chichón	Mexico	3,478	1,060	Eruption in 1982 killed 187 people and released a cloud of dust and sulfur dioxide gas high into the atmosphere.
Hibokhibok	Philippines	4,363	1,330	In 1951, red-hot cloud of gas and dust killed about 500 people.
Kilauea	Hawaii	changing	changing	Destroyed 200 homes as of 1991 changing radically the surrounding area.
Krakatoa	Indonesia	2,667	813	Great eruption in 1883 heard about 3,000 miles (4,800 kilometers) away; produced sea waves almost 130 feet (40 meters) high that drowned about 36,000 people on nearby islands.
Lassen Peak	California	10,457	3,187	One of several volcanoes in the Cascade Range; last erupted in 1921.
Mauna Loa	Hawaii	13,677	4,169	World's largest volcano; rises almost 30,000 feet (9,100 meters) from ocean floor and is about 60 miles (97 kilometers) wide at its base.
Mont Pelée	Martinique	4,583	1,397	Glowing cloud from 1902 eruption destroyed town of St. Pierre, killing about 38,000 people in minutes.
Mount Etna	Sicily	11,122	3,390	About 20,000 people killed in 1669 eruption.

Table 10-4. Some Famous Volcanoes (continued)

Name	Location	Height above sea level		Interesting facts
		In feet	In meters	
Mount Katmai	Alaska	6,715	2,047	Eruption in 1912 produced glowing flood of hot ash that traveled about 15 miles (24 kilometers) and formed Valley of Ten Thousand Smokes.
Mount St. Helens	Washington	8,364	2,549	In 1980, violent eruptions released large amounts of molten rock and hot ash; killed 57 people.
Mount Tamboro	Indonesia	9,350	2,850	In 1815, eruption released 6 million times more energy than that of an atomic bomb; killed over 12,000 people.
Nevado del Ruiz	Colombia	17,717	5,400	Eruption in 1985 triggered mud slides and floods; destroyed city of Armero and killed about 25,000 people.
Paricutín	Mexico	9,213	2,808	Began in farmer's field in 1943; built cinder cone over 500 feet (150 meters) high in six days.
Stromboli	Mediterranean Sea	3,031	924	Active since ancient times; erupts constantly for months or even years.
Surtsey	North Atlantic Ocean	568	173	In 1963, underwater eruption began forming island of Surtsey; after last eruption of lava in 1967, island covered more than 1 square mile (2.6 square kilometers).
Thira (formerly Santorin)	Mediterranean Sea	1,850	564	Eruption in about 1500 B.C. may have destroyed Minoan civilization on Crete; legend of lost continent of Atlantis may be based on this eruption.
Vesuvius	Italy	4,190	1,277	In A.D. 79, produced history's most famous eruption, which destroyed towns of Herculaneum, Pompeii, and Stabiae.

Source: World Book

Kilauea Crater. It measures the earthquake activity, long-term tilt and bulging (deformation), makes regular temperature measurements and chemical analysis of the gases and lava of the volcanoes, and assesses the activity to establish normal patterns. The detection of any unusual activity can then be the basis for preparing predictions and warnings.

According to the OEP Report, the study of volcanoes in the Cascade Range of the Pacific Northwest is largely concentrated on Mt. Ranier and Mount St. Helens in Washington, and Lassen Peak in California, and has consisted mostly of mapping and evaluating existing hazards, with brief periods of micro-earthquake monitoring of these three volcanoes.(62)

No one can be sure a volcano is extinct. Some thought to be extinct have come to life. In 1973, a volcano that had been dormant for more than 5,000 years erupted on the island of Heimaey near Iceland. It poured so much ash

Molten lava flowing from an erupting volcano careens across the countryside obliterating everything in its path. (Photo courtesy of Ron Buckingham.)

over the island's only town of Vestmaneayjar that all residents were forced to escape.(63)

In 1990, the *Los Angeles Times* reported most persons are willing to take a fatalistic view of dormant volcanoes assuming they will not erupt. In the Mount Shasta area, including the communities of Weed, Mt. Shasta, McCloud, and Dunsmuir, the residents paid little heed to a United States Geological Survey (USGS) report that stated that, "Future eruptions of Mount Shasta are certain to occur eventually and can be neither prevented nor stopped."(64)

Damage Control—Volcanoes

Since the eruption of volcanoes cannot be prevented there will always be hazard to persons who live in the vicinity. Two types of damage limitation action can be taken; plot the areas that are likely to be in the path of molten lava flows and institute zoning restrictions that control use of land in such areas.

Emergency Action

Fire departments in areas where volcano eruptions pose a threat must keep in communication with agencies which provide warning of impending danger and activate the disaster plan that deals with the volcano danger. Such a plan should provide for rescue of persons in structures in the path of the lava flows, and containment of fires started by molten rock and hot cinders. Breathing apparatus should be used in areas where the air is heavily contaminated with sulfur dioxide, hydrogen sulfide, and acid vapors while conducting rescue and fire suppression operations.

If a volcano "blows its top" with no warning and showers the city with burning cinders and suffocating ash, the fire department will encounter many simultaneous fires. If the eruption is accompanied by strong earthquakes, which is likely, the problems will be compounded.

An aerial survey might be hazardous due to flying rocks and cinders, and the sky would be so filled with ash that visibility would be greatly restricted. Debris fallout can make surface travel hazardous and eventually might bury the community completely. If this prospect appears likely, the fire department has no alternative to assisting and joining in the evacuation of the populace.

LANDSLIDES

One secondary effect of an earthquake or volcanic eruption is a landslide. Independent landslides, large and small, also occur. Fire departments sometimes respond to emergencies involving the burial of people, autos, and even buildings under debris that slides off hills and mountains. Sometimes structures slide and collapse, or a mass of earth may break from a mountain and carry structures along. Sometimes land falls away from under a building, leaving it teetering on the edge of a cliff. As states become more populated there is a tendency to subdivide hillsides and build terraced residences without thorough investigation of geological conditions of the region. To add to the problem, natural vegetation is often cleared away to such an extent that soil erosion takes place, and the burden on the land is increased by many tons in the form of swimming pools or other structures built near the edge of the slope or cliff. A ground cover may be planted to replace brush and trees that have been removed (sometimes to reduce the fire hazard) and then watered so frequently as to saturate the soil, with the result that the risk of landslide is increased.

It has been estimated that costs of landslides every year in the United States amount to hundreds of millions of dollars including damage to highways, railroads, utility company structures, public works, industrial property, residences, and personal property.(65)

Causes of Landslides

Landslides occur when a mass of earth or rocks slips down the side of a slope. They may be falls, slides, slumps, or flows. Water seepage, cracks in rocks, earthquakes and volcanic action are common causes of landslides. Heavy rains, snow accumulation, and erosion are additional causes. Action of waves on a coastline may start a landslide.

Falls of earth or rock occur from natural cliffs and over steepened river banks, or from cuts or excavations made for highways, railroad roadbeds, and construction sites.

Earth blocks that slide on fracture surfaces may be a few hundred to several thousand feet across, and the fracture surfaces may be a considerable distance below the surface of the earth. Such slides usually involve a slow downward movement of an entire slope due to a subsurface shearing; when it moves as a unit it is called "slumping." If it breaks into small units it is called "avalanching." These landslides move a few inches to feet per day, some considerably faster.

Flows may be dry or wet and involve greater masses of material that continue downhill far beyond the base of the slope from which they originate. Mudflows usually begin on steep slopes when shallow soil layers are transformed into a liquid state, usually by rainfall, though leaking water systems might be the water source. When this happens, liquefied soil flows like streams of soft plastic down canyons, arroyos, and valleys, spreading on the plain below. Large flows may move as fast as 20 mph, carrying along houses, barns, fences, trees—everything on top of the ground and in the ground, including sewer pipes, gas lines and other utilities.

Following a brush fire that denuded California hillsides, and a subsequent long period of steady rain, huge flows of jelly-like mud flowed through windows and doors of beautiful homes at the foot of a hill; this river of mud continued into the street, leaving a deposit several feet deep that later congealed into a hard mass of clay, encasing the furniture like cement. Houses were pushed from foundations, causing them to topple, collapse, break, or become buried completely.

Slumps occur when the strength of the ground is exceeded by forces that

Massive landslides can wreak havoc with the local landscape and with transportation. (Photo courtesy of Ron Buckingham.)

tend to move the earth downhill. "These forces arise from the weight of earth materials and from ground water seepage. Landslides are triggered by both natural and manmade process that reduce the strength of ground or increase the forces associated with weight and seepage. Such processes include rainfall, erosion, earthquakes, and various modifications such as cutting and filling that are made to ground by man."(66)

An OEP Report lists five causes for landslides.

(1) Slow erosion, which results in rock falls due to gradual wearing away from supporting materials and avalanches that sweep down a mountainside gathering momentum and masses of loose debris.
(2) Water, causing marginally stable slopes to give way in avalanches, slumps or mudflows.
(3) Glacial action, melting and removing the support that glacier ice gives to canyon walls, resulting in landslides and mudflows.
(4) Earthquakes and volcanic action, shaking material loose and forming an avalanche, such as the one that formed Earthquake Lake in 1959 at Hebgen, Montana. Earthquakes create vibrations that cause soil to act like liquid, with resulting ground failure and slumping over wide areas; this type of action took place in the Van Norman Dam during the 1971 earthquake in Sylmar, California, necessitating evacuation of 80,000 residents for several days while the reservoir was being drained.
(5) Engineering defects, such as miscalculating slope stability when

excavating cuts through a slide area for dams, bridge abutments, and buildings.(67)

Effects of Landslides

The primary effect of landslides is removal of a vast amount of earth accompanied by destruction of everything in the path. On gradual slopes where soil liquefaction occurs, foundation material can lose strength and an entire town can gradually move downward.

A secondary effect is air blast. When a large amount of material slides into a canyon it pushes air forward and to the side with great force. There are reports of persons having clothes torn off and receiving serious injuries from air blast.

Another effect of landslides is the huge water wave (seiche) formed when a slide hits a lake or bay. When this happened in Lituga Bay, Alaska, following the 1958 earthquake, a wave was formed more than 1,700 feet high.

Landslides affect the economy. The loss to the U.S. and Canadian railroad systems is $4 million annually without considering the cost of repairing the equipment, damage to the freight, or loss of revenue. Another example is the $20 million spent in a 20-year period for Grand Coulee Dam to correct damage from landslides and avoid further damage. Many states spend between $100,000 and a million dollars each year to handle the damage from slides to highways.

Property damage is not the only effect of landslides. They have caused many deaths in the United States and elsewhere.

Location of Landslides

Table 10-5 shows the major areas in the United States where landslides occur. Landslides are widespread in central and western United States, and are most prevalent in the Colorado Plateau, Wyoming, Montana, southern California, Oregon, and Washington. These regions average about 100 landslides per year per 40,000 square miles in hill areas, and about ten per year per 40,000 square miles in plateau areas. The losses in these regions have been about 325 million dollars.

Examples of Landslide Disasters

On May 1, 1991, a Soviet town vanished from earth and 81 people died when the village of Khakhieti was obliterated by a mountainslide. The village was buried so completely that not one of its two-story houses was left visible. This slide, triggered by an earthquake, left no ruins, no bodies, and no signs of the farming village which once was there. It was described as "just dirt, rocks, and fallen oak trees."(68)

Japan's Ministry of Construction pinpointed 35,000 sites where rapid land development had upset the balance of nature and could endanger lives. In July 1976 torrential rains on the overbuilt Izu Peninsula, a popular weekend resort near Tokyo, triggered 487 landslides, leaving 9 dead and 9,794 persons homeless.(69)

Table 10-5. Landslides of the United States

Type Slide	*Major Areas*	*Number of Historical Slides*	*Approx. Frequency per 100 sq. mi.*	*Approx. Frequency per 40,000 sq. mi.*	*Est. Prop. Damage (millions $ adjusted to 1971 values)*	*Recorded Deaths*
Rockslide & rockfall	White, Blue Ridge, Great Smoky, Rocky Mtns., & Appalachian Plateau	Several hundred	—	1 per 10 yrs.	30	42
Rockslump & rockfall	Widespread in Central & West U.S.; prevalent in Colo. Plateau, Wyo., Mont., Southern Calif., Oreg., & Wash.	Several thousand	Av. 10 per yr. hill areas; 1 per yr. plateaus	100 per yr. hill areas; 10 per yr. plateau areas	325	188
	Appalachian Plateau	Several thousand	1 per 10 yrs.	70 per yr.	350 (mainly in highway & railroad damage)	20
	Calif. Coast Ranges, Northern Rockies	Several hundred	1 per 10 yrs.	10 per yr.	30	—
Slump	Maine, Conn. Riv. Valley, Hudson Valley, Chicago, Red Riv., Puget Sound, Mont. glac. lakes, Alaska	About 70	1 per 100 yrs.	1 per yr.	140	103
	Long Island, Md., Va., Ala., S. Dak., Wyo., Mont., Colo.	Several hundred	1 per 50 yrs.	1 per yr.	30 (mainly to highways & foundations)	—
	Mississippi & Missouri Valleys, eastern Washington & southern Idaho	Several hundred	1 per 10 yrs.	1 per yr.	2	—
	Appalachian Piedmont	About 100	—	1 per yr.	less than 1	—
Debris flow & mudflow	White, Adirondack, & Appalachian Mountains	Several hundred	1 group slides, 10+ per group, per 100 yrs. in White Mtns. & North Carolina	1 group slides, 10+ per group, per 15 yrs.	100	89

When Peru experienced its 7.75 magnitude earthquake in 1970, most of the 70,000 persons who lost their lives died as a result of landslides. Yungay, a town of 25,000 people, was completely destroyed when the earthquake loosened the northern face of Mt. Huascaran and an avalanche of water, rocks, and mud roared down upon it. "A few people reached high ground in time. Rescuers found them huddled in a cemetery. Everyone and everything else had been obliterated."(70)

Four years later, in April 1974, Peru had other massive landslides which killed 250 persons; an entire village was buried, and its 500 residents were presumed dead.(71)

On July 31, 1972, in the Philippines, more than two weeks of incessant rains caused widespread flooding and landslides on the island of Luzon, killing 427 persons and leaving about 2 million homeless.(72)

In the middle of the night on August 17, 1959, an eight-state area felt the first heavy jolt of the Hebgen Lake earthquake, and a huge landslide consisting of about 38 million cubic yards of rock slid into the Madison River Canyon at a speed of about 100 mph. People, cars, and trailers were crushed and buried. Some of the survivors were engulfed by a huge wave of water from the displaced Madison River. The slide blocked the canyon so that hundreds of vacationers asleep in their tents, campers, trailers, and lodges near Hebgen Lake and along the river had their escape routes cut off.

By dawn, rescue units were on the way from many agencies in Idaho and Montana. Aerial reconnaissance was made and the Forest Service parachuted smokejumpers into the canyon to give first aid and set up communications. Others came in on horseback and by helicopter to assist the survivors. By evening, bulldozer roads had been built around highway slipouts, and all who wished to leave were able to get out of the canyon. Twenty-eight men, women, and children lost their lives in the disaster.(73)

The slide was about a mile long, and three-quarters of a mile wide, and it formed a dam about 225 feet thick. After the valley was blocked by the mass of displaced earth, the waters that poured over the dam as a result of the seiche, plus the waters normally released through the spillway at the dam, began to fill the canyon and form Earthquake Lake.

Landslide Prevention

Although there is little that can be done to prevent earthquakes or heavy rains, which are the cause of many landslides, the danger of landsliding can be greatly lessened in hillside developments if construction practices are regulated by grading ordinances and building codes and if geologic and engineering advice is followed. Also, the laying of rights-of-way for dams, bridge abutments, and other structures should be done with special consideration for the landslide characteristics of the area.

"If unavoidable risk must be taken, good engineering practice can do much to reduce the danger. Methods of slide prevention include removal of unstable materials, selection of a safe slope factor in excavation, provision for both surface drainage and subsurface drainage between the overburdened material and bedrock, and the installation of retaining walls, bulkheads, pilings, and tie rods.(74)

Once a landslide begins, little can halt its progress. In the Palo Verdes movement near San Pedro, California, residents of homes below the main scarp attempted to continue living in the houses by connecting flexible pipe and hoses to utility lines before they became severed by the movement; tunnels were dug back into the hillside to drain ground water, and gas heaters were installed in them to dry out the soil; steel rods and retaining walls proved useless, as did the other measures tried, and eventually the homes had to be moved, disassembled, or abandoned, as the subdivision gradually slid down to the ocean below.

Although it might be very costly, highways, railways, or aqueducts that must traverse unstable areas where control of slumps is not possible can sometimes be bridged over the area, allowing movement to continue without disturbing the construction project.

Emergency Action—Landslides

From the fire department's standpoint, landslides that involve the eventual destruction of buildings through the slow process of slump are not problems calling for emergency response unless a gas line is sheared off during the movement, fallen power lines start a brush fire, or some similar emergency dictates an immediate need.

When part of an embankment fell away from beneath a residential area in the Pacific Palisades a number of years ago, Los Angeles fire fighters assisted persons to escape from homes left precariously perched on the edge of the bluff. In some cases it was possible to bridge a gap with ladders and shore up the overhanging building sufficiently to permit the occupants to remove some of their personal property before the house fell or slid any further.

In another instance where a series of homes were situated on the top of a bluff in the Santa Monica Mountains that started to slump, many homes had swimming pools. Fire department pumpers were called and using submersible pumps and siphon ejectors, the pools were emptied quickly to lighten the burden on the affected terrain. In a few weeks the homes were beyond saving but the occupants had ample time to remove possessions and find other quarters.

In mud flows that struck many homes following brush fires in California and western states in the early 1990s, fire departments dispatched companies to assist in filling sand bags to use in diversion of flowing mud and water. An attempt was made to keep homes from being inundated with the putty-like effluent that was oozing out of the gullies and arroyos as a result of the heavy rains. The efforts of many fire fighters were concentrated on the digging of channels to provide a bypass for homes being crushed, pushed, and infiltrated by thick mud flows slowly moving down the slopes, but it was often a losing battle. Bulldozers and skiploaders were put to work in an attempt to divert the mud streams to the streets, where similar equipment would keep it moving. Between rains, hastily bulldozed bulwarks were constructed in gullies that posed the greatest threat from mud flows, but these were soon filled after the next rain commenced and provided only temporary relief.

CONCLUSION

In this chapter, we have covered the terrain-based types of natural disasters that are awesome and devastating in their effects. Earthquakes which strike without warning have the power to level even the most earthquake-resistant structures and wreak havoc on an unsuspecting community. Volcanoes on the west coast, Hawaii, and Alaska have potential atmospheric repercussions that could stretch across the United States and Canada, and economic repercussions that could touch every part of the North American continent.

Finally, landslides because they occur independently are found everywhere. A brief study of these three major types of disasters would easily lead one to believe that the term *terra firma* is a conflict in terminology. In the next chapter, we turn our attention away from natural disaster and toward those caused by the mobility of mankind—transportation disasters.

CHAPTER 10 REFERENCES

1. Associated Press reports in *The Cincinnati Enquirer*, January 3, 1991, p. A-2, and June 30, 1991, p. A-8.
2. Webre, A. L., and P. H. Liss. *The Age of Cataclysm*, Berkeley Publishing Corporation, distributed by G. P. Putnam's Sons, New York, 1974, p. 11.
3. *The Cincinnati Enquirer* April 28, 1990.
4. *Disaster Preparedness*, Report to the Congress, Executive Office of the President, Office of Emergency Preparedness, January, 1972, Washington, D.C. U.S. Government Printing Office, Vol. 1, p. 73.
5. Gutenberg and Richter, *Seismicity of the Earth*, Princeton University Press, 1949.
6. Ibid.
7. Canby, T. "Can We Predict Quakes?" in *National Geographic*, Volume 149, No. 6 (June 1976), p. 830.
8. *UCLA Monthly*, Vol. 17, No. 3 (Jan./Feb. 1987), pp. 10–11.
9. Gwynne, P., with S. Liu. "Thunder Out of China," *Newsweek*, June 16, 1975, p. 45; also see No. 5, p. 834.
10. Carrol, R., et al. "China's Killer Quake," *Newsweek*, August 9, 1976, p. 30.
11. See note 7, p. 831.
12. The modest 1986 earthquake which struck East Los Angeles *may* have fulfilled Dr. Whitcomb's prophecy, but this is a quake-prone region, regardless.
13. Alexander, G. "Experts Won't Accept Quake Prediction," *Los Angeles Times*, May 1, 1976, Part I, p. 1.
14. "The Big Seismic Scare," in *Eastsider*, Cincinnati, Winter 1991.
15. Alexander, G. "Forecast Role Seen for 'Silent Quakes'," *Los Angeles Times*, March 31, 1976, Part II, p. 2.
16. Alexander, G. "Concern Voiced Over Bulge in Land Along San Andreas Fault," *Los Angeles Times*, March 12, 1975, Part II, p. 1; and "Cracks in Quake Forecast Seen," *Los Angeles Times*, December 2, 1975, Part II, p. 1.
17. Paddock, R. "Senate OK's Bill to Perfect U.S. Quake Warnings," *Los Angeles Times*, May 25, 1976, Part I, p. 1. at p. 3; also, Senator Alan Cranston, *Report to Californians*, No. 34, July 1976, p. 3.

18. "Quake Forecasts: How Will We React?" *The UCLA Monthly*, Vol. 6, No. 2 (November–December, 1975).
19. Cited in note 1, p. 13; *Nature*, Vol. 241, January 12, 1973, p. 85.
20. "Mexico City Rocked by Quake" in *The Cincinnati Enquirer*, June 1, 1990, p. A6.
21. Alexander, G. "Possible Link Between Great Quakes Studied," *Los Angeles Times*, April 7, 1975, Part II, p. 1.
22. See note 4, Vol. 3, p. 72.
23. ______, *Los Angeles Times*, April 7, 1975.
24. *The World Book Year Book*, 1971, Chicago, IL: Field Enterprises Educational Corporation, p. 294.
25. Heck, Nicholas. *Earthquakes*, Princeton NJ: Princeton University Press, 1936.
26. "Quake Kills Thousands," *The Cincinnati Enquirer*, June 22, 1990, p. A-1.
27. *Report of the Fire and Rescue Problem—Earthquake Disaster, Managua, Nicaragua, December 23, 1972*, Office of Emergency Services, Fire and Rescue Division, State of California, February 20, 1973, p. 12.
28. "Large Region of Soviet Asia Hit by Strong Quake," *Los Angeles Times*, May 18, 1976, Part I, p. 1.
29. "Death Toll Climbs in Italian Quake," *Los Angeles Times*, May 8, 1976, Part I, p. 1.
30. "At Least 9,000 Feared Dead in Aftermath of Earthquake in Indonesian New Guinea," *Los Angeles Times*, July 9, 1976, Part I, p. 12.
31. "2 Major Earthquakes Strike Asia During 2-Hour Period," *Los Angeles Times*, August 17, 1976, Part I, p. 1; "Phillipines Quake Toll Tops 4,000," *Los Angeles Times*, August 19, Part I, p. 6.
32. McDowell B. "Earthquake In Guatemala" in *National Geographic*, Vol. 149, No. 6, June 1976, p. 810.
33. Earthquakes of the 1980s taken from the respective yearbooks of the *World Encyclopedia*.
34. Blackburn, Frank T., Charles Scathorn, and Jack Neil. "Earthquake Preparedness for Fire Departments" in *Firehouse*, Vol. 14, No. 12 (December 1989, p. 37.
35. Scathorn, Charles. "First Report on the Quake" in *Firehouse*, Vol. 14, No. 12 (December 1989), p. 31.
36. *Rapid Visual Screening of Buildings for Potential Seismic Hazards*, a handbook, FEMA-154 Earthquake Hazards Reduction Series 41, July 1988."
37. Hume, E. "Unsafe Buildings Endanger 200,000," *Los Angeles Times*, April 7, 1975, Part II, p. 1.
38. Baker, E. "Panel OK's Shift in Quake Safety Policy," *Los Angeles Times*, August 28, 1976.
39. Blackburn, Frank T., et al. 1991, supra. p. 48.
40. Ibid. p. 52.
41. *Report of the Los Angeles County Earthquake Commission—San Fernando Earthquake, February 9, 1971*, Los Angeles County Board of Supervisors, November 1971, p. 20.
42. Blackburn, Frank T., et al. 1991, supra. p. 48.
43. Dektar, Cliff "Extensive Rescue Operations Follow Los Angeles Quake" in *Fire Engineering*, June 1971, p. 38–40.
44. Long, K. "The Fire Department's Fire Problem (Resulting From the Unexpected Structural Damage)," *Earthquake Fire Seminar*, p. 9.
45. Blackburn, Frank T., et al. 1991, supra. p. 37.
46. Ibid. p. 49.
47. Connick, W. B. "An Isolated Fire Department's Approach to Earthquake Caused Problems," *Earthquake Fire Seminar*, June 5, 1975, cited in note 34, p. 11.
48. *The Press Democrat*, Santa Rosa, CA April 19, 1986.
49. Enquirer News Services, *The Cincinnati Enquirer*, June 23, 1991, p. A-6.
50. *USA Today*, May 1, 1990.

51. Webre, A., and P. Liss. *Age of Cataclysm*, Berkeley Publishing Company, distributed by G. P. Putnam's Sons, N.Y., 1974, citing C. F. Richter, *Elementary Seismology*, San Francisco: W. H. Freeman and Company, 1958, p. 407.
52. Office of Emergency Preparedness, Report to Congress, *Disaster Preparedness*, 1972, Volume 3, Washington, D.C.: U.S. Government Printing Office, p. 121.
53. ———, *Disaster Preparedness*, U.S. Government Printing Office.
54. *USA Today*, May 1, 1990.
55. *Newsweek*, June 2, 1980, p. 25.
56. *World Book Encyclopedia*, 1991.
57. See note 50, Vol. III, p. 119.
58. The OEP Report (No. 4 above) cites Cliff Ollier, *Volcanoes*, Cambridge, MA, and London: The MIT Press, 1969, p. 154.
59. *Newsweek*, June 2, 1980, p. 25.
60. "Volcano Erupts on Hawaii Island," *Los Angeles Times*, July 7, 1975, Part I, p. 8.
61. Panati, C., with M. Kellogg. "Thunder Mountain," *Newsweek*, August 18, 1975, p. 74.
62. See note 50, Volume I, p. 103.
63. *World Book Encyclopedia*, 1976, Vol. 10, p. 18.
64. *Los Angeles Times*, circa May 1990.
65. *Disaster Preparedness*, Report to Congress, Office of Emergency Preparedness, 1972, Washington, D.C.: U.S. Government Printing Office, Vol. I, p. 87.
66. *Landslides*, U.S. Department of Interior, Geological Survey, 1974, Washington, D.C.: U.S. Government Printing Office, p. 1.
67. See note 63, Volume III, p. 89.
68. Associated Press report in *The Cincinnati Enquirer*, May 2, 1991, p. A-4.
69. "Lost Weekend," *Newsweek*, July 26, 1976, p. 54
70. *The World Book Year Book*, 1971, Chicago, IL: Field Enterprises Educational Corporation, p. 294.
71. *The World Book Year Book*, Chicago, IL: Field Enterprises Educational Corporation, 1975, p. 301.
72. *The World Book Year Book*, Chicago, Il: Field Enterprises Educational Corporation. 1973, p. 307.
73. *The Night the Earth Shook*, U.S. Department of Agriculture Forest Service, Miscel. publication No. 907 August, 1972, U.S. Government Printing Office, Washington, D.C.
74. See note 1, Volume III, p. 97.

CHAPTER 11

TRANSPORTATION DISASTERS

"ABOUT 4,000 PERSONS, almost a third of the loss of life in all fires, die in transportation fires each year, most of them on the nation's highways. The NFPA estimates that about 500,000 transportation fires each year result in property losses of more than $300 million." The fact that more than 50,000 people may lose their lives annually in auto accidents is not commonly viewed as a disaster, because these deaths occur a few at a time; but when 27 persons, mostly children, died in a bus crash on Interstate 71 near Carrollton, Kentucky, on May 14, 1988, the tragedy was acknowledged as a "disaster."(1)

Regardless of how many are killed at a time, or whatever the form of transportation involved, the observation made by Dr. Anne Wright Phillips, M.D., as a member of the National Commission on Fire Prevention and Control, will be applicable: "The public will look to the fire chief and fire fighter for knowledge and direction and perhaps to blame. Simply because one is a fire chief, one is expected to know how to deal with emergencies."(2)

Transportation accidents can be divided into three general classes: those in the air (landing and takeoff of aircraft), those on the ground (highway and railroad disasters), and those on water (rivers, lakes, and seas). These kinds of accidents may be the result of an impact or collision; or of the ignition of hazardous materials; or the very threatening potential of a massive fire or an explosion. In many instances a transportation disaster becomes simultaneously a hazardous materials incident with additional associated hazmat ramifications to be discussed in the next chapter.

On responding to a transportation accident, the fire officer can assume that people will be endangered and that rescue operations might be the first necessary action. Transportation accidents almost always involve human casualties, if only a driver, a few passengers, or railroad or airline crew. The second important consideration for the fire officer will be to estimate how much water, foam, or other extinguishing agent will be needed for the task. This reality, plus the availability of fire fighting and rescue crews, determines how quickly the fire officer must radio for assisting fire companies or other support.

AVIATION DISASTERS

Most of us can easily recall the terrible destruction that occurred when a Pan American Boeing 747 exploded over Lockerbie, Scotland. Two hundred and fifty-nine passengers and crew on board died, as well as 11 on the ground. This tragedy which occurred on December 21, 1988, was traced to a terrorist bomb, signifying the international aspect of disasters and indicating how all of us are touched, at least indirectly, by international terrorism.(3) On May 27, 1991, in a similar incident, 223 persons died when an Austrian airliner in northern Thailand exploded in flight. A team from the National Transportation Safety Board (NTSB) assisted in the investigation, which again linked this crash to a terrorist bomb. Prior to this incident, western intelligence had pinpointed Bangkok, Thailand, as a possible staging sight for terrorism.

Aviation disasters, however, are not new. Looking backward through timo, we are reminded that many of the 119 German Zeppelins met with accidents or disappeared from sight. The Hindenburg ended in a disastrous

Aviation fire fighters must be trained in "high heat" and "swift response" tactics needed in their profession. (Photo courtesy of Salt Lake City Fire Department.)

fire on May 6, 1937, while landing at Lakehurst, New Jersey, at the end of one of her transatlantic flights. In 32 seconds her flimsy bag containing over seven million cubic feet of hydrogen went up in flames, along with 32 passengers.(4)

Modern aircraft carry hundreds of passengers and the equivalent of several tankcar-loads of fuel, with the result that the disaster potential for this mode of transportation has increased in the more than five decades since the Hindenburg tragedy.

The possibility of disaster is not limited to passenger aircraft, but includes military and civilian cargo planes; nor is it confined to fixed-wing aircraft, but includes cargo and passenger helicopters, the capacity of which is increasing.

Types of Disasters

Aircraft accidents have occurred because of collisions with other aircraft in the air or on the ground; impact with tall structures, or with the ground; and from involvement with fire, in the air, or on the ground. Impact may be on the ocean, against a mountain or structure. The aircraft may be disabled because of other reasons, such as wind shear, ice on the wings, sabotage, ingestion of birds by the jet engines, loss of a door or engine, chemical spill, or bomb explosion in the cargo compartment, or by fire.

Causes of Aircraft Disasters

Airport hazards have caused a number of accidents. These include inadequate clearance of operational runways; insufficient navigational aids; runways too short, or obstructed by hills, bays, power lines, blast fences, dikes, or unimproved terrain. In poor visibility conditions, aircraft have struck blast fences and dikes in the undershoot areas off runways.

Overrun portions of runways create another area where accidents occur, owing to inability of aircraft to brake sufficiently to avoid hitting boundary fences, or ending up in a swamp, bay, or against a hill. On September 20, 1989, USAir Flight #5050 crashed into the river at the end of runway 31 at New York's LaGuardia Airport.(5)

In some cases, a community experiences rapid population growth, its airport, surrounding and hemming in a long-existing area airport. Such is the case in North Perry, south Florida. This has been the site of numerous small aircraft accidents resulting in a clamoring by local residents to have the airport shut down or moved. When a multiple fatality incident occurred in late 1989, the residents of Pembroke Pines, Florida, took their fight to the air waves. On an *American Heat* video journal segment, they argued the merits and demerits of moving the airport.(6) While some residents were rather vocal about wanting the airport moved, others were more resigned and philosophical, pointing out that the airport had been there first and that it was essential for maintaining the local economy. Certainly, the field of aviation in general, despite its inherent dangers and occasional disasters, has been of tremendous benefit to mankind, locally, regionally, and nationally.

Prevention of Aviation Disasters

While aircraft accidents are noteworthy, especially of commercial carriers involving multiple fatalities, the excellent safety record of the nation's airlines must be kept in perspective. As the number of flights continues to increase dramatically from all major airports, and as the number of passenger miles flown escalates by the millions, the accident rate per million passenger miles flown actually continues to decrease through time. In 1991 about 35,000 commercial flights took off daily and that number grows each year.

One reason why air transportation has achieved such a remarkable safety record is that safety on aircraft starts on the drawing board. Fire detection and extinguishment systems are normally installed to cover areas that have inherent fire potential; flammable liquid systems are separated from ignition sources; fire-resistant and noncombustible materials are used wherever possible; fire hazardous areas are compartmented from control systems and structural components of the aircraft.(7)

In the crash of United Airlines #232 in Sioux City, Iowa, in 1989 fire-resistant seat upholstery installed at the insistence of the National Transportation Safety Board was one of many factors credited with a high survival rate. Many consumer safty groups continue to argue for even more fire resistant materials and greater fire safety in aircraft.(8)

In addition to providing "non-jammable" exits from the aircraft interior, to permit crash survivors an opportunity to escape, considerable work is being done in improving methods for reducing post crash fire hazards. These include better fuel containment; reduction of ignition sources by de-energizing electrical systems prior to impact; cooling hot engines and inerting the air around them at time of impact; preventing backfire flame emission by cutting off engines prior to touchdown. Experiments have been conducted in the use of fuels that can be "gelled" at time of impact to prevent "misting" of spilled fuel when the tanks rupture.

A disaster may be averted in an aircraft ground impact accident if properly equipped and trained fire fighters can rescue passengers before they are overcome with combustion products or trapped in flames. The ability to evacuate large aircraft rapidly is critical in disaster prevention. This was well illustrated back in 1965 when a Boeing 727 had both main landing gears separate from the plane while landing at Salt Lake City's airport. About 37 seconds after touchdown, ramp crews were responding to the accident site, doors and window exits were being opened, and notwithstanding the heavy, dense smoke that was pouring into the cabin, 50 of the 91 occupants successfully escaped before flames enveloped the whole interior of the aircraft.(9)

In July 1989, a United Airlines DC 10 was enroute from Denver to Chicago when its #2 engine exploded with shrapnel disabling the entire hydraulic system. The skilled pilot managed to steer the crippled jet for 41 minutes into an emergency landing at Sioux City, Iowa, Airport. With no hydraulic control, and capable only of right turns, the aircraft was within yards of a successful touchdown when the right wing tip touched the ground and the jetliner broke up on impact in a large fireball.

Miraculously, 186 of the passengers survived, many walking away; 110 persons were killed.

In a conversation with author William Kramer, Fire Chief James Hathaway of the Sioux City Gateway Airport stated that many lessons were learned about rescue and about aircraft safety in general. For example, he stated "the children who sat lower buckled into their seats were often survivors, the adults whose heads extended above the back of the seats often suffered massive head injuries and did not survive." One suggestion might be higher backed, stronger seats to assist with survival upon impact. All news accounts of this aircraft disaster, however, indicated that it could have been much worse.

Fire departments and rescue organizations in Sioux City and in the surrounding Woodbury County had run through a drill two years previously in which a large plane was assumed to have crashed at the airport with 150 survivors needing help. The area was prepared. Even before Flight #232 came into view, Dr. David Greco, Head of the Medical Disaster Teams, was hovering overhead in a helicopter. Numerous ambulances and medivac choppers were ready to take survivors to two local hospitals. Fire, police, and a National Guard Unit were rushing to the scene.(10)

It was incredible that anyone could have survived the tremendous fireball that enveloped Flight #232, but because the jetliner broke into two large sections which remained intact, these became tunnels to survival into the bright Iowa sunlight. Few of the passengers had understood the real danger when the rear engine exploded crippling all three hydraulic systems in the aircraft. The danger became increasingly pronounced, however, as the situation deteriorated and they braced for survival. Fortunately, because Sioux City was well prepared for, and had rehearsed, a major aircraft disaster, these survivors received the best possible attention once they hit the ground. According to *Newsweek*, "Rarely in the history of U.S. Aviation has an airborne emergency gone so horribly wrong, and so blessedly right."(11)

Examples of Air Disasters

Our complete listing of disasters in Chapter 1 is punctuated with regular examples of multiple fatality air disasters. Here are a few of the more noteworthy over the past several decades.

On March 3, 1974, near Paris, France, a Turkish DC-10 jumbo jet exploded and crashed in a forest when a cargo door blew off soon after takeoff from Orly Airport. The 346 persons aboard were killed, making it the worst disaster in aviation history at that time. That record was broken in 1977 when two Boeing 747 planes collided at the airport in Tenerife, Canary Islands; 577 persons died as a result.(12)

On June 24, 1975, 113 persons were killed in the worst U.S. air accident involving a single plane at that time when a passenger jet crashed while attempting to land at Kennedy International Airport in New York during a rainstorm.(13) A collision of airliners over New York City in 1960 resulted in 134 fatalities.

Another tragic air disaster occurred in Saigon, South Vietnam, on April 4, 1975, when a U.S. Air Force Transport plane crashed during takeoff, killing at least 155 persons. The plane was carrying about 247 Vietnamese orphans to safety just prior to the collapse of the South Vietnamese government.(14)

On August 16, 1987, a Northwest Airlines jet crashed during an aborted take-off from the Detroit Metropolitan Airport in Romulus, Michigan. This crash was noteworthy for the fact that the lone survivor, a four-year-old girl aboard the jet, endeared herself to the nation. One hundred fifty-six people, including two on the ground, were killed in the crash of this DC-9. Twenty-one fire departments responded to the disaster and described the destruction as "beyond comprehension."(15)

In November 1987 a snow storm raged and the temperature dropped below freezing as Continental Flight 1713, bound for Boise, took off from Denver's Stapleton Airport. Shortly after taking off, it clipped the ground with its wing and cartwheeled down the runway, breaking into three pieces. Twenty-eight persons died, including the pilot and copilot, and 81 people were injured. Investigators from the National Transportation Safety Board focused on the build-up of ice on the plane's wings during a 23-minute wait before take-off and the inexperience of the pilot and copilot.(16)

On September 20, 1989, at 11:26 P.M., USAir Flight #5050, a Boeing 737/N400 crashed into the river at the end of runway 31 at New York's LaGuardia Airport. Immediately, numerous vehicles from various agencies responded as shown in Table 11-1. This large contingent of responding persons and vehicles resulted in a successful rescue of 55 passengers and 6 crew members. There were only 2 fatalities.

Table 11-1. USAir, Flight 5050, Response Apparatus & Personnel.

Agency	*Vehicles*	*Boats*	*Helicopters*	*Personnel*
Port Authority of New York & New Jersey	26	1	1	150
New York Fire Department	25	1	-	150
New York Police Department	70	28	-	700
Emergency Medical Services	60	-	-	100
U.S. Coast Guard	-	10	3	200
Department of Correction	10	1	-	100

On January 25, 1990, Long Island's exclusive North Shore was visited with tragedy when Avianca Flight #52 from Bogotá and Medellin, Colombia, ran out of fuel attempting a second approach to Kennedy Airport. The relatively small Oyster Bay Fire Protection District was suddenly faced with its largest ever emergency operation. What followed was the largest "mutual aid" response in the history of the Nassau County Fire Service. A heroic rescue effort resulted in 89 passengers surviving and being sent to area hospitals. Sixty-nine persons died upon impact. This was the incident that became newsworthy when drugs from Colombia were found on and inside the bodies of many of the victims.

The fact that more than half of the 161 people aboard survived was credited in part to the fact that there was no fire or explosion due to the lack

TOP: Avianca Flight No. 52 ran out of fuel and crashed on Long Island's exclusive north shore on January 25, 1990. BOTTOM Fire fighters ponder how to extricate a victim trapped between the fuselage of the Avianca aircraft and a tree. (Photo courtesy of Ron Buckingham.)

of fuel, but also to the efforts of fire fighters, police, and medical personnel from dozens of Long Island communities who assisted in the rescue effort. The most seriously injured were flown out by helicopters while others waited for ambulances, sometimes up to two hours due to the narrow accessibility along a two-lane highway.(17)

On August 29, 1990, shortly after Iraq invaded Kuwait, causing the international Middle East crisis that sent the United States into the Persian Gulf War, a giant U.S. cargo jet was headed from Germany to Saudi Arabia loaded with supplies for U.S. forces. It crashed and burned in West Germany, killing 13 persons aboard. These were the "first casualties" of the war with Iraq.(18)

On October 3, 1990, a Chinese hijacker caused the crash of a Boeing 737, which sliced into two parked jets killing 127 people of numerous nationalities, including one American. This worst reported air disaster in Chinese

Figure 11-1.

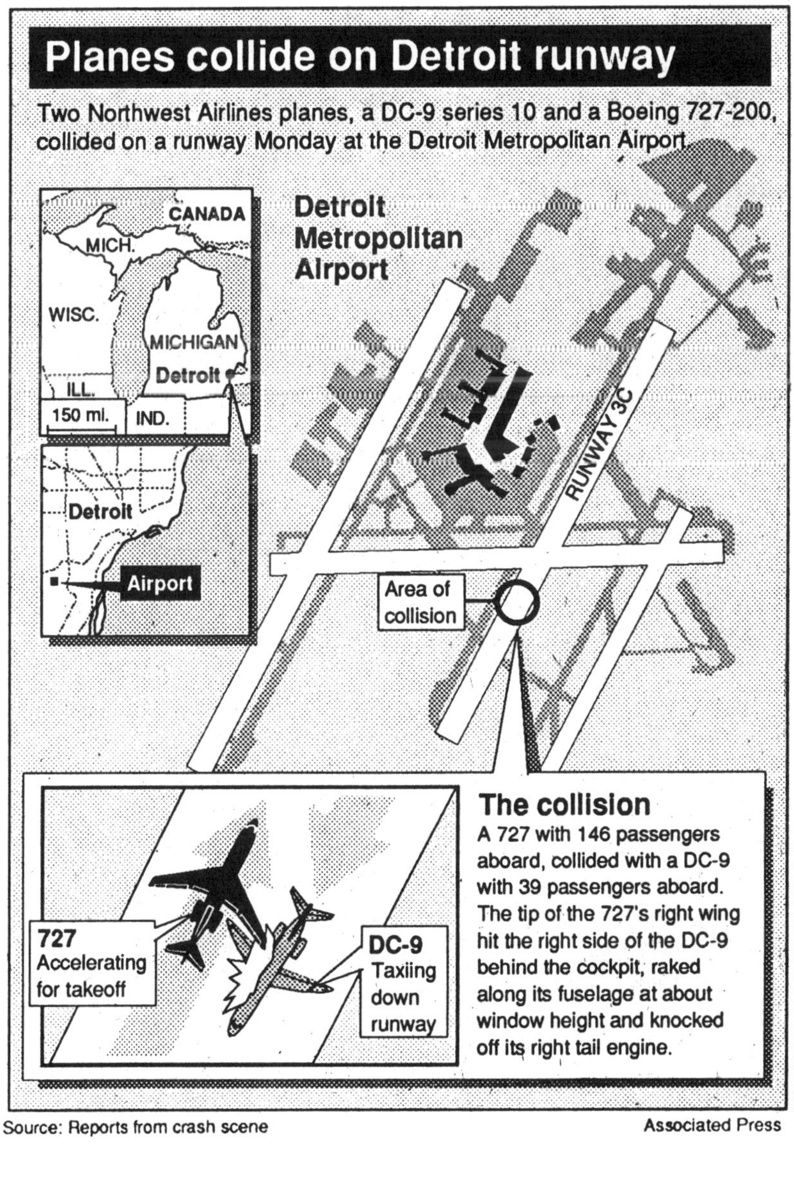

Source: Reports from crash scene

Associated Press

history reflects a problem that has emerged over the last two decades—terrorism. Air traffic controllers stated that the hijacker pirate appeared confused about where he wanted to go and the investigation of this crash was shrouded in mystery.(19)

On November 14, 1990, an Alitalia DC-9 crashed in Zurich, Switzerland, arriving after a 50-minute flight from Milan, Italy. Since the plane was travelling approximately 1000 feet below normal altitude, pilot error is suspected as the cause of the fiery crash which killed all 46 people aboard, including six Americans. About 300 police and rescuers worked to retrieve charred bodies well into the following day.(20)

On Monday, December 3, 1990, a Northwest Airlines jet strayed into the path of another jet that was screaming toward takeoff. Eight people were killed and at least 20 injured as the two collided in heavy fog at the Detroit Metropolitan Airport. The fatalities were all aboard Northwest Flight 1482, a DC-9 bound for Pittsburgh. Significant in this event were the weather conditions, the fact that two moving planes were involved, the fact that there was an ensuing fire which caused the fatalities, and the fact that an on-duty, well-equipped airport fire department, despite its best efforts, was unable to save some of the passengers. See Figure 11-2 for a description of this incident.(21)

A contributing factor in this disaster was a rear exit handle which malfunctioned. On January 9, 1991, the National Transportation Safety Board urged the Federal Aviation Administration to require an immediate inspection of all DC-9 jets to inspect the rear exit handles.(22)

On February 1, 1991, a USAir jet landed on top of a commuter plane waiting to take off from Los Angeles airport. Although the primary cause was initially considered air controller error, contributing factors included outdated and unworkable radar equipment, and poor visibility from the control tower. Thirty-four persons were killed including all in the commuter plane.(23)

On February 17, 1991, an Indianapolis-bound DC-9 carrying U.S. mail flipped upside down onto a snow-covered runway at Cleveland Hopkins International Airport demolishing the cockpit. Although the two crew members perished in the accident, the ensuing fire was brought swiftly under control and fire fighters and postal service agents succeeded in salvaging much of its cargo of mail.(24)

Colorado Air Crash

On March 3, 1991, a United Airlines Boeing 737 from Peoria, Illinois, to Colorado Springs disintegrated as it was on final approach to the runway. Flight #585 appeared to be on a normal approach when it veered sharply and went vertically into the ground. All 25 persons aboard died, as the plane was pulverized beyond recognition. This was the third major accident in four months involving U.S. airlines, one of the deadliest periods in American history. The pilot was credited with maneuvering the plane to avoid nearby homes and apartment buildings, thus averting what would have been a much more serious disaster.(25)

On May 27, 1991, 223 persons died when an Austrian Airliner in northern Thailand exploded in flight. A team from the U.S. National Transportation Safety Board assisted in the investigation, which linked the crash to a terrorist bomb. Prior to this incident, western intelligence had pinpointed Bangkok, Thailand, as a possible staging site for terrorism.

Between the time this book goes to press and the time when most persons will read these paragraphs, there will unfortunately be additional noteworthy air disasters to add to the list. The occurrence of airline disasters may seem to be as frequent as always. However because the number of daily flights and passenger miles flown continue to increase dramatically, and the number of aircraft accidents is relatively stable, a far smaller percentage of passengers are involved in aircraft disasters on an average yearly basis today than in the past.

One of the reasons that some people attribute to contemporary aircraft disasters is that technology has raced ahead of the human element. An example is the highly automated Airbus A-320 Jetliner, which is operated in the U.S. only by Northwest Airlines. After two overseas crashes occurred within a short span of time, Federal Aviation Administration and U.S. airline officials were forced to take a closer look. During a 1988 air show in France one of the first models plunged into a thicket killing three passengers and injuring about 50. Then in February 1989 an Indian Airlines A-320 crashed near the city of Bangalore, killing 93. Although the official cause of both crashes was ruled as pilot error, some pilots and air safety experts have asked whether the plane's auto thrust system which controls speed may have been a contributing factor. Ken Plunkett of the Aviation Safety Institute, a nonprofit research group states, "People may be becoming overconfident with the Airbus."(26) Northwest issued a bulletin to alert its pilots that there may be possible glitches in the plane's cockpit computer.

Other critics have singled out the DC-10 and possible hydraulic problems as a source of air disasters. Several incidents in the 1970s and the later Sioux City crash, all involving hydraulic difficulties, have raised questions about technological design.(27) The fact is that no major airliner has been immune to crashes for one reason or another and all of them have an amazingly safe record in light of the millions of hours flown and billions of passenger miles.

Emergency Action in Air Disasters

Aircraft crashes that result in a large number of fatalities are likely to involve passenger rather than cargo planes. They could involve helicopters of large capacity that shuttle persons between airports and the roofs of highrise hotels and office buildings or between suburban heliports and major urban air terminals. Such crashes may occur on or off airfields.

Cargo aircraft accidents may present a hazard to fire fighters from leaks, spills, or fires involving radioactive materials and other dangerous chemicals. Hazardous cargo might also be responsible for the crash of a transport plane, as was the case in Boston, Massachusetts, on November 3, 1973, when a Boeing 707-321C crashed at Logan International Airport, destroying the craft and killing its three crew members. The National Transportation Safety Board found that spontaneous chemical reaction between leaking nitric acid, improperly packed and stowed, and the improper sawdust packing surrounding the acid's package, caused the dense smoke that entered the cockpit and resulted in the loss of control during final approach.(28)

Maintenance of good communications is essential in any disaster control operation, but is especially important when aircraft approaching your airport have malfunction problems. FAA (Federal Aviation Administration) personnel assigned to control towers at airports are responsible for directing traffic in the immediate vicinity of the airport as well as movement of aircraft on the ground.

"Ground Control" is one of several frequencies used by the tower; it is used mainly for controlling traffic on the ground and is shared by other agencies in addition to the tower and the aircraft, such as, emergency crews, airport operations, and airline operations. To obtain clearance for fire department vehicles to cross runways, to acquire information relating to emergencies, and to inform the tower of conditions at the scene of an emergency, it is important that the fire department communicate directly with the tower. This can be accomplished by providing fire units with a radio having "ground control" frequency or providing the tower with a radio having fire department frequencies.

Although the tower does not dispatch municipal fire apparatus, there should be a direct telephone "hot line" to fire stations at the airport. This way the tower can provide instantaneous information regarding crash or emergency conditions to all emergency agencies at the airport and the fire department. Upon receiving this information, the fire department dispatcher

Fire fighters scramble as aviation fuel ignites during their search and rescue operation. (Photos courtesy of the Federal Emergency Management Agency.)

can send units to handle the emergency without having to repeat the information the tower has given to fire companies. At Cincinnati's Fire Station #18, which serves the Lunken commuter airfield as well as the surrounding residential area, a red phone connected directly to the airport control tower sits in the alarm room. Affectionately dubbed "the Bat Phone" by the members of this station, it has provided early alert for numerous airport emergencies over the years.

When the tower gives notification of an emergency, it describes the type of plane, its location, and nature of the trouble; it may supply more information by radio while the companies are responding or are on standby. While keeping the fire department informed, the tower will divert air traffic, close runways, and relay messages from the plane, as may be required. Information that may be available includes exact time of arrival, wind direction and velocity, number of persons on board, type and quantity of fuel remaining, change in runways, and specific facts relating to the plane's problems. Likewise, the tower may relay from the fire department to the aircraft pilot observed hazards such as blown tires, locked wheels, broken struts, fuel leaks, and fire.

Fire department and airport personnel should establish an agreement (preferably in writing) concerning duties and responsibilities among all agencies that would be involved in an aircraft incident. An annex to the fire department's disaster plan might be an "Emergency Operating Procedure for Airport Emergencies."

Chicago Air Crash Plan

The City of Chicago has adopted a disaster plan that incorporates an example of action to be taken if a jetliner crashes in a congested area of the city. It prefaces its advice with the observation that, "In a disaster of this type, the Chief Fire Officer will have a terrific fire fighting problem on his hands that could develop into a spreading fire and exposure hazards and serious loss of life." The example starts with the question, "What action is necessary?" and then advises size-up of the situation: spreading fire—exposure hazards; life hazard—plane occupants, building occupants; pedestrians in street; jet fuel spillage; emergency help and multiple alarm response that might be needed. Also it advises that the Fire Alarm Office may be requested for necessary response and be given the Disaster Plan Number to be placed into operation. (e.g., Plan 1—five to fifteen victims, Plan 2—fifteen to thirty victims, and Plan 3—thirty or more victims.)(29)

It is of interest that the U.S. Air Force found a helicopter downwash could extend passenger escape time and provide a considerable reduction in the radiant heat and air temperature in a rescue path. The record showed that 96.6 percent of all air crashes occurred within 61 miles of their air bases (72 percent within 10 miles), and 46 percent of these accidents resulted in fire; it was imperative that a fast and reliable method be developed to rescue crash victims that might be located in places inaccessible to standard ground fire fighting equipment.(30)

Large helicopters with counter-rotating main rotors produce large volumes of low velocity air with a longitudinal forward thrust, thereby opening

The remnants of a jumbo jet after a crash and burn. (Photo courtesy of the National Fire Academy.)

a cooling corridor through the fire to the burning fuselage. Fire fighters have been able to walk through this corridor without protective clothing and experienced no burns. The H-43B Huskie fire fighting helicopter, developed during the Vietnam war, was used by the Air Force crash rescue teams. They were able to gain entry to a burning fuselage in less than a minute after arriving on the scene. The turbine engine required no warm-up time, thus permitting rescue to be accomplished in seconds while utilizing the Huskie's syncropter cooling downwash.

On September 24, 1961, in North Carolina a C-123 crashed on takeoff at the Wilmington Airport show. An H-43B hovered twenty minutes over the nose of the large plane, supplying air to the trapped personnel while crash crews worked inside the burning fuselage to extricate the injured; twelve persons inside were rescued. The Huskie then became an aerial ambulance to deliver the injured to the hospital.(31) While potential for the Huskie was great, actual use was limited. The end of the Huskie's career is now dignified by its inclusion in the Air Force Museum near Wright Patterson Air Force Base, Fairborn, Ohio.(32)

The Systems Research and Development Service of the Federal Aviation Administration, in its final report on tests conducted by burning C-97 aircraft under post-crash conditions, found that the helicopter downwash extended escape time when fire existed solely on the upwind side of a C-97 fuselage, but reduced the escape time when fire was on both sides, or downwind of the fuselage. Ability of crews to improve escape time in standard fire conditions depended on pre-burn time, integrity of the fuselage, and whether emergency doors were open or closed. An escape time of 50 seconds was computed for a C-97 with emergency doors open as compared to 138 seconds with doors closed, stressing the need for quick control of the fire.(33)

HIGHWAY DISASTERS

Road transportation accidents may involve autos, buses, and trucks, and require prompt notification of emergency services so rescue and fire fighting can begin at the earliest possible moment. With the increasing use of cellular car phones and a continuing interest in CB radios by the motoring public, the summoning of assistance to an accident is faster than ever. Unfortunately, CB radios may bring swift appearance of spectators, most of whom will hinder rather than help, and may even become casualties in event of explosion. Some CB users have also caused "false" summons for "apparent" accidents or emergencies.

Causes of Highway Accidents

Mechanical failure, adverse weather, and driver incompetence account for a large percent of accidents on the highway. Multiple collisions involving trains, tank trucks, buses, street cars, and other vehicles (with each other or with stationary objects alongside the road) are not uncommon, and frequently end in fire and a loss of life.

Collapsing bridges, landslides, and flash floods have also resulted in multiple deaths to persons traversing the highways. Explosion of dangerous chemical cargos in fires have resulted in fatalities among the fire fighters, spectators, and nearby building occupants.

Examples of Motor Vehicle Disasters

Not all motor vehicle disasters are the result of fire, but considering that responses by fire departments to motor vehicle fires throughout the United States average over 500,000 calls a year, it should not be surprising that a few of these result in very large monetary loss or a large loss of life.

Auto Disasters

Dense fog that blankets a highway for several miles can be the case of a chain-reaction pileup of motor vehicles of disaster proportions. Such fogs are not uncommon on highways bordering the ocean, but they also develop in inland valleys in certain weather conditions. Without warning, motorists suddenly find themselves enveloped in a medium known as "tule fog," cutting off vision beyond the hoods of their vehicles.

On January 2, 1991, 70 automobiles, tractor trailers, and other vehicles were involved in a deadly chain reaction accident that killed three people and injured more than 19 on Interstate 215 in Salt Lake City. Like most chain reaction accidents fog was blamed which in this case had reduced visibility to zero.(34)

On Wednesday, December 12, 1990, 13 persons died in an 83-car wreck on Interstate 75 in Calhoun, Tennessee. Fifty people were also injured in the pile-up, which was caused by two separate chain reaction crashes in the same spot of the interstate. As with so many similar incidents, dense fog was blamed for the massive destruction. One driver, Virgil Metts, was quoted as saying, "for three minutes after I stopped, cars kept hitting each other. They kept driving into the fog at 60 miles per hour. It was deadly silent. You couldn't see anything because of the fog, and all you heard were exploding gas tanks." In this case, history repeated itself because back in 1978, 62 vehicles were involved in a similar accident.(35)

In February 1990, a highway accident in Flagstaff, Arizona, involved 58 vehicles. All available fire and rescue personnel from the Flagstaff and surrounding fire departments were called to action. Access was difficult with slopes on both sides of the accident site. It was foggy and icy. Manpower and equipment were stretched to the limit as several extrications were being attempted at once. One extrication involved a 14-year-old girl who was trapped for five and a half hours.(36)

On October 6, 1985, eight persons were killed and 45 injured on Interstate 5 north of Sacramento, California. The crash was triggered by dense smoke from a nearby grass fire. Survivors lay in the wreckage listening to car after car slam into the chain-reaction pile-up, which went on for 5 minutes, demolishing 39 vehicles.(37)

A record-setting accident occurred on April 30, 1975, when nearly 200 cars and trucks were trapped in a series of chain reaction traffic accidents on the Riverside Freeway near Corona, California. This was one of a series of fog blankets that plagued freeway motorists in southern California during the early morning hours, reducing visibility to almost zero. On this morning, rear end collisions of cars and truck pile-ups extended several miles, and the highway had to be closed in both directions for several hours while fire fighters fought the resulting fires and helped trapped victims out of the wreckage. No one was killed, but eighteen persons had to be hospitalized and property damage was inestimable. "The cars and a tremendous number of trucks were piled on top of each other. There were people walking around in panic and in shock just yelling."(38)

This area had been the site of previous mass collisions because of the Santa Ana River bed fog, and highway patrol officers had been escorting some motorists minutes before the pileup started, but the heavy fog moved in suddenly and the crashes began. The officers had posted temporary fog warning signs and radio stations were broadcasting traffic advisories, but these measures were not enough.

A few days after this accident a warning plan was initiated by the California Department of Transportation working with the California Highway Patrol (CHP); it consisted of pickup trucks mounted with illuminated, changeable advisory signs that flashed speed warnings and other messages to motorists before they hit the fog area. These messages could be varied from "Dense Fog Ahead" and "Delay Ahead" to "Stop Ahead" and

"Wreck Ahead." The messages are programmed in solid state modules, inserted by the truck driver in units in the cab. The CHP officers working in the fog areas radio truck operators outside the fog bank with the message to be displayed. By making the motorist aware of the fog and advising him of a safe speed for the conditions, such pileup accidents can be reduced. These vehicle-mounted, illuminated signs are quite commonplace now in highway construction zones.(39)

Bus Disasters

When a chain-reaction pileup involves a loaded bus, the loss of life is apt to be larger. In Barrie, Canada, a collision involving a bus, a truck, and 30 automobiles killed 11 persons and injured 43.

On August 1, 1991, a chartered yellow school bus apparently lost its brakes careening off a mountain road and rolled over several times down a rocky slope. The bus was transporting a group of Girl Scouts and adult supervisors from an outing in Palm Springs, California. The driver, two adult chaperones, and four scouts were killed, and 31 others were hospitalized.(40)

On July 19, 1990, two American high school students died and more than 50 were injured when a double-decker bus went off the road and flipped over near Oxford, England.(41)

In Mexico City on the Tlahuac-Chalco Highway, a bus driver lost control

A flammable liquid carrier and school bus collide in upstate New York, but tragedy is averted when neither is carrying its intended cargo. (Photo courtesy of Ron Buckingham.)

in dense fog in November 1987. The bus drove off the road. Fifty-nine were killed.

One streak of bad luck involving buses occurred back in 1974. On July 28, at Belem, Brazil, a bus and truck collided head-on, killing 69 persons; the next day, 29 were killed when a Mexican bus ran into a bridge near San Mateo Atenco; and two weeks after that, two buses collided on a highway linking Ankara and Istanbul in Western Turkey, killing 21 and injuring 41 persons.(42)

Buses by their nature transport numerous potential victims and deserve special consideration and training by fire and rescue personnel.

Truck Disasters

A consideration of highway disasters must include trucks, and among the heaviest of trucks are those used by fire department personnel. One of the more dangerous aspects of the fire fighting profession is the response to emergencies along highways by fire fighters riding vehicles whose weight is measured in tons. They do not always arrive. In June 1990, two fire fighters were killed and another two injured seriously when the brakes failed on a reserve pumper in Waterbury, Connecticut, and crashed into a tree as they were responding to an automatic alarm at a commercial establishment. Most highway disasters, however, involve other vehicles, creating an array of bizarre emergencies.(43)

On March 20, 1990, U.S. Highway 101 in California was shut down when a tanker truck jackknifed and rolled onto its side. The truck contained five thousand gallons of tetrahydrofuran (THF) and burst into flames. Approximately 150 personnel from 18 different agencies were used to control this 18-hour incident, which was noteworthy because it illustrated a changing attitude in fire service personnel. Instead of rushing in to apply water, a system of cautiously planned action resulted in a successful mitigation with minor injuries to the driver only.(44)

Consider the situation confronted by the Memphis Fire Department on Friday, December 23, 1988, when a propane truck accident killed eight persons and injured more than a dozen others. Occurring on Interstate 240, just north of the Interstate 40 interchange in Memphis at approximately 10 o'clock in the morning a 10,000-gallon semitrailer carrying propane overturned on the northbound ramp leading from I-40 to northbound I-240. This was the third fatal incident involving a semitrailer at the interchange. As the tractor trailer went around an inclined ramp, it slid approximately 200 feet and struck a bridge abutment. When the front end of the tank ruptured it released liquid propane forming a large vapor cloud carried in a northwesterly direction. It travelled 1,500 feet up the expressway before igniting, forming a huge fireball. Vehicles in both north and south bound lanes were caught underneath the cloud. For the most part occupants died instantly. Occupants of some other vehicles were able to escape, some burned severely, before their vehicles ignited. The incident grew in complexity as the tank shell rocketed and ricocheted from building to building adjacent to the interstate, igniting several structures at different locations.

Suddenly the Memphis Fire Department was faced with four different incidents at the accident site: three structure fires, and the numerous automobiles on the highway. This required four separate command posts under an extended incident command system, with difficulties in communication as fire fighters scrambled to understand which commander or which incident command post was communicating.(45)

This was reminiscent of a similar incident that occurred 16 years previously, when a tractor-semitrailer (tank) carrying liquefied propylene gas sideswiped a bus (carrying no passengers) in the southbound lane of the New Jersey Turnpike about a mile north of Exit 8. After impact, the bus, while rotating clockwise and sliding across the highway, was struck by a southbound automobile. The tractor-semitrailer scraped, then straddled the turnpike's median guardrail, jackknifed, spun into the northbound lanes, and overturned when struck by a northbound auto.

When this tractor-semitrailer hit the guardrail, fire broke out, and after burning about 25 minutes, the cargo tank exploded in a ball of flame. Segments of the tank rocketed more than 1,300 feet northeast and 500 feet southwest of the tractor-semitrailer. Two persons were killed and 28 others, including seven police officers, were injured by the explosion of the cargo tank.(46)

Three years later in 1975 a tractor-tank-semitrailer struck a concrete head wall on Route 277 near Eagle Pass, Texas, after swerving to avoid an auto that had slowed ahead to make a left turn. The tank-semitrailer, having separated from the tractor, ruptured upon impact with the head wall (where the road crossed a canal) and the ensuing fire and explosion of LP-Gas destroyed a building and 51 vehicles. Fifty-one persons in the area were burned and 16, including the truck driver, were killed.

"After the tank broke into two pieces, the large forward section of the tank was propelled into the air, where it struck an elevated sign; traveled 1,029 feet and struck the ground; bounced into the air and traveled 278 feet; struck and demolished a mobile home and came to rest 1,654 feet from its point of departure."(47) The aft end of the tank split into three parts which rocketed about 800 feet through the air to the east and northwest and fell to the ground. In his letter to the Secretary of the U.S. Department of Transportation the Chairman of the National Transportation Safety Board declared that this was the worst highway accident involving hazardous material that the Safety Board had investigated. This accident was especially severe because the large quantity of LPG vaporized so rapidly and the vapor cloud engulfed the area so quickly that victims were unable to respond to the hazard when the cloud ignited moments after the crash.

The catastrophic potential of trucks transporting extra-hazardous materials including explosives, oxidizing materials, flammable liquids and gases, poisonous substances, and radioactive materials is growing. Millions of trucks are on the highways, and about one-fifth of their cargos by volume consists of one or more of almost 1,000 items listed by the Department of Transportation as explosives or other dangerous articles requiring special identification and labeling.(48)

Another gas responsible for disasters has been ammonia. It is transported in great quantities in tank trucks, and though it is flammable and toxic, the Department of Transportation was slow to require a label or placard to indicate either of these hazards. In May 1976, a tanker loaded with 5,000 gallons of ammonia crashed through a guardrail at the busiest highway interchange in Texas, near Houston, and plunged into another lane of traffic twenty feet below, violently rupturing on impact. Clouds of ammonia gas billowed over the area, causing motorists to flee from their cars. As a result of the accident, four persons died and 119 were treated at 14 hospitals.(49)

When a truckload of dynamite and blasting agent was rolling along the highway near Unionville, Ohio, many years ago, its tarpaulin cover caught fire from its exhaust pipe, and two hours after the driver pulled off the road, it exploded. Thanks to the presence of mind of the driver, no one was killed. However, around the same time, another truck carrying dynamite and ammonium nitrate blasting agent was parked by a lumber and supply warehouse in Roseburg, Oregon. The driver had just completed a 290-mile run from the State of Washington. Having arrived too late to make deliveries at two points outside the city, he locked the truck and checked into a downtown hotel about three blocks from the warehouse. At about 1 A.M., fire broke out in the warehouse, and after it spread to the truck, an explosion followed. Thirteen persons were killed and more than 125 injured. The blast leveled the immediate 12 blocks area, tore a crater 52 feet in diameter and 20 feet deep at the site, caused extensive building damage over a 50-block area, and glass breakage as far as 9 miles.(50) Although more commonly the proximate cause, it does not require a fire to trigger a cargo explosion. A gun can suffice. On September 30, 1970, a sniper's bullet exploded 21 tons of dynamite aboard a truck being operated on a highway near Springfield, Missouri, killing the driver and causing extensive damage to nearby structures.(51) Less than a year later, on June 4, 1971, a dynamite cargo played a more damaging role in an accident on U.S. Highway 78, near Waco, Georgia. An automobile crossed over the centerline of a two-lane highway and collided with a tractor-semitrailer transporting 25,414 pounds of explosives. The auto driver apparently was killed and fire broke out instantly along the left side of the tractor and in front of the trailer. Fire fighters arrived at the scene shortly thereafter and while they were trying to extinguish the fire, the truck driver was doing his best to persuade the crowd of bystanders to move away from the burning wreckage. About ten or fifteen minutes after the collision the cargo detonated, killing two fire fighters, a wrecker driver, and two bystanders; 33 others were injured and the property damage was estimated to be in excess of one million dollars.(52)

Fire fighters can be expected to recognize the danger of fighting a fire involving explosives, unless the warning placards have been removed from the truck, as was the case in Marshall's Creek, Pennsylvania, where four of them were killed in an explosives truck blast.(53)

It is unreasonable to expect fire fighters to recognize trade names of all the pesticides and other toxic products transported in trucks that carry no placards to warn of danger. About 9 P.M. on April 10, 1975, Los Angeles City fire fighters responded to a well-involved truck fire in a diesel tractor and two 27-foot cargo trailers on the Hollywood Freeway. The front trailer contained 24,000 pounds of Methomyl, an insecticide. The product was in a carrier of methyl alcohol in one- and five-gallon plastic containers in cardboard boxes, clearly labeled for toxicity and flammability. As the insecticide solution burned it released highly toxic vapors. The rear trailer held 10,000 pounds of tires.

The truck driver was unaware of the nature of the load, and fire fighters were already applying aqueous film-forming foam on the fire when someone noticed the warning label on one of the fallen cartons and reported it to the battalion chief. Orders were issued to don breathing apparatus, but more than 100 persons had already been exposed while fighting this fire. Fifteen of the 43 stricken fire fighters taken to the hospital had to remain there; 59 police officers and other persons were also taken to the hospital for treatment; the fire department's chemist was able to give the doctors information that he had obtained from the chemical manufacturer's biochemist in Washington, D.C., as to the best antidote to use.(54)

Six months after that accident, near Yuba City, California, a fiery collision involving a fertilizer truck and a pickup towing a trailer hospitalized 21 persons as a result of exposure to vapors from a spill of telone from a ruptured tank that had fallen off the back of the fertilizer truck. The victims included five California Highway Patrol officers, three Sutter County fire fighters, passersby, and the occupants of the vehicles.(55)

Emergency Action for Highway Disasters

A highway crash involving autos, buses, or carriers of hazardous cargos can require considerable rescue, fire fighting, and evacuation operations. To determine the nature and extent of the emergency, it may be necessary to reconnoiter by helicopter, especially if there is apt to be delay in getting through traffic during peak rush hours. An aerial survey should help to supplement information already obtained from the dispatcher. In any case, it will be necessary to alert all public service agencies as to the nature of the problem, exact location, and extent of damage.

Initial size-up should disclose whether mutual aid forces will be needed. The scene of the accident may be on the outskirts of your community, and since your emergency vehicles will probably all be responding from one direction, it may be desirable to have the next community beyond the accident scene respond from the opposite direction. This is especially important if the emergency is on a divided expressway where emergency response needs to be made from both directions, owing to the infrequency of crossover ramps providing access to the other side.

Auto and Bus Emergency Action

Special equipment may be required, depending upon the type of emergency. If a bus load of people has plunged into a lake, a scuba team could be brought in by helicopter; boats, cranes, heavy duty tow trucks with powerful winches may be useful in retrieving the bus.

An auto or bus having its engine in the front is apt to take a nose-dive, with the forward momentum carrying it to an upside down position on the bottom of the lake or reservoir. If the bottom is muddy, the vehicle will sink somewhat, making it difficult to open doors from either side. Some air probably will be trapped inside a tightly closed vehicle, permitting uninjured persons to survive a short while. Soon the pressure equalizes from inflow, but the chances that anyone will have presence of mind to wait until the condition stabilizes and then open a window and exhale while he escapes to the surface is not too great. (If the swimmer fails to let out his breath on the way up to the surface, he is likely to suffer from lung rupture and embolism.) If a scuba team (or team of skin divers, if the water is shallow) is readily available, its job is apt to consist of attaching hoisting or tow cables to the sunken vehicle, or extricating the dead victims through windows and escape hatches.

For the same reason as given before, a bus that falls off an elevated highway is apt to land upside down, crushing some passengers to death. There will be immediate need for fire apparatus to safeguard against a fuel spill fire, and heavy rescue vehicles with hoists, hydraulic prying tools, power saws, comealongs, cutting tools, cranes, ambulances, and law enforcement personnel to provide traffic and spectator control.(56)

Hazardous Cargo Vehicles—Emergency Action

In the next chapter we will cover the topic of hazardous materials in detail. If a highway emergency involves trucks or trailers loaded with hazardous materials, considerable caution is required and a number of questions need to be answered, if possible. The first question "who is in charge?" should be resolved under your disaster plan. Does the responsibility for directing operations at a highway emergency fall upon the local police department if the location is within the city limits? Does it fall upon the state highway patrol (or troopers) if outside the city limits? In any case, the decision of how fire fighters should be deployed will rest with the fire chief in that jurisdiction.

The action that the fire chief calls for will depend upon whether there is a leak, spill, fire, or a combination of these hazards; whether materials are in the form of gas, liquid, or solid; whether terrain is level or slopes into a river, flood control basin, or storm drains; types of exposure (inhabited or not) that may be subjected to heat, poison gas, or toxic products of combustion; availability of mutual aid and back-up water supplies; availability of police assistance to stop traffic, remove spectators, notify the occupants of exposed structures; and the probable amount of time available before an explosion is apt to occur.

Difficult for fire fighters is the identification of hazardous materials that may be involved. There is no guarantee that required identification for hazardous cargos will be on the carrier or will be readily visible. The U.S. Department of Transportation continues to improve its identification system for hazardous materials. Recent improvements include special classifications for some poisons as chlorine and ammonia, of which the latter is also flammable. Caution is in order, however, because placarding is often missing, or if present, inadequate. For one class of explosives, no warning placard is required unless the cargo exceeds a large quantity.(57)

Even if the vehicle were labeled so as to give fire fighters and the public adequate warning of the nature of the hazard, there are situations where the sign or placard may not be visible in dense smoke. The truck driver, if not killed in the accident, may have no knowledge of the hazards involved. But a fire chief can take certain practical steps under emergency conditions involving hazardous materials. Suggested procedures for handling tank truck fires, accidents involving explosives, rocket propellants, cryogenic materials, pesticides, chlorine leaks, ammonia spills, flammable liquid and oil spills on the highway, and LP-Gas tank truck emergencies are described in many current hazardous material texts.58 A bulletin on "Highway Emergency Training on the Transportation of Extra-Hazardous Cargo" has been issued by the American Insurance Association to assist members of police and fire departments.(59)

A 1973 incident in Los Angeles City illustrates coordination needed among diverse units. The chief in charge could observe from his helicopter that access roads were completely blocked with traffic, so first water on the fire came from two two-and-a-half-inch lines from a parallel surface street just north of the freeway. The first line had been laid by an engine company in readiness for a second helicopter to pick it up and carry it 1,500 feet to the incident. A crash truck arrived from a nearby airport station to help the engine companies; by applying AFFF (aqueous film-forming foam) from its two turrets it was able to control the ground fire while the aluminum tanker burned down to the liquid surface and was also blanketed with AFFF. However, due to the varying levels of gasoline in the compartmentalized tanks, heat caused the gasoline to continue to bubble, to break through the blanket formed by the AFFF, and become reignited.

The Air Pollution Control agency wanted the fire put out immediately to stop pollution of the atmosphere. The California Highway Patrol wanted the fire extinguished at once to alleviate the traffic problem. The State Division of Highways wanted the fire ended as fast as possible to prevent further destruction of the roadway. "On the other hand, the Bureau of Environmental Conservation was not as concerned with the fire as they were with the gasoline spill going into the flood control basin. The Department of Fish and Game did not want any foam or gasoline going into the flood control basin because it would kill fish and frogs. The Department of Water and Power did not want the gasoline in the flood control basin fearing an explosion of a rubber dam they constructed at Forest Lawn Drive."(60)

The fire department considered all these interests and used controlled burning to reduce the quantity of exposed fuel while extinguishing ground fires and diluting fuel runoff with water until extinguishment could be effected by AFFF from the crash truck's turrets. That would be a prudent operation today.

RAILROAD DISASTERS

Devastating accidents have destroyed freight and passenger trains. As with highway transportation, these may be caused by human error, mechanical failure, sabotage, or extraneous factors, such as landslides, blizzards and other storms. Collisions, derailments, and pileups, often followed by fire and explosions, are important factors in railroad disasters especially when hazardous materials are involved.

Freight Train Disasters

For five days, July 8 to July 12, 1986, the community of Miamisburg, nestled in the scenic Miami River Valley of southwest Ohio, was in a state of emergency. A derailment caused tank cars of phosphorous to ignite triggering an evacuation of Miamisburg and several surrounding cities. In an attempt to recover its expenses in handling the incident, the city of Miamisburg negotiated for several years with the railroad to recover more than $433,000 of expenses incurred by the city of Miamisburg and responding mutual aid fire departments.(61)

In April 1990 a freight train derailed in Worthington, Pennsylvania, just outside of Pittsburgh. This incident rapidly became a major hazmat fire as six crude oil cars, two coal cars, and one car loaded with sodium hydroxide began to burn and explode. Hundreds of fire fighters converged on the accident and over a two-day period brought it under control. A major consideration was the environment, and indeed at least a thousand fish lost their lives from water runoff that entered the Buffalo Creek. This indicates yet another dimension on the tactical complexities imposed on the modern fire service. Often it must choose between air pollution from not extinguishing

The city of Miamisburg, Ohio, and surrounding communities were rocked by a series of explosions as a white phosphorus tank car burned. (Photo courtesy of the Miamisburg Fire Department.)

a hazardous fire and water pollution from the ensuing runoff if extinguishment is attempted.(62)

Two decades earlier, a 15-car derailment in Crescent City, Illinois, resulted in rupture of one of the nine tank cars containing liquefied petroleum gas. Escaping propane ignited and when the fire fighters attempted to battle the fire, tank car explosions occurred at various intervals for several hours, injuring more than 60 persons and destroying most of the business district. Large pieces of the tank cars rocketed as far as 1700 feet from the derailment site.(63)

A year later, tank cars were involved in another derailment at Houston, Texas. One fire fighter was killed and 50 fire fighters, reporters, photographers, and spectators were injured. In this accident, six tanks cars of vinyl chloride, three of fuel oil, and one each of acetone, butadiene, and formaldehyde were among the 16 cars that piled up in the derailment area.

Vinyl chloride escaped from a 48,000 gallon tank car and immediately ignited. Attempts to control the fire with water did not prevent the vinyl chloride explosion in a huge fire ball about 45 minutes later, causing the death and injuries.(64)

The fact that such accidents happen should not be surprising in view of the vast amount of dangerous chemicals that are shipped by rail. The Southern Pacific Railroad each year moves over 100,000 car loads of hazardous materials in California alone. This includes cargo in box cars, in trailers on flat cars, and in tank cars. Probably the greatest problem area today, according to the Superintendent of the Hazardous Materials Control for the Southern Pacific Transportation Company, is with liquefied petroleum gases and flammable liquids in transit.(65)

The Southern Pacific Railroad might well revise the preceding opinion to include the moving of tank cars of toxic chemicals in its greatest problem areas. One of them fell into the Sacramento River on July 14, 1991, near the northern California town of Dunsmuir, as the result of a train derailment, releasing the deadly herbicide metasodium. Over 100,000 trout were killed in a 45-mile stretch of the Sacramento River as the 20,000-gallon spill moved downstream.(66)

Passenger Train Disasters

On January 4, 1987, one of the most memorable train disasters in recent memory occurred. A Conrail locomotive operated by personnel who tested positive for drug usage skidded through a switch right into the path of a speeding Amtrak train. The wreck of train No. 94—the worst in Amtrak's history—killed 16 people and injured 184, and left a serious question regarding rail safety in America. Rescuers from throughout Baltimore County converged on the scene, including teams flown in from the renowned Shock Trauma Center 15 miles south of the accident. Thirty-seven medic units and 10 helicopters were used to treat the 184 injured passengers. Within hours, the disaster scene was compounded by the expected influx of federal railroad administration officials and numerous law enforcement agencies. Fire service personnel and first responding rescue units operated under a command complicated by the

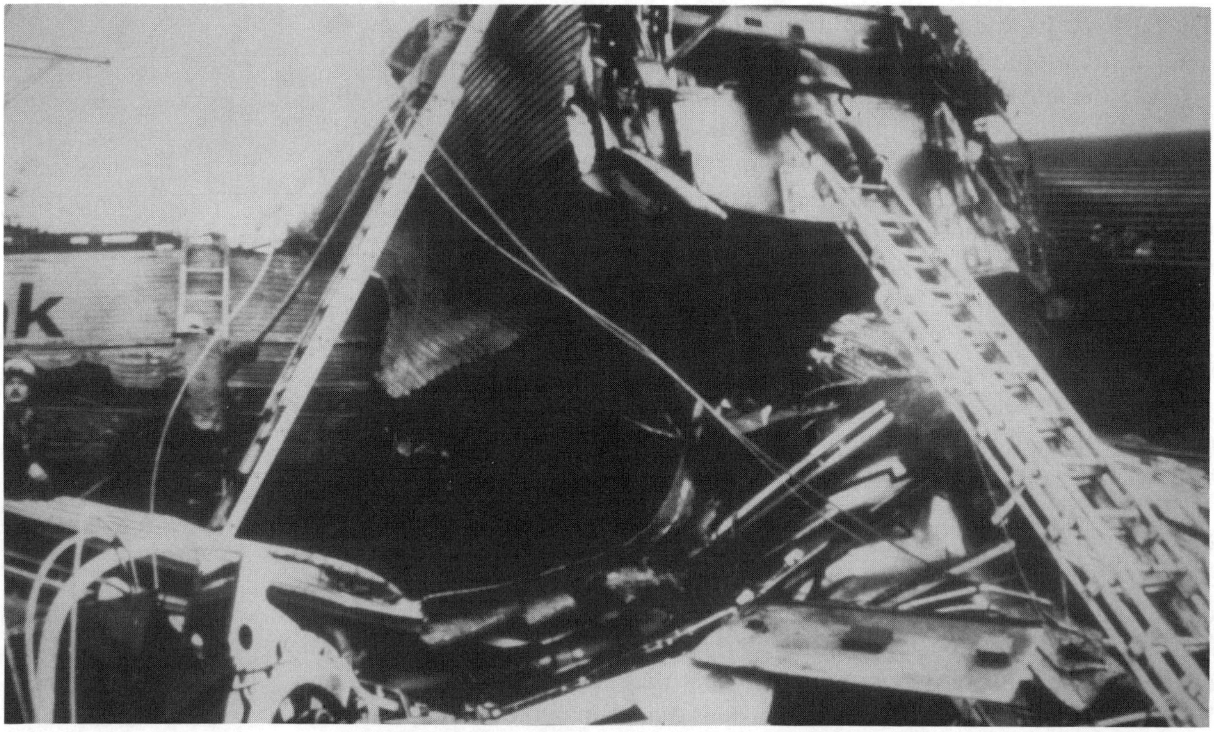

The twisted wreckage of an Amtrak train derailment and crash presents major extrication and life-saving demands on the fire service. (Photo courtesy of Ron Buckingham.)

additional agencies involved. As disaster and mitigation become more complex, it can be expected that disaster command centers will include many new agencies and require new dimensions of cooperation and coordination.(67)

On December 13, 1990, an Amtrak train was reportedly speeding when it jumped the tracks in a station and slammed into a rush hour commuter train. Amtrak denied that inexperience caused this crash, but an apprentice engineer had begun operating on the Washington-Boston line on the Monday prior to the accident. The Boston train wreck, which injured 264 passengers—10 critically—was so violent that cars were hurled into the tunnel ceiling buckling the road above. As the lights went out and the tunnel filled with smoke from igniting diesel fuel, emergency workers were hard pressed to safely evacuate all of the passengers.(68)

On May 13, 1991, two passenger trains collided head on near Shigaraki in western Japan, killing 42 passengers and injuring 415. Japanese fire fighters and rescuers were called upon to exercise their knowledge of extrication and mass-casualty skills.(69)

Like the airlines, however, millions of passenger miles are traveled for each fatality. For example, high-speed Amtrak trains carry 60,000 people between Washington and Boston alone, on a daily basis.

Although Amtrak does have an occasional derailment or accident involving fatalities, its record is relatively safe. There are times, however, when its safety record is questioned. For example, on May 4, 1990, *U.S.A. Today* reported that Amtrak had its third accident in ten days.

The incident triggering this news report was an Amtrak passenger train carrying about 200 passengers which collided with a logging truck near Durant, Mississippi. Only minor injuries were recorded. The second incident was a Chicago to Indianapolis Amtrak train which rammed a line of parked freight cars in Crawfordsville, Indiana, injuring about 30 people. The third was Amtrak's California Zephyr carrying 420 people which derailed near Batavia, Iowa, injuring 97. Twenty-seven of these passengers were hospitalized.(70)

Emergency Action for Railroad Wrecks

The initial action and size-up for a railroad accident is about the same as that for a highway accident. However, if the incident occurs in a remote location, an aerial survey by helicopter may be of greater importance than most roadway emergencies where access, though sometimes difficult, is usually obtainable by motor vehicle.

Passenger Trains—Emergency Action

When a train wrecks, the train dispatcher should be notified immediately and be told the accident location and tracks involved. It will be necessary to watch for other trains, and it is practical to put out flag men in both directions until you are notified that trains have been stopped. Immediately request that all electric power be cut off and electrified rails be de-energized, as might be needed in a subway or elevated railway. Get confirmation that this has been done.(71)

Even though the fire department may share control of such emergencies under your community's disaster plan, establish your command post nearby; you probably will be requested to supply ladder equipment for gaining access to the top of overturned cars, forcible entry tools, acetylene cutting torches, heavy duty jacks, hydraulic prying equipment, power saws, lighting equipment, rescue companies for extrication of victims, and engine companies to handle outbreak of fire.

It may be necessary to disconnect individual cars and get the train moved to perform rescues; elevating platforms, aerial ladders, and cranes may be of assistance under these circumstances. When ambulances are summoned, notify the hospitals where the victims will be taken so preparations can be made for their reception.

A triage system may be used to sort and classify the victims in accordance with the priority needed for treatment. This objective should not be defeated by permitting less seriously injured persons to be whisked off to the hospital by private citizens before the critically injured have been loaded into ambulances and transported to these hospitals; otherwise, the nearest hospitals will be filled with noncritical cases, and the seriously wounded may have to be taken to more distant ones.

The coroner may have to handle victims whose lives cannot be saved. The police department may need to control the spectators and establish barriers to vehicular traffic in the vicinity.

Freight Trains—Emergency Action

As with truck accidents, until the nature of the cargo is ascertained, a highly defensive posture should be maintained in freight train accidents. Manifests or waybills, usually carried in the locomotive (previously in cabooses when they were widely used) should indicate whether hazardous materials are being transported. Not all hazardous materials require a warning placard; not all placards are indicative of every dangerous reaction that a product can create, and such placards are often difficult to locate in twisted wreckage and dense smoke.

If the name of the chemical can be determined, but your hazardous materials references do not list it, do not assume it is harmless. Have your dispatcher call CHEMTREC at 800-424-9300 (toll free) if there is any doubt about its toxicity or instability and try to obtain more information before you commit fire fighters to a close attack. Due to the large volumes of hazardous materials that can be carried by train, and possible vapor clouds of corresponding large magnitude, a railroad incident provides an ideal application for a CAMEO program.(72)

If the accident has occurred in a nonpopulated area where no buildings are exposed, withdraw all personnel and notify the railroad that you will standby a half-mile away until given assurance by its safety personnel (not the brakeman, conductor, or engineer) that the materials

A deadly tank car of hydrogen cyanide distinctively painted to warn emergency responders of its deadly cargo. (Photo courtesy of the National Fire Academy.)

involved do not present an unusual hazard to your fire fighters. Do not ever risk lives of personnel merely to protect cargo and railroad equipment.

If the accident occurs in a business district, such as derailment of tank cars and box cars piled on top of one another, and one of the ruptured tank cars is releasing a large volume of liquefied gas, do not assume that it is nontoxic or nonflammable. Have everyone responding to the emergency don self-contained breathing apparatus and full protective clothing while laying lines to protect exposures and getting occupants of nearby structures evacuated to a distance of at least 2,000 feet. Where such defense action is taken promptly, a disaster can be averted, as was the case in a Dow Chemical Plant in Freeport, Texas. About ten minutes before the tank car of ethylene oxide exploded, employees saw the tank venting and immediately had the plant evacuation alarm sounded; plant personnel were able to clear out of the area and find protection behind remote buildings and structures.(73)

It is not safe to assume that you will have time to prepare for an explosion. The box car of bagged ammonium nitrate did not detonate in the Traskwood, Arkansas, train derailment until 40 minutes after the accident, but the tank car of propane at Kingman, Arizona, exploded about 15 to 20 minutes after vapors escaping from its relief valve became ignited. Thirteen fire fighters were within 150 feet of the tank car when it "let go." Twelve died from thermal burns—their turnout coats and whatever street clothes they had been wearing burned off their bodies. A thirteenth man died and 95 spectators who had gathered along the highway about 1,000 feet from the explosion were injured.(74)

In this instance, the fire department was attempting to keep the tank car cool with a one-inch booster line until larger backup lines could be laid from a water supply 1,200 feet away; as demonstrated here, a small hose stream is apt to be insufficient to prevent a BLEVE (Boiling Liquid Expanding Vapor Explosion). Even 500 gpm directed from portable or fixed turrets cannot prevent such an explosion unless the streams are able to bathe the area where impinging flames are causing the metal shell to heat and weaken from internal pressure of the heated product.

You cannot rely upon suddenly increased sound intensity or flare-up of the burning product to give sufficient warning to withdraw your forces prior to the explosion. Therefore, unless it is necessary to take a calculated risk to save lives, do not attack such fires at all where flames are impinging upon the shell of any tank car; even if the product is nontoxic and nonflammable, a violent rupture can take place from vapor expansion that relief valves cannot handle, especially if the car is inverted, and the valves are in the liquid space.

If cars are piled up, and you cannot see if flames are impinging on tank car shells, assume they are, and do not allow anyone but your reconnaissance team to approach the wreck until the product is identified and you realize that you have the personnel and resources available to handle the situation. While making this determination, keep all fire companies and everyone else clear of the fire to a distance of at least 2,000 feet.(75)

If nursing homes or similar institutions are nearby that will take considerable time to evacuate, it may be possible to set up portable monitors, left unattended, with large tips to direct heavy streams on the exposed tank shells. If a good water supply is not readily available, even by relaying from distant hydrants or ponds, the railroad may be able to bring in tank cars of water, along with cranes and other emergency equipment.

If the disaster consists of exploding bombs, as was the case in the railroad yards at Antelope, California, you may have to protect homes that are not heavily involved in fire, and disregard buildings that are beyond help. In this incident, almost two million pounds of bombs exploded over a period of several hours, leveling the town of Antelope, destroying a Citrus Heights fire station and 205 railroad cars, injuring about fifty, and causing millions of dollars in damage to buildings within a wide radius. Thousands of residents in a two-mile area in western Citrus Heights and Roseville were evacuated. Some of the fire fighting forces, not learning until after the emergency was over that the county sheriff was legally in charge of the disaster, disregarded orders of his deputies to stay out of the area where exploding bomb fragments were igniting homes. It took about two hours to get the Emergency Operations Center functioning following the initial explosion at 7:52 A.M. Unexploded bombs which flew through the air, along with great quantities of flying shrapnel, landed great distances from the railroad yards. It was considered a miracle that no one was killed.(76)

The fire chief of Wenatchee, Washington, learned the value of having a mutual aid plan when a tank car of monomethylamine nitrate solution detonated in the railroad yard, obliterating 75 railroad cars, demolishing buildings in a three block radius, killing two persons and injuring 77 others. His request for aid brought 182 fire fighters on twenty-three pumpers, nine brush fire rigs, eight tankers, plus several ambulances and rescue rigs. "One of the major problems that occurred at the explosion was communications between the time (Fire Chief) Harris found out about the box cars of ammonium nitrate and the time it took fire officials to locate the proper railroad personnel."(77)

MARINE DISASTERS

Disasters on the water have involved passenger and cargo vessels. Passengers are transported on ferry boats and cruise ships; ferry boats that carry autos are rather large ships of typical hull design; some boats on major rivers in the U.S. are large vessels having stern or side paddle wheels and are practically floating hotels on inland waters. For relatively short trips, such as between Copenhagen, Denmark, and Malmo, Sweden, hydrofoil ferry boats are used. In inland waters, such as San Francisco Bay, ground effect ferry boats have been used to move passengers from one port to another.

A limited number of passengers are carried on some freighters, but the cargo vessels apt to be involved in a disaster are more likely to be those used for transporting hazardous materials.

Causes of Marine Disasters

Shipwrecks can be caused by a vessel going aground, colliding with another ship, a rock, or an iceberg. Other reasons for voyages ending in disaster include boiler explosions, fire, storms, capsizing and sinking.

Fire and explosion have been leading causes of disasters originating with vessels in harbors or alongside piers and wharfs. Usually the barge or ship has a cargo of hazardous materials that becomes ignited and results in a detonation before the fire can be suppressed. Barges loaded with a highly toxic liquefied gas have broken loose in a storm and drifted miles downstream in a navigable river, finally sinking in an unknown location.

Passenger Ships

Despite tremendous strides in passenger ship safety, disasters with large loss of life have occurred through time and will continue as long as people continue to use the waterways.

On April 10, 1991, the Associated Press reported that an Italian ferry crashed into an oil tanker causing a floating inferno that killed 139 people. That same day 13 sailors were killed and numerous others injured when a ship tanker fire broke out not far from the original disaster. This two-fisted calamity that hit the Italian waterways was an eerie coincidence that showed how suddenly simultaneous disasters can strike an area.(78)

Twenty-one U.S. sailors became additional casualties of the Persian Gulf War when they were among those who drowned when a ferry overturned in Haifa, Israel. The sailors were returning to the S.S. *Saratoga* after an "R&R" break.(79)

On March 6, 1987, the water rushed through the open bow doors of the British ferry, *Harold of Free Enterprise*, in the Belgian port of Zeebrugge, causing it to capsize; 189 people drowned.(80)

Passenger ships are growing both in popularity and in numbers of passengers carried. When many states legalized gambling on the inland waterways in the early 1990s, many of these vessels became floating casinos, increasing their popularity. As gambling loses the stigma once attached to it and becomes both respectable and affordable to ever increasing numbers of persons, these "floating casinos" will become increasingly serious objects for prefire planning.

One of the most deadly of the historic ship disasters involved the *General Slocum*, an excursion steamer. It had left the 13th Street Pier on the East River in New York City at 9 A.M. June 15, 1940, with 1,400 persons aboard, most on a church picnic. Just about an hour after this wooden sidewheeler had commenced its cruise, a fire was discovered in the forward deck. The ship's fire hoses broke when the water was turned on; its life preservers tore apart when handled; lifeboats were improperly loaded and overturned before they reached the water, and many of the passengers drowned. A strong breeze fanned the flames and many women and children had their clothing ignited and died from the burns. By the time the ship was beached, 1,030 persons had died from the flames or from drowning.(81)

There have been other ship disasters with great losses of life, such as the sinking of the *Titanic* in 1912 as the result of striking an iceberg, wherein 1,517 lives were lost, and the collision of the *Mont Blanc* in Halifax Harbor in 1917, where 1,635 perished. However, it was the *Morro Castle* disaster in 1934, where 125 died as a result of its burning off the New Jersey coast, that prompted the drafting of fire safety standards for passenger vessels.(82)

The International Convention for the Safety to Life at Sea (SOLAS '60), which became effective on May 26, 1965, regulates construction and operation of merchant ships of all signatory countries. Following a number of disastrous passenger ship fires in the 1930s, an international convention was adopted in 1948, a predecessor to SOLAS '60. Both conventions used three principles for increasing fire safety: (1) separation of the passenger accommodation spaces from the remainder of the ship by thermal and structural boundaries; (2) containment, extinction, or detection of the fire in the space of origin; and (3) protection of the means of escape. U.S. flag passenger ships add a very important fourth principle: minimum use of combustible materials in construction of the vessel.(83)

Cargo Ships

The same factors responsible for mishaps on passenger ships, such as fire, collision, and going aground, are applicable to cargo ships. While the loss of life potential is not so great on board freighters, they may be loaded with hazardous materials that could explode while in a harbor and cause great destruction and loss of life on shore.

The most devastating explosion that ever occurred in the United States originated in a ship. When longshoremen were loading the hold of the S.S. *Grandcamp* with ammonium nitrate fertilizer about 8:00 A.M. on April 16, 1947, a fire was discovered in the No. 4 hold. When the workmen were unable to extinguish the fire with fire extinguishers, the hatches were closed, steam was tried, and the volunteer fire department of Texas City was called. While hose streams were being applied to the fire, which involved wooden dunnage and paper-bagged nitrate, a tremendous explosion disintegrated the ship and pier, killing 26 of the 27 fire fighters, and demolishing their four fire trucks. Four hundred sixty-eight persons were killed in the blast. More than 1,000 were injured, and 128 were missing.

Flying metal weighing many tons shattered nearby structures; pieces of the ship penetrated oil storage tanks in a refinery a considerable distance from the blast, starting fires; other fragments started fires in the S.S. *High Flyer* cargo ship loaded with ammonium nitrate berthed at a nearby pier. No attempt was made to extinguish these fires because of the great amount of rescue work that needed to be accomplished in the shattered reinforced concrete buildings of a nearby chemical plant. A later attempt to tow the burning *High Flyer* away from the pier was abandoned about 1:00 A.M. and everyone withdrew from the operation; at 1:15 A.M. its cargo detonated. The property damage was estimated at fifty million dollars.(84)

The date of June 11, 1990, was not a good one for the shipping industry in the United States. First, in Galveston, Texas, fire fighters were struggling to keep a burning supertanker with 38 million gallons of oil—more than three times what the *Exxon Valdez* spilled in Alaska—from sinking in the Gulf of Mexico. After powerful explosions sent flaming debris rocketing into the sky, the tanker's aft deck was just five feet above the surface of the Gulf of Mexico when fire fighters from Galveston and the supertanker experts—Williams, Boots & Coots—succeeded in quenching the flames. On the same day, a 90-foot tear in the bottom of a cruise ship threatened to spill 110,000 gallons of diesel fuel off the coast of Cape Cod. Clean-up workers manned booms around the 617-foot ship and succeeded in corralling the fuel.(85)

On New Year's Day, 1991, a British cargo ship *The Maersk Yare* ran aground and was capsized during an attempt to refloat her. In this case the potential disaster was averted when all 13 crew members were rescued.(86)

Emergency Action—Marine Disasters

It is sometimes possible to lend assistance to vessels heading for your port with a fire on board by obtaining via radio a complete description of the conditions, type of ship, nature of fire; its exact location, description of contents and compartments on all sides, above and below; type of extinguishing system or fire equipment being used; condition of the ship's stability; ability to maneuver and take advantage of wind for smoke and flame deflection, and progress being made toward containment.

If the ship is not too far away and the seas are not too high, it may be possible to dispatch a large fire boat to meet it and render assistance while

Marine fire fighters can often play only a limited role in disaster mitigation due to their distance from land. (Photo courtesy of the Federal Emergency Management Agency.)

it is making port. If special extinguishing agents and technical assistance are requested, a shipboard fire fighting team, equipped with an extinguishing agents package, such as dry chemical and aqueous film-forming foam (similar to the kits used in crash fire fighting) could be flown by helicopter to help control the fire, pending the arrival of the fire boat or the ship's arrival in the harbor.

Generally, fire departments with port facilities are equipped and trained to handle ship fires, and they are aware that fighting a fire in the lower reaches of a ship is comparable to coping with a basement fire with the additional problem of stability.

It is of utmost importance that a disaster plan identifies the chain of command in harbor emergencies, to avoid repetition of the S.S. *Normandy* fiasco. There the burning liner capsized in the New York slip because each agency working at the fire thought that another was in charge, and no one ordered the water streams stopped when the ship developed a critical list. The loss of 53 million dollars was only the monetary value; the real loss was the badly needed troop-carrying capacity that the ship would have provided in World War II upon completion of re-outfitting.

A harbor disaster plan should indicate explicitly the respective authority of the port warden (city officials), state harbor fire marshal, captain of the port (Coast Guard), and the commander of any military base or shipyard in which an emergency may occur. The fire chief must know in advance which agency has full authority to order a burning vessel moved away from a pier or wharf (private or military); to order a burning vessel towed to a designated location to be anchored or scuttled; whether or not the chief will be in charge of fire fighting aboard a military vessel alongside a municipal pier or in a naval dry dock; and whether the fire chief of a naval facility where the ship fire occurs understands that the municipal fire chief in whose jurisdiction the facility is located (unless otherwise agreed upon) has overall responsibility for suppression of fires within the city, including the harbor and vessels in port. In the event of an explosion like that in Texas City, affecting a wide area, the authority of the agencies likely to be involved is not immediately relevant, but should eventually be delineated.

Rescue may have to be conducted under conditions of intense heat and dense smoke issuing from the passageways and hatches, necessitating a large supply of spare oxygen or air cylinders for breathing apparatus. The cargo storage plan should be obtained from the ship's captain or duty officer to ascertain the nature of materials involved. If large quantities of explosives, poison gases, unstable chemicals, or other extremely hazardous materials are threatened by fire, the same considerations will have to be weighed as in highway and railroad cargo fires, including the need for evacuation of inhabited structures on shore, limiting reconnaissance and rescue teams to the minimum number required to handle the situation, and keeping the rest of the companies two to three thousand feet away from the ship to fight secondary fires caused by an explosion.

If size-up reveals that the fire is one that you should attack, then hose lines that have already been laid out upon arrival can be brought aboard by helicopter, over aerial ladders, elevating platforms, through side cargo hatches and fixed gangways. In closing down ventilation systems and trimming ventilators against the wind, consider the effects of premature ventilation on tightly sealed spaces where combustible products of combustion may have accumulated.

Although maximum use of the ship's extinguishing systems is desirable, use your fire department's portable fire fighting equipment (hose, ladders) wherever possible. In using hose streams or master streams from water towers, aerial ladder pipes, or portable monitors, keep in mind the need for ship stability. Simultaneous de-watering operations may have to be carried out in flooded compartments on the listing side of the ship to prevent its capsizing or foundering. Confine fire to the space involved if possible, using no more water than necessary.

If the ship has already sounded the national fire signal (five prolonged blasts of whistle or siren), other agencies may also be responding to the scene, such as tug boats, Navy fire boats, Coast Guard emergency personnel, police boats, and the harbor emergency force. If not, and their services are needed, your dispatcher can summon any needed assistance, including longshore crews and lighters if off-loading is required for overhauling burned cargo.

One lesson learned in the port of Charlotte Amalie, Virgin Islands, is that an entire crew should not be permitted to go ashore at one time. Here a passenger ship burned to the water at the dock.

CONCLUSION

We as a society have become very mobile, and as transportation modes have increased in number and type, the number of transportation disasters has increased in direct proportion. We have seen how serious multiple-fatality incidents have occurred on land, both by rail and highway, in the air, and at sea. Many of the incidents chronicled in this chapter tested the tactical capabilities and resources of the local fire service agencies to the limit and the situation is not expected to improve. In the decade of the 1990s, except for the period of the Persian Gulf War, people showed the tendency to become even more mobile, increasing the frequency with which they traveled for leisure and business. Likewise, as our society has become more technologically complex, the number of physical goods that must be transported continues to grow dramatically. All of this leads to an increased probability that a given fire department will be called upon to handle a major transportation disaster. These incidents are compounded where multiple fatalities or hazardous materials, or some combination of the two, requires maximum utilization of incident command. In the following chapter we will cover the problem of hazardous materials in greater detail.

CHAPTER 11 REFERENCES

1. "Carrollton, Ky. Bus Crash," in *The Cincinnati Enquirer*, May 14, 1990, p. A-2.
2. Reiser, J. "Fire Department Classroom" in *Western Fire Journal*, July 1973, p. 43.
3. "Bomb Destroys Pan Am 747 In Blast Over Scotland" in *Aviation Week & Space Technology*, January 2, 1989, pp. 28–32.
4. Bahme, Charles W. *Fire Officer's Guide to Disaster Control*, 1st ed., Boston, MA: NFPA, p. 150.
5. US Air Flight 5050, LaGuardia Airport, 9/20/89.
6. Kramer, William M. *American Heat* video journal, St. Louis, MO, December 1989.
7. *Fire Protection Handbook*, Boston, MA: NFPA.
8. *Time*, July 31, 1989, p. 15.
9. Tryon, George H. "The Salt Lake City Airport Crash" in *Fire Journal*, September 1966, p. 5.
10. *Time*, July 31, 1989, p. 12.
11. *Newsweek*, July 31, 1989.
12. *The World Book Year Book*, Chicago, IL: Field Enterprises Educational Corporation, 1975, p. 298.
13. *The World Book Year Book*, Chicago, IL: Field Enterprises Educational Corporation, 1975.
14. *The World Book Year Book*, Chicago, IL: Field Enterprises Educational Corporation, 1976, p. 282.
15. "Romulus Air Crash" in *Firehouse*, December 1987, p. 42.
16. *Time*, November 30, 1987.
17. *Newsweek*, February 5, 1990, p. 22.
18. *Cincinnati Post*, August 29, 1990.
19. *The Cincinnati Enquirer*, October 4, 1990, p. A2.
20. *Cincinnati Post*, November 15, 1990, p. 2A.
21. *USA Today*, December 4, 1990.
22. *The Cincinnati Enquirer*, December 4, 1990, p. A-8.
23. *USA Today*, February 5, 1991, p. 6-A.
24. *The Cincinnati Enquirer*, February 18, 1991, p. A-2.
25. *USA Today*, March 4, 1991, p. 1-A.
26. *Newsweek*, May 21, 1990, p. 70.
27. *Time*, July 31, 1989, p. 15.
28. "Aircraft Accident Report," National Transportation Safety Board Report #NTSB-AAR-74-16, December 2, 1974.
29. Bahme, Charles W. *Fire Officer's Guide To Dangerous Chemicals*, Boston, MA: NFPA, 1968.
30. Bahme, Charles W. *Fire Officer's Guide To Emergency Action*, Boston, MA: NFPA, 1976.
31. "U.S. Navy Aircraft Firefighting and Rescue Manual," NAVAIR 00-80R-14, January 1968, Naval Air Systems Command, Code 1416C, Washington, D.C.
32. "Aircraft Emergency Rescue Information, Technical Manual" T.O. 00-105E-9, July 1972, Hq. NRAMA-MMSTD, Robbins Air Force Base, Georgia.
33. "Aircraft Fire Protection and Rescue Procedure," 1st Edition, International Fire Service Training Association, Oklahoma State University, Stillwater, OK, 1970.
34. *USA Today*, January 3, 1991, p. 3A.
35. *The Cincinnati Enquirer*, December 13, 1990.
36. Kramer, William. *American Heat* video journal, St. Louis, MO, March 1990.
37. *U.S. News & World Report*, October 21, 1985.
38. "Highway Accident Report," NTSB-HAR 76–4, National Transportation Safety Board, May 5, 1976, p. 2.

39. Hebert, R. "Warning Plan to Avert Fog Pile Up Starts," *Los Angeles Times*, March 13, 1975, Part II, p. 1.
40. Associated Press Report in *The Cincinnati Enquirer*, August 2, 1991, p. A-4.
41. CNN broadcast, July 19, 1990.
42. Kramer, William. *American Heat* video journal, St. Louis, MO, July 1990.
43. *Fire Engineering*, July 1990.
44. Kramer, William. *American Heat* video journal, St. Louis, MO, January 1989.
45. Ibid.
46. "Highway Accident Report," NTSB-HAR 73–4, National Transportation Safety Board, October 17, 1973.
47. "Highway Accident Report," NTSB-HAR 76–4, National Transportation Safety Board, May 5, 1976, p. 2.
48. *Firehouse*, April 1, 1989, p. 58.
49. "Ammonia Truck Crashes in Houston; 4 Die, 119 Hurt," *Los Angeles Times*, May 12, 1976, Part I, p. 5.
50. Vincent, L.A. "Hazardous Materials on the Road," State Government, Autumn, 1961, p. 241.
51. "Sniper Bullet Explodes Dynamite Laden Truck," *Los Angeles Times*, October 1, 1970. Part I, p. 25.
52. "Highway Accident Report," NTSB-HAR 72–5, National Transportation Safety Board, September 21, 1972.
53. "Marshall's Creek Explosion" in *NFPA Quarterly*, October 1960.
54. "Insecticide Fires Severe Hazard to 54 Firefighters" in *Western Fire Journal*, July 1975, p. 9. Dektar, Cliff, "Insecticide Fumes Fell 94 at Truck Fire" in *Fire Engineering*, 1975, p. 166.
55. "Toxic Fumes in Collision Send 21 to Hospital," *Los Angeles Times*, October 10, 1975, Part 2, p. 1.
56. Bahme, Charles W. *Fire Officer's Guide To Emergency Action*, Boston, MA: NFPA, 1969.
57. "Hazardous Materials Regulations of Department of Transportation DOT Proposals for Identification of Hazardous Materials," Parts 290–297 of Title 49 of the Code of Federal Regulations.
58. Bahme, Charles W. *Fire Officer's Guide To Dangerous Chemicals*, Boston, MA: NFPA, 1969.
59. Engineering Safety Service, *Police and Fire Interest Bulletins*, No. 1 to 11 inclusive, American Insurance Association, June 1973.
60. "Freeway Tanker Fire" in *Firemen's Grapevine*, November 1973, p. 7.
61. Woodhouse, Linda R. "Miamisburg Revisited," in *Nation's Cities Weekly*, July 20, 1987, p. 3.
62. *American Heat* video journal, St. Louis, MO, June 1990, Vol. 4, Prog. 11.
63. Burgess, W.V. "Crescent City Disaster," FDIC Kansas City Conference Report, 1971, p. 161.
64. Hayes, Terry. "Houston Train Derailment and Subsequent Hearing" in *Fire Journal*, March 1972, p. 13–17; "Railroad Department Accident Report," NTSB-R AR 72-G, NTSB, December 13, 1972.
65. "Communications, Rail Emergencies Topics at Recent California Chiefs Conference" in *Western Fire Journal*, July 1975, p. 39.
66. Data supplied by Sandi Frey, Lakeport, CA, Librarian.
67. CNN broadcast, January 4, 1987.
68. *Cincinnati Post*, December 13, 1990, p. 2A.
69. Associated Press Report in *The Cincinnati Enquirer*, May 15, 1991, p. A-7.
70. *USA Today*, May 4, 1990.
71. Bahme, Charles W. *Fire Officer's Guide To Disaster Control*, 1st ed. Boston, MA: NFPA, 1978.
72. *Cameo Demo*, by National Safety Council, Chicago, 1991.

73. Bahme, Charles W. *Fire Officer's Guide To Disaster Control*, 1st ed. Boston, MA: NFPA, 1978, p. 176.
74. Ibid. p. 176.
75. Ibid. p. 177.
76. Ibid. p. 178.
77. Ibid. p. 178.
78. Associated Press Report, April 11, 1991.
79. *Cincinnati Post*, January 3, 1991, pg. 2A.
80. *The Cincinnati Enquirer*, March 6, 1991, pg. A-2.
81. "Tragic Trip of the '*General Slocum*,'" *Firemen*, June 1964, p. 19.
82. *The World Book Encyclopedia*, Field Enterprises Educational Corporation, Volume 17, p. 346, 1969.
83. *Fire Protection Handbook*, NFPA, Edition.
84. "Kansas City Fatal Explosion Report" in *American Heat* video journal, St. Louis, MO, May 1989.
85. *Cincinnati Post*, June 11, 1990, p.12-A.
86. CBS News broadcast, January 1, 1991.

CHAPTER 12

HAZARDOUS MATERIALS INCIDENTS

CONTRARY TO SOME misconception, hazardous materials disasters are not a recent phenomenon, but have been a serious problem for public safety and industrial personnel for more than a half century. When Chuck Bahme was researching the subject of explosions in 1948 in preparation for writing the Los Angeles *Dangerous Chemicals* code,(1) he listed every chemical that had produced an explosive reaction during the period of 1900 to 1948, as recorded in *Chemical Abstracts*.(2) He found that the number exceeded 3,000. He had been assigned to this task and that of heading up a new Dangerous Chemicals Detail by his department's fire chief after his response, as an engine company commander, to a devastating electroplating plant explosion. All the buildings in the neighborhood had been leveled and the fragments of seventeen bodies gathered up in gunny sacks by the fire fighters.

It was in the early 1900s that the detonation of a huge pile of ammonium nitrate in Oppau, Germany, almost wiped out a whole town, killing about a thousand of its residents; and it was back in 1929 that an explosion of X-ray film killed 125 occupants of a clinic in Cleveland, Ohio. In the more than 60 years that have elapsed since then, several thousand new chemical products have been developed annually, out of which possibly a thousand or more enter into commercial production and transportation each year.

SCOPE OF THIS DISCUSSION

In this chapter we will not deal with the ordinary incidents involving chemicals and other products that fire departments handle almost daily, such as acid carboy spills, chlorine and ammonia leaks, fires in acetylene, oxygen, butane, and other liquefied petroleum gas cylinder storage areas.

Our focus in this chapter will be on the hazardous material incidents that cause, or have the potential for causing, a major disaster. Chapter 11 on transportation disasters discusses those arising from shipboard fires and train derailments, as well as truck and plane crashes where hazardous materials are involved. Our attention in this chapter will be directed toward those occurring off the seas, channels, highways, rails, and runways.

HAZARDOUS MATERIALS DISASTERS DEFINED

Aside from the transportation element, there is still much to discuss about hazardous materials disasters. First, let us define our subject in its broadest terms. A "*hazardous materials disaster*" could be described as a large-scale interruption of normal activities by an unplanned release, spill, fire, or explosion involving hazardous materials that cause, or are likely to cause, widespread injury or damage to people, property, or the environment. For our purposes, an "explosion" is defined as the "sudden going away of things from where they have been, accompanied by a loud noise."(3) We will not try to distinguish among the reactions called deflagration, decomposition, dissociation, detonation, polymerization, or pressure vessel rupture, where the violent destructive results are apt to be similar. If we learn what we need to know about these dangerous reactions, we should be able to cope with them without becoming a target for things that "suddenly go away with a loud noise."

WHAT ARE THE HAZARDOUS MATERIALS?

Having arrived at a broad definition of a hazardous materials disaster, let us take a look at some of these materials we are so concerned with. Their principal components are often various chemicals and their compounds, as well as products in which they are used. One hazardous materials incident plan defines them as "any substance capable of posing an unnecessary risk to health, safety, property, and the environment." These include but are not limited to materials that are highly explosive or flammable, poisonous and toxic chemicals, radioactive products, corrosive materials, and biological agents.(4) The plan also defines "deleterious materials" as products which in themselves do not necessarily constitute a threat or danger to human life and property but when allowed to enter water sources, such as sewers, streams, ditches, creeks, and rivers, may be a threat to aquatic life.

To simplify our discussion, we are going to use the commonly employed abbreviation of "hazmats" in the balance of the chapter, and stipulate that they frequently turn out to be dangerous chemicals. You do not have to be a chemist to know that just about everything in the universe of known composition is comprised of chemicals. Nor do you need a Ph.D. degree to know that practically every chemical can be dangerous under certain conditions. Even the air we breathe if liquefied and absorbed in combustible materials becomes a high explosive. The word "hazardous" may apply to a condition or circumstance in which the chemical may be placed, as well as its properties.

Chemicals are usually classed as hazardous because they may explode, react violently, become unstable, ignite, develop heat, intensify combustion, rupture their containers, corrode, or yield toxic or flammable gases, all with possible injury to life and property. Some explosions are caused by materials that we do not ordinarily consider as chemicals. Powdered wood or sugar are examples. A dust explosion on March 23, 1976, at a plywood plant in northern California killed seven persons, injured nine others, and started a $9 million fire that took nine fire departments with 118 fire fighters to bring under control.(5)

Examples of some of the commonly encountered hazmats that are not usually thought of as chemicals include, gasoline, natural gas, cleaning fluids, paints, plastics, pesticides, fertilizers, ammunition, and drain cleaners.

CAUSES OF HAZMAT DISASTERS

There are a multitude of reasons why leaks, spills, fires, explosions, and so on can occur, and other reasons why a minor mishap may evolve into

55 gallon drums of methyl ethyl ketone are protected by Dayton, Ohio, fire fighters from the large Sherwin-Williams warehouse fire. (Photo courtesy of Dennis McCloskey.)

a major disaster In general, however, a great many may be traced to "The Four Blind I's." They are Ignorance, Incompetence, Indifference, and Inertia. People who are burdened with these handicaps may not face the wrath of a farmer's wife, but where hazmats are concerned, could end up very seriously dead! Lack of vision is not found only in the ranks of the people who encounter hazmats in their business, but also in front lines of the emergency personnel who are called upon to mitigate their mistakes.

Ignorance

Lack of knowledge on the part of the formulators, manufacturers, packers, shippers, warehousemen, and consumers who use hazmats has resulted in disasters such as the one described earlier in the electroplating plant, where apparently the chemist who combined acetic anhydride and perchloric acid in the 200-gallon electropolishing bath was unaware that a power failure to the tank's refrigerator would result in its detonation. As one of the victims, he never survived to learn the lesson of that mistake. Even more appalling was the statement made to Chief Bahme by an executive of a large chemical company in the presence of the company's chief chemist that there was no need for drafting regulations for liquid oxygen, since it was "so cold that it would freeze out any fire in which it might become involved!" He quickly dispelled that fallacy with a burning cigarette demonstration using liquid oxygen.

On June 17, 1987, fire fighters in Memphis, Tennessee, found the body of a 39-year-old tool worker inside a fuel oil tank that exploded. It is believed the worker's torch and buffer may have ignited the two-hour blaze. Four others, including a fire fighter, were injured.(6)

Fire fighters who respond to hazmat emergencies without the requisite knowledge of latent dangers and proper procedures for handling hazmat incidents have also had to pay with their lives. A costly example occurred in Texas City, Texas, where 28 fire fighters who battened down the hatches on the *Grandcamp*'s cargo of ammonium nitrate fertilizer as fire was spreading in its hold, unaware of the need for maximum ventilation, all were killed along with almost 500 others when the detonation let go.(7)

Incompetence

Ignorance may be responsible for not recognizing dangerous potentials of some hazmats, but incompetence is often pinpointed for the failure to provide adequate safeguards that might prevent or mitigate a disaster. Chief Bahme witnessed a shocking example of incompetence on the part of a fumigator who loaded a building with highly flammable and toxic hydrogen cyanide to rid it of termites without first shutting off the pilot light to the basement water heater. The resulting explosion not only demolished the multistory building but killed the operator and his assistant (and, no doubt, the termites).

Just imagine what incompetence on the part of city councils, county supervisors, state legislators, and federal regulators might produce in the way of ordinances, laws, and regulations for safeguarding hazmats. For example, if dangerous substances are not required to be tested, properly classified, put in safe packages, conspicuously and correctly labeled, stored in fire resistive structures having automatic fire detection and extinguishing systems especially designed for the particular hazards, and transported in properly equipped and suitable placarded vehicles, the possibilities for disaster are manifold.

If testing is inadequate to determine all possible dangerous reactions prior to marketing a product, disaster can result. If there is wrong classification of chemicals that have long been known to be dangerous, the public can be exposed to greater risks that would be the case if such chemicals were properly classified.

If a chemical, known to be dangerous, is omitted from a list of regulated chemicals, the absence of adequate regulation can lead to disastrous results.

If the effect of requirements for labeling hazardous materials is to conceal the most dangerous aspects of chemicals from emergency forces and the public, disaster can result.

If adequate regulations of hazardous materials can be circumvented by issuance of special permits that exempt chemical handlers from complying with safeguards relating to container design, for example, disaster can result.

If adequate regulations are not well understood by persons to whom they apply, nor adequately enforced by designated agencies; if the promulgation and enforcement of regulations are in a large measure dependent upon industries being regulated; if the reporting of facts relating to chemical accidents—facts that would help substantiate the need for more stringent regulation—are withheld from regulatory agencies, there is little prospect that chemical disasters can be eliminated. On the contrary, as we proceed from test tube to tank car quantities of dangerous chemicals, the prospect of larger chemical calamities appears to be greater than ever before. Let us consider one example of danger in the expanding production, shipping, and storage of liquefied natural gas. Natural gas incidents are frequent and occur in all parts of North America. As we were preparing this chapter, for example, on January 2, 1991, approximately 650 homes in Novi, Michigan, were left without heat after a portion of a natural gas main exploded, sending flames 50 feet into the air. Six power company employees were injured.(8) One month later, on February 4, 1991, in Atlanta, Georgia, approximately 20,000 people were evacuated after a six-inch gas main mysteriously ruptured, spreading noxious fumes throughout a four-block, downtown area. Somewhere in the United States there is a serious natural gas emergency on an average monthly basis.(9)

One news article from the west coast stated: "Despite industry assurances that the risk of accident from importing liquefied natural gas into southern California is extremely small, new estimates indicate that serious hazards will exist, jeopardizing more than 100,000 lives and property worth hundreds of millions of dollars."(10)

A report cited in the above article issued by the Rand Corporation regarding the proposal to construct liquefied natural gas (LNG) off-loading facilities within two populated areas (Los Angeles Harbor and Oxnard, California) points out that spills could occur either on land in port facilities, at sea, or in the harbor, especially during transfer procedures. Such spills could be caused by ship collisions or groundings, sabotage, natural disasters, design faults, construction or fabrication weakness, or human error. The Socio-Economic Systems, Inc. released a report about the same time as the Rand report, in which it estimated that an LNG tanker collision 4.5 miles off shore and a subsequent fire could kill 70,000 persons and create a total property loss of $325 million.

It is estimated that a spill of 100,000 cubic meters would create a flammable cloud ranging from 17,000 to 670,000 feet in length. One group of consultants estimated that a 25,000 cubic meter gas cloud would travel 1.9 miles, but the Coast Guard predicted that the same size spill would produce a cloud 14.4 miles long.(11)

Incompetence that leads to disaster is not just limited to legislators and administrators who fail to require and put into effect the necessary safeguards for hazmats. Too often, it is found among the ranks of the everyday handlers. Some examples are illustrated by the LP gas tank truck driver who drove away without disconnecting his filler hose, resulting in the destruction of many buildings in a Massachusetts military base, and the sea captain who let his tanker run aground off Alaska where it spilled millions of gallons of oil to produce an environmental disaster, or the backhoe operator who dug up a high pressure gasoline pipe line in Culver City, California, resulting in a conflagration that burned down much of the city.(12)

Indifference

Indifference on the part of hazmat handlers who have the requisite knowledge and competence to avoid accidents, but who choose to take a chance for various reasons, can result in disastrous consequences. Perhaps the old adage that "Familiarity breeds contempt" led to the continued use of thin metal containers for liquid hydrocyanic acid in the adjacent county, after their use in Los Angeles was forbidden upon finding that they would rupture if dropped from the bed of a truck. The desire to save time and money could also lead to the type of indifference shown by the truck driver who illegally took his cargo of carbon bisulfide, a highly toxic and flammable liquid, through a New York tunnel in which it caught fire, menacing the lives of hundreds of motorists. In another instance, the driver of a truck load of blasting agent hid the "Explosives" placard under the seat in his cab, and when fire ensued, drove off in the detached cab leaving the cargo with no warning sign for the responding fire companies. This resulted in the death of several fire fighters and the destruction of their fire apparatus when the cargo detonated.

Indifference on the part of emergency response teams is not generally a problem today. However there was a time when the hardy "smoke-eaters" would show their contempt for danger by not wearing breathing apparatus to hazardous areas or by rejecting the advice of the plant manager that the fire they were about to attack might be unsafe to approach. Such apparently was the case when the arriving fire officers at a Norwich, Connecticut, plant were warned that the fire in the benzoyl peroxide cargo might explode.

When it finally did, several fire fighters were killed and their fire trucks demolished.(13)

Inertia

Lastly, in our discussion of the principal causes of hazmat disasters, there are those people who are fully aware of most of the risks involved and genuinely desire to provide adequate safeguards, but who just never get around to doing anything about it. They are the victims of the "blind eye of inertia." Some examples of this are clearly evident in the inaction on the part of federal, state, and local officials, both legislative and administrative, and also on the part of industrial and public safety managers.

This inertia not only extends to the adoption of needed uniform laws and regulations to provide improved safeguards for hazmats, but also to observing and enforcing the existing standards of care, however deficient they may be in some respects.

While it is true that some industries have been slow in establishing training programs for their employees relating to hazmats, and in promulgating plans for emergency action for coping with disasters, this inertia has not been entirely absent from the fire service. In the matter of providing personnel, equipment, and training for hazmat problems in their response districts, or in establishing a target hazard inspection program, hazmat reporting and identification requirements, and a disaster plan, the fire service still has much to accomplish. Fortunately, this inertia is being overcome, slowly, in almost every aspect of the problem, with the result that we have a much better chance of preventing more hazmat disasters in the future. How can we improve the odds?

PREVENTION OF HAZMAT DISASTERS

Most would agree that the best way to obviate the need for controlling, managing, and mitigating disasters is to prevent them from happening in the first place. It has long been recognized that the elimination of unnecessary risks in almost every field of endeavor can be achieved through the application of the "The Three E's of Safety," namely, Engineering, Education, and Enforcement. These same three E's can be valuable tools in the continuing goal of preventing disasters of any kind, but especially those involving hazmats.

Since it is not within the scope of this study to provide a detailed discussion of two aspects of disaster prevention, namely, Education and Enforcement, we will comment upon them only briefly.

Engineering for Hazmat Safety

Probably the best place to commence any safety endeavor is at the "drawing board." In the field of hazmat safety it might well start with the chemical engineer's test tube and proceed to the design of the processing, storage, and handling equipment. Engineers can do much to reduce hazmat hazards arising from exothermic reactions, spontaneous polymerization, decomposition of unstable chemicals, etc., through the provision of relief valves, soft plugs, automatic dumping or flooding devices, special walls and roofs for explosion venting, backflow and check valves, dikes, drains, sumps, high-speed detection and extinguishing systems, automatic fire door, venting and alarm systems. Helpful as some of these systems may be in preventing disaster-causing accidents, they do not install themselves, nor do all plant operators see the wisdom of spending the money to have them provided, even though national standards-setting organizations may recommend them. Hence, in many instances it becomes advisable to adopt laws, ordinances, and regulations that make their installation mandatory. But merely having laws on the books is not very helpful unless they are enforced.

Enforcement

Before you commence a conscientious enforcement program, make sure that your local requirements are adequate to obtain the desirable safeguards for hazmats in your jurisdiction. If they are not, try to have ones adopted that will meet the needs. In any case, you will have to train a sufficient number of personnel to carry out the enforcement responsibility, and this is not limited to your own local ordinances, but also to state and federally mandated hazmat requirements. Before we take a look at some of the national legislation that relates to hazmats, you may justifiably ask why so much responsibility is imposed upon fire departments. Why not the public health departments, or the law enforcement agencies? The answer lies in the fact that the fire service has traditionally been the one that the people look to in time of disaster, whether a personal one, such as a hand caught in a bicycle sprocket, a person stuck in a bathtub, a child locked in the bathroom, or people trapped in an elevator or a collapsed building—any one of a hundred emergencies that fire departments are called upon to handle that do not involve fires of any kind. Because the fire service responds to these disasters, why not trust them to mitigate the major disasters, including those arising from hazmats? And since fire departments have always been given the duty of enforcing fire codes, why not hazmat laws also? Legislators and political leaders at every level of government have come to realize that hazmat problems can erupt anywhere, from the smallest hamlet to the largest metropolis. If there is one service agency that they all have in common it is the fire department.

GOOD LAWS ARE NEEDED

Enforcement of poorly drafted laws will not prevent hazmat disasters. And if there are industrial managers who are unwilling to voluntarily adopt safe practices in the absence of regulatory coercion, there also are some who might urge legislators to adopt rules that have the appearance of establishing needed safeguards for hazmats, but because of monetary considerations, end up being less than satisfactory from a public safety standpoint. For example, suppose you can hire a stevedore to handle a product labeled as an "oxidizer" for half the wage rate he would demand if it were more accurately labeled an "explosive" because, even though it may only be intended for use as a fertilizer, it has long been known to detonate when involved in boxcar, structure, and shipboard fires. Now you can understand why Chief Bahme's objections to this type of labeling were rejected in Washington, D.C., when he met with regulators following the Texas City shipboard explosions on the *Grandcamp* and *High Flyer* many years ago. (See section on Cargo Ships in Chapter 11 for a description of this disaster.) Lack of enforcement was not the culprit in that disaster, nor the lack of knowledge of the hazards, but the lack of useful regulations that would have required a label sufficiently descriptive of the product's latent danger to warn the fire fighters and hundreds of spectators of their impending sudden demise.

CONSENSUS STANDARDS

Perhaps it is not always desirable to leave the drafting of regulations to the industrial organizations being regulated and then pass laws forbidding state or local governments from adopting any rules that would require more effective safeguards. Furthermore enforcement is not of great value unless the rules being enforced are drawn up by persons of integrity who truly seek to obtain adequate safeguards for hazmat handlers, the public, and the emergency responders. Codes drafted by a consensus of committee members, many of whom represent the industries to be regulated, do not always contain the best standards that could be applied to a given problem, instead they will be watered-down standards that a majority of the committee could agree upon. If these turn out to be mediocre, then a requirement that they be uniformly enforced throughout the nation, barely results in the uniform adoption of mediocre rules at every level of government, and may preclude the enforcement of really good safeguards at the local level. This may not be a pleasant fact to contemplate, but it is a fairly accurate one. As someone once said, "You shall know the truth and the truth shall make you flinch before it makes you free."

To sum it up, then, enforcement will not be of great benefit in preventing hazmat disasters unless the regulatory and legislative bodies adopt good regulations that can be uniformly applied within their respective jurisdictions. For example, there is little doubt that hazmat disasters are more likely to be avoided if use, storage, manufacturing, and processing of hazardous materials are required by zoning laws to be restricted to relatively uninhabited areas. Nor would anyone deny that risks might be reduced by prohibiting their transport through tunnels, congested business districts, past schools, hospitals, and densely populated residential areas. However, basic to such solutions is the determination of which hazmats are so dangerous, and in what quantities, as to justify such restrictions. And that is where the third "E" becomes applicable. This requires the "education of our legislators, hazmat handlers, and the public as to the need for reasonable restrictions and safeguards that can be helpful in preventing hazmat disasters.

Education

To eliminate, or at least reduce, the catastrophic consequences of a hazmat disaster, education is an indispensable tool. Obviously, the educators must acquire a thorough understanding of the potential risks that hazmats pose, especially if they are called upon to assist the regulators in drafting rules for safe handling or in devising a warning system to provide ready identification of such hazards for everyone exposed to them.

LEARNING THE RISKS

Hazmat education must start with the people who produce them. They must learn the scope of the risks that may be generated under all of the conditions likely to be encountered from the time the hazmats are manufactured until packaged, transported, stored, used, and ultimately discarded. There are various ways of determining a product's hazards. A way that definitely should be discouraged is to put it into production, ship it around the country, and wait to see how many fires and explosions it will cause. A better way is to test it thoroughly according to recognized standards *before* it is put into commercial use. If tests show that a chemical is hazardous in one or more respects, such as fire, health, or instability, then it should be listed in a suitable classification in regulatory publications, and the laws and rules applicable to its packaging, labeling, and handling should be rigidly enforced. For example, if a product is tested and found to be susceptible of detonation under conditions that can be reasonably anticipated to occur during normal handling in rail shipment, could there ever be any justification for sending tank cars of that material around the country where they might be bumped or shocked during switching operations? Was any testing done before the monomethylamine nitrate solutions were shipped in the tank car that detonated in Wenatchee, Washington, causing the disaster described in Chapter 11 on transportation disasters?(14) At the time of the accident, this chemical could not be found in any listing of the national organizations that promulgate safety standards or transportation regulations. Why?

PROVIDING A WARNING

A practical system must be devised and promulgated for warning not only the people who manufacture, transport, store, use, and dispose of hazardous chemicals and products, but also members of the public and emergency personnel who may be on the scene of a mishap where these materials may be involved. Several systems have been proposed, but the two that are currently used most extensively in the United States are the "704M" of the National Fire Protection Association(15) and the "Hazard Information System" of the U.S. Department of Transportation (DOT).(16) There is also a United Nations Classification System for use with the transportation of products in foreign trade. There has been considerable improvement made in the classification, labeling, and placarding requirements in the DOT system in recent years, but there are still some chemicals assigned to a classification that fails to disclose their most serious risks. For example,

Markings. Stencils and placards are essential if first responders are to know whether or not the contents of rail cars are hazardous.

nitromethane was known to be a nonpropellant capable of detonation when the first tank car containing this flammable liquid exploded in 1958, near Niagara Falls, New York, injuring 180 people and causing damage in excess of $1 million to buildings within a 3.5 mile radius. Many of the injured were children in eight of the damaged schools. Even if regulatory agencies had previously been ignorant of the explosive properties of this rocket fuel when the first detonation occurred, what is the explanation for a second tank car of the same chemical detonating only four months later at a siding near Mt. Pulaski, Illinois, causing the same amount of loss, killing two trainmen, and injuring forty other persons?

"This tank car was also involved in switching operations, but like the previous one, had no 'Explosives' or 'Dangerous' placard on it to indicate that it was susceptible to detonation, nor has the chemical ever been subsequently required to be identified as other then 'Flammable'." The DOT's 1987 *Emergency Response Guidebook*, in describing fire and fire explosion hazards of this material (Guide 26) omits any reference to its susceptibility to detonation by shock. Fortunately the 1990 Guidebook was more definitive.(17)

Classifications of chemicals in the safety standards and regulatory publications should reflect their most hazardous properties. How could the people who needed to know the dangerous properties of such poisonous gases as chlorine, sulfur dioxide, nitrosyl chloride, and ammonia be alerted to their lethal potentiality if the national list of regulated hazardous commodities merely listed them as "nonflammable compressed gases?"

Hazmats should be required to be labeled under an identification system that will provide adequate warning of their most dangerous properties. Any set of regulations that does not require materials capable of deflagration, such as organic peroxides, or detonation, such as ammonium nitrate-fuel oil mixtures, under fire conditions, to be labeled as explosive materials fails to impart the true hazards to the public and the emergency services. To permit explosive materials to be exempt from being labeled as explosives merely because they are not primarily intended to be used as explosives cannot be justified in the judgment of fire fighters who may not always be able to distinguish between "oxidizers" that are not explosive, such as sodium nitrate, and those that are, such as ammonium perchlorate.

Likewise, any set of regulations that deliberately defined a "flammable gas" in such a devious manner as to exclude ammonia, a gas that has been responsible for many explosions, was not fair to the public; to permit this gas to be labeled with the same label that was required for compressed air did nothing to prevent further disasters caused by this chemical. Ammonia's capability of causing a disaster was well-illustrated in Crete, Nebraska, on February 19, 1969, when a freight train derailment struck a tank car of anhydrous ammonia standing on a siding resulting in a blinding flash and release of dense clouds of ammonia gas which drifted through the town, killing eight persons and putting eleven others in the hospital.(18)

It is impossible to devise an identification system that will indicate the true hazard of every chemical, compound, or cargo, whether mixed or homogeneous, by trade name, chemical name, signal word, degree of hazard, and by code number to a reference supplying information for emergency procedures in a number of languages. However, it is reasonable to require that hazmats be identified with basic warning symbols for protection of the public and emergency services. It is also reasonable to expect that the emergency services will obtain the knowledge and training to cope with hazmat emergencies even though all of the desired information is not instantly available. Industrial personnel who produce, ship, store, and use hazmats should also learn their risks and become familiar with the applicable labels and placards of the identification system currently in use. First responders of the emergency services need to be able to quickly assess the risks at the scene of an incident through their knowledge of the hazard information system, by means of labels, placards, waybills, or data cards that clearly indicate the hazards or at least give them clues that will enable them to seek needed information from other sources. These other sources will be discussed later on in this chapter.

TYPES OF HAZARDOUS MATERIALS DISASTERS AND SOME EXAMPLES

Hazmat disasters have taken many forms. The most serious precipitate extensive pollution of the air we breathe, the water we drink; they poison the wildlife in oceans, lakes, and streams, sometimes making vast land areas unfit for human habitation; and sometimes they explosively shatter a neighborhood of businesses and dwellings, killing occupants and often the fire fighters who attempt to rescue them. Unlike an earthquake or a tornado, these disasters do not "just happen" in most cases but can be traced to oversight, neglect, or downright stupidity on someone's part. While it is true that hazmat disasters can grow out of earthquakes, floods, and violent storms that dislodge tanks, overturn ships, trains and trucks, uproot factories and warehouses, where there has been good engineering foresight and disaster control planning, damage mitigation can often be accomplished. Let us review some major hazmat disasters that occurred around the world to better understand how to prevent, control, and mitigate future disasters. The fact that some of these incidents occurred many years ago does not render the experience less valuable. When you recall that "experience" has been defined as "that which enables you to recognize a mistake when you make it again," it would seem well worth while to periodically review hazmat catastrophic experiences whenever they may have occurred. Let us start with those that have affected the air we breathe, as we commence our survey of some environmental disasters; we can then proceed to disasters that have polluted our water and made our land uninhabitable.

Environmental Disasters

Air pollution can arise from many sources, including factory smokestacks, chemical spills and releases, ruptured pipelines, combustion products from hazmat fires, decomposition of buried or submerged materials that come to the surface, and radioactive emissions from nuclear accidents.

Industrial Emissions

Many years ago, in a little town in Pennsylvania called Donora, a thick, smoggy blanket of factory fumes and fog choked its 14,000 inhabitants for four consecutive days. That was in October 1948. "Almost 6,000 people had coughing, sore throats and breathing problems, smarting, watering eyes and nausea and vomiting. People with heart disease and bronchitis were crippled by the filth in the air; many people had to be given emergency oxygen treatment. In the four days before the wind and rain came, twenty people died."(19)

In the 1952 London disaster, industrial pollutants choked the city for five days, seeping into homes and causing people to grope blindly through the muggy darkness. Of the many thousands who became ill, over 4,000 died.

The smog that blanketed the City of New York in 1963 for only a brief time was directly responsible for the deaths of between 200 and 400 persons.

Chemical Leaks, Spills, and Releases

On December 1, 1984, a drum of sodium hydroxide (drain cleaner) in a dumpster began to give off a corrosive vapor cloud that caused the evacuation of several thousand weekend shoppers from a mall in Hicksville, New York. The first arriving fire units sealed off the rubbish chutes and stairways leading from the area under the mall (where the dumpster was) with tape and salvage covers and reversed the ventilation system, creating a positive pressure that assisted police and fire units in the evacuation of the four floors of the mall building. While fire fighters handled the dumpster fire, the Hazmat Unit set up a "Hot Zone" around the tunnel entrance, a medical triage and treatment area, and began construction of a decontamination site near the hazmat command post. A staging area was set up near the primary command post for arriving apparatus from nearby towns; these brought additional floodlight and air charging units, along with additional ventilation fans and tarps. Using a dumpster pick-up unit, the Hazmat Unit brought the dumpster out in the open where it removed the lye drum and packed it in a recovery drum. The burning material was covered with an absorbent-dryer compound and then sand. After thorough ventilation, the mall was reopened in about three-and-a-half hours.

In a somewhat similar incident which occurred on January 9, 1991 in Broken Arrow, Oklahoma, 22 oil tool company workers were injured when a container filled with lithium batteries exploded during pressure and heat testing. The building was evacuated until it was properly ventilated and any toxic fumes eliminated.(20) We do not always get by so easily, however. Many recorded incidents have affected large numbers of persons.

About 6,000 residents fled from their homes in Auckland, New Zealand, in 1973, to escape the deadly vapors of twelve leaking drums of a chemical defoliant that had been unloaded in that port city. While 350 persons were being treated in various hospitals, hundreds of fire fighters, police officers, ambulance operators, and troops tried to neutralize the lethal gas and the liquid in the drums. With this type of chemical, death comes from respiratory failure either by inhalation or through skin penetration.(21)

Another highly toxic chemical defoliant known as dioxin, or ICDD, escaped in a cloud of gas from a malfunctioning safety valve in an Italian factory in Seveso, on July 10, 1976, commencing a medical disaster. For more than a week, none of the 100,000 residents of the area grasped the magnitude of the disaster. When animals began dying by the thousands, trees began to wither, and doctors' offices were deluged with hundreds of patients complaining of skin infections, the health minister realized that something was wrong and had the area evacuated. To halt the spread of contamination, an army of veterinarians in protective suits began to kill all surviving animals in the Seveso area; corn fields and vegetable crops had to be destroyed, for the fine vapor had been washed down into the soil by a heavy rain. Small doses of this chemical can not only damage human kidneys, livers, and lungs, but can cause deformed fetuses in pregnant women. As a result, therapeutic abortions were commenced on August 13 at the request of women from the most contaminated region. About 5,000 persons had to abandon their homes permanently. A government scientific commission recommended that all vegetation in the locale be destroyed in special furnaces and that all buildings, including the factory (Icmesa plant) and the evacuated homes be razed within three years.(22)

A large leak of silicon tetrachloride occurred in Chicago at 122nd Street and Stony Island on April 26, 1974, about noontime. The emergency was not over until May 3rd. A malfunctioning valve at the bottom of a 1,250,000-gallon storage tank containing 800,000 gallons of this water-reactive chemical caused a spill of about 300,000 gallons into a diked area. Mixing with air, this chemical formed a dense white cloud of hydrogen chloride, an acid vapor that has a toxic threshold of about 5 ppm. By 6 A.M., a five-mile long, half-mile wide cloud of toxic fumes had spread over the south side of the city, causing cessation of all port activities and evacuation of a residential area known as Altgeld Gardens, three miles southwest of the tank terminal. The area affected by the leak included at least 30,000 persons. After flying over it, Michigan's governor ordered mobilization of three National Guard battalions to assist in evacuation of 16,000 persons who had been ordered to leave the area.

The fire department was called at 2 P.M. to furnish all available ambulances and to assist in other operations that continued for eight days. The first evacuation headquarters was set up at the Washington Carver Hospital, but it had to be abandoned when fumes seeped in and persons began to collapse. A high school was then selected as a new center and about 2,500 persons had already gathered there by 11 P.M.(23)

The fire department transported victims to the hospitals, administered oxygen inhalation to others, and assisted in activities aimed at getting the leak stopped, the liquid transferred, the spill neutralized and covered, and provided aerial reconnaissance of the cloud path. Attempts to stop the leak included the use of transfer pumps, hot-tap equipment, and special quick-drying cement. Attempts to neutralize the product in the diked area included the use of lime, sand, oil, covering it with vinyl and rubber sheeting, and various diking techniques. In trying to stop the leak in the pipe between the tank and the valve, fire fighters wearing self-contained breathing apparatus and chemical handler's suits tried shredded lead wool, compounds, and shrouds, but none of them were effective. The chemical suits ripped and tore easily and nine fire fighters had to be treated for acid burns caused by the liquid and vapors. When 200 gpm chemical pumps were used in an attempt to pump out the tank, the pumps burned out. They tried to get a "hot tap" into the tank; they used fire boat heavy stream appliances to throw up a water curtain to contain vapor spread.

The diked area was first covered with lime but then it rained, causing further vapor emission. The area was then covered with vinyl sheets to keep it dry and sand was put on top to hold them down, but chemical action caused the vapors to penetrate the vinyl. It was then paved with asphalt and tons of lime spread on top, but this was not a complete vapor barrier. A sight scope was used to determine the depth of the liquid in the tank and the location of the hot spots under the vinyl covers. A fast-drying cement was pumped over the pipes and the leak, but it still leaked, though not as much, and the tank was finally pumped out. The acid vapors coated the lenses of eyeglasses and damaged fire vehicles.

Technical assistance was supplied by the Department of Army Operations Center (AOC), U.S. Army Technical Escort Center (TEC) Edgewood Arsenal, Aberdeen Proving Ground, Maryland. However, the Army chemical equipment did not arrive on the site until 11 A.M., April 27.

Chicago's Fire Chief, Bill Foley, told author Chuck Bahme that if he had to do it over again, he would "flash" the vapors of the spill right at the start, and get rid of it, instead of trying for days to cover it up.

The leaking of a poisonous gas, methyl isocyanate, from a Union Carbide Corporation pesticide plant in Bhopal, India, on December 3, 1984, created a continuing catastrophe that has no foreseeable end. About 2,500 people who lived around the plant died immediately(24) and by 1987, 8,000 more had died and 300,000 more would suffer debilitating injuries.(25) In 1987 the Supreme Court of India ordered Union Carbide to pay the victims $470 million, apparently based upon the Indian government's contention that the plant was poorly designed and maintained, and rejected the argument that it had been sabotaged by disgruntled employees.(26)

Fire-Caused Pollution

Air pollution can also be caused by fires, sometimes with serious consequences. On January 14, 1976, Los Angeles City fire fighters were called to a fire in an industrial building in Chatsworth, a northern suburb of the city, where containers of hydrochloric acid, cylinders of chlorine, and large

Members of the Milwaukee, Wisconsin, hazardous materials team train under realistic but simulated conditions. (Photo courtesy of Scott Franken.)

quantities of a plastic derivative of polyvinyl chloride were stored. Fire fighters wearing breathing apparatus entered and quickly knocked down the fire, unaware that there were chemicals in the building. "Soon the effects of the fumes were felt. Fire fighters dropped to their knees, vomiting, and with burning sensations in their mouths, throats, and lungs, and on exposed areas of skin. Their masks provided no protection from the fumes and fire fighters in the building, unable to breathe, managed to stumble out or be dragged to safety."(27)

Forty fire fighters, 11 police officers, and 21 civilians were rushed to eight hospitals for medical treatment. Helicopters broadcast the alarm. Over 2,000 persons had to be evacuated from the area, including 1,400 students from two schools. The fire was brought under control in 32 minutes, but the fumes of the gases and vapors could be detected 17 miles away. Students were taken to a high school in a nearby community, and other evacuees were taken to a center established by the Red Cross in the Chatsworth High School gymnasium. More than five hours after the fire was controlled, policemen were still making a house-to-house search of the area to see whether any persons had been overcome in their homes.

Despite modern fire codes, which provide detailed regulation for chemical occupancies among thousands of businesses and warehouses where dangerous chemicals are to be found, fires are occurring daily. Occasionally one results in a near disaster because of release of toxic gases from chemicals. Such was the case on June 15, 1975, when fire broke out in a swimming pool chemical firm in the San Fernando Valley, where 250,000 pounds of dry chemicals used for chlorinating water, packed in 200- and 80-pound drums, were stored. The pall of poisonous chlorine gas released from these chemicals, due to heat and the application of hose streams, spread over four square miles of the valley, forcing evacuation of about 5,000 people and hospitalizing 50. After the fire was checked, fire fighters stopped the application of water to reduce the chlorine emission, and let the fire burn out over a period of nine hours. At times the dense white smoke billowed as high as 500 feet above the scene, forcing police helicopters to withdraw. (Two flying officers were among those treated for fume inhalation.) But for the most part, the smoke settled close to the ground and was then spread by the wind. The Red Cross established two evacuation centers at gymnasiums of secondary schools where about 2,000 persons including 100 elderly people and invalids from a nearby convalescent home were provided cots and mats to spend the night.(28)

A wide-scale evacuation of residents in Darien, Wisconsin, was necessary on February 3, 1991, when a 48,000 square foot fertilizer warehouse was destroyed. This disruption to the community could likely have proven to be disastrous without the evacuation.(29)

Sometimes chemicals are mistakenly dumped into a vat of highly reactive materials that produce great volumes of toxic vapors and gases; however, some chemicals can poison the air by their own reaction to it.

More than 1,000 persons had to be evacuated from a twenty square block area in Hawthorne, California, when a chemical reacting in a detergent manufacturing plant generated clouds of irritating sulfuric acid fumes. Eleven persons had to be given hospital treatment for the effects of vapors which caused extreme breathing problems. Policemen cordoned off streets leading into the area while fire fighters, wearing breathing apparatus, washed down the area where sulfur trioxide had been in the process of being transferred to a storage tank at the time of the mishap. Upon hitting the air, the sulfur trioxide converted to sulfuric acid vapor, and the acid fumes billowed through the plant and out into the streets.

This same chemical put another city's disaster plan for hazmats to a severe test. On January 4, 1978, a leak developed in the discharge pipe of a Baltimore, Maryland, soap company's sulfur trioxide tank, thus producing a sulfuric acid cloud that spread over the city. More than one hundred persons were treated for respiratory injuries during the four and a half hour incident.

The DOT's guidebook suggests that you evacuate an area two miles long and a mile wide if you have an 800-square-foot spill of this toxic and corrosive liquefied gas. However, no one could advise the chief how far he would need to evacuate the 300,000 inhabitants affected by a leak from a four-by-ten-foot tank of sulfur trioxide.

The cloud of vapor was tracked by Maryland State Police in a medevac helicopter as it moved into Pennsylvania. Radio and television messages warned people to get off the streets and stay indoors; fire fighters assisted the staffs of three nursing homes to keep the fumes from entering the buildings and used smoke ejectors to move fumes out of buildings.

A field headquarters command was established to direct three command posts where spare air cylinders were brought for replacement of depleted ones on breathing apparatus. The contaminated air precluded recharging the empty bottles with the fire department's air compressor.

The fire chief's disaster plan contemplated localizing this type of chemical emergency and evacuating persons in danger, but it did not envision the need for evacuating hundreds of thousands of residents.(31)

Air pollution from vaporized irritating toxic liquids at least gives some warning of the hazard; such is the case with the nonflammable sulfur dioxides and nitric oxides. However, if a tank car of liquid hydrogen cyanide were to rupture and vaporize, untold numbers of people could drop dead without even suspecting that the mild almond odor was highly poisonous. Like hydrogen cyanide, styrene is both flammable and unstable, but it does give off an unpleasant odor. The following incident, taking considerable time to control, also illustrates the ignorance of some high ranking plant officials.

Two tanks containing 52,000 gallons of styrene monomer ruptured about 6:00 A.M. on August 14, 1978, in a plastics plant in Orange, California. Styrene monomer is a flammable liquid that must be inhibited to prevent polymerization. Fire fighters sprayed water on the two tanks as the liquid spewed out, vaporized, and blanketed the surrounding area with an irritating gas. About a hundred members of five fire departments engaged in this operation while 25 policemen went throughout the community with sirens and loudspeakers ordering approximately 2,500 people to evacuate a one-and-a-half square mile area.

After about five hours, the vapor-producing reaction terminated, and fortunately the flammable vapors that are heavier than air did not ignite during their dispersal, though many persons were sickened by the irritating odor and dozens of businesses could not open because of the explosion hazard in the event the vapors ignited. Despite the fact that styrene was known to be flammable, irritating, and unstable, the vice president of the plastics plant was quoted as saying that he couldn't understand why people were evacuated, for "there wasn't any danger."(32)

Ten days later 500 people from 14 plants had to be evacuated because of a leaking tank car of styrene in the City of Industry, another community not far from the scene of the preceding incident. Fire fighters had the tank car moved to isolate it from other railroad cars and sprayed it with heavy stream appliances; the addition of chemical inhabiters to the tank car did not stop its heating. Two oil company experts were flown in from Houston, Texas, and took samples from the bottom of the tank car to help determine when the vaporizing process would end; they also removed a one-foot diameter lid on the car to relieve pressure caused by the vapors expanding inside the car.(33)

Water Pollution

Industrial wastes are a principal cause of drinking water pollution. The Mississippi River, which serves as the source of raw water for 1.5 million people along the 258 miles from St. Francisville to Venice, Louisiana, had about ten metal pollutants discharged into it at one time.(34) Although water pollution of this type may have disastrous effects over a long period of time, it is not the kind that fire departments are apt to be concerned with in emergency situations. Flammable liquid spills from overturned tank trucks or railroad tank cars or from leaking storage tanks that end up in rivers and harbors, whether directly or by means of storm drains and flood channels, can create potentially disastrous situations.

In one memorable incident, a tank truck apparently sheared off a pipe leading to a 30,000 gallon tank of vinyl acetate at Los Angeles Harbor, starting a fire that immediately exposed other nearby tanks of acetate solvents. Within minutes the fire spread to other tanks, causing their rupture, blowing off their covers, and sending balls of fire into the air.

Notwithstanding the use of 27 engine companies, 5 truck companies, 2 foam units, 2 paramedic units, and a large crash truck brought in especially to apply its aqueous-film-forming-foam (AFFF), 21 of the tanks were eventually destroyed with damage estimated at a half million dollars. Two of the city's fire boats were used, one pumping its full 10,000 gpm capacity through several lines, including a 3-½ inch line.

The chief flew to the fire by department helicopter. One of his first concerns was to prevent burning liquids from reaching the harbor waters, since dikes around the tanks were only 30 inches high and were being deluged with foam and filled with water being applied to cool exposed tanks. Three pumpers were ordered to start drafting water from the dikes, and one of the airport's crash trucks was specially called to apply 750 gpm to AFFF from its main turret nozzle to kill the ground fire. With the assistance of Los Angeles County fire crews, the fire was under complete control in about three and a half hours, and pollution of the harbor was prevented.(35)

Pollution from Radioactivity

Radioactivity is a source of environmental pollution that probably has the potential for more widespread and longer lasting disasters than any hazmat known so far. The authors of a *Newsweek* article on the Three Mile Island nuclear plant episode said that "it began with a brief but terrible bump in the night and grew into the worst commercial nuclear accident in history."(36) They would have added, "in the United States," if they could have foreseen that six years later on April 26, 1986, one of four reactors at Chernobyl would catch fire and explode in the USSR. Unofficial reports say that 250 died in the blast that spewed cesium, strontium, and plutonium across much of the Ukraine, Byelorussia, and Russia and sent a cloud of radioactive gas around the world. About 116,000 people were evacuated shortly after the accident. Hundreds of thousands will suffer long-term health effects.(37) Later 135,000 people were evacuated from Ukrainian villages where the delayed reaction is killing or maiming thousands of humans and animals. For example, on a collective farm about 40 miles south of the blast site, another calf was born without any eyes, and the children "have a lot of throat trouble and illness that nobody understands," probably caused by the unendurably high levels of radiation. Soviet leaders agreed in April 1990, that $26 billion dollars should be spent to resettle people in the affected areas.(38) In a 1987 trial of the Chernobyl plant officials, the chief engineer and his deputy got a sentence of 10 years in a labor camp.(39)

Cooling towers at Three Mile Island are a stark memorial to the dangers posed by nuclear energy. (Photo courtesy of Ron Buckingham.)

As a result of the Three Mile Island accident near Harrisburg, Pennsylvania, where mechanical and human failures resulted in a breakdown of a reactor's core cooling system, radioactive steam leaks developed that continued to spread for several days out over the community to a distance of twenty miles. The spread of radiation might have been catastrophic had the technicians not succeeded in preventing a core meltdown.

Although nuclear plants do not release solid or chemical pollutants into the atmosphere, they do threaten the environment in other ways. They release small amounts of radioactive gas into the air. In water-cooled plants a small amount of tritium remains in the water when it is returned to the river or lake. Moreover, this heated water causes thermal pollution that can harm animals and plants that live at lower temperatures. One of the largest unsolved problems to date is the disposal of radioactive wastes produced by reactors. These wastes remain radioactive for thousands of years and can cause serious radiation damage to all living things. In the United States these wastes are now stored on the plant grounds. Most scientists believe that storage in underground salt deposits would be the best solution to the problem. These storage areas might have to be guarded for thousands of years to eliminate the risk that radioactivity might be released by inadvertent digging into them by our descendents.

Radiological Considerations

Accidents involving radioactive materials cause a great deal of concern among both responders and the public because of relatively poor understanding of the physical properties and effects of radioactivity and because it can be detected only with special equipment such as a Geiger counter. This high anxiety, in many cases, will have a much larger impact on the medical response than will the actual cases of exposure. A great many people may believe they have received large doses, despite assurances from authorities, and will seek medical assistance.

A major consideration in radiological incidents is whether or not there has been dispersion of radioactive contaminants in particulate, liquid, or gaseous form. Such conditions will require decontamination procedures similar to those required with other hazardous materials. Many radioactive materials, such as industrial radiographic cameras, employ sources sealed in welded, stainless-steel capsules. Accidents with these sealed sources may still involve large-dose radiation exposures to those within a few yards of the source. Unless the capsule is damaged, however, contamination will not be a factor.

Most accidents in transportation or typical industrial settings involve just a single radioactive material. Nuclear power plant accidents, however, can release a variety of radionuclides (radioactive chemical elements) which may escape primarily in particulate or gaseous forms. This may complicate medical response strategies. Some materials may be highly radioactive, but may decay to background levels in just a few days. Others may be less radioactive, but may have very long half lives and cause long-term problems by entering the food chain.

A National Radiological Response Plan, developed by the federal gov-

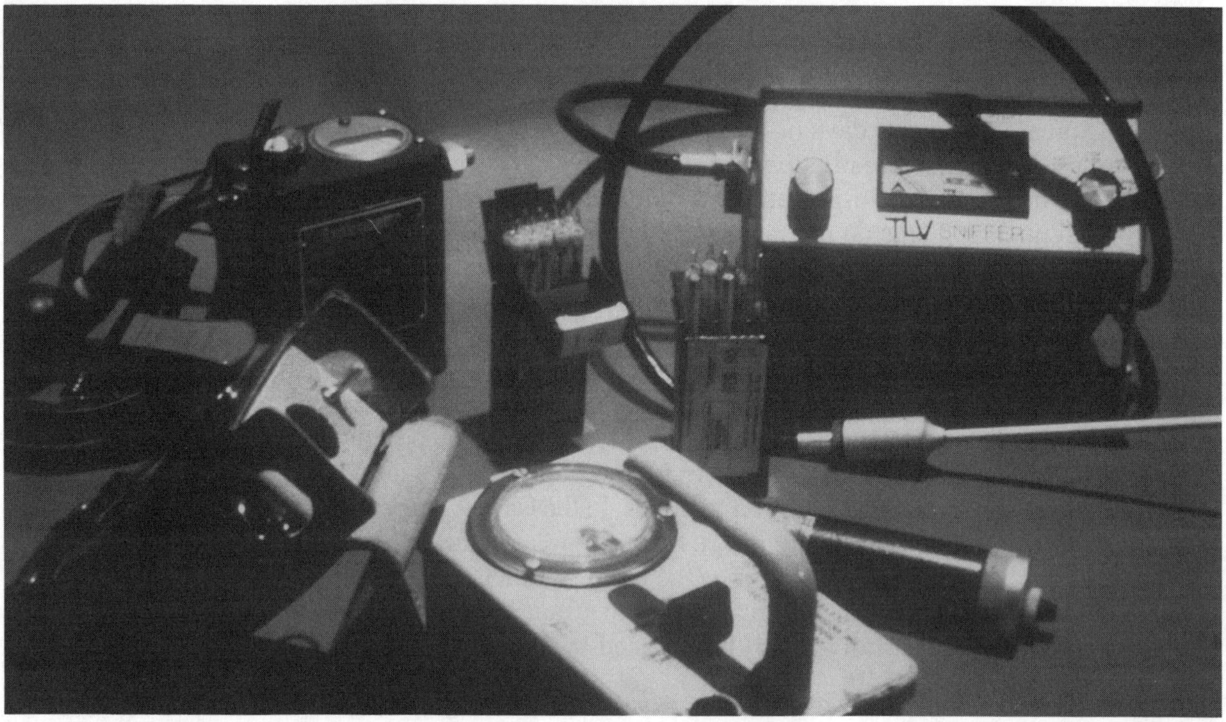

Fire fighters must be familiar with radiological monitoring equipment if they are to handle radiation emergencies. (Photo courtesy of the National Civil Defense Administration.)

ernment in the aftermath of the Three Mile Island incident, describes how federal agencies will coordinate their activities and interact with local and state government. Principal federal agencies include the Nuclear Regulatory Commission, Department of Energy, Federal Emergency Management Agency, Department of Defense, and the Federal Bureau of Investigation. Fire departments that are within nuclear power plant emergency planning zones or could otherwise encounter a major radiological incident should be familiar with this plan.

Other Nuclear Accidents

On March 22, 1975, fire was started accidentally by an employee in the Brown's Ferry, Alabama, nuclear plant when he was using a candle to test for air leaks in the combustible fire wall seal. The fire was not extinguished by the plant's brigade, though its members discharged portable carbon dioxide and dry chemical fire extinguishers on the polyurethane sealing material burning in the cable orifice of the division wall between the two units. The fixed carbon dioxide system did not extinguish the fire which extended into the cable tray system, consuming insulating and jacket materials, though the system discharged three times. The Athens Fire Department was called to help, and after surveying the fire, the chief recommended that water be used to extinguish the blaze. However, use of portable extinguishers was continued because the plant superintendent feared that water might further degrade plant operation.

About seven hours after the start of the fire, water was used for extinguishment. As a result of the fire damage, two nuclear generating units were shut down for more than a year. Property damage was estimated at $10 million, and cost of replacement power was about $120 million. There was no release of radioactivity as a result of this fire, although products of combustion of the cable jacket and insulation materials were toxic and required use of breathing apparatus during fire fighting. This incident affirmed the necessity for using water on cable fires to prevent the loss of circuits not already affected. Water would have diluted corrosive vapors and limited the extent of necessary cleanup. The fire emphasized the need for plant fire protection systems.

Shortly before the Brown's Ferry incident, the Vermont Yankee reactor was shut down after simulated tests showed that a steam explosion might not be adequately damped by its protective steel "doughnut."(40) In 1980 it was reported that near Miami, Florida, "large leaks of radioactive water continued at the Turkey Point Power Plant, whose two nuclear reactors provide Florida with about 15 percent of the state's electricity."(41)

There is some concern that sabotage might be used against nuclear power installations. The Nuclear Regulatory Commission (NRC) reported that 99 threats of violence were directed against nuclear facilities in a two-year period, while 76 other threats were aimed at federal nuclear power generators.(42)

The NRC is said to have accepted a probable casualty figure of about 3,300 if the worst possible meltdown should occur, though this figure would vary up or down depending upon the population density and the wind. However, Director Henry Kendall of the Union of Concerned Scientists estimates that between 23,000 and 36,000 persons might die after exposure to the radiation of a meltdown.(43) Experts today, in view of the Chernobyl accident, are revising casualty estimates upward.

Chemical Explosions Causing Disasters

Chapter 11 on transportation disasters discusses how certain chemicals have been responsible for disastrous explosions in trucks, tank cars, trailers, and vessels in every conceivable location, including railroad yards, business districts, harbors, highways, rivers, and on the high seas. The largest ship ever sunk—the 207,000 ton tanker *Marpessa*—sank in December 1969, after an explosion ripped open its hull.(44)

There have been explosions unrelated to transportation that are memorable for their great destructiveness or large loss of lives. Most were not caused by "explosives" as defined in transportation regulations but by chemicals never intended to be used for producing an explosion.

Cellulose nitrate has not generally been considered to be an explosive, though it is the main ingredient in smokeless gun powder. However, its ability to produce an explosion has been often demonstrated.(45)

Smokeless powder was the cause of the second and major explosion in the basement of a hardware store in Richmond, Indiana, on April 6, 1968. Forty-one people were killed and over 100 injured. The first and lesser explosion may have been of natural gas escaping from the ancient and corroded street main.(46)

Cellulose nitrate produces lethal nitric oxides when ignited or undergoing decomposition. As noted earlier, these toxic gases killed 125 persons at the Cleveland, (Ohio,) Clinic fire on May 15, 1929; in this disaster the hazmat was in the form of X-ray film.(47)

Though not commonly classified as explosives, powerful oxidizing chemicals have been responsible for disasters. The explosive power of ammonium nitrate was amply demonstrated at Texas City in the two ship detonations discussed previously. It was later demonstrated ashore at Traskwood, Arkansas, in the box car detonation following the train derailment and fire, and in the fertilizer plant detonation at Mount Vernon, Missouri.(48)

In May of 1988, the city of Henderson, Nevada, just south of Las Vegas, was rocked by a series of mighty explosions at the Pacific Engineering and Production Company "PEPCON" plant, which manufactured rocket propellant fuel used for the space program.(49)

Hot slag from a welding job touched off a fire and sent employees scrambling. When the fire spread to a large supply of ammonium perchlorate, the oxidizer used to make the rocket fuel, the first of seven major explosions occurred. One registered 3.6 on the Richter scale. The major explosion rocked the entire Las Vegas valley, throwing many people into a panic as a giant mushroom cloud spiraled a mile in the air.

While the quick evacuation limited the death toll to two, scores were injured including the responding Henderson Fire Department. The windshield was blown from the first responding pumper and the chief's car was immobilized.(50)

The explosions in Henderson, Nevada, would have been somewhat reminiscent of a similar devastating incident that had occurred in England almost two decades earlier, except that most of the PEPCON employees had time to escape. At the English chemical plant in Flixborough, at 4:53 P.M. on Saturday, June 1, 1974, a huge vapor cloud consisting of mostly cyclohexane began to rise. When the base of the cloud reached a source of ignition, a devastating explosion occurred that killed 28 persons and injured over 100 others. In addition to the damage caused to 2,488 surrounding homes, shops, and factories within an eight-mile radius, the plant itself was nearly totally destroyed at a loss of about $48 million. Because of the fear of further explosions and possible release of 20,000 tons of stored toxic ammonia, about 20,000 people within a five-mile radius of the plant evacuated the area. It took about two and a half days for the 250 fire fighters to bring the fire under control, and some units continued patrol and cooling operations for an additional six days.(51)

A natural gas explosion demolished a modern two-story high school building in London, Texas, on March 18, 1937, killing 204 students and teachers. Gas leaking from pipes under the building was ignited by an arc caused by a teacher plugging in an electric sander in the woodworking shop. In the flash and explosion that followed, the destructive violence was so great that the heavy slab floor in this concrete, steel, and tile building was blown upward from below, carrying bodies, desks, and lockers with it 75 feet into the air; one two-ton slab of concrete flew through the air and flattened an auto 200 feet away. Because of the noncombustible construction, there was very little fire following the explosion.(52)

"Combination" Hazmat Incidents

Some hazmat incidents do not lend themselves to neat categorization. The BASF plant explosion in Cincinnati discussed in Chapter 1, for example, included all the elements of a hazmat operation, major fire, mass-casualty incident, and large-scale evacuation.(53) Many disasters may be multifaceted and involve several types at once.

At least 155 persons were treated at three Memphis, Tennessee, hospitals for respiratory problems, eye irritation, and chemical poisoning resulting from exposure to smoke issuing from a burning pesticide plant. About 40 employees were in the plant when an 8,000 gallon tank of methyl parathion exploded, forming a mushroom cloud of toxic smoke that rose over one thousand feet in the air.

Fire fighters wearing breathing apparatus dodged exploding drums of organic phosphates while directing large-stream appliances on the burning structure. Police officers in helicopters covered a five-square mile area broadcasting evacuation orders issued by civil defense officials. All boat traffic on the Mississippi River was halted seventy miles south of Memphis, and railroad cars were stopped from entering the industrial area around the plant.(54)

The usual hazmat references advise of the health danger of methyl parathion, which is commercially found formulated in a 20 percent solution of xylene, a toxic flammable liquid, but they do not mention that it is also an explosive that can be detonated in a fire. However, the warning can usually be found that it "decomposes at temperatures above ambient and may develop sufficient internal pressure to cause the container to rupture violently."(55)

On February 22, 1980, the *Los Angeles Times* reported "A big refrigeration compressor exploded with earth shaking force at the Knudsen Corporation dairy products plant in Central Los Angeles Thursday night, sending up a huge cloud of ammonia gas that overcame or seared scores of persons."(56)

Twenty-two of the injured from the plant and nearby apartment house were taken to six hospitals in seven ambulances for treatment on February 21, 1980. Four fire fighters were among the victims; they had suffered inhalation and corrosive injuries while trying to effect a shutoff of the ruptured high-pressure pipe in the plant's engine room. Other fire fighters stripped off the injured fire fighters' clothes and washed the victims with streams of water before loading them into the ambulances.

Eleven fire companies attempted to keep the ammonia gas from reaching a high enough concentration to enter its flammable range (16–25 percent) through the application of water spray on the escaping gas which is quite soluble and readily converts from anhydrous to aqueous ammonia. The compressor blew up at 6:00 P.M. and it was 8:00 P.M. before the gas flow could be stopped and the ammonia cloud began to dissipate.

Pipeline Rupture Disasters

Pipelines carrying flammable and toxic gases and liquids sometimes rupture because of overpressure, corroded and weakened walls, and from being dug up by excavators. Lines can be found in the business districts of large cities, such as the 2-inch high-pressure line of ammonia that ruptured in the center of Los Angeles a few years back, causing evacuation of several multistory buildings that were constructed before the era of nonopenable windows. This line supplied refrigeration equipment from a cold-storage plant on the east side of the city.

Pipelines also go through industrial districts, such as the one that ruptured near Los Angeles in 1973 and permitted more than 100,000 gallons of oleum to escape and produce a huge cloud of sulfur trioxide over a 25 square-mile area. Thirty-eight persons went to hospitals for treatment.(57)

Underground transmission lines of natural gas, gasoline, LPG, coal slurry, and ammonia are in rural areas; some extend through several states. There are over 700,000 miles of natural gas pipelines in the United States.

Examples of Pipeline Rupture

In June of 1989, lightning struck twice in San Bernardino when a gasoline pipeline ruptured and exploded just thirteen days after a runaway train had careened into a group of homes at the same site. Both calamities claimed fatalities, injuries, and severe damage to homes.(58)

It was in a rural area near Natchitoches, Louisiana, in March 1965, when a pipeline failure resulted from corrosion in a 24-inch natural gas pipeline. It had been installed in 1944 and operated at 750 pounds per square inch pressure. When the rupture occurred it took about 30 minutes to bleed down 17 miles of pipe—the distance between isolation valves. The rupture tore a huge crater in the ground, and the escaping gas ignited immediately, acting as a gigantic torch to ignite seven homes, several autos and outbuildings. The heat was so great that all buildings were completely consumed and the automobiles melted to the ground. Seventeen persons were killed by the fire and at least nine others injured.(59)

In a Hutchinson, Kansas, suburb on August 13, 1974, an eight-inch underground pipeline carrying liquefied anhydrous ammonia ruptured and sent a huge white cloud up from the crater. About 200 residents of the Willowbrook area and nearby farms were evacuated until the next day, and four sheriff's deputies required medical treatment for respiratory problems caused by the ammonia gas. The gas continued to escape long after the isolation valves were closed, which took about an hour, because six miles of pipeline had to be bled off through the break. Fortunately no one was killed by this toxic and flammable gas.

Another example of a pipeline rupture disaster occurred in a business district on the edge of the Los Angeles suburb of Culver City on June 16, 1976. An excavation machine struck and ruptured a buried eight-inch high pressure line on Venice Boulevard. The 600 pounds of pressure caused gasoline to shoot up in the air like a fountain, forming a huge cloud of mist that seemed to hang in the sky for about ninety seconds. Ignited by some unknown source, a tremendous explosion leveled the north side of the street, demolishing seven buildings, several autos, and sending a wall of fire along the street. Six persons were either killed outright or died from injuries, and about twenty-six others suffered severe burns and other injuries. About 16,000 gallons of gasoline spewed from the broken pipe during the approximate two minutes that it took for an alarm system to cause the pipeline to be shut off. Three hundred miles of such gasoline-carrying pipelines run under the City of Los Angeles.

The Los Angeles Fire Department had 150 fire fighters on the scene within minutes after the blast, and they were assisted by 25 members of the Culver City Fire Department. The Los Angeles Police Department dispatched 100 officers and Culver City sent 50. This disaster occurred while Los Angeles fire fighters were cleaning up after the San Fernando Valley chemical fire discussed earlier in this chapter. At one time, 250 fire fighters from 40 companies were deployed to various sections of the city.(61)

Disaster struck the Soviet Union's Trans Siberian Railroad on June 4, 1989, when 460 passengers were killed in the blast of liquefied petroleum gas that had escaped from a ruptured pipe line and became ignited by sparks from one of two passing passenger trains. This gas explosion caused the second-worst train wreck in world history, exceeded only by a 1981 crash that killed 800 people.(62)

CONTROL AND MITIGATION OF HAZMAT DISASTERS

After viewing the many instances in which hazmats have brought about incalculable harm to the inhabitants of many communities and their environment, it is appropriate at this point to consider what we can do about such catastrophes beyond taking every possible preventative measure. It is especially important to do this in view of the fact that it is the fire departments that are usually called upon to handle such emergencies. If you happen to be the incident commander who responds to the scene you can almost certainly predict that not every built-in safeguard designed to control the situation will function as it was designed to do. As the old familiar "Murphy's Law" pointed out long ago, "if anything at all can go wrong, it will," despite all the precautions that may have been taken to avoid it. It would appear that Murphy was right, for instance, in the 1986 explosion of the space shuttle *Challenger* that killed the seven astronauts on board shortly after it was launched. Probably, no vehicle ever has been constructed with more carefully designed safeguards for its solid rocket booster, and yet a single design flaw in the O-rings spelled disaster.

So, if we can't prevent them, it would be prudent that we make the best possible plans for controlling and mitigating the effects of hazmat disasters. Who should make these plans? What laws need to be considered in their preparation? What agencies should be consulted? What data needs to be assembled for the task? What should their content be? Who has to approve them? How do you test them? Many of these questions have already been considered in an earlier chapter on planning, so it might be well to focus on those aspects that have a particular bearing on the hazmat problem. Since they have been outlined by Robert Obermayer in his essay "The Intricacies of Hazardous Materials Planning," which he prepared for the Open Learning Fire Service Program,(63) we will accept this kind offer to let us include some of the material here under the heading of "Planning."

PLANNING

The achievement of most desirable goals is only possible through planning.

Reasons for Planning

To protect its personnel, the fire department must recognize the dangers involved in the mitigation of a hazardous materials incident and plan accordingly. Planning takes on a life and safety connotation and involves an intensive effort that begins with recognition of the need to plan, extends to identification of hazards, continues with identification of needed and on-hand resources, and culminates in successful mitigation of an incident.

Reasons why the fire service must look to pre-incident planning as the most appropriate means of facing the hazardous materials challenge include these:

- magnitude of the danger
- need for specialized training, equipment, and expertise
- number of agencies that will become involved in an incident
- need for compliance with federal and state legislation

Federal Legislation and the Hazardous Materials Issue

This section discusses several pieces of federal legislation that apply to hazardous materials and must be considered in the development of a pre-incident hazardous materials plan. Indeed, in many communities the local planning effort may have been initiated by one or more of these federal laws. Limited space precludes a discussion of relevant legislation in each of the fifty states; however, you should be aware of your own state laws concerning hazardous materials. Copies may be obtained from the local public library.

GENERAL FEDERAL AUTHORITY

Article I, Section B of the United States Constitution grants the federal government the authority to regulate interstate commerce. Although this federal authority has been used to develop regulations concerning the transportation of hazmats, the federal government's intent is to support and abet local hazmats mitigation efforts. The legislation that is discussed below exemplifies this intent.

NATIONAL CONTINGENCY PLAN

The National Contingency Plan (NCP) required by Section 105 of the Comprehensive Environmental Response, Compensation, and Liability Act (CERCLA) calls for preparedness and training. One important feature of the act was the establishment of the National Response Team (NRT), which is the coordinating federal agency for emergency planning. It consists of representatives of 14 federal agencies with significant responsibilities in the areas of worker safety, transportation, the environment, public health, and emergency management.

On-Scene Coordinators, employed within the Coast Guard and the Environmental Protection Agency, are federal officials predesignated to coordinate and direct federal responses and hazmat removals under the National Contingency Plan.

Federal resources may be triggered by a report to the National Response Center (NRC), which is operated by the Coast Guard and relays the initial report to the appropriate Coast Guard or EPA on-scene coordinator, depending on the location of the incident. Provision of several federal laws, including the Federal Water Pollution Control Act and CERCLA, require the immediate notification to the NRC of any discharge or release of hazmats. Based upon an evaluation of the initial report, the decision to send an on-scene coordinator to the scene is made. The on-scene coordinator will monitor the response of the responsible party, and the state or local government. He or she may also initiate federal action.

Regional Response Teams (RRT) are composed of representatives of federal agencies and a representative from each state within the region. During the mitigation of a major incident, the on-scene coordinator may request a meeting of the Regional Response Team to provide advice and recommendations on specific mitigation issues.

TITLE III, SUPERFUND AMENDMENTS AND REAUTHORIZATION ACT (SARA)

SARA contains planning requirements relative to hazmats including the establishment of state emergency response commissions, emergency planning districts, and local emergency planning committees. Appointed by the state governor, the state emergency response commissions have the following responsibilities:

- designation of emergency planning districts
- appointment of local emergency planning committees for each planning district
- review of emergency plans
- establishment of procedures for processing public requests for information about emergency response plans, material safety data sheets, and the list of extremely hazardous substances

Emergency planning districts are intended to facilitate the preparation and implementation of emergency plans. Such districts may follow existing political subdivisions or may create multijurisdictional planning organizations. The local emergency planning committee for each district must include, among others, members of state and local government, emergency services, community groups, the media, and owner/operators who are subject to the provisions of SARA Title III. Each planning committee was expected to have a plan in place by October 1988 and the plan is to be reviewed annually. The plan must address these items:

- facilities and transportation routes relative to specific chemicals
- response procedures
- EMS
- training Programs
- evacuation plans
- means of identifying populations and areas that may be at risk
- means of detecting a spill or release

Also known as Emergency Planning and Right to Know Act of 1986, SARA Title III requires all manufacturing employers to provide Material Safety Data Sheets (MSDS) to state and local authorities. Listing the properties, hazards, and other necessary information about every chemical on the site, the MSDS will be an indispensable aid to your planning process.

RESOURCE CONSERVATION AND RECOVERY ACT

The Resource Conservation and Recovery Act establishes a framework for the proper management and disposal of all hazardous wastes. The Hazardous and Solid Waste Amendments of 1984 expanded the law's scope and placed increased emphasis on waste reduction, corrective action, and treatment. This act authorizes the EPA to issue standards and regulations for the management of hazardous wastes by the generator and the transporter, which are implemented through permits issued by the EPA. The permit application process requires the operator wishing to store, treat, transport, or dispose of hazardous wastes to characterize the wastes to be handled, demonstrate compliance with the appropriate regulations, and have a contingency plan. Your emergency planning process must interface with the operators within your area who are subject to the Resource Conservation and Recovery Act, and your plan should be integrated with their plans.

OSHA REGULATIONS: THE RIGHT TO KNOW

Occupational Safety and Health Administration (OSHA) regulations require employers to develop and implement response plans. Specifically, employers whose employees will be responding to hazmat incidents from their regular duty station must have an emergency response plan. The plan should include a mechanism to recognize emergencies, methods to alert employees, evacuation procedures, EMS, line of authority, on-site decontamination, site control means, and a method for the evaluation of the plan.

An important OSHA regulation is the Hazard Communication Standard issued in 1982 and known as the Right-to-Know law. Again, the Material Safety Data Sheet is the mechanism through which the employee receives needed information on the specific hazmats contained on the site.

The student may wish to read the *Hazardous Materials Emergency Planning Guide*, which is available from the National Response Center at 1-800-424-8802.

The International Society of Fire Service Instructors has prepared a helpful guide for fire departments for implementing Title III and the OSHA Hazardous Material Standards. For information about this 1987 loose leaf publication, prepared by William H. Stringfield, write to the ISFSI at 20 Main Street, Ashland, Massachusetts 01721.

ADDITIONAL FEDERAL REGULATIONS CONCERNING HAZMATS

Under the provisions of the Hazardous Materials Transportation Act, the Department of Transportation (DOT) has the responsibility for protecting the nation against the risks to life and property that are inherent in the commercial transportation of hazmats. DOT's authority ranges from issuing placards and labeling regulations to developing specifications for highway and railway hazmat transport vehicles. Several agencies within DOT have a regulatory responsibility relative to hazmats.

- Federal Highway Administration—regulates vehicular transport
- Federal Railroad Administration—regulates rail transportation
- Federal Aviation Administration—regulates civil aviation
- U.S. Coast Guard—regulates marine transportation
- Materials Transportation Bureau—promulgates hazardous materials regulations for other DOT agencies

NATIONAL TRANSPORTATION SAFETY BOARD

The National Transportation Safety Board (NTSB) is a part of DOT but reports directly to the Congress. The five-member board attempts to determine the cause of major transportation accidents and recommends measures to prevent recurrences. It is authorized by the Independent Safety Board Act of 1974 to assess the effectiveness of other government agencies in preventing transportation accidents and to publish its findings for public inspection. Although this agency has proven to be an effective monitor of other regulatory agencies, it has no authority to enforce its recommendations.

DEVELOPING THE HAZARDOUS MATERIALS PLAN

Pre-incident planning for hazmats involves a number of important considerations for proper development of a good plan. Because of the ever present threat of hazardous materials, this topic deserves special consideration and should definitely constitute an appendix to the disaster planning process introduced in Chapter 1. Here are a few guidelines that can help ensure that disaster planning will adequately treat the hazmat threat.

The Planning Team

As the planning authority, the planning team must respect all legitimate interests; the team must attempt to balance the needs of the emergency responders with the needs of the site and vehicle owners or operator and the community at large.

The success of the planning effort will depend on the qualities of the team members. Needed qualities include the ability, commitment, authority, and resources to accomplish the planning task. Further, team members should have a wide range of experience and expertise relative to the specific aspects of the community and the community's hazmat problem. Team members must agree on their purpose, be capable of working together, and faithfully represent all interest within the community. The following is a list of potential members of an Emergency Planning Team.

POTENTIAL MEMBERS OF AN EMERGENCY PLANNING TEAM

Part A

Experience shows that the following individuals, groups, and agencies should participate in order for a successful plan to be developed.(64)

- ○ Mayor/city manager (or representative)
- ○ County executive (or representative)/board of supervisors
- ○ State elected officials (or representatives)
- ○ Fire department (paid and volunteer)
- ○ Police department
- ○ Emergency management or civil defense agency
- ○ Environmental agency (e.g., air and/or water pollution control agency)
- ○ Health department
- ○ Hospitals, emergency medical service, veterinarians, medical community
- ○ Transportation agency (e.g., DOT, port authority, transit authority, bus company, truck or rail companies)
- ○ Industry (e.g., chemical and transportation)
- ○ Coast Guard/EPA representative (e.g., agency response program personnel)
- ○ Technical experts (e.g., chemists, engineers)
- ○ Community group representative
- ○ Public information representative (e.g., local radio, TV, press)

Part B

Other group/agencies that can be included in the planning process, depending on the community's individual priorities.

- ○ Agriculture agency
- ○ Indian tribes within or adjacent to the affected jurisdiction
- ○ Public works (e.g., waste disposal, water, sanitation, and roads)
- ○ Planning department
- ○ Other agencies (e.g., welfare, parks, and utilities)
- ○ Municipal/county legal counsel
- ○ Workers in local facilities
- ○ Labor union representatives (e.g., chemical and transportation, industrial health units)
- ○ Local business community
- ○ Representatives from volunteer organizations (e.g., Red Cross)
- ○ Public interest and citizens groups, environmental organizations, and representative of affected neighborhoods
- ○ Schools or school districts
- ○ Key representatives from bordering cities and counties.
- ○ State representatives (Governor, legislator's office, state agencies)
- ○ Federal agency representatives (e.g., FEMA, DOT/RSPA, (Research and Special Programs Administration) OSHA)

TASKS OF THE PLANNING TEAM

During the planning effort, three tasks will confront the fire department's hazmat planning team:

- ○ review existing plans
- ○ perform a hazard analysis
- ○ assess available capabilities and resources

The plan review function may encompass a review of hazmat response plans of other agencies as well as your own. In addition to providing the planning team with new ideas, the review will help prevent overlap and inconsistency while fostering interagency cooperation.

The hazard analysis will yield the necessary documentation for the support of the hazmat planning effort and the factual basis for the establishment of planning priorities. It allows for the following:

- ○ examine your community's potential for an incident
- ○ estimate the potential magnitude of the incident
- ○ identify the most probable location of the incident
- ○ identify the material most likely to be involved

The hazard analysis may be simple or complex and may focus on the most common hazard, such as the highway transportation of gasoline, or the most potentially destructive hazmat that is either stored, utilized, manufactured, or transported through the community. In any case, it is the task of the planning team to determine the appropriate level of detail for the hazard analysis in view of local resources.

Hazard analysis employs several terms that should be mastered.

- *A hazard* is a situation that has the potential for causing damage to life, property, and the environment.
- *Vulnerability* is the susceptibility of damage to life, property, and the environment if a hazard reaches its full potential.
- *Risk* is the probability that an injury or damage will occur.

To illustrate these terms, consider an example, an elementary school at the bottom of a steep hill on a highway that is used for the transportation of hazardous materials. How does this situation represent a hazard? Where is the concept of vulnerability? Is risk involved? Here, the potential for damage exists simply because a hazmat accident may occur in close proximity to the school. Why does the potential for an accident exist? Because of its proximity to the highway, the school is vulnerable. Risk is the chance that an accident will occur. What factors may affect the level of risk that is present? Risk can be moderated or exacerbated by such factors as regulations that govern the transport of hazmats and road conditions.

Hazards Analysis: Three Steps

As you recall from Chapter 1, the first step in the performance of a hazard analysis is hazard identification. Here we seek to identify the facility and transportation situations that have the potential for causing injury to life, property, and the environment. Factors to be considered include the types and quantities of hazmats, their location and routes of transport, and the nature of the hazard that is most likely to accompany an incident.

The second step, the vulnerability analysis, identified what in the community is susceptible to damage should an incident occur. The analysis includes the extent of the vulnerable zone, the size and characteristics of the population in the affected area, the property that might be affected, as well as the environmental issues of damage to sensitive areas and endangered species.

Risk analysis, the third step, assesses the probability of damage occurring due to the release of a hazmat. Planners may assume a worst-case scenario, a best-case scenario, or take a middle-of-the-road approach. The hazard analysis summarizes all three steps and forms the basis for a hazmat plan.

Capability Assessment

After the hazard analysis, the planning team must determine the amount and composition of the resources that will be required in the mitigation effort. Following resource identification, the team must assess the availability of the required resources. Paramount considerations are the resources that are available from the facility or transporter management, other local private sector sources, and the public sector including local, state, and federal governments.

A Community-Based Plan

Planning is certainly a local responsibility because each community is the best judge of the local hazards, the conditions of mitigation, and the availability of resources. The community, including the local fire department has the duty to determine what level of response commitment is appropriate. *The Hazardous Materials Emergency Planning Guide* has summed up this sentiment in the following terms, "Each community must choose the level of planning that is appropriate for it, based upon the types of hazards found in the community."(65)

As the planning process unfolds, the fire department may take the multihazard or the single hazard approach.

- The multihazard approach seeks a common planning ground for all emergencies, such as initial response, command and control needs, and communications. Following the construction of the planning framework, the multihazard approach then looks at each hazard type individually via a dedicated plan annex or appendix that contains specific information and action provisions. In addition to providing the fire department or the community with an integrated emergency management system, the multihazard approach also avoids unnecessary confusion and redundancy.
- On the other hand, the single-hazard approach, requiring a less intensive planning effort, is useful where the fire department has limited time, resources, and personnel. The single hazard planning approach assumes that any plan is better than no plan at all. This planning option may be acceptable where the hazard analysis has indicated that the community may face a small number of quite specific hazards. However, the single-hazard planning approach becomes counterproductive when the number of hazards escalates so that instead of having a general disaster plan with several appendices, the fire department has many disaster plans, each addressing a specific hazard.

Specific Hazmat Planning Issues

A number of special issues must be addressed in the hazmat planning process—training, interagency cooperation, unique health and environmental concerns of hazmats, and monetary issues.

TRAINING

Training should acquaint department members with their role within the plan, train them in the skills and equipment that their role requires, and keep them informed about plan changes and alterations.

- First, the training effort must convey to all plan participants the social value of the role they undertake. To provide motivation, plan participants should see themselves as important in the solution of a significant community problem.
- Second, the participant should have an appreciation of the importance of each role within the plan, and the affect that each individual can have on the implementation of an effective emergency response.
- Third, the training program must include classroom work, hands-on skills, and equipment familiarization sessions that should be presented to the plan participants immediately after the adoption of the plan and on a continuing basis.

Training assistance is available from the EPA, FEMA, OSHA, DOT's Research and Special Programs Administration (DOT/RSPA), and the Coast Guard. SARA Title III, Section 305, mandated FEMA to provide $5 million for training grants within fiscal years 1987, 1988, 1989, and 1990. The FEMA training grants were designed to improve local emergency planning, preparedness, mitigation, response, and recovery capabilities.

LIAISON WITH OTHER AGENCIES

Hazmat incidents, because of their potentially destructive effects, require a concerted effort on the part of many agencies for a successful mitigation.

A hazmat team practices. Note the helicopter poised for medical evacuation purposes. (Photo courtesy of the Federal Emergency Management Agency.)

A series of containment booms are stretched across the Miami River to collect a dangerous spill of toluene diisocyanate. (Photo courtesy of the Kettering, Ohio, Fire Department.)

Your plan must recognize the reality that your fire department will not possess all of the required equipment or expertise; therefore, you must seek the resources and cooperation of other agencies. The fire department that attempts to address the issue of hazmats solely with in-house resources is doing a disservice to its community as well as its members.

HEALTH AND ENVIRONMENTAL ISSUES

The planning document must look beyond the immediate mitigation of the incident and prepare for the health and environmental impacts that may accompany any release of hazmats or products of their decomposition. A review of the November 1, 1986, fire at the Sandoz Chemical Plant in Schweizerhalle, Switzerland, highlights this concern.

The one-and-a-half-hour blaze occurred in a warehouse that contained insecticides, various herbicides, and over four tons of flammable solvents. The smoke emitted by the blaze contained compounds of sulphur dioxide, nitric oxides, and sulphuric hydrocarbons, and forced area residents to remain indoors and avoid opening their windows and doors. Further, the water run-off from the fire reached the Rhine River killing more than 500,000 fish and rendering 11 miles of the river biologically dead, incapable of supporting fish and other forms of wildlife. Domestic water supplies were also disrupted for scores of communities along the river.

This incident began as a fire. However, a plan that required fire control resource exclusively would have been insufficient. Although fire was involved, the incident was a hazmat release and its impact went beyond the immediate destruction of the structure and its contents. Consequently, planners should not limit themselves to the problem of fire extinguishment.

Incident management and the underlying planning effort must consider the effects of the incident beyond the immediate damage that is evident at the scene. In environmental terms as well as in the health of the community and the operating forces, a "poorly chosen response to a hazardous materials incident can have effects that last for years."(66) Therefore, certain planning team activities, including the compilation of the results of basic chemical tests of major waterways (for example, the pH and dissolved oxygen concentration levels) will, by providing the incident commander with baseline environmental data, greatly facilitate the mitigation of environmental damage.

MONETARY ISSUES

As with other types of disasters, the hazardous materials annex to your plan must consider monetary issues such as the responsibility for extraordinary expenses; for example, overtime for personnel, meals and lodging, large quantities of special extinguishing agents such as dry powder and aqueous film forming foam, and large quantities of consumable supplies such as sodium bicarbonate or dry ice. Further, the value of the material released, the container from which it escaped, and the potential for damage to surrounding valuable property, make it imperative that your planning process, which involves representatives of the private sector, adopts a policy to deal with such monetary concerns. We live in litigious times, and expenses resulting from the incident, such as loss of business and damage to property, will often result in litigation. Having a plan does not exempt your department from being sued, but it will simplify the assessment of possible liability and free your incident commander from this concern.

By guaranteeing vendors' payments for supplies and services rendered during an emergency, your plan can help speed delivery of needed items.

Another issue is compensation. Who will pay for damaged and lost equipment, lost business, and consumable supplies? It is far better to answer these questions in the planning process than in the courts after the event. Litigation can be costly and uncertain; you may be unable to recover all of your expenditures and payment for your losses.

The planning process may be viewed as a consensus since it involves representatives from many interests in the community. In the area of monetary issues, this consensus process enables members of both the government and the business community to consult and agree on a monetary policy that can be incorporated into the planning document.

LEGAL LIABILITY ISSUES

Another important subject that should be covered in a plan annex is the possible legal considerations related to liability of a hazmat instigator. (The possible civil and criminal liability of the fire department and its officers is discussed in Chief Bahme's book *Fire Service and the Law*, which Timothy Callahan has recently assisted in revising.)(67) It is not the possible negligence of response teams that will be mentioned in the annex, but that of the persons or corporations responsible for the hazmat incident. Under the laws of most states a number of different agencies have the authority to initiate legal sanctions against the responsible party. Prosecution, if appropriate, would be coordinated between such agencies, to avoid duplication of the process and avoid unnecessary hardship on the defendants.

When a spill or release of a hazmat occurs, the responsible person(s) (e.g., driver, carrier, processor) may be criminally liable, civilly liable, or both; in some cases there may be joint liability.

Abatement of toxic spills or releases can result in tremendous costs to public entities. The incident commander should be familiar with civil and criminal statutes applicable to such incidents to expedite necessary cost recovery procedures. Questions regarding liability of either kind should be directed to your district attorney, county counsel, city attorney, or other available legal advisor.

The question of civil penalties is usually dependent upon the actual or potential damage to roadways, water resources, agricultural land, domestic livestock, air quality, wildlife and other environmental resources. Criminal liability for misdemeanor or felony violations of the various codes (e.g., Vehicle, Penal, Fish and Game, and Health and Safety) can arise out of hazmat spills, and any questions regarding the applicable laws in this area should be directed to the district attorney's office, specifically the deputy in charge of toxics.

It might be well to include in the plan's annex a list of the various section numbers of each state code that can result in the imposition in civil and criminal liability arising out of hazmat accidents.

ENFORCEMENT OF TOXIC RELEASE LAWS

In situations where the city or county team directs the on-scene operations, the district attorney's office should be requested immediately to determine whether any criminal or civil action is appropriate on its part. If the response teams are from federal or state agencies then their leaders are usually responsible for enforcement of their respective laws and regulations.

RECOVERY OF DAMAGES

Recovery of damages to public property and expenses incurred by a city or county department (or other agency involved) is usually the responsibility of that department or agency; it has to institute an action against the spiller through its respective government attorney office.

In states such as California incident commanders of local response teams must obtain prior approval for the use of state funds from the state's Health Department before incurring any expenses at a hazmat incident.

Extensive documentation of agency costs, as well as those incurred by private contractors, is required if reimbursement is to be considered from state or federal agencies.

If a state or federal agency coordinator has directed response operations, then any claim for reimbursement of costs incurred by a local jurisdiction should be submitted to that agency; and such agencies usually will only approve such expenses as have been incurred after their arrival on the scene.

MEDIA RELATIONS

The ever-increasing number of hazmat incidents has caused the press and television media to consider them a source of newsworthy information. It is not uncommon for their representatives to monitor the radio channels of the public safety agencies and respond to the scene of hazmat spills in a matter of minutes, looking for the incident commander in the desire for detailed information which they can televise, radio transmit, or write up for their newspapers.

At the command post there should be an information officer who can perform the very important function of providing factual and timely information. To appear disinterested or provide inaccurate data to media representatives would be a disservice and counterproductive. The right of these people to reasonable access to the scene should be respected, and unless it would jeopardize their safety or unduly impede the handling of the emergency, every reasonable attempt should be made to help them perform their tasks.

Only the incident commander or persons designated by him should be permitted to release news concerning the overall operations. Of course, if there are other agencies assisting the fire department in its operations, their leaders should be consulted in preparation for news releases and answering questions related to their operations.

State laws frequently require that properly identified members of the news media be permitted to enter disaster scenes, including the scenes of hazmat spills. These same laws also require that they be warned before they enter that it may be hazardous to their safety and that they be given a recommendation to stay at a safe distance to avoid injury. If they disregard the warning and enter into a contaminated area, then their equipment and personnel will be deemed contaminated and be subsequently subjected to decontamination measures.

SUPPORTING SYSTEMS AND RESOURCES

Communications

In Chapter 6 on Communications, we saw how cellular telephones are frequently found in emergency vehicles in metropolitan areas. The radio, however, is the primary means of communication for emergency services whether city, county, or fire district; special hazmat vans are often equipped with teletype machines, computers with printout capability, and even equipment to receive fax readable data. For guidance in maintaining and upgrading your communications, there is a book now available from *Fire Engineering* called *Fire Service Radio Communications*.(68)

Volunteers

In a great many fire departments hazmat responses are augmented by volunteer fire fighters, not all of whom have received special training in coping with the risks involved. In California and some other states, volunteers who are duly enrolled and assigned duties are considered employees of the state for worker's compensation purposes under its labor code. It is the policy of the state to only use such volunteers when their services will materially contribute to the abatement operations and to assign them only to tasks that have minimal safety risks. Volunteers who spontaneously engage in life saving operations, usually in the initial phase of an incident, are afforded some protection provided under the state's government code. Most states afford similar protection to volunteers.

Resources

No hazmat disaster plan would be complete without an annex showing the fire service resources, and another annex listing those of the public works departments and private contractors.

PLAN REVIEW AND UPDATE

As in all disaster plans, one for hazmat control should have a requirement that it be submitted for review to the appropriate agencies not less than annually to make sure that its procedures and data are up to date. Agencies responsible for hazmat incident response or its support should advise the designated agency, such as the county health department, at any time there is a contemplated change in procedures or notification telephone numbers.

ANNEXES

Typical annexes that are commonly included in a hazardous materials incidents response plan are illustrated in the plan of Lake County, California, having a population of about 40,000 with two small incorporated cities and a dozen villages served by mostly volunteer fire departments. Among the annexes listed are the following.

1. Incident Classification by Type and Location
2. Extracts of Pertinent Legislation
3. Notification Sequence/Responsibility
4. Hazmat Incident Notification

5. Hazmat Incident Report
6. Role Definitions (for the 23 agencies listed)
7. On Scene Manager Checklist
8. Technical Reference Sources and Emergency Telephone Numbers
9. Services (Private, Industrial, and Specialized)
10. Activation

PRE-INCIDENT RESPONSE PLANS

Commencing in the 1990s the state legislatures began to question the effectiveness of hazmat response on the part of the various emergency agencies that would be called upon to handle chemical leaks and spills wherever they occurred. For example, author Bahme has served on a committee of the California State Firemen's Association in recent years that had as its main objectives the education of fire service members throughout the state and the adoption of legislation that would establish training facilities to teach and equip responding emergency forces in handling hazmat incidents in the safest possible manner. Fire departments everywhere in the nation should be involved in preparing such pre-incident planning efforts.

In the ensuing discussion we will resume our outline of some of the materials that Robert Obermayer prepared for use in the National Fire Academy's Open Learning Fire Service Program, in which he stresses the importance of replacing quick action with a deliberate, thoughtful decision-making system based on a risk-benefit analysis. He asserts that contemporary fire departments are faced with at least two important responsibilities relative to hazardous materials:

- planning—participation in the development of effective pre-incident response plans for hazmat incidents, and
- mitigation—service at the scene of the incident.

Since the planning initiatives of SARA Title III are a relatively new development at the beginning of the 1990s, fire officers were not too well versed in the preparation of comprehensive hazmat response plans, though they were quite familiar with prefire planning for target hazards.

RESPONSE DIFFERENCES—HAZMAT VS. TYPICAL

The general comparison of fire incidents with various types of disasters, as outlined in the first chapter, could also be applied to major hazmat incidents. However, with respect to the nature of the destructive impact, the decision-making system utilized, and the services required, the hazmat response differs from the typical fire department response in most cases.

Destructive Impact

Whether hazmat response planning is designed as a separate annex to a comprehensive major disaster plan or is undertaken independently, fire service officers assigned to the hazmat planning process must keep in mind the destructive impact that the uncontrolled release of a hazmat may exert on the community's property, inhabitants, domestic animals, and the environment. Remember, it is the curtailment or mitigation of the harmful forces unleashed in a hazmat mishap that justifies the existence of the plan. If the plan does not recognize the potential danger inherent in hazmat incidents, not only will it be ineffective, but it may actually increase the risk of personal injury and property damage.

The planning team must be aware that it is not planning for fire control operation but for an incident potentially far more destructive. For this reason, the hazmat control operation represents a radical departure from the traditional fire department mission. When confronted with a residential fire problem, the preferred means of operation is a quick, aggressive attack on the fire. The destructive impact is usually limited to the structure involved, its occupants, and the operating forces. On the other hand, in the hazmat incident the destructive potential includes immediate, long-term, and recurring risks of damage, and injury to the community and the environment, as well as to the operating forces.

Deliberate Decision Making

Planners must recognize that the traditional incident command decision-making process, usually calling for quick aggressive action, will not always be appropriate for hazmat incidents; it must often be modified to suit the circumstances. Some fire officers may find it difficult at first to adopt a more deliberate approach, but after reviewing the records of several catastrophic hazmat disasters, some of which have been mentioned previously, they may become convinced of the need for a more restrained approach in the decision-making process. Our references to these disasters will not dwell on the monetary losses, though often tremendous, but on the impact to the affected people and their environment. That will be a primary first consideration. See Table 12-1.

These incidents could have occurred in many communities. Knowledge and planning are imperative on the part of all emergency services, and especially on the part of fire departments. The incident in Switzerland was precipitated by the failure to control water runoff from a fire control operation. Sometimes, in the effort to meet the competing demands on the incident commander to heed the concerns of control strategy, tactical alternatives, and monitoring suppression and containment progress, environmental problems can be overlooked.

As mentioned earlier, the Bhopal plant in India has been considered the site of the "world's worst disaster"(69). However, the full effects of the 1986 Chernobyl meltdown cannot be calculated even yet,thus it may surpass Bhopal. In any case, the design of the Union Carbide plant at Bhopal may not be too dissimilar from pesticide chemical production plants elsewhere.

As a matter of fact, one postincident investigator described the plant as exhibiting the "same design, same equipment, same everything" as operations in the United States.(70)

The Mexico City incident was a BLEVE (boiling liquid expanding vapor explosion). Knowledgeable fire officers can no doubt recall several similar instances in our own nation. Names of such locations as Crescent City, Illinois, Kingman, Arizona, and Waverly, Tennessee, have a special significance to the fire service. Since this type of hazmat incident poses an especially severe threat to fire fighters, only a deliberate decision-making command procedure will be suitable.

Table 12-1. Destructive Impact of Selected Incidents

Date	*Location*	*Hazmat*	*Cause*	*Impact*
1986	Chernobyl	Radioactive	nuclear reactor fire & explosion	200 killed 200,000 injured
1986	Switzerland	1. disuloton 2. thioneton 3. ethoxyethyl-mercury-hydroxide	warehouse	500,000 fish dead, water supplies disrupted
1984	Bhopal, India	methyl-isocyanate	human error failure of safety system	10,500 dead (as of 1987) 300,000 injuries
1984	Mexico City	liquefied petroleum gas	BLEVE	472 dead 4,000 injuries

Services Required

Another important difference between a hazmat response and the typical fire department incident is in the type of services usually required for successful mitigation. The average response requires fire suppression services or medical care that fire departments have traditionally provided with varying levels of proficiency, depending upon the amount of training and experience they have had. In contrast, a hazmat incident may demand the provision of any number of technically complex and distinctly different services, such as, the plugging and diking of leaking tanks, monitoring of the water and air supplies, and efficient systems for determining the chemical properties and potential risks of the hazmat involved. Though still unfamiliar to some fire departments, such services have been supplied by some fire departments for a great many years; the Los Angeles City Fire Department, for example, started to respond to hazmat incidents in 1950, following the creation of its Dangerous Chemicals Detail of five members with a captain in charge (Author Bahme). Today, such response units in large departments may be several in number and be part of a separate command headed by a chief officer. In the Los Angeles City Fire Department, the Hazardous Materials Section is the largest section in the Fire Prevention Bureau and as such is broken down into five units, each with its own specialty. In addition, it operates four Hazardous Materials Squads.

ORGANIZING THE HAZMAT RESPONSE

During the planning process, the fire department must determine how it will respond to the hazmat challenge. This involves the consideration of at least two issues; the first response and the hazmat response team.

First Response

The objective of a disaster plan annex or hazmat plan is to organize the hazmat incident mitigation process. This begins with the initial response of your fire department. Your plan must organize the first response by assigning to it a mitigation role, address the problems that will affect the actions of the first response, and develop operational priorities to facilitate a good size-up of the situation.

Role of the First Response

A most important function of the first response commander is the recognition of the fact that it is a hazmat incident and the identification of the hazmat involved. Early recognition and identification will allow the fire department to determine the magnitude of danger and the resources and precautions that will be required. Toward this end, your plan must train all unit commanders and other department personnel in the following:

- potential location of hazmat incidents
- chemicals that may be involved
- hazards of such products
- potential for human, monetary, and environmental damage
- equipment and supplies that will be required to prevent or moderate such damage.

Source of Hazmat Data

In addition to the information provided by the hazard analysis, your plan should catalogue the following sources of information:

- the Department of Transportation Hazmat Identification System
- the National Fire Protection Association 704 Marking System
- CHEMTREC
- shipping papers
- consultation with the facility operator or transporter's personnel.

First Response Issues

Any tasks that you assign to the first response units should recognize that they will probably operate under the following constraints.

- They may lack specialized hazmat training.
- They may not have access to all relevant information on the situation.
- They may be unable to obtain the full departmental inventory of hazmat control equipment.
- To save life or property, they may be required to make split second decisions that have life or death consequences with additional potential for catastrophic monetary losses.
- They may commit or omit actions that affect the outcome of the event.

Your plan must also recognize that the first response unit may experience pressure from the people who summoned it. Civilians on the scene are probably unaware of the danger of hazmats and will expect fast action, not understanding the need for deliberation. Imagine yourself at a flammable liquids fire. You realize that you have insufficient quantities of foam available to effect extinguishment; you have summoned an additional supply and are awaiting its arrival. Meanwhile, the transporter or fixed facility operator (realizing that each gallon of liquid consumed by the fire increases the financial loss) may attempt to induce you to start extinguishment as soon as possible. Your personnel may also want to take quick action. You must have the presence of mind to resist an action that your training tells you is inappropriate.

Team Issues

If your plan has created a hazmat team, you have increased your planning and management responsibilities and must now decide the following issues.

- Identification of the Team's Mission
  Will the team attempt to prevent as well as contain, leaks and spills? prevent chemical fires and explosions? Will the team engage in clean-up activities once the incident has been controlled? In view of the fact that hazmat incidents may require costly clean-up measures for streets, buildings, streams, vehicles, etc., private sector interests may urge the planning team to make this residual clean-up a responsibility of response agencies.
- Personnel Issues
  How many members will be assigned to the team? What qualifications or level of training will they have? Will the unit be staffed with on- or off-duty personnel, or both?
- Management and Resource Issues
  Where will the unit be located within the department? How will it be alerted? How will the plan provide for periods of personnel shortages? What level of equipment and supply inventory should be maintained?

○ Health Issues
What provisions will be made for the continual assessment of the team's health? Since many hazardous substances have long-term effects, the physical condition of team members must be carefully monitored.

○ Team Response Issues
What units will respond with the team? How will the first response units be integrated into the overall hazmat response plan? What are the means through which field commanders will be appraised of the capabilities of the team.

○ Level of Protection
Which of the EPA levels of protection is appropriate? See Table 12-2. First check the results of hazard analysis which will tell you just what is required to handle your hazmat problems.

Table 12-2. Environmental Protection Agency Level of Protection (71)

Level	*Type of Clothing*	*When Used*
A	Totally Encapsulating Chemical Protective Suit Pressure Demand SCBA or Air Supplied Respirator with Escape SCBA Chemical-Resistant Gloves and Boots	Severe Skin, Eye, Respiratory Hazard
B	Chemical Resistant Coverall, One or Two Piece Splash Suit, Gloves and Boots Pressure Demand SCBA or Supplied Air Respirator with Escape SCBA	Severe Respiratory Hazard or Moderate Skin Hazard
C	Chemical Resistant Coveralls, One or Two Piece Splash Suit, Gloves and Boots Full Facepiece, Air Purifying Canister Equipped with Respirator	Moderate Skin and Respiratory Hazard
D	Coveralls, Safety Boots, Safety Glasses, and Hard Hat	No Respiratory Hazard, Mild Skin Hazard

Decontamination Planning

Since hazmats often have a destructive effect upon the environment, unlike ordinary fire debris, they must usually be removed from the scene of the incident after the damage control measures have been taken. The physical removal of hazmats must often be preceded by decontamination through chemical degradation or neutralization.(71) To ensure that these measures will be carried out with minimum risk to operating forces, the planning team must recognize the need for protective clothing(72 and 73) and establish procedures to do the following:

○ procure decontamination equipment
○ identify the hazmats for which decontamination will be required
○ establish the safe level of contamination
○ determine the best means for carrying it out.

Failure to solve the decontamination problem could result in long-term health and environmental dangers arising from the released materials.

OPERATIONAL GUIDELINES

Your plan should provide departmental officers with operational guidelines that will ensure the achievement of personnel safety and incident control objectives outlined in the planning document. To attain these objectives the incident commander must make a number of determinations.

Determine If the Incident Is a Fire Service Problem

Upon arrival at the scene, the commanding officer must determine whether the incident is actually a fire service problem requiring emergency action, or a cleanup problem that should be handled by employees of the plant or transport carrier, or by a professional hazmat disposal contractor. In some instances the situation may call for drastic measures that no fire department could be expected to provide. For example, when a chemical firm in England was contaminated by dioxin in 1968, the only solution for getting rid of the toxic chemical was to level the factory and bury the wreckage in an abandoned mine shaft.(74)

Fire personnel should not be committed to performing cleanup operations unless there is an imminent threat to life and property. The safety of your personnel should never be jeopardized in the effort to save product, property, or the environment. Be aware that huge sums of money could be at stake and transport or facility operators may seek to minimize their costs at the expense of your personnel.(75)

Determine If It Is Your Problem

Is the incident your responsibility, another jurisdiction's, or a private sector problem? Often, the person reporting the incident, or a nonfire department first responder, may not be familiar with jurisdictional boundaries, the hazard present, or other operational considerations, and may unknowingly request the response of your hazmat team. Your planning document should not assume that just because your resources are summoned to the scene of the incident, they must be used. There are valid situations where no action is the best alternative. Using the risk-benefit analysis, if no benefit is to be derived, then why commit resources and undertake a level of risk? Also, do not assume that your neighbors automatically want your team at the scene of their hazmat incidents.

Determine Who Is in Command of the Incident

If you have ever attended an incident at which incident command was disputed among rival agencies, you would not want to commit the welfare of your personnel to such chaotic command structure. Each agency may feel that its role is indispensable and that its leader should be in overall command. Incident command should be determined during the planning effort in order to alleviate conflicts.

In spite of the plan, command disputes may arise. In a worst-case scenario, you should report to the individual who appears to be running the operation and then assess the situation. As the fire department commander, you are responsible for fire department personnel and equipment resources. Accordingly, never place your personnel directly under the command of officers from another service; retain command of your forces because you alone understand their capabilities, limitations, training, equipment inventory, and prior experience.

Evaluate the Situation

Effective mitigation requires that the incident commander identify the type of situation that exists, the extent of the incident, and the relevant hazards before determining objectives, identifying resource needs, and embarking upon a specific plan.

Various identification systems and procedures will furnish the basic identity of the hazmat and an elementary understanding of its properties. However, during the mitigation effort, additional up-to-date, detailed information will be required. Consequently, the planning effort must establish a means through which such information is readily available to operating forces.

Until recently, the fire service was restricted to hard copy data sources. Unfortunately the contemporary proliferation of hazmats tends to make some reference books out of date. However, new technologies including portable computers and fax machines have dramatically increased the team's ability to quickly obtain fairly accurate data. Your planning document should address the problem of hazmat proper identification through use of computerized data bases.

Chapter 6 introduced you to CAMEO (Computer Assisted Management of Emergency Operations) as an excellent system, developed in 1986 through the joint efforts of the Seattle Fire Department and the National Oceanic and Atmospheric Administration (NOAA). In addition to containing information on over 2,600 chemicals, CAMEO can predict vapor cloud paths and through its Seattle Map capability, relate such information to a map of the jurisdiction in question. It does this with its ALOHA component (Aerial Location of Hazardous Atmospheres).

Operational Priorities

To give guidance to the first response commander, your plan should include the following operational priorities:

1. Preservation of human life, both civilian and operating forces
2. Containment of the hazmat release to the area of origin
3. Preservation of property. (Note that this is the third priority; the safety

of civilians and operating personnel should not be jeopardized for the preservation of the environment or property.)

Hazmat Response Team

Because hazmat incidents require special skills, knowledge, attitudes, and equipment, many fire departments have recognized the value of a hazmat response team. The hazmat response team is defined as a group of individuals whose abilities, training, and equipment inventory represent an asset in handling hazmat incidents. Let us look at some questions that should be answered in the organizing process.

Is There a Need?

The first question for the planning body is, Does your department really need a hazmat response team? Considering the cost of training, equipment, personnel, and the potential number of incidents, perhaps it may be advisable to rely on the services of an adjacent fire department or to participate in a regional team. Your hazard analysis will help you decide if the severity of your problem is sufficient to require the formation of a local hazmat response team. If your department opts to join a regional team, or to utilize an adjacent team, your document must address joint operational procedures and resource commitment issues.

HAZMAT APPARATUS AND EQUIPMENT

An increasing number of fire departments are obtaining vans for carrying not only special equipment useful in handling chemical leaks, spills, toxic gas releases, etc. but also for transporting communications gear, computers, fax machines, and lab equipment, along with reference books, files, report forms, check-off sheets, notification and resource lists, and other items contained in the annexes of your disaster plan. Some have a desk for the hazmat incident commander, with duplicating equipment for reproducing maps, material safety data sheets, and instruction guides to give members of his command and on-scene managers of other agencies.

Using the hazard analysis, the planning team will identify any special apparatus or equipment needed and will establish how the equipment is brought to the incident. Workable ideas include departmental purchase of some equipment, donations from facilities that contain hazmats, and possibly the establishment of multijurisdictional equipment pools.

Hazmat Apparatus

Some fire departments, such as Boston, have converted beverage delivery trucks into hazmat response units to suit their needs and some have rigs that more closely resemble motor homes, complete with toilet and shower, coffee makers, etc. Others have come up with a unique design that makes for easy access to the various types of equipment it carries. Such is the case with the New Zealand Fire Service. In an excellent *Fire Engineering* article (October, 1990), the Western Division Commander of the City of Auckland describes the design of its new hazmat unit. He presents several photographs depicting its horizontal-sliding "concertina" doors that run the full length of the vehicle, permitting easy access to either end on both sides. It was built on a standard commercial nine-ton chassis with a stepped floor design into a low-loading deck. In loading the rig, the heavy and important items are stored near the deck, and the less frequently used items higher up. In the two years that it had been in service, its good load carrying capability and ease of access to all equipment was working out well at incidents.(76)

Hazmat Equipment

High technology equipment for communications, data storing and retrieval, spectrometers for chemical analysis, combustible gas analyzers, radiation detectors, and similar devices are being employed in hazmat emergencies more than ever before.

Equipment commonly carried in a hazmat van include protective clothing, such as encapsulating suits, self-contained breathing apparatus, absorbents, neutralizing materials, liquid transfer pumps, monitoring equipment for gases and radiation, inflatable shelters, showers, fans, portable generators, hand tools, tarps, placards, barricade items, decontamination showers, soaps, detergents, temporary clothing, and full recording and monitoring

Specialized hazardous materials teams such as Squad 5 in Atlanta, Georgia with unique equipment and apparatus are available in most areas to handle immediate hazmat needs. (Photo courtesy of Atlanta Fire Chief David Chamberlin.)

items for cleanup action on 100 or more fire service personnel or other victims.

Equipment pools can be a convenient procurement alternative to the purchase of expensive specialized items such as backhoes, skip-loaders, and bulldozers. Where such pools can be arranged by a regional planning team, each participating agency could submit a list of the equipment that it could afford to provide by a mutually agreeable date. The regional planning committee would also establish the rules and policies under which the pool would operate.

Consumable Supplies

Many hazmat incidents require the use of special supplies and extinguishing agents that are easily obtainable any time of the day or night every day in the week. Moreover, when used in large quantities in control operations the expense could be considerable. Unlike ordinary fire incidents where the fire department consumes mostly water and air for breathing apparatus, commodities that are not difficult to replenish, a hazmat incident may require the following.

- Large quantities of various foaming agents, dry chemical extinguishing agents for both flammable liquids and metals.
- Dry ice, sodium bicarbonate, limestone, soda ash, and other agents to retard or neutralize reactions.
- Sand, sawdust, absorbent materials, etc., for diking and absorbing spills.

Your planning committee should see that the plan provides for the procurement of large quantities of such materials on short notice. The preplan analysis of the principal hazards in your response area should help you determine the need for the type and amount of such supplies that may be required for your particular conditions. For example, if oil storage and refineries are of major concern, arrangements could be made with the local petroleum companies and neighboring fire departments to set up a bank for various foam agents, diking materials, containment drums, absorbent compounds, etc. that could be readily transported to a flammable liquids incident. It may also be possible to make similar arrangements with local industries and commercial suppliers of such commodities, whose day and night telephone numbers would be noted in your plan's resource annex.

ACTION—NO ACTION DECISION, GO—NO GO!

Your plan should recognize that operating forces may arrive on the scene of a hazmat incident just before an explosion or other form of material release occurs, with resources that are woefully insufficient to reverse the course of events. Since your primary objective is to save lives, your plan must give the incident commander the authority to defer any action if an unacceptably high level of risk is present. No immediate action or withdrawal are viable tactical alternatives, and incident commanders should be urged to err on the side of caution in such decisions.

To encourage the exercise of this tactical option, the planning document should include the criteria upon which to base the ACTION—NO ACTION decision. Because fire fighters are accustomed to aggressive tactics, the plan should contain a list of specific conditions that would indicate the need for immediate withdrawal or the noncommitment of department personnel. These might include the following:

- flame impingement on a pressure vessel or tank and inability to deliver the required volumes of water at the site of the incident
- inability to obtain special extinguishing agents
- lack of protective equipment that is commensurate with the hazard
- fire involving an explosive material
- inability to adequately identify the product or its properties.

The ACTION option of the ACTION—NO ACTION decision must not be made before completion of a risk-benefit analysis. An incident commander should first consider the level of risk to operating forces and then compare it with the expected mitigation benefit. Of course, this rule should be followed in all fire department tactical decisions whether involving hazmats or not; however, a review of many fire fighter fatalities reveals that it has often been ignored. Sadly there have been hazmat incidents that resulted in fire fighter fatalities, most notably the death of 26 Texas City, Texas, fire fighters in 1947 and the death of 13 Kingman, Arizona, fire fighters in 1973.

If an explosion or widespread release of toxic gases has already occurred prior to your arrival on the scene, then a decision to not take any action is not a viable option. Your evaluation, sometimes with the aid of binoculars, may result in activating your evacuation plan for persons in the endangered area exposed to gas clouds, or in calling for the specialized equipment required for rescuing persons from collapsed structures, suppressing incipient fires, closing valves on ruptured pipes, and instituting such other damage control and medical measures as needed. Actually you may be fortunate if the hazmat explosion takes place while you are responding to the emergency; Chuck Bahme was fortunate twice in this respect. The first time was when he was a young Naval officer and an ammunition magazine blew up while his response team was still 1,000 yards away from the fire. Again he was fortunate when as a battalion chief leading his first alarm assignment to a fire in a Hollywood gun shop a gunpowder explosion blew out the front of the building before they all got in front of it.

In any case, your plan should contemplate the possibility that your fire department's first-in-officer, whether a company commander or a chief officer, may be the first to arrive on the scene of the incident and will be expected to notify the dispatcher where the interim command post will be located pending the arrival of whoever assumed the responsibility of the incident commander. If your first response commander even suspects that the hazmat situation might (if it has not already) become a disaster, there should be no hesitation to request that your hazmat disaster plan be activated immediately so that your dispatcher can make the required notifications and get your hazmat unit on the scene as soon as possible. If the situation fails to become as dangerous as the rapid size-up anticipated, and the special team is not essential, there is no harm done. Just put it down as an unplanned drill.

POSTOPERATIONS PLANNING

Among the more famous aphorisms, Yogi Berra's "It ain't over until it's over" is a classic. It certainly applies to hazmat incidents. They are not over when the leak is plugged, the spill removed or neutralized, the fire extinguished, the exposed inhabitants evacuated, the decontamination of personnel and equipment accomplished, the injured cared for, and the dead carted off. There may be still much to do that your plan should take into consideration, for instance, the demobilization of your units deployed, including that of your mutual aid responders; the gathering of data for the preparation of required reports, e.g., the exposure of your personnel to toxic substances for their medical records, the amount of supplies used, expenses incurred, agencies that rendered assistance, injuries to personnel and damage to equipment.

We have already discussed the need for postoperation consideration of legal liabilities on the part of the disaster instigator for injuries, deaths, and property damage, and the fire department's possible liability for negligent conduct. There could be a situation where the city could be fined by OSHA, as was Waterbury, Connecticut, for failure to inform the fire fighters that hazmats were stored at the fire scene, resulting in cyanide poisoning of 28 of them. "While Waterbury fire fighters fought the chemical blaze, a 38-page hazardous materials report the company had filed with the city in 1984 was locked in the fire marshal's office."(77)

Recovery

When the disaster has been brought under control, there may be homeless people and animals to be fed and housed in temporary shelters, priorities established for the resumption of services, arrangements made for salvage and cleanup operations, restoration of critical services, e.g., water, electricity and sewer, and, eventually, the eviction of the out-of-town volunteers from the shelters where they have been fed and housed (sometimes for weeks), ever since their arrival.

Postincident Analysis

Your disaster plan for hazmats should call for some "Monday-morning quarterbacking"—a postincident critique—in which all the major participants sit down and analyze what went right and what went wrong. Assuming that it could happen again, how could the operations be improved the next time around?

To lessen the psychological impact of disasters, there should be some planning for the emotional support of rescuers and victims; and since man-made spills of oils, chemicals, and similar hazmats could possibly undermine public confidence, especially if there have been serious injuries or deaths resulting from the incident, the plan should consider how public confidence and your credibility as an effective emergency service can be restored.

HAZMAT INCIDENT ACTION

Now that we have completed the discussion of the planning process and your hazmat plan has been drawn up and approved, it might be well to set forth a brief outline of the actions that the incident commander will usually need to take upon approaching and arriving at the scene of a major hazmat problem. In the preceding chapter suggested steps for evaluation, control, containment, mitigation, and disaster management for transportation disasters were presented in some detail. Therefore, at this point we will only outline disaster management options in general. Remember, though hundreds of tasks may have to be carried out by numerous agencies—fire, police, medical, utility, etc.—your primary job as incident commander is to set priorities, to decide what should be done first and by whom. If you have done your "homework" on hazmat study and planning, you stand a good chance of making most of your decisions right the first time. At least that should be your goal, for if you do not, there may never be a next time! Make it your aim to get all the knowledge and training for hazmat responses that you will need to cope with any kind of hazmat disaster that may confront you. *Knowledge* overcomes most fear of making mistakes, so aim to get it. He who aims at nothing is almost sure to hit it!

HAZMAT INCIDENT RESPONSE

I. *Approaching the Scene*
When rolling up to the scene you may observe a cloud of yellow gas or smoke in the distance; ask your dispatcher if any further information has been called in about the nature of the incident, e.g., fire in fertilizer or pesticide storage, ruptured acid drum, etc.
   - A. **Size-up carefully:**
      1. Use binoculars if indicated, and don't get too close.
      2. With the information available, give responding units a minimum safe distance to advance.
      3. Give appropriate dispatcher your size-up report, and update as needed.
   - B. **Take initial action** to effect control, e.g., call for activation of Hazmat Disaster Plan, deploy arriving forces, call for additional help as may be required, etc.

II. *Arrival on Scene*

Park clear of magazines, flammable liquid spills, and buildings that may blow up from gas ignitions, ammunition and gun powder, etc., and establish needed controls:

1. Locate command post where easily accessible and recognizable.
2. Designate staging areas for incoming special-called units.
3. Designate "Hot Zones" if needed, and mark off their peripheries with barrier tape, rope, signs, etc. Prohibit anyone from entry into contaminated areas unless
   a. properly protected, and
   b. decontamination equipment and procedures in a designated area are being provided.

III. *Evaluating the Hazards*

A. Identifying the product

1. Check the warning signs: acrid odors, colored smoke, clouds of vapors, pools of sizzling liquids, etc.
2. Check the container type and shape
3. NFPA identifying system labels
4. DOT identifying systems labels, placards, HI System numbers
5. Markings—chemical names on containers
6. Shipping papers

B. Assessing the risks

1. Access available information sources to determine risks to health, property, and environment.
2. Coordinate the evaluation process with appropriate agencies, e.g. health department, CHEMTREC, EPA, etc.

IV. *Securing the Scene*

Security and access control are very important in the disaster zone. It may become necessary to implement and enforce an emergency security plan in at least two respects:

A. Access to the general area, to keep out sightseers and permit legitimate media and business persons to enter

B. Access to "HOT" areas, to permit emergency forces to enter who are properly equipped, and to keep all others out.

V. *Managing Emergency Medical Services*

Integrating your EMS system into your disaster control operations for:

A. Triaging victims

B. Providing emergency medical services to emergency forces and victims

C. Removing victims for transport to designated medical centers.

VI. *Monitoring and Decontamination*

A. General guidelines for personnel

1. Do not commence decon operations until the product has been identified, if possible.
2. Decon of personnel should preferably take place inside the "HOT" area.
3. Proper levels of protection should be used by decon personnel.
4. Try to contain the materials used in decon procedures.
5. The decon team should be properly decontaminated.

6. Where possible, exposed personnel should be monitored and checked by appropriate health agencies before being released from the area.
7. Follow-up checks by such agencies should be performed as necessary.

B. General guidelines for equipment decontamination

1. Decon of equipment should await product identification.
2. Decon of equipment should take place inside "HOT" area only.
3. Decon teams should wear proper gear for the hazard level.
4. Decon materials should be contained as far as possible.
5. Decon teams should be properly decontaminated.
6. Equipment should be checked by appropriate health agencies before release from the area.

C. Coordination of decontamination

1. The incident commander should coordinate any other than primary emergency decontamination.
2. The appropriate health agency should assure that proper decontamination takes place.
3. Additional support units, e.g., fire, hazmat, public works, etc., may be requested to assist in decon operations.
4. Normally, decon should be attempted only after technical expertise on the material has been obtained and proper safety precautions have been taken.

VII. *Servicing the Forces*

A. Logistical support

1. For Hazmat unit—replenish supplies of dry ice, foam, neutralizers, absorbents, compressed air or oxygen for SCBA, fuel for generators, clothing to replace torn or contaminated clothing (that can't be decontaminated).
2. For supporting agencies assisting in the operations

B. Personnel needs

1. Medical attention for the injured and exhausted
2. Portable toilets for extended operations
3. Food and beverages for extended operations
4. Relief teams to permit rotation for sleep and feeding

C. Accounting needs

1. At the staging area to record arrival and departure of assisting departments' apparatus and personnel
2. At the Command Post to record time and services of private contractors, off-duty reporting personnel, volunteers' overtime, persons hospitalized, equipment damaged, etc.

VIII. *Leveraging Resources*

Review the human, equipment, and information sources that are available to assist you in controlling and recovering actions. (Resource list should be in your hazmat plan annex.)

XI. *Caring for Evacuees*

A. Determine to what extent agencies such as the Red Cross can provide food, shelter, blankets, cots, etc., if evacuees will not be permitted to return to their homes for an extended period.

B. Arrange for separated family members a means of notifying their parents, spouses, and children of their condition and location.
C. Show some concern for the care of pets of evacuees.
D. Arrange for their homes or businesses to be checked for safety of occupancy, and advise when it is considered safe for them to return.
E. Advise them of whatever security measures will be provided for their property pending their return to it.

X. *Restoring Critical Services*—water, power, sewer, etc.
Assist public utility companies in determining when their services can be safely restored.

XI. *Managing Clean-Up and Salvage Operations*
Determine to what extent this will be done by the plant management, property owner, persons or corporations responsible for causing the hazmat incident, or other agency, and to what extent your services will be feasible and affordable.

XII. *Assessing the Damages*
Seek professional assistance in determining the monetary and property loss sustained in the disaster (insurance adjusters may be of assistance in this); request the medical services manager to provide a report of the deaths and injuries sustained by people and livestock at the incident.

XIII. *Demobilizing the Forces*
In major hazmat disasters it may be several days before the emergency forces can all be returned to their quarters. It is the hazmat incident commander's responsibility to advise the overall incident commander when the situation has been sufficiently controlled or decontaminated to permit removal of the security barriers, and permit security forces to be released. Management's hiring of private security guards to keep the area secured over an extended period of time may obviate the need for department personnel to stay behind on standby watch, thereby speeding up total forces demobilization.

XIV. *Postdisaster Operations*
These matters are not part of your hazmat disaster plan, for the situation has been handled, the troops have left the scene, the media have begun to sing your praises, and your mayor is so pleased with reports that he has been receiving of your department's efficient managing of the whole incident, that he may be about to ask if there is anything your department could use to make your hazmat plan function even better, e.g. more personnel, more equipment, more budget appropriations for training programs, etc. To be fully prepared to justify an affirmative answer, it might be well to lay the groundwork.

A. Preparing your reports

1. If you can get your facts, figures, and photos on paper showing what you accomplished with the resources now at your disposal, and what more you might have done, in a shorter time, if they had been augmented in certain respects, this report could support an argument for upgrading your response capability.

2. Data showing how injuries occurred, and how they might be more easily avoided with improved equipment or training, should be part of your report.

B. Critiquing your performance

1. Assembling the major participants in the operation for a post-incident analysis should help in pinpointing the weaknesses in your plan or in the manner in which it was carried out. It is not unusual to find that a plan is workable in most respects, but the communications between the participants and the coordination of their activities at the scene could be improved. Discuss ways to correct these problems.
2. If the plan needs changing in any way, submit the modifications for approval to the appropriate authorities and get them adopted right away.
3. Try out the modified plan in your next joint-agency drill, and expect to find that your activities function more smoothly.

CONCLUSION

It is not contended that the suggestions offered in this chapter represent the final solutions to the many complex problems in a hazmat disaster. At best they offer only partial guidance to fire fighters facing perilous situations. Notwithstanding years of study and training, if you are the first to respond to the scene of a widespread hazmat incident you may face difficult choices of courses to pursue and directions to take. Remember Lucy of the Peanuts comic strip and her reluctance to play on Charlie Brown's ball team? Standing out there in tall grass over her head she exclaimed, "I don't mind playing left field if they would only tell me which way to face!" Well, firefighters do not mind taking the risks they have to face in handling hazmat disasters if only they would be given the right directions to take. This chapter will not give you *all* the right directions, but perhaps it can show you some signposts that will help you find the right course of action.

CHAPTER 12 REFERENCES

1. Los Angeles Fire Department, Bureau of Fire Prevention: "*Dangerous Chemicals Code*," Parker and Co., Los Angeles, CA, 1951.
2. *Chemical Abstracts*, American Chemical Society, Baltimore, MD.
3. Davis, Tenny L. *Chemistry of Powder and Explosives*, Wiley and Sons, NY, and Chapman and Hall, London, 1991.
4. Lake County, *Hazardous Materials Incident Response Plan*, 1986.
5. Kemp, D.L. "Plywood Plant Dust Explosion Kills Seven; Nine Fire Departments Contain $9 Million Dollar Fire" in *Western Fire Journal*, August 1976, p. 25.
6. "Fire Wire" in *Firehouse*, Vol. 12, No. 8 (August 1987), p. 27.

7. Bahme, Charles W. *Fire Officer's Guide to Dangerous Chemicals*, 2nd ed., Boston, MA: NFPA, 1978.
8. "Fire Wire" in *Firehouse*, Vol. 16, No. 3 (March 1991), p. 85.
9. "Fire Wire" in *Firehouse*, Vol. 16, No. 4 (April 1991), p. 72.
10. Pryor, L. "Liquefied Natural Gas Accident Risks Looming Larger," *Los Angeles Times*, August 16, 1966, Part II, p. 1.
11. See Note 10, at p. 2.
12. Culver City, CA, June 16, 1976; nine persons killed.
13. See Note 7, at p. 114.
14. "Devastating Blast of Railroad Cargo" in *Fire Command*, April 1975, pp. 32–34.
15. NFPA's *Fire Protection Guide for Hazardous Materials* includes 704M, *Recommended System for the Identification of Fire Hazards of Materials*, along with 325A, 325M, 481M, and 49M.
16. "Hazardous Materials Regulations of Department of Transportation," Title 49 of the Code of Federal Regulations, 1991.
17. U.S. Department of Transportation, *Emergency Response Guidebook*, 1990.
18. See Note 7, 1st ed., p. 215.
19. *Your Chance to Live*, Defense Civil Preparedness Agency, SM 3-12, 1972, p. 71.
20. "Fire Wire" in *Firehouse*, Vol. 16, No. 3 (March 1991), p. 85.
21. "Killer Gas Peril Hits Auckland," *Los Angeles Times*, March 1, 1973, Part II, p. 1.
22. "Abortions for Italy Poisoning Begins Today," *Los Angeles Times*, August 13, 1976, Part II, p. 1.
23. "Chemical Mist Envelops Chicago's Far South Side; 16,000 Told to Flee," *Los Angeles Times*, April 27, 1974, Part I, p. 8.
24. *Newsweek*, August 10, 1987, p. 47.
25. *USA Today*, April 26, 1990, p. 4A. See also Kurzman, Dan, *A Killing Wind*, McGraw-Hill, 1987, which updated the casualty figures given here.
26. *Newsweek*, February 27, 1989, p. 54. See also, *Time*, February 27, 1989, p. 53.
27. "Phosgene Fumes Hospitalize 40 Fire Fighters" in *Los Angeles Firefighter*, February and March 1976, Vol. 14, No. 7, p. 1; *Los Angeles Times*, May 16, 1976, Part I, p. 2.
28. Belcher, J., and K. Roderick. "Chemical Fire Burns 9 Hours; 50 Hospitalized, 5000 Flee," *Los Angeles Times*, June 17, 1976, Pt. II, p. 1.
29. "Fire Wire" in *Firehouse*, Vol. 16, No. 4 (April 1991), p. 72.
30. Chemical Fumes Force Evacuation of Hawthorne Area," *Los Angeles Times*, February 10, 1973.
31. Peige, John D. "Sulfuric Acid Over Baltimore" in *Fire Engineering*, May 1978, pp. 22–23.
32. "California City Flees Chemical," *Las Vegas Sun*, August 15, 1978; also, Kennedy and Turner, "Chemical Tanks Rupture; Orange Area Evacuated," *Los Angeles Times*, August 15, 1980, Pt. I, p. 1.
33. Oliver, Myrna. "500 Evacuate 14 Plants Near Leaking Fumes," *Los Angeles Times*, August 25, 1978, Pt. II, p. 5.
34. See Note 19, p. 73.
35. Dektar, Cliff. "L.A. Crash Truck with AFFF Halts Chemical Storage Fire" in *Fire Engineering*, October 1972, p. 40.
36. Matthews, Tom, Susan Agrest, Gloria Borger. "Nuclear Accident," *Newsweek*, April 1979, p. 24.
37. *U.S. News and World Report*, July 20, 1987, p. 14.
38. Kirk, Don. "In The Shadow of Chernobyl, Fear and Deformity," *USA Today*, April 26, 1990, p. 4A.
39. *Newsweek*, August 10, 1987, p. 47.
40. Sawyer, R. G., and J. A. Elsner. "Cable Fire at Brown's Ferry Nuclear Plant" in *Fire Journal*, July 1976, pp. 5–10; Gwynne, P. and J. Bishop, Jr. "How Safe is Nuclear?" in *Newsweek*, April 12, 1976, pp. 70–75, at p. 71.

41. *Newsweek*, April 12, 1976, at p. 70. in *Los Angeles Times*, March 16, 1976, quotes the *Miami Herald* as follows: "Highly radioactive material has been leaking from a nuclear generator storage site near Miami into Biscayne Bay for more than 2 years . . . at a rate of 90 gallons per hour."
42. See Note 41, in *Newsweek*, at p. 73.
43. *The World Book Year Book*, 1970, Chicago, IL, Field Enterprises Educational Corporation, p. 318.
44. *Year Book, 1969, World Book Encyclopedia*, Chicago, IL, Field Enterprises Educational Corporation, 1969; 1975, p. 300.
45. See Note 7, at p. 109.
46. See Note 7, at p. 143.
47. *Hospitals—Occupancy Fire Records*, NFPA Fire Records Bulletin No. FR61-1, 1961, p. 4.
48. See Note 7, at pp. 68–69.
49. "Heat Sheet," Training Guide; *American Heat* video Journal, Vol. 3, Prog. 3. (September 1988) p. 1.
50. Kramer, William M. "PEPCON Blasts," Narration in *American Heat* video Journal, Vol. 3, Prog. 3. (September 1988).
51. Di Meo, Michael. "Explosion at Flixborough" in *Fire Journal*, November 1975, pp. 58–67.
52. "Schools," Occupancy Fire Records No. FR 57-1A, NFPA, 1965; A detailed description of this London, Texas, school disaster is presented in Lyons, Paul R. *Fire in America*, Boston, MA: NFPA, 1976, pp. 157–159.
53. Uebelacker, Carl. "Explosion & Fire in Cincinnati Kills Two, Injures 83," *Firehouse*, October 1990, pp. 43–51.
54. "Burning Cloud From Burning Pesticide Routs 2,000 at Memphis," *Los Angeles Times*, July 6, 1979, Pt. I, p. 4.
55. NFPA *Hazardous Chemicals Data*, 1975, p. 202.
56. West, Richard, and Sue Goldfarb. "Dairy Plant Releases Ammonia Cloud: 22 Hurt," *Los Angeles Times*, February 22, 1980, Pt. I, p. 3.
57. "Pipe Line Rupture Blamed For Toxic Gas Leak at Plant in Carson," *Los Angeles Times*, August 20, 1973, Part II, p. 1.
58. "Heat Sheet," Training Guide; *American Heat* video Journal, Vol. 4, Prog. 1. (July 1989).
59. "Bimonthly Fire Record" in *Fire Journal*, Vol. 59, No. 4 (July 1969) p. 49.
60. Culver City, CA: June 16, 1976; nine persons killed.
61. Hurst J., and D. Rosenweig. "Blast and Fire Level City Blocks,"*Los Angeles Times*, June 17, 1976. Pt. I, p. 1.
62. *World Book Encyclopedia*, 1988 Edition; *Year Book 1990*, p. 275.
63. Prepared by Robert Obermayer for the National Fire Academy's Open Fire Service Program, and partially based on the writings of Chief Bahme for the same program.
64. National Response Team of the National Oil and Hazardous Substances Contingency Plan, *Hazardous Materials Emergency Planning Guide*, Washington, DC. Government Printing Office, p. 13.
65. Ibid. p. 35.
66. Orvos, David, and John Cairns. "Hazmat and Habit" in *Fire Engineering*, July 1987.
67. Bahme, Charles W., and Timothy Callahan. *Fire Service and the Law*, 2nd ed., Boston, MA: NFPA 1987, Chapters 6 and 13.
68. Spahn, Edwin J. *Fire Service Radio Communications*, Saddle Brook, NJ: Fire Engineering Books.
69. See Notes 25 and 26.
70. Ibid.
71. Maslansky, Steven P. "Decontaminate: Persepctive for Responders" in *Fire Engineering*, February 1985, pp. 16–20.

72. Stull, Jeffrey O. "Making Sure That Protective Clothing Protects" in *Fire Engineering*, September 1987, p. 27.
73. Lesak, David M. "Protective Clothing: Overcome the Shortcomings" in *Fire Engineering*, February 1985, p 45.
74. Carrol, R., and L. Jenkins. "Italy 'Our Own Hiroshima'," *Newsweek*, August 16, 1976, p. 49.
75. Fire, Frank L. "Handling Hazardous Material Incidents: An Emergency Guide, Part I" in *Fire Engineering*, February 1985, pp. 32–34.
76. O'Sullivan, Kevin G. "A Soft-sided HAZ-MAT Unit" in *Fire Engineering*, October 1990, pp. 55–57.
77. "City Fined for Failure to Provide Hazmat Information" in *Fire Command!*, March 1986, p. 10.

CHAPTER 13

ENEMY ATTACK AND UFO POTENTIAL

FEW RESIDENTS of the United States, except for those in Hawaii, have experienced an enemy attack on their hometown in this century; some think they have. The Great Los Angeles Air Raid of February 26, 1942, began at 2:25 A.M. when the U.S. Army announced the approach of hostile aircraft and the city's air raid warning system went into effect for the first time in World War II. "Suddenly the night was rent by sirens. Searchlights began to sweep the sky. Minutes later gun crews at Army forts along the coast line began pumping the first of 1,433 rounds of ack-ack into the moonlight. Thousands of volunteer air raid wardens tumbled from their beds and grabbed their boots and helmets. Citizens awakened to the screech of sirens and, heedless of the blackout warning, began snapping on their lights . . . The din continued for two hours. Finally the guns fell silent. The enemy, evidently, had been routed. Los Angeles began to taste the exhilaration of its first military victory."(1)

WHAT IS WAR?

"War is a man-made disaster. Earthquakes and floods *happen* to mankind, but man makes war himself."(2)

War is a struggle in which two or more large groups of people try to destroy or conquer each other. There are "hot" wars and there are "cold" wars. When overt acts of aggression involving force are employed, it is a "hot" war, and there was a time in history when such wars were actually declared, but not in the past half-century. However, the United States in each decade since World War II has found itself involved in at least limited military actions. Since the 1940s there have been the Korean "police" action in the 1950s, the Vietnam conflict in the 1960s and 1970s; Panama and

Grenada skirmishes in the 1980s, along with the shelling of Beirut by our U.S. Navy that accompanied our U.S. Marines "peacekeeping" attempts in Lebanon, also in the 1980s.

In 1991 the United States led a multinational coalition of forces in the Persian Gulf, involving the greatest deployment of military forces ever assembled in our nation's history.(3) Without a declaration of "war," though with approval of the Congress, the President of the United States launched an attack on Iraq on January 16, 1991, with the objective of carrying out a United Nations Resolution requiring the withdrawal of Iraq's occupational forces from Kuwait. In the first ten days of this "Pushbutton" war, more bombs, missiles, rockets, and Patriot missile interceptors were launched by the allied combatant nations against the enemy than in all of World War II. When the air attacks had accomplished their objectives, the ground forces, numbering over a million altogether, met in combat to decide the question of who would rule Kuwait.(4) In a swift and decisive victory the U.S. led forces drove Iraq from Kuwait suffering minimal casualties.

Meanwhile, the "cold" war with the Soviet Union had thawed considerably with the removal of the Berlin Wall, the reunification of Germany, the initiation of *Glasnost* and *Perestroika* by President Mikhail Gorbachev. It took a turn for the worse when some of the Soviet Republics, encouraged by the trend toward more democratic policies, sought greater independence and their actions resulted in deployment of military forces to repress overexuberance. The efforts of the Soviet President to restrain a possible dissolution of his Union gave rise to some concern over the survival of the democratic movement.(5) A "chill" seemed to set in in 1991, hovever before the end of one year the Soviet Union had ceased to exist. It was unclear what forms of unrest would prevail on the new republics.

CAUSES OF WAR

There can be a difference between the *causes* of war and the *reasons* for war. The unstated cause of a war may be selfish, but the reasons given usually are lofty and noble. "Both sides in a war may show reasons which they consider valid."(6) For example, in the Persian Gulf War, President Bush stated that he had sent a half million troops to Saudi Arabia to protect it from invasion by Iraqi forces and to drive them out of Kuwait so as to restore that country's Emir as head of its government. However, there were a few prominent skeptics who had the temerity to suggest that the real underlying *cause* of the war was the need to protect the vast oil resources of the Middle East for future needs of the western nations. Still others asserted that the most important cause of the war was the need to curtail future aggression of President Saddam Hussein and curtail his desire to increase his nuclear, chemical, and biological warfare capability. In any event, there is little likelihood that we will ever eliminate all the causes of war, nor ever eliminate statesmen who find good reasons to justify them.

Chemical Warfare

The arsenal of modern warfare includes chemical and biological weapons. Chemical weapons are agents (persistent and nonpersistent) formulated to harass the enemy or kill him. In World War I, the French were the first to use a poison agent, bromine; this brought retaliatory use of chlorine by the Germans, resulting in hospitalization of two million people, with two out of every hundred casualties being fatal.

When the United States and its allied forces engaged in combat in the Middle East, one of the most fearsome threats it faced was the use of chemical and biological weapons by Saddam Hussein; he had already demonstrated his willingness to use chemicals against Iran and rebels in his own country, with an untold loss of thousands of lives. While they were not used against the United States, they were used to suppress Kurdish rebels within Iraq.(7)

Biological Warfare

Although Iraq and some other countries have developed deadly biological agents capable of transmitting communicable disease quickly, there are no public records of their deliberate dissemination. However, if some person should cultivate and distribute bacteria, fungi, viruses, rickettsia, and similar deadly agents from living organisms, great suffering

Both fire and police departments are involved in bomb disarmament and bomb disposal. Seen here is a bomb unit used in New York City. (Photo courtesy of F.D.N.Y.)

could be caused among the people of the world. The most widespread and catastrophic epidemic of bubonic plague was the "Black Death" (from the dark color of the body after death), which began in Constantinople in 1347 and devastated much of Europe; mortality was estimated at three-fourths of the population.

Conventional Bombs

"Brush-fire" wars still require use of conventional bombs of demolition, fragmentation, anti-personnel, and other special types. During the Persian Gulf War, for example, cruise missiles were launched from U.S. naval vessels at Iraqi targets, SCUD missiles were directed by Iraq at Israel and Saudi Arabia. (8) Their likelihood of being used in attack against targets in the United States is not considered great.

However, there is always the possibility that the era of limited wars may someday be terminated by the launching of an all-out attack by mankind's most devastating weapon, the thermonuclear bomb.

The Hydrogen Bomb

The H-bomb has an explosive force thousands of times more powerful than the atomic bomb that destroyed the Japanese cities of Hiroshima and Nagasaki in 1945. Its energy comes from fusion of atoms of hydrogen to form atoms of helium, an element heavier than hydrogen. This thermonuclear, or fusion, reaction releases energy in the form of heat. The heavy isotopes of hydrogen (deuterium and tritium) are used in this reaction.

Not all hydrogen bombs produce radioactive fallout, for unlike the fission process used in the atom bomb, fusion does not produce radioactive products. The earlier three-stage H-bombs employed a Uranium-238 jacket to produce radioactive fission products. The newer weapons are called "clean bombs" for they derive only a small part of their energy from fission.

U.S. scientists detonated a hydrogen bomb at Bikini Atoll in 1954. It exploded with the force of 15 megatons, or the equivalent of 15 million tons of TNT. It formed a fireball about 3 ½ miles across and sucked up pulverized coral to a height of about 25 miles in a cloud that spread out for 100 miles. The radioactive fallout from the bomb fell over an area of 7,000 square miles.

The Russians exploded their first megaton-class hydrogen bomb in 1953; in 1961 they exploded one with a power of 58 megatons. China and France each exploded their first H-bomb in 1967. By the time the first nuclear test ban treaty was signed in 1963, more than 500 nuclear weapons had been exploded. Since then, all nations except France and China have detonated their nuclear bombs below ground. A United Nations treaty went into effect in 1970 to stop the spread of nuclear weapons to nations that did not already have them; forty nations in addition to the United States, USSR, and Great Britain ratified this treaty.(9)

The greatest nuclear threat today may well be from third world nations and unstable regimes in the Mideast.

A typical terrorist bomb capable of being hidden in a suitcase, as seen at an anti-terrorist training program in Suffolk County, New York. (Photo courtesy of Ron Buckingham.)

How Many Nuclear Bombs Are in the Hands of Terrorists?

In Chapter 8 we saw how civil unrest and international political dissatisfactions can lead to acts of terrorism. Targets most likely will be major cities in air terminals, government or embassy buildings, or other high-visibility areas.(10)

Hence, when we think of "enemy attack" we should not limit our consideration to attack by another nation, but should recognize that the enemy could be a terrorist group possessing atomic weapons capable of causing great devastation in any community. Dr. Theodore B. Taylor, who formerly designed atomic bombs for the federal government at Los Alamos, stated that, "Present U.S. safeguards applied to commercial plutonium and highly enriched uranium are not adequate to prevent theft by heavily-armed groups with resources and motivation similar to the Brinks gang and other groups of professional criminals that carried out successful major robberies in the past. Given roughly 10 kilograms (22 pounds) of reactor-grade plutonium oxide, or about 20 kilograms of highly enriched uranium oxide, and using information that is widely published and materials and equipment available from commercial sources, it is quite conceivable that a criminal or terrorist group or even one person working alone could design a crude fission bomb."(11)

Such bombs could be carried in a small automobile and could be capable of producing an explosion that in a densely populated area such as lower Manhattan, could kill 100,000 people. When such threats first became real a *Nuclear Blackmail Emergency Response Plan* was drawn up for the State of California. It defines procedures to obtain technical assistance in determining credibility of the threat and outlines responsibilities and actions in response to the threat.(12)

EFFECTS OF A NUCLEAR ATTACK

In modern times, or at least since World War II, only "limited" wars have been waged, not "total" war. A limited war is one in which the warring nations limit the weapons they use, the targets they attack, or the areas involved. The Korean and Vietnam wars are examples of this, for neither side chose to use nuclear weapons, and limits were voluntarily placed upon the targets selected and areas involved. Although such target limitations prevailed in the 1991 Persian Gulf War, it was militarily successful because the military leaders were not strapped with equipment limitations or tactical meddling by politicians.(13)

However, it is conceivable that a war, intended to be limited, could escalate into total war through use of tactical nuclear weapons, at the outset, and then when the losing side seeks to prevent further losses, it could decide to use strategic nuclear strikes as a defensive measure.

No nation has ever suffered a full-scale attack employing modern thermonuclear weapons, and therefore it is impossible to appraise the damage that the United States would sustain under such attack. However, based upon data obtained from the bombing of two Japanese cities in 1945 and the tests conducted since then, we can estimate the effects that would result from a single nuclear explosion, and extrapolate the effects of bursts, with certain assumptions: size of overall attack, size of individual weapons, types of burst, and nature of target.

If we assume that an attack is made on the United States with 10-megaton weapons giving a 50 percent fission yield, half surface bursts and half air bursts, an attack on all military targets probably would result in blast damage and thermal effects (2nd-degree-burn level) over 42 percent of the nation; 95 percent of the country would be affected by fallout; about 30 percent of the country's industrial facilities would be made unusable; of the 130 million people killed, 15 percent would die of blast and thermal injuries, and 57 percent from fallout.(14)

Physical Effects of Above Ground Nuclear Explosions

Following a blinding flash of light a fireball forms; in about two seconds it reaches a diameter of about 3.4 miles for a 10-megaton (mt.) bomb, and 4.6 miles for 20-mt. A low burst, in which the fireball touches the earth, would vaporize all structures not made of heavy concrete inside the fire. At the same time there would be a release of energy—5 percent in the form of nuclear radiation, 35 percent as thermal radiation, and 50 percent in blast waves and ground shock. The remaining 10 percent of energy release becomes radiation fallout.

The initial burst of nuclear radiation is lethal to everyone within the area of the diameter of the fallout.

The thermal radiation from a surface burst of a 10-mt weapon can cause first-degree burns on unprotected people and start fires up to 21 miles. In an air burst, this distance would be increased to 35 miles.

It has been estimated that the typical American city has about five to twenty-five points per acre—places where easily ignitible materials can be found in which the several-seconds-duration of initial heat can start fires. If the conditions are conducive to formation of many simultaneous fires over a wide area, these fires could develop into a fire storm.

The heat rays and initial nuclear radiation are followed by a blast wave which starts at more than 2,000 miles per hour, but loses much of its damaging force about ten miles out. It starts as a high-pressure shock front, but after a few seconds, a negative pressure phase follows. Drag forces of the winds created would cause much damage to buildings and most of the blast injuries to people, and would extend much further out than the area affected by shock pressures.(15)

A 10-mt bomb would knock down frame houses and most trees at a distance of 8.5 miles in a surface burst and 11.3 miles in an air burst.

There would be injury to persons struck by debris or being thrown about a distance of 15.5 miles in a surface burst, and 18.5 miles in an air burst. These distances would be 33 and 40 miles if the bomb were 100-mt.

Ruptured gas pipes and fuel storage tanks, splintered structures and debris in the blast damage area would provide vast amounts of fuel for fires that start from overturned heating appliances, arcs and sparks from broken electrical wires, and other sources of ignition.

The fourth major effect of the explosion of a nuclear weapon (after radiation, heat, and blast) is radioactive fallout. The amount of fallout is enormously greater if a weapon detonates on or near the surface than if it explodes high in the air. Radioactive elements are incorporated into earth particles which are scattered by winds and eventually fall to the ground. The early fallout—that which returns to the earth during the first day—is highly radioactive, but radioactive decay sets in at once; it is rapid at first and then slows down as time passes. After two days the radioactive emission is about one-hundredth of that in the early fallout. Information has been gathered regarding the effects of fallout after the Chernobyl nuclear power plant disaster in the USSR. It was discovered in the spring of 1991 that in the regions of the Soviet Union hardest hit by the nuclear fallout, abnormally high numbers of children were dying and large numbers of adults were showing varying degrees of symptoms from the long lasting effects of radiation sickness.(16)

Global Fallout

About 20 percent of fission products from a surface burst go into the atmosphere and become global fallout. While being carried by upper level winds the fine particles will decay over a long period of time, but heavier particles which finally drift down to earth contain long-lived isotopes, such as strontium-90. This type of fallout increases radiation levels that can have genetic effects and can enter the food chain and be selectively absorbed by the human body to cause bone cancer.

Local Fallout

Local fallout that results from a surface burst contains about 80 percent of the fission products and it is much quicker to take effect. Though lighter particles may take several hours to fall to earth, heavier particles descend in the first hour. Radiation from this fallout can create a hazard over hundreds of square miles from a single 5-mt. weapon. In a nuclear attack involving thousands of megatons, local fallout would affect the entire nation and global fallout probably would exceed that produced by all weapons tested thus far.

Fallout Radiation

Fallout particles emit several kinds of radiation. Most neutrons have a short range, about 1,500 feet, for lethal effect. Although fallout consists mostly of beta particles, these are not dangerous unless inhaled or swallowed. Beta rays have a range of only about 40 feet, for they are quickly absorbed by the air. The same is true of alpha particles, which have a range of about 7 inches.

Gamma rays, like X-rays, are very penetrating and can be lethal to unprotected personnel for a distance of about one-half mile. A sheet of aluminum foil will stop beta particles, but it takes about a half mile of air or 2.5 feet of earth to attenuate gamma rays by a factor of 1,000.

Fission products of fallout are generally so dispersed in the rising column of air that these radioactive elements are not hazardous after a detonation. But only a small part of the bomb is fissioned, and the unfissioned material can be around for a long time. The half-life of uranium is 713,000,000 years and plutonium, 24,000 years, so that these radioactive residues will take a long time to decay to stable elements. However, in an air burst of a single weapon, these residues would be so scattered and dispersed by the upper level winds that by the time they fall back to earth, dangerous concentrations are not likely to be found in any one location. In an all-out attack against U.S. military, industrial, and population centers, radioactive fallout could be expected to spread over about 75 percent of the land area of the United States, in varying intensities, contaminating many places for periods ranging from hours to weeks, with potentially disabling effects on the population.(17)

Radiation Effects upon People

Radiation exposure is measured in roentgens (R). The effect upon humans varies with their age and general health. Although some people might not survive a dose of 300 R received in a twenty-four-hour period, a few might recover from a dose twice that amount. The time period in which a given dose is accumulated has a bearing on the effects. Recent studies indicate that it may be possible to survive an otherwise lethal dose if it is accumulated in small doses over a long period of time.

Effects of less than fatal doses can produce radiation sickness; it is caused primarily by damage to the blood-producing function of the bone marrow, and to the lymph glands. During the first day after exposure the symptoms are apt to be nausea, vomiting, diarrhea, and general debility. In more severe cases, depilation (loss of hair) and skin ulcers may follow. The most sensitive cells to radiation are the white blood cells, reproductive cells, and membrane lining cells. "There is no specific treatment, but antibiotics and blood transfusions may help. Recovery is slow, involving weeks or months. The sickness is not communicable."(18)

The table on the next page shows some of the effects following exposure to radioactive fallout and body accumulation of roentgens.

The maximum dosage for safety is .3 R per week and .1 R per day; in an emergency, 50 R may be permissible, but basically, it is considered advisable that roentgen exposure to the body not exceed 25 per day, with a total accumulation dose of not over 100 in a four-week period. Another way of estimating the effect of radiation exposure is the likelihood that 100 R

The nuclear threat poses a danger in peace time as well as war time due to potential radiation emergencies." (Photo courtesy of Atomic Energy Commission.)

received on one day, or 150 R in one week, or 300 R in one month (on a one time basis only), will produce drowsiness and an undetectable shortening of life. By the way of comparison, the dose received from a chest X-ray is about 1 to 2 ½ R; fluoroscopy, 40 R; dental X-ray, 30 R on tongue; cosmic rays, 2 R at sea level and 20 R on a mountain top.

Roentgens	Effects
25	No effects
50	Least detectable dose; minor blood changes
100	10 percent sick; no deaths
200	50 percent sick; 4 percent deaths
400	100 percent sick; 50 percent deaths
600	100 percent fatalities

Radiation Effects on Communications

One effect of high-altitude bursts is the blackout of high frequency radio communications. "Disruption of the ionosphere (which reflects radio signals back to earth) by nuclear bursts over the Pacific has wiped out long-distance radio communications for hours at distances up to 600 miles from the burst point."(19)

An electromagnetic pulse (EMP), sometimes called "radioflash," can affect a large area. Most of the EMP energy lies within the radio frequency spectrum (ranging from power line frequencies to radar system frequencies). EMP is electrical in nature and crudely similar to waves radiated by local lightning or a broadcast station. "This pulse is easily picked up by any conductive material, damaging any electrical or electronic equipment connected to this material."(20)

Radiation Effects Upon the Environment

In the 1980s we became aware of the danger from aerosol sprays to the ozone layer, which helps protect life on earth from the sun's ultraviolet radiation. In the 1990s, this danger became more measurable. In April 1991, a National Academy of Sciences (NAS) committee concluded that global temperatures could soar 9° F in the near future and urged steps to limit global warming. The largest danger comes from fluorocarbons, inefficient automobiles, and industrial atmospheric discharges. But while these are having a gradual negative effect on the environment, the effects of large-scale nuclear detonations can rapidly accelerate this detrimental effect. (21) If an attack and counterattack were to result in detonation of 10,000 megatons, a National Academy of Science study estimates that as much as 30 to 70 percent of the ozone might be eliminated from the northern hemisphere and 20 to 40 percent from the southern hemisphere. Though long-term changes cannot be ruled out, the Academy's study indicates that recovery would take probably three to ten years; the resultant increase in ultraviolet rays would cause incapacitating cases of sunburn in the temperate zones and snow blindness in the northern countries. There could also be a drop in the average surface temperature of the world that could have an adverse effect on agriculture in areas where the number of frost-free days in raising crops is critical. The cooling of temperature by one degree centigrade would eliminate commercial growing of wheat in Canada.(22)

"Probably man's most extensive modification of the global environment to date occurred three decades ago, back in September 1962, when a nuclear device was detonated 250 miles above Johnson Island. The 1.4-megaton burst produced an artificial belt of charged particles trapped in the earth's magnetic field. Though 98 percent of these particles were removed by natural processes after the first year, traces could be detected six or seven years later. A number of satellites in low earth orbit at the time of the burst suffered severe electronic damage resulting in malfunctions and early failure."(23)

Summary of Effects of Enemy Attack

Other effects would also follow an enemy attack with thermonuclear weapons. These include the effects of discussed in Chapter 10: earthquakes: deaths, injuries, collapsed buildings and dams, power failure, streets blocked, all forms of transportation disrupted, fires, with possible fire storm or conflagration, communications facilities knocked out, release of toxic gases, spills of toxic and flammable liquids and solids, water mains broken, water supply contaminated. And as in the case of an earthquake disaster, where you never know when or how large an aftershock to expect, there would be no way of telling when, or with what size weapon, the enemy might follow the initial strike with another on the same targets.

Decades of research, along with some empirical studies of Chernobyl and Three Mile Island, indicate that a massive attack with many large-scale nuclear detonations could cause such widespread and long-lasting environmental damage that the aggressor country might suffer serious physiological, economic, and environmental effects even without a nuclear response by the country attacked.(24)

After many years of development and study of the effects of nuclear weapons, there is much that we still do not know. Some of the most disastrous effects are unpredictable, particularly the global effects of a large-scale nuclear war.

WORLD WAR II NUCLEAR ATTACKS

Hiroshima, Japan

On May 9, 1991, American and Japanese survivors of the Pearl Harbor attack assembled in Austin, Texas, for a three-day symposium to mark the fiftieth anniversary of the event that initiated U.S. involvement in World War II. This war had been brought to a speedy close with the first use of nuclear weaponry.(25)

This use of a nuclear weapon against a military target took place on August 6, 1945, at 8:15 A.M. Over half of Hiroshima's population became casualties. It was estimated later that about 80 to 100 thousand Japanese were killed and thousands of others were injured. In addition to causing terrific blast damage, the explosion started thousands of fires. Within twenty minutes after the detonation, the city was a mass of flames. About 68,000 of the city's 75,000 buildings were either destroyed or seriously damaged. Nearly five square miles of the city were reduced to ruins.

The Hiroshima bomb weighed only half as much as the largest aerial bomb used before 1945, but it produced as much energy as a stack of TNT the size of the Washington Monument (equal to about 19,000 tons of explosive). The bomb exploded about 2,000 feet above the ground and demolished all buildings except those of earthquake-resistant construction out to a distance of one-half mile from ground zero. The fireball created was hotter than the center of the sun (70,000,000 degrees C.), and on the ground below the temperature was about 3,500 degrees C. Flash burns killed persons exposed to the heat as far as a mile away from ground zero and started primary fires in the structures of flimsy construction. Secondary fires resulted from overturned stoves, braziers and lamps, broken gas lines, interrupted industrial processes, and other damage. Fire breaks did not prove effective because fires started in so many places.(26)

"Fire fighting in Hiroshima was effective only at the outer perimeter of the fire-storm area and in four of the fifty-eight fire-resistive buildings which were damaged by fire. The public fire department and rescue units played a minor role in extinguishing the fires, as 80 percent of the fire fighters on duty had been killed or critically injured, 60 percent of the public fire stations totally damaged, and 68 percent of the fire trucks destroyed. As a result, only sixteen pieces of fire equipment were available."(27)

Because of a combination of wind, fire, fighting, streets acting as fire breaks, and low building density, the fire stopped at the perimeter; by early evening of the day of the attack most of the fire had burned itself out or was extinguished. Many dwellings on the outskirts of the city were saved by the use of hand fire equipment. Documented reports assert that some people survived within the fire storm area in fire-resistant buildings."(28)

Nagasaki, Japan

Nagasaki, with a population of more than 400,000, is the Japanese city with which Westerners have had the longest contact, for its harbor was opened to foreign trade in 1556. It is on the west coast of the island of Kyushu, and became the target for an atomic bomb attack in World War II only because the intended target city happened to be obscured by a cloud cover: Nagasaki was a second choice.

On August 9, 1945, three days after the atomic bomb attack on Hiroshima, the citizens of Nagasaki, although they had read vague references to the Hiroshima attack in the newspaper, were scarcely better prepared to withstand the nuclear detonation that occurred at 11:02 A.M. on their own city. The bomb exploded about 1,700 feet above ground, and destroyed 1.8 square miles in the heart of the city. It injured 40,000 persons, and another 40,000 were killed or missing.(29) As in Hiroshima, primary fires were started by the initial flash, and secondary fires resulted from the combination of disrupted electrical circuits, broken fuel lines, overturned hibachis, demolished light and flammable structural materials, and other blast effects. However, unlike Hiroshima, these fires did not create a fire storm. These cities had nearly the same population at that time—around 235,000 for Nagasaki and 245,000 for Hiroshima—but Hiroshima had a built-up area of about thirteen square miles, while Nagasaki had a densely built-up area of four square miles in the center of its metropolitan area.

Direct damage to the fire department in the Nagasaki attack was much less than at Hiroshima. Only two fire stations and one truck were lost by the paid fire department. The auxiliary police and fire units lost four fire stations and fire trucks. Though the fire fighter casualties were not nearly as great as those sustained in Hiroshima, the effectiveness of the fire suppression efforts was no better, mainly because of the failure of the public water supply. Not only were there breaks in the main distribution lines, often at bridges, but the collapse of 5,000 dwellings (from blast, fire, or both) severed most water service connections. Mutual aid from neighboring villages brought only manpower, for no motorized fire apparatus was available, and most of the efforts were directed at rescue work instead of fire suppression.

"The prefecture had made elaborate plans to cope with a possible incendiary attack. Instruction was given the people, and the volunteer organization was mobilized and in readiness. The havoc created by the atomic bomb completely disorganized all plans, not only for control of fires, but for rescue work, care of wounded, and burial of the dead."(30)

PREPARATION FOR ENEMY ATTACK

Since there may not be much warning that an enemy is about to launch a nuclear attack on any given city or nearby military installation, it is well to consider what preparation will enable your community, as well as your fire department, to survive.

Community Preparation

In 1961, the Secretary of Defense established the Office of Civil Defense (OCD) to develop and direct civil defense programs. That same year the Office of Civil and Defense Mobilization (OCDM) was renamed the Office of Emergency Planning. Its name was changed again in 1968 to Office of Emergency Preparedness. In 1972, the Secretary of Defense abolished OCD and established the Defense Civil Preparedness Agency (DCPA) to take over its civil defense programs. The Office of Emergency Planning was abolished in 1973.

In the 1980s and 1990s it was the Federal Emergency Management Agency (FEMA) that was charged with centralization and consolidation of national disaster preparation including nuclear attack.(31) This agency maintains a visible image and while many can criticize its cumbersome bureaucracy, few doubt the need for a central federal disaster agency.

Currently, it is the responsibility of FEMA to develop and direct a defense program against chemical, biological, and radiological warfare; and warning and communications systems. The agency has programs to provide aid to state and local governments after attack, and some publications to assist communities to meet an attack-caused emergency.(32)

FIRE DEPARTMENT PREPARATION—GENERAL

Consider what would be necessary to prepare an annex to your major disaster plan outlining actions your fire department would take if an enemy attack were imminent. This "Nuclear Attack Bill" should list actions to be taken before an attack (if a warning has been received from official sources) and after the attack (assuming there is capability to take action). Although a nuclear or enemy attack may be remote, the exercise would provide "spin-off" benefits that could help with other types of disasters.

Request the local civil defense organization to supply you with a list of your fire stations rated as acceptable for fallout protection. If any have inadequate protection, obtain a list of nearby buildings having an accept-

Refineries are likely targets during times of enemy attack. In this picture of a Unicol refinery fire, tanks are seen being launched like rockets as their fuel ignites. (Photo courtesy of Mary Paluszek.)

able rating where temporary quarters might be established for each of such stations.

If your community has no civil defense personnel who can make a survey, request your fire officers to appraise shelter locations within reasonable distance of inadequately protected stations and provide a plan for relocating these companies on a warning alert. Government publications describe the relative protection of different types of construction.(33)

The ability to operate portable radios in shelter structures should be considered. When the division commander in the San Pedro area of Los Angeles made a survey for his command, he found that an unused tunnel in a nearby militery fort would provide excellent fallout protection for one of his companies, and radio transmission was possible near the opening.

After you draft tentative pre-attack and post-attack plans discuss them with all officers, and review these plans in training sessions for all members of your department.

Fire Department Nuclear Attack Plans

If there is a possibility of enemy attack, your state will receive notification from the national warning system "Readiness Conditions," namely: Condition 4—Normal; Condition 3—Tense International Situation; Condition 2—Risk of War; and Condition 1—War Inevitable (Declaration of National Emergency). There are three "Warning Conditions"; Strategic (A Blue Alert under the Bell-and-Light Warning System); Tactical (Yellow Alert), and Attack—No Warning (Red Alert).

The state warning center will transmit this information to designated recipients in government and industry, and the public will receive notification over the Emergency Broadcasting System and by sirens, horns, or whistles.

If your Emergency Operations Center is to receive notification from the state warning point, it might be well to draft your nuclear attack plan with four major subdivisions, based on actions your fire department will take under the three warning conditions, and your post-attack plan. If your fire department is responsible for sounding the warning sirens, you will receive at least two notifications, whether or not your communications center is a recipient of the Readiness Conditions notifications: you may be asked to sound the siren in a long steady blast, as an alert to the public that radios should be turned on to receive official announcements relating to the imminent risk of attack (or natural disaster), and you may receive a request to sound the sirens in an undulating tone for several minutes to warn the public that the enemy has attacked (or bombs are on the way). The relationship between the National Defense System and local fire service organizations is being strengthened, in large measure by mutual understanding, mutual use of FEMA training, and shared disaster planning.(34) These should prove to be helpful in the event of war threats in the future.

Irrespective of the number of plan subdivisions you may decide upon using, the following suggestions may be incorporated in your plan.

Preattack Plan—Strategic Warning Condition One (Blue Alert)

Your plan might be prefaced with a statement such as this:

"When advised through official sources that a state of National Emergency has been declared or that international relations have deteriorated to the point where an enemy attack may be launched within a short time (i.e., a matter of hours or a few days), in the absence of other orders issued by competent authority, the fire department commander will order the Emergency Operations Center to be activated and direct the department's communications officer to transmit a message to all stations directing that the Department's Condition Blue Plan be put into effect at once."

Your plan may then provide this information. The following actions will be taken:

Blue Alert

1. All members will be notified that a Condition Blue has been declared and that all leaves except military and sick leave have been cancelled; recall telephone numbers will be used to notify off-duty members to provide for protection of their families and to hold themselves ready for recall.
2. Nonemergency activities will be cancelled. All companies not engaged in emergency operations will return to quarters.
3. Division commanders will retain sufficient manpower from the off-duty platoon to six-member companies using reserve apparatus.
4. The Department's Supply Bureau will obtain and deliver to each fire station and department operation facility sufficient food to provide a four-day reserve food supply for twice the complement of personnel normally on duty in those locations.
5. Supply Bureau will fill all department gasoline and oil tanks. Subsequently tanks will be filled when down to one third capacity or less.
6. Division commanders will require that reserve engines (not already loaded) be equipped with spare hose from hose towers or drying racks; fittings from other apparatus may be used pending replacement from the department's store room.
7. Supply Bureau will deliver spare hose and fittings to stations having unloaded reserve apparatus, and to auxiliary supply centers designated in the appendix to this plan.
8. Additional companies created by recalled personnel and reserve apparatus, designated for relocation at peripheral stations, will be moved; companies not having adequate shelter for fallout protection will move to alternate locations.
9. The department's operations officer will report to EOC and provide the augmenting communications equipment as necessary.
10. Key personnel will report to assigned duty locations and carry out assigned missions.

11. Company commanders will prepare to improvise decontamination vestibules for fire stations and to minimize flash and fallout hazard to personnel.
12. Training sessions will cover use and operation of radiological instruments and disaster control procedures, with emphasis on radioactive emergencies.

Tactical Warning Condition Two (Yellow Alert)

This Warning Condition (of imminent enemy attack) is announced to the public by the long steady siren signal, and if your community is equipped with a civil defense warning system, all personnel, on and off duty should follow procedures set forth in the emergency plan. This signal may only give fifteen to thirty minutes warning, and may or may not follow a prior Strategic Warning Condition One (Blue Alert).

The following suggestions might be modified to suit your local requirements:

Yellow Alert

1. Complete steps being taken under Blue Alert.
2. If no Blue Alert is given, carry out priority missions:
   a. Move designated companies (not having adequate shelter) to designated locations.
   b. Move loaded reserve and split companies to designated locations in peripheral areas.
   c. Activate EOC; all key personnel to report to duty locations and advise operations officer when they are on the job.
3. All on- and off-duty personnel take survival precautions consistent with duty requirements.
4. Emergency response will be handled as quickly as possible and nonemergency activities will cease.
5. All personnel listen to news broadcasts of EBS to be informed of enemy attack.

Attack with No Warning—Condition Three (Red Alert)

This means that one or more cities and/or military targets are under attack, and your community may be next. If you are not in a target area, you may be in a fallout pattern. Your plan needs one simple instruction for this condition:

Red Alert (Wavering siren signal)
All personnel, on or off duty, take cover immediately.

Postattack Plan—Nuclear War

Your postattack plan can be in two sections for off-duty and on-duty personnel. If your community survives initial attack; you may have so much damage that options for your personnel may be limited to staying under cover until radiation fallout decays sufficiently to go outside. This might take a week or two, depending upon your location with respect to ground zero and fallout pattern.

The following suggestions might be adapted to your postattack plan.

Off-Duty Personnel—Postattack

1. Follow survival instructions given in radio and television broadcasts. Depending upon where you are, the fallout pattern, and protection, you may leave the area, stay put, or report for duty.
2. If radio and television stations are not operating (and no other source of advice is available), stay where you are if your home provides adequate fallout protection. If not, and fallout has begun, arrange to move your family to the nearest available shelter. Report to your fire station or to the designated location to which you have been assigned for disaster duty.
3. If advised by radio or television that you will be in a hazardous fallout area, and your home affords inadequate protection, proceed to the nearest designated shelter or building, or to a point beyond the fallout pattern, in accordance with radio instructions. As soon as fallout has abated, report to your duty assignment.
4. If unable to reach your station or assigned post of duty, and fallout conditions do not require that you remain indoors, help organize rescue and fire fighting operations in peripheral areas.
5. When conditions permit, report to a public building (fire station, police station, post office, school) for transportation to the nearest fire service assembly area. Manpower pools will be established in parks, playgrounds, high schools, colleges, churches, and campgrounds after fallout pattern has been determined.

On-Duty Personnel—PostAttack Plan

1. Company commanders will appraise the availability of personnel (casualties); damage to apparatus and equipment; adequacy of water, food, medical supplies, fuel, etc.; stability of station, accessibility of roads, and hazards from fire, radiation and gas leaks.
2. Officers will make a status report to the Emergency Operations Center when conditions permit. Portable and vehicular radios may be the only means of communication; to conserve batteries, use only one radio (with greatest range) for monitoring.
3. Carry out orders in the chain of command, in accordance with provisions in the appendix to this plan. (Note: this should show the ascending order of command under a State of Extreme Emergency, such as, from the fire chief, to the civil defense director of the city, to the area commander, to the region commander, to the state disaster office director. If martial law is declared, everyone would be subject to orders of the area's military commander.)
4. All members will do whatever possible to safeguard personnel, restore the department to condition of readiness, insure continuity

of operations, improve communications, and augment the department's fire fighting and rescue capability.

5. Commanders of facilities, including fire stations, that qualify as being adequate for fallout protection, will make room for any other company instructed to share quarters. Post a fire look-out near the windows; post a radiation watch with suitable instruments; improvise a decontamination vestibule for personnel; the hose tower might be a place where personnel can be showered (while in turnout gear) by a hose rigged outside the door; turnouts can be removed in the tower and clean clothes put on inside the station. (Suggested plans for rigging such vestibules could be included in the appendix of the plan.)
6. Upon return from a run under fallout conditions, leave apparatus outside, and wash it; cover with salvage covers or tarps between runs.
7. If in a station which does not meet fallout protection standards, or in one having impaired stability, load spare hose and needed supplies and move to another location that will provide better protection; if the designated location has been damaged, keep moving (if possible) toward the periphery and beyond predicted fallout pattern, or to a structure that will provide adequate fallout protection. (It is not necessary to find a place to get fire apparatus inside the building.)
8. If you must relocate your base of operations, notify your commanding officer as soon as practical. (Ionizing radiation may cause radio blackout for several hours following nuclear attack.)
9. You may be on your own for a considerable period, so screen all requests for help (not received through the chain of command) very carefully.
10. Limit exposure of personnel:
    a. Limit response to fires and emergencies where life is threatened, or key installations are endangered, or where spread of fire to the rest of the city can be prevented.
    b. Send a few people on limited response to small fires, and heavy assignment to large fires (to get a quick knockdown).
    c. Rotate fire fighters on responses, to keep individual exposure to a minimum. (Note exposure limits provided earlier in this chapter.)
    d. Curtail overhauling to bare requirements to prevent rekindling.
    e. Eliminate salvage operations not absolutely necessary.
11. Be prepared to move out to an assembly area along the perimeter in case a fire storm develops and you are likely to be within it, or the fallout pattern makes your location untenable.
12. If you have to leave, keep personnel and apparatus covered to protect from fallout; have them wear breathing apparatus if radioactive dust and smoke are present.

Guard Against Complacency

It is easy for those of us in North America to consider the possibility of enemy attack as being quite remote. While it is true that this continent has not been visited with the death and destruction accompanying past wars, we have found in each generation that wars are all too common and could indeed at some future date involve our country. Certainly many portions of North America are within easy range of enemy nuclear submarines and we could at the least expected time be suddenly involved in a conflict.

The Persian Gulf War served as an example of how quickly the United States can find itself in military conflict. Hence, there is always a need to guard against the complacency that has evolved as our country has enjoyed relative insulation from previous wars.

THE UFO THREAT—A FACT

In this chapter we will now turn our attention to the very real threat posed by Unidentified Flying Objects (UFOs), whether they exist or not. The well-documented and highly publicized *War of the Worlds* radio drama by Orson Welles shows how even a perceived existence to alien creatures can cause very real disaster-like conditions and panic among a given populace. In addition, if the apparent visits by alien beings and their space vehicles should pose any type of threat, it will, as always, be the fire service that is called upon to provide the first line of life-saving defense and disaster mitigation.

On April 25, 1991, radio station KSHE in St. Louis, Missouri, was fined $25,000 by the Federal Communications Commission for broadcasting a mock warning of a nuclear attack during the Persian Gulf War. The seriousness with which the FCC treated this case is indicative of the very real panic that can be created from even illusionary or fictional phenomena. Certainly if these unexplainable events become more prevalent, the possibility of panic could be even greater; and again, the fire department will be the agency called upon to handle the situation.(35) Hence, as we near the year 2000 and move beyond, any comprehensive disaster plan should address the potential for panic and other deleterious effects that might befall a populated area when unexplainable phenomena occur. We will see, as we continue our discussion in this chapter, that widespread blackouts, communication disruptions, and other potentially disastrous conditions have been linked directly to UFO sightings. Hence, fire service leaders who want to ensure that their disaster planning is complete will not neglect an appendix to outline those things that could be done in preparation for the occurrence of such phenomena.

Throughout this book, many of the references to actual events are based on the experiences of both of the authors. However, in this area of UFOs and their potential, we are relying largely on the research and experiences of Charles Bahme. Chuck has made a considerable study of this subject and is acquiring many publications and VCR tapes to augment his library on this and related phenomena. His interest in UFOs was greatly heightened when Congress in 1969 adopted a law (14 CFR Ch. V Part 1211—Extraterrestrial Exposure) which gave the NASA Administrator the arbitrary discretion to quarantine under armed guard any object, person, or other form of life which has been extraterrestrially exposed. The very fact that our congressmen believed there was a necessity for such drastic authority made Chuck wonder if they had only *our* astronauts in mind when they adopted it. Could it be applied to anyone who has had a UFO encounter? Whether it has or not is not likely to be a topic for public dissemination.

UFO Discussion—Why Now?

The subject of UFOs (Unidentified Flying Objects) was not included in previous editions of this book. The first edition was the *Handbook of Disaster Control* which Chuck personally published in 1952 following his release from active naval duty in the Korean War. Although his services in the conflict as Security Coordinator for the Chief of Naval Operations involved the creation of a worldwide disaster control organization for the protection of the physical properties of the Navy, it must be admitted that the directives approved for this new organization did not reflect any significant concern for a flying saucer threat to its shore establishment. That was in the 1950s. Now that we are in the 1990s it is doubtful that the UFO potential would be brushed off so lightly by our military security forces. This change of attitude was evidenced as far back as December 24, 1959, when the Inspector General of the Air Force issued the following Operations and Training Order: "Unidentified Flying Objects—sometimes treated lightly by the press and referred to as 'Flying Saucers'—must be rapidly and accurately identified as serious Air Force business. . . ."(36)

There is no uncertainty about the reality of the war between nations on our planet and the disastrous effects of military actions. The 200 sorties flown every hour against Iraq in the Persian Gulf provided ample evidence of global war's destructive power. On the other hand, there are many persons who may believe that a discussion of the theoretical harm that could be caused by a real or imaginary invasion of UFOs would be "far out!" But this is not so for the thousands of witnesses of unexplained aerial phenomena. To them it is also serious business.

Chuck's interest in UFOs commenced during the early morning hours of August 26, 1942, while he was roller skating from his house to the nearest fire station a few blocks away; the wail of sirens had signaled his recall to fire duty, and with the stringent blackout orders in effect, driving was not wise; besides, it was much more exciting to be out in the open where he could see the spectacular aerial "fireworks" that filled the heavens all around him. Few residents of the U.S. had ever experienced a real or imaginary invasion of UFOs like that which occurred in what has become known as "The Los Angeles Air Raid of 1942." The Army announced the approach of hostile aircraft and the city's air raid warning system went into effect for the first time in World War II. The defense to this "attack" is described in dramatic terms in the opening paragraph of this chapter.

But what enemy had been routed? No one ever knew. All the fire fighters saw in the sky were the 15 or 20 moving "things" which seemed to change course at great speed apparently unaffected by the flak from bursting shells all around them. Rumors that one had been shot down were never verified, nor was the explanation that these zig-zagging invaders were weather balloons ever taken seriously. In any event, for Chuck, that unforgettable episode aroused a continuing interest in UFOs, rivalling his professional fields of law and fire protection. The fact that he subsequently was a member of a group whose sighting of a flight of UFOs was authenticated by airport radar helped to sustain that interest.

UFO Background Information

With no intention of trying to prove or disprove the authenticity of the numerous UFO encounters often related by very credible witnesses including airline and military pilots, astronauts, police officers, fire fighters, members of Congress, and even a U.S. President, the balance of this chapter will present a brief history and nature of UFOs and their alleged occupants; their widespread sightings over the globe since ancient times; their appearance, propulsion origin, and possible motives for continuing reconnaissance.

A quick look at some of the classic accounts of encounters documented in numerous foreign and U.S. publications might help us judge the magnitude of their threat, if any, to social stability, and, if deemed desirable, propose a fire service plan for coping with some of the conceivable catastrophic effects that UFOs could produce on cities and densely populated areas.

For readers who already have made up their minds that there is no such thing as a UFO notwithstanding the overwhelming evidence to the contrary, it should be pointed out that there is circumstantial evidence that disastrous effects have already been attributed to UFO activity in more than one nation, including the United States.

UFOs—What Are They?

William Shakespeare put a fitting observation in the mouth of Hamlet, the Prince of Denmark, that went like this: "There are more things in heaven and earth, Horatio, than are dreamed of in your philosophy." Whether Hamlet was referring to those strange lights or objects that appear in the sky or near the ground and have no known cause, we will never know, but the *World Book Encyclopedia* defines such things as UFOs.(37)

Several theories have been propounded as to what they might be. Some scientists believe that they are of extraterrestrial origin—coming from other planets. Military officers conjecture that they might be alien aircraft. Some attribute them all to natural causes, such as meteors, comets, sun dogs, light reflections, marsh gas, ball lightning, even though they must admit that scientists cannot explain all UFO reports in that manner. Still others are inclined to believe that they may be forms from other dimensions which can materialize and dematerialize at will perhaps by making a wavelength or frequency transition so as to become invisible to humans. Some believe they are time travelers from the future.

UFO Classification System

Dr. J. Allen Hynek, Northern University Professor Emeritus of Astronomy and an advisor to the Air Force's Blue Book Project, adopted a very simple classification system based solely upon the manner of observation:

1. Nocturnal lights
2. Daylight disks
3. Close encounters (day or night)
4. Radar readings.

He concluded that this system tells us nothing about the nature of the UFOs, but can suggest a means for gathering data.(38) He found that while a large number of such reports were readily identifiable by trained investigators as misconceptions of known objects or events, a small residue (about 1,000) were not. These came from credible witnesses from such widely separated places as Canada, Australia, South America, and Antarctica. He concludes with: "Although I know of no hypothesis that adequately covers the mountainous evidence, this should not and must not deter us from following the advice of Schroedinger: to be curious, capable of being astonished, and eager to find out."(39) Dr. Hynek has an excellent, well-illustrated article on UFOs in a 1982 book which gives a detailed history of the UFO sightings, together with the reports of some well-known people who made them, including President Jimmy Carter while governor of Georgia.(40)

Shapes of UFOs

Witnesses have described the shapes of UFOs as anything varying from a sphere to a boomerang. Some have resembled flying saucers with a lid; others a glowing tube; some as semi-spherical with colored apertures; some with reddish-orange glows, or fire-like or sparking discharges. Incredible speed and maneuverabilities not attainable by aircraft of any kind are commonly observed. Many of the books and articles in Appendix H have excellent photographs of these unexplained visitors—photos that have been checked by experts for their authenticity.

History of UFOs

For hundreds of years mysterious objects in the sky and strange moving lights have been reported by many people, including the military pilots in World War II who called them *foo fighters*, ("Where there's Foo, there's Fire"). In the middle of the 1900s flying saucers were increasingly observed in the United States and other countries.

Scientists at the University of Colorado hired by the Air Force from 1966 to 1968 to study this type of aerial phenomena could explain most of the UFO reports as a star (Venus), meteor, planet, balloon, rocket, artificial satellite, etc. Sometimes atmospheric conditions, aircraft exhaust trails, or unusual lighting conditions may produce optical illusions that observers thought were UFOs. After investigating more than 12,000 reports, the U.S. Air Force was unable to explain where the unexplained UFOs come from, but apparently concluded that the national security was not threatened by them.(41) The emphasis of the university's team, headed by Edward U. Condon, seemed to be more concerned with the establishment of the emotional stability or instability of those who reported the sightings than with other evidence.

Psychiatrists have examined the witnesses who claimed to have encountered UFOs and even been taken aboard their craft, such as the two shipyard workers in Mississippi, and found that they are not unbalanced people.(42) "They're not crackpots. There was definitely something here that was not terrestrial."(43)

Dr. J. Allen Hynek agreed, and added, "Where they are coming from and why they are here is a matter of conjecture, but the fact that they were here on this planet is beyond a reasonable doubt."(44)

The Air Force, after 20 years of being deluged with UFO sightings and spending millions of dollars on their investigation, decided to drop the inquiry business and turned the project over to a Kensington, Maryland, group called NICAP (National Investigation Committee on Aerial Phenomena). This left NASA (National Aeronautics and Space Administration) with part of the task of trying to run UFO sighting reports, including many by its own *Apollo* and *Skylab* astronauts. By 1974 over a score of astronauts saw and photographed UFOs during their flights beyond the earth's atmosphere.

Early in the *Apollo II* mission, which culminated in the moon walk, astronauts Neil Armstrong, Edwin Aldrin, and Michael Collins reported sightings of what seemed to be a UFO during the first half of their flight to the lunar surface. There were many more sightings by U.S. and Soviet Astronauts. On November 11, 1966, *Gemini XII* astronauts Jim Lovell and Edwin Aldrin said that they saw four UFOs linked together, and on October 12, 1964, three Russian astronauts aboard *Voskod* reported that they were surrounded by a "formation of fast-moving disc-shaped objects."(45)

UFO Organizations

In addition to NICAP, some of the other organizations that study UFO phenomena are MUFON (Mutual UFO Networks), CAUS (Citizens Against UFO Secrecy), GSW (Ground Saucer Watch), CUFOS (the Center for UFO Studies), and APRO (Aerial Phenomena Research Organization), an Arizona nonprofit scientific and educational organization, founded in 1952.(46)

Why the Secrecy?

In their book *UFOs Over America*, authors Jim and Carol Lorenzo charge that the CIA (Central Intelligence Agency) has been closely involved in the collection and suppression of UFO information. "Witnesses to the phenomena have been bribed, coerced, and threatened by the CIA, who wanted valuable evidence given to them alone."(47) One reason given is that military intelligence may view the UFOs as a tool of either a known or unknown potential enemy. "If these vehicles prove evasive and surreptitious, all the more reason to suspect them. . . . the probability looms large that the minds behind these vehicles may well be gathering intelligence of their own."(48)

Another reason for secrecy may lie in the hope of obtaining knowledge relating to advanced propulsion methods and anti-gravity systems before other potential enemies on earth may acquire it. Hence, though many nations are secretly investigating UFOs, they are reluctant to share their findings. Robert Lofton, in his book *Identified Flying Saucers*, claims that the Air Force became the "goat" in the effort of the CIA to debunk many sightings by pilots, radar technicians, and reliable civilian observers. He thinks that the suppression of information about how dangerous UFOs can be is wrong. After citing a case where a child was burned over 50 to 60 percent of her body by a low flying UFO and then taken to an Air Force hospital, no one would explain why her clothes were not burned at the same time. He also describes another burn case in New Mexico and another man who recently received a sledge-hammer like blow that knocked him unconscious by the force field of a 100-foot diameter UFO. "The public ought to be told the danger! . . . Nothing helps rumors and panic more than ignorance."(49)

Major Donald Kehoe describes in his book *Aliens from Space, The Real Story of Unidentified Flying Objects* the difficulties he had in 1957 in trying to get the truth from government agencies after he was director of NICAP, the world's largest UFO research organization with over 30 subcommittees in the U.S. and abroad.(50)

According to some UFOlogists the attempts at cover-up by the CIA extend to destruction of evidence that it could not confiscate. Apparently some of our nation's important leaders have been denied access to some UFO secrets in the possession of an agency of the United States, the very existence of which is classified above top secret.(51) Senator Barry Goldwater, a retired Air Force Reserve Brigadier General and pilot with many decades of flying experience, was quoted as saying "I certainly believe in aliens in space. They may not look like us, but I have very strong feelings that they have advanced beyond our mental capabilities." He said he was refused permission to check the Air Force files on UFOs and added, "I think some highly secret government UFO investigations are going on that we don't know about—and probably never will unless the Air Force discloses them."(52) He said that he put faith in the reports of the Air Force, Navy, and commercial pilots who reported instances where a UFO would fly near them—right off their plane's wing—and then just zoom away at incredible speeds. "I remember the case in Georgia in the 1950s of a National Guard plane going after a UFO and never returning. And I recall the case in Franklin, Kentucky, when four military planes investigated a UFO. One of them exploded in midair and no one knows why."(53)

Unleashed by the policy of *Glasnost* (greater openness) the Soviet media felt free to include accounts of UFO sightings. A *Tass* report of October 10, 1989, reported a large shiny ball or disk hovering over a Voronezh park; residents saw the UFO land and three creatures similar to human beings emerged, accompanied by a robot.(54)

Apparently the Russians felt no need to suppress this report which was poked fun at in *Newsweek* and *Time* magazines(55) but not in *U.S. News and World Report*: "A scant few decades ago, both the U.S. government and the media treated flying objects as no laughing matter—which even Congress looked into. In 1966, Representative Ford responded to a rash of sightings in his home state of Michigan by calling for, and getting, a House hearing on UFOs."(56)

UFO Missions

Many reasons have been advanced for the purpose of the UFOs visits to our planet. Although some of the persons who apparently have been the subjects of genetic investigation, such as the family of Whitley Streiber may not agree, the majority of those who have studied possible UFO visitors feel that they are friendly. Mr. Streiber described his experience as terrifying, and believes that these "little figures with eyes that seem to stare into the deepest core of being are asking for something. Whatever it is, it is more than simple information. The goal does not seem to be a sort of clear and open exchange that we might expect; whatever may be surfacing, it wants far more than that. It seems to me that it seeks the very depth of soul; it seeks communion."(57)

From the thousands of reports he has studied, William Spaulding, aerospace engineer and head of the Arizona-based Ground Saucer Watch, believes that a pattern indicates that UFOs are here on a surveillance mission; the fact that a majority of sightings occur around our military installations, research and development areas leads to the conclusion that a methodical study is being made of the earth and its defensive and offensive capabilities. "The phenomena is not unlike our own space explorations: scout ship survey; soil samples; landing."(58)

In his book *Incident at Exeter*, John Fuller discusses the seeming affinity of UFOs for electrical power lines in the northeastern part of the United States. In a later section of this chapter dealing with the effects of UFOs on our terrestrial activities, we will see how this affinity may have been responsible for causing 36 million people to lose power over an area of 8,000 square miles.(59)

Because of our recent adventures into space, there are some who speculate that UFOs are more concerned with what we will do there, than in settling here. In any event, the Air Force's official publication (issued by the Government Printing Office 1968) called *Flying Objects* says that "No UFO has been determined to represent a threat to our national security." That conclusion, however, should not rule out less disastrous consequences than the overthrow of our government.

ADVERSE POTENTIAL OF UFOS

Regardless of its past evaluations, the Air Force could be wrong about a number of things. "It can't even guess within a couple of billion dollars what one of its planes is going to cost; maybe, despite the skepticism of the scientists and other investigators, the UFOs sent from other planets do exist and have visited earth."(60) And maybe they have exhibited some destructive effects, whether or not intentionally in every instance, which we need to consider when drafting a plan for coping with an emergency situation where UFOs are involved. Some of these documented effects are as follows.

UFO Hazards

The two principal hazards noted with relation to UFOs have been attributed to powerful electrical fields which they can project in a general or localized area and the psychological effects they have produced on the general populace or individual contacts.

Force Field Impact

The disruption of air and ground travel has often been reported in the presence of UFOs. The ignition systems of auto and aircraft engines have apparently been affected by energized force fields to such an extent as to stop their operation; the headlights and radios have also ceased to function. Here are a couple of examples. In Buenos Aires, on March 29, 1978, "A strange force shut off their engine and headlights of their Citroen CG, lifted it 15 feet off the road, then set it down a minute later and 75 miles to the north." The driver had noticed a yellow and violet light shining in his rear view mirror while driving the last leg of a long stock car race, and he realized that it was approaching too fast to be a competitor. A month later a Colombian bank manager and a navy officer had their car headlights go off when buzzed by a UFO, with the navy man suffering temporary paralysis. Other South American countries in which similar actions were reported around that time included Argentina, Chile, Peru, and Uruguay.(61)

These effects have also been noted to influence the controls and instruments of aircraft, e.g., the pilot of a Piper PH-24 reported that his controls became inoperable when he was approached by three disk-shaped objects, 10 to 12 feet in diameter, over Mexico City on May 3, 1975.(62) Similar cases have been reported by military pilots, illustrated by the classic case of the near mid-air collision of an army helicopter with a UFO on October 18, 1973, over Ohio, where not only did both the UHF and VHF radio wavelengths go dead temporarily, but the downward movement of the helicopter with its four occupants was levitated upward by a green beam from the UFO in time to prevent its crash into the ground.(63)

Communications Disruption

In addition to the impedance of radio transmissions and reception, such as that described in the preceding incident, telephone interference has occurred, illustrated by the chagrin of President Lyndon Johnson in having his conversation from the Texas White House cut off while talking to assistants in Washington, D.C.(64) The ability to render inoperable all electronic forms of communications, including those that control the launching of defense weapons systems, has been considered within the range of UFO capability. Whether this could extend to the erasing of recorded computer data such as bank records, personnel data, FBI, CIA, and NSA files, along with critical information of every kind, is not beyond the realm of possibility.

Regional Power Blackouts

It has long been suspected that UFOs have the capability of blacking out a city, state, or many states by exerting a force field sufficient to overload the circuits of public and private utility installations. "Few things are more disturbing than to be plunged into pitch darkness without warning; it is dangerous for masses of people. It paralyzes cities, blocks highways, stops trains, leaves elevators suspended between floors. In general it simply plays hell with the modern way of life."(65) You would think that the power companies would have achieved sufficient reliability in their high tech systems that a mass failure such as that which blacked out New York and New England in 1965 would never happen—but it did. Although, as we have mentioned before, it was known that UFO activity was associated with disturbances with compasses, instruments, ignition systems, radios, etc., it was inconceivable that it could also interfere with generation and distribution of electrical power. Such a connection was also inferred in November 1953, when a glowing red object went over a residential area of New Haven, Connecticut, causing lights to dim out on both sides of the object's path and then come on when it went out of sight.

Power failures were also reported in association with UFOs in Brazil in 1957 to 1959, Rome, Italy, in 1958; and Mexico in 1965. Likewise, in Uberlandia, where the power station operators promptly closed the circuits when the UFO apparently caused them to open, it did no good, and they were unable to restore the power until the UFO departed.

"The Granddaddy of all blackouts to date was the stygian blanket that fell over 30 million people in the northeastern corner of the U.S. during the early evening rush hour period on November 9, 1965."(66) Relay services that were supposed to automatically transfer the load in case of failure in one area to an alternate source malfunctioned. Military communications relying on public power without alternate backup systems also failed, but communications were operable to make a quick public announcement that there was no military emergency. Though it was largely over by the next morning, the official explanation about a malfunctioning small device in a Canadian hydroelectric generating plant never accounted for the failure of millions of dollars worth of electronic devices to shift the load when the breakdown occurred.

Fireballs Over Syracuse—The Blackout Connection

Airplane pilots reported that UFOs were being chased across Pennsylvania about 4:30 P.M., and electronics and construction engineers who were driving in the area of the Syracuse airport saw UFOs moving about 5:30 P.M., just prior to the Great Blackout. A veteran flight instructor who had been flying over Syracuse on a training flight saw a glowing globe over the power lines leading to the Niagara Falls generating plant. Hundreds of others saw the glowing object in the sky on the night of the big power failure.

That was on November 9th. On December 2nd, about 700,000 persons in Texas, New Mexico, and Mexico also had their power fail. It was said to have originated in a regular failure in El Paso; then on December 5th, three nights later, 40,000 homes plus military installations in the area of East Texas were also blacked out—overloaded! Missile grounds (White Sands), Fort Bliss, Holloman Air Force Base, and numerous airports were all blacked out (with no emergency power backup), and this was when President Johnson's telephone call to the White House in Washington, D.C. was cut off. In response to his request for an explanation, President Johnson was told that his calls were fed into a cable system that went dead when the surge of power caused by the El Paso regulator blow-up hit it, and the backup batteries didn't work. Though it may be debatable whether the above cases of electrical transmission failure were merely coincidences with UFO activity, an incident on April 18, 1962, involving a UFO that had been tracked from New York, through Kansas to Eureka, Utah, was well documented. The Air Force spokesman admitted that the object had landed, and during the 42 minutes that it was on the ground near the power station there was no power, but it was restored when the UFO left. The object was pursued by jet interceptors summoned from Phoenix and Stead Field in Reno until it exploded over the Mesquite Range in Nevada in a brilliant glare that was visible over five states.(67)

UFOs—The Panic Hazard

The second major disastrous effect that UFO activity, real or imagined, can have on the populace, is the creation of fear, panic, flight, and all kinds of irrational behavior. We have mentioned already the rather well-documented case of hysterical contagion and mass hysteria created by *War of Worlds*, the radio drama by Orson Welles about an invasion of Martians. It was broadcast on Halloween of 1938 during the period of the invasions of Germany into Austria and Japan into China. "The drama, realistically presented in the form of news bulletins and interviews concerning an alien spaceship landing in New Jersey, resulted in many kinds of hysterical actions, including thousands of panic-stricken phone calls, wildly fleeing automobiles, and impromptu shot gun brigades."(68)

Though most persons reporting UFOs do not interpret them as personal threats, it is possible that some of the large volume of reports may be attributable to hysterical contagion. In any case, one of the reasons often cited for the tight secrecy on government UFO research findings is the need to prevent the possible panic that a revelation of the truth might arouse. Rumors that people were being abducted, dematerialized, burned, made radioactive, rounded up and impounded, liquidated with ray guns and lasers or shipped off to Mars or Venus might well give rise to fear—fear of the unknown. Hysteria could cause frightened persons to imagine that their water was poisoned, the air contaminated with undetectable but lethal aerosols or nerve gases. With hundreds of UFO squadrons zooming across the landscape from California to New York, Toronto to Mexico City, communications disrupted, widespread power failures, airports and railroads paralyzed, highways turned into giant parking lots of immobilized vehicles full of terrified motorists, the problem of restoring order and sanity would be a tremendous challenge to all of the emergency services, assuming their personnel would remain calm, detached, and able to resist the human impulse to put the safety and well being of their own families ahead of the public's. To make matters worse, some of the more excitable gun owners might be tempted to rush out Rambo fashion, and in utter disregard of the damage that falling bullets might cause innocent residents below their fallout, start firing at the evasive objects regardless of the range.

Personal Hazards—Physiological

The force field affects on the physical environment—communication, transportation, illumination, and computerized data storage—have already been considered. We might have added that some physical effects have been observed at locations where UFOs have landed—circular patterns of crops destroyed by heat or radiation and baking or sterilization of the soil at the site.

On a more practical basis there may be grounds for concern that more than just the environment can be adversely affected by UFO actions. While pursuing UFOs, military aircraft have disappeared in mid-air, exploded, and suffered harassment. Persons on the ground have sustained serious burns, paralysis, and "blows" from a force field, radiated emissions, or rays and beams that have been described like that of a "stun-gun"(69)

In 1980, three witnesses saw a red ball of light hovering above houses in Kuala Lumpur, Malaysia, when suddenly a bolt of blue light shot down and two houses burst into flames.(70) An Indiana man saw a bright light flash pass his window; the power went off in his house so he went outside to investigate and found a brilliantly lit object hovering above him; when he started to walk toward it his body commenced to tingle and he was unable to move until the object disappeared. A similar tingling sensation swept over another man in Lynn, Massachusetts, one night when he approached a domed object with a red glowing cone rising from a parking lot. He too was immobile until the object moved out of sight. Some believe that even animals may be at risk by UFOs; in trying to account for the death of 15 ponies, the leader of an investigative team believed they were crushed by the anti-gravity field of a flying saucer as it took off.(71)

Thus, UFOs may not only have the power to control some of our military and industrial establishment's highly technical scientific hardware, they may also possess the ability to impose pain and control over people who attempt to attack them, even to the extent of "liquidating" them in one way or another.

UFOS—EMERGENCY ACTION

In view of the fact that many UFOlogists believe that we are fast approaching a time when overt landings of UFOs will become less remarkable, and in the absence of our knowing whether their visits are friendly or hostile, it would not be remiss to give some thought to the part that fire departments might play in the event of the unexpected arrival of UFOs in their communities. For example, what would be your course of action as an incident commander at the scene of a school ground where a UFO has crashed into the boiler room, rupturing a fuel line, and ignition has occurred in the spilling oil, endangering the occupants of the craft who are trapped in the wreckage? If your rescue attempts are successful, and two of the five small alien creatures are injured but still alive, how do you dispose of the dead and treat the survivors? How would the presence of children on the school grounds affect your actions? What persons and agencies would be notified?

The authors have never read any advice on these matters. The following admonition was printed on the inside front jacket of Frank Edward's book on flying saucers:

> WARNING
>
> "Near approaches of UFOs can be harmful to human beings. Do not stand under a UFO that is hovering at low altitude. Do not touch or attempt to touch a UFO that has landed. In either case the safe thing to do is to get away from there very quickly and let the military take over. There is a possibility of radiation danger and there are known cases where persons have been burned by rays emanating from UFOs. Don't take chances with UFOs!"

In view of the federal law (cited earlier) empowering NASA's administrator to impound, without a hearing, anyone who touches a UFO or its occupants, it would be inadvisable to make personal contact unless you are willing to submit to NASA's quarantine requirements, should the law be invoked.

Besides the possible physical effects of approaching a UFO, e.g., burns, radiation, etc., there may be psychological effects produced by force fields that could induce a hypnotic state in the viewer, loss of consciousness, memory relapse, and submission to the occupants. Jacques Vallee, author of *The Invisible College* cautions that we should consider psychic effects, such as space-time distortions experienced by percipients of craft-like devices which appear to fade away—dematerialize—and then reappear; of alien, strange voices or thoughts that may effect involuntary changes in the manner in which witnesses may react in such circumstances.(72)

Perhaps the above warnings of Edwards and Vallee are a little too cautious and apprehensive to adopt as a general pattern of conduct in every situation. In the absence of overt acts indicating hostility, there may be no danger in approaching a landing (or landed) UFO with a positive, solicitous

Some combination of air and ground units should be available to treat potential victims of unexplained aerial phenomena. (Photo courtesy of Ron Buckingham.)

attitude of wanting to be of service. This nonaggressive mental state may be telepathically sensed by those aboard or emerging from the craft; a form of nonvocal communication is a possibility. It goes without saying that any display of firearms or other weapons on your part could be construed as unfriendly and likely to thwart your intention of conveying a helpful attitude.

In a best case scenario, you may be able to obtain guidance as to the appropriate actions to take, whether of a life-saving nature, e.g., in quelling a fire, abating a spill, and of preservation of property, or even in the reduction of apprehension on the part of your response team and the spectators.

In a less optimistic scenario, you may have engine trouble upon approaching the scene, and radio contact could be lost with your dispatcher. If at night, your headlights could go out, the city could be blacked out, and your portable generators may malfunction when you attempt to use them for fans and portable lights. It would certainly be an inopportune time for your comrades to announce that they had decided to take their pensions, effective immediately.

In any event, the incident could provide invaluable experience for further training in coping with rare and difficult emergencies. Whatever "inside" information you are able to pass along to your fellow officers and citizens of the world might help to alleviate unreasonable fear, so that there would be less likelihood that we would ever again experience the panic and hysteria that was created by *War of the Worlds* a half century ago. Truth is the best cure for the unknown. A list of some of the available books on Unidentified Flying Objects (UFOs) is found in Appendix H.

CONCLUSION

Some fire chiefs have little confidence in disaster plans, especially those dealing with UFOs or enemy attack. If you develop a plan that sets forth your responsibilities, resources, organizations, supplies information, telephone numbers, and special data that will be useful in obtaining help and fulfilling your role in disaster control, commit it to an electronic medium, a computer with a capability for continuous updating through modern word processing. Bring it forth when the need requires. With a good plan, good leadership, and adequate resources, you may save many lives in any disaster, including attack from possible enemies.

CHAPTER 13 REFERENCES

1. Smith, Jack. "The Night L.A. Bombed," *Los Angeles Times*, September 9, 1975, Part I, p. 1.
2. *World Book Encyclopedia*, Volume 21, p. 21, Chicago, IL Field Enterprises Educational Corporation, 1976.
3. Cloud, Stanley W. "Gathering Storm" in *Time*, September 3, 1990, pp. 24–28.
4. "Secret History of the War" in *Newsweek*, March 18, 1991, p. 28.
5. "Snubbing People Power" in *U.S. News & World Report*, April 8, 1991, p. 38.
6. *World Book Encyclopedia*, Vol. 21, p. 22. Field Enterprises Educational Corporation, Chicago, IL 1976.
7. "Preparing for Iraqi Chemical Warfare," photo and story in *Time*, September 3, 1990, p. 26.
8. "Steel Rain" in *Newsweek*, March 18, 1991, p. 31.
9. See note 2, volume 9, p. 416.
10. See Chapter 8 for a more in-depth discussion of terrorism.
11. Gilliam, J. "A-Bomb Materials Can Be Stolen, Expert Says," *Los Angeles Times*, December 11, 1975, Part II, p. 1.
12. *Nuclear Blackmail Emergency Response Plan for the State of California*, Officer of Emergency Service, State of California, June 1976.
13. "After the Storm" in *Newsweek*, March 11, 1991, pp. 26–29.
14. *Nuclear Attack and Industrial Survival*, McGraw-Hill Publishing Company Special Report, 1962, p. 52.
15. Willenson, K., and L. Norman. "Missiles on the Move" in *Newsweek*, February 16, 1976, p. 42.
16. CBS News broadcast, April 1991.
17. *Disaster Planning Guide for Business and Industry*, Defense Civil Preparedness Agency, 1974, p. 7, U.S. Government Printing Office, Washington, D.C.
18. L. A. Postal Report, Volume 10, No. 9, April 23, 1965, p. 2, Los Angeles, CA.
19. *Worldwide Effects of Nuclear War—Some Perspectives*, a Report of the U.S. Arms Control and Disarmament Agency, 1975, p. 5, U.S. Government Printing Office, Washington, D.C.
20. Ibid. p. 6.

21. "Turning Up the Heat on the Greenhouse" in *Newsweek*, April 22, 1991, p. 69.
22. See note 19, p. 6.
23. See note 19, p. 5.
24. "Deadly Meltdown" in *Time*, May 12, 1986, p. 39.
25. "A 'Big 50' for Pearl Harbor in *USA Today*, Thursday, May 9, 1991, p. 3A.
26. Bahme, Charles W. *Fire Officer's Guide To Disaster Control*, 1st ed., Boston, MA: NFPA 1978, p. 340.
27. *Fire Effects of Bombing Attacks*, Technical Manual 9-2, October 1959, Office of Civil Defense Mobilization, U.S. Government Printing Office, Washington, D.C.
28. *Fire Aspects of Civil Defense*, TR-25, Office of Civil Defense, July 1968, p. 4, U.S. Government Printing Office, Washington, D.C.
29. *World Book Encyclopedia*, 1991.
30. *Disaster Operations*, Defense Civil Preparedness Agency, 1972, p. 29, U.S. Government Printing Office, Washington, D.C.
31. Bruno, Hal. "The Wait May Be Over at FEMA" in *Firehouse*, Vol. 15, No. 5 (May 1990) p. 10.
32. See note 30, p. 40.
33. The U.S. Government Printing Office, Washington, D.C. has several publications available as of 1991 which suggest measures that can be taken to safeguard dwellings and other buildings, in preparation for a nuclear attack; they illustrate the relative protection afforded for fallout radiation by various types of construction and in various locations within a building. Two available from the U.S. Government Printing Office, are *Fallout Protection* and *In Time of Emergency*, both Office of Civil Defense.
34. Weldon, Curt. "The Fight for Fire Protection" in *Firehouse*, Vol. 16, No. 4 (April 1991), p. 20.
35. Radio Broadcast on station KSHE, St. Louis, MO, reported on April 25, 1991.
36. Edwards, Frank. *Flying Saucers—Serious Business*, NY: Lyle Stuart, 965, p. 315.
37. *World Book Encyclopedia*, World Book Inc., 1988, Vol. 20, p. 19.
38. Sagan, Carl, and Thornton Page. *UFOs—A Scientific Debate*, Cornell Univ., 1972, p. 44.
39. Ibid. p. 51.
40. Readers Digest. *Mysteries Of The Unexplained*, p. 219.
41. Steiger, Brad, Editor. *Project Blue Book*, NY: Ballantine Books, 1976, p. 170.
42. Uphoff, Walter and Mary Jo, *New Psychic Frontiers*, Colin Smyth Ltd., 1975, p. 152.
43. Ibid., quoting Dr. James Harder, University of California.
44. Ibid. p. 152.
45. Macomber, Frank. "UFOs Spotted by Astronauts Still Haven't Been Identified," *Santa Cruz Sentinel*, April 17, 1974, p. 30.
46. APRO's address was given as 3910 E. Kleindale Rd., Tucson, Arizona, 85716.
47. Lorenzen, Jim and Coral. *UFOs Over America*, NY: Signet, 1968, pp. 182 et seq.
48. Ibid. p. 186.
49. Lofton, Robert. *Identified Flying Saucers*, NY: David McKay Co., 1968, p. 86.
50. Kehoe, Donald, Major. *Aliens From Space, The Real Story of the UFOs*, New York: Doubleday, 1972.
51. Friedman, Stanton. *Cosmic Watergate, New Realities*, 1979; Hynek, J. Allen, "UFOs," *This World*, Aug. 30, 1981, p. 13.
52. Zullo, Allan A. "I Believe That Earth Has Been Visited By Creatures From Outer Space," *National Enquirer*, December 1973.
53. Ibid. p. 2.
54. *Press Democrat*, October 10, 1989, p. A-5; also reported in weekly news magazines; see notes 22 and 23.
55. *Time*, October 23, 1989; *Newsweek*, October 30, 1989; UFO update, *Omni*, January 1990.
56. *U.S. News & World Report*, "UFOs in Uncle Sam's Closet," October 23, 1989, p. 19.

57. Streiber, Whitley. *Communion*, NY: William Morrow, 1987, p. 15.
58. Adamski, George, *Inside the Flying Saucers*, NY: Paperback Library, 1967, p. 11.
59. Fuller, John. *Incident at Exeter*, cited in UFO Update, in *New Realities*, 1978, p. 52.
60. "Shooting Down The Flying Saucers," *Los Angeles Times*, December 15, 1959, Pt. II, p. 5.
61. Boudreaux, Richard. "South Americans Take UFOs Seriously," *Los Angeles Times*, November 29, 1978, Pt. VI, p. 7.
62. See note 6, p. 224.
63. Randles, Jenny. *The UFO Conspiracy*, NY: Sterling Pub. Co., 1990, p. 105.
64. See note 1, p. 267.
65. See note 1, p. 255.
66. Ibid. p. 259.
67. Ibid. p. 269.
68. See note 4, p. 216.
69. Blundell, Nigel, and Roger Boar. *The World's Greatest UFO Mysteries*, NY: Berkeley Books, 1990, p. 175.
70. Ibid. p. 176.
71. Ibid. p. 179.
72. Vallee, Jacques. *The Invisible College*, NY: E. P. Dutton, 1975 p. 6.

CHAPTER 14

MASS CASUALTIES AND MASS EVACUATION

DISASTERS USUALLY HAVE a serious impact on a large number of persons. As we saw in the preceding chapter on war, human losses that are often staggering in ordinary disasters become incomprehensible in activities such as a major earthquake or nuclear attack.

In this chapter we discuss the role of the fire service in dealing with large numbers of persons, both the injured, or mass casualties, and the uninjured, or mass evacuees.

In many disaster situations, the fire service is confronted with numerous fatalities and injuries, that is, mass casualties, and also numerous people who must be moved from harm's way. A simultaneous need is created to treat the mass casualties and to effect orderly removal of the uninjured—a mass evacuation. In the opening chapter, we saw how Cincinnati fire fighters were confronted with fatalities, scores of major and minor injuries, and also the need to evacuate more than a square mile area, when the BASF Chemical Plant exploded and erupted into flames on July 19, 1990.(1)

MASS CASUALTIES

The first part of this chapter deals with mass casualties. The fire service has become a key player in managing mass casualty incidents in the decade of the 1990s for two major reasons. First, it has moved from simply providing first responder rescue services to providing transportation and advanced life support or paramedic services. Second, it has developed a comprehensive incident command system that can be effectively applied to managing large-scale emergencies involving multiple casualties and fatalities.

Causes of Mass Casualty Disasters

Before discussing planning and response considerations in multiple casualty incidents, it is helpful to review the causes and effects of disasters that can generate large numbers of injuries or fatalities.

Serious students of mass casualty incidents at the National Fire Academy are quite familiar with W. Steve Collier, adjunct academy instructor and EMS director for Indianapolis, Indiana. Steve has conducted extensive research regarding mass casualties and has documented his findings in the National Fire Academy course "Disaster and Fire Defense Planning."(2) Steve has graciously agreed to provide much of the information in this portion of the chapter.

Mass casualty disasters can be natural, technological, and human-caused. We will review emergencies in each category, some of which have been covered in preceding chapters, and discuss the nature of the medical problems they cause and some common response problems.

Natural Disasters

Natural disasters include all of those covered in chapters 9 and 10. These are caused by one or more natural phenomena: earthquakes, tornadoes, windstorms, dust storms, floods, hurricanes, blizzards, prolonged cold, prolonged heat, drought, ice storms, hail storms, insect infestation, volcanic eruptions and lava flows, subsidence (sink holes), mud slides, avalanches,

Here police and fire officials account for bodies from an airline disaster. (Photo courtesy of the National Fire Academy.)

tsunamis, and tidal waves. There have even been rare occurrences of "exploding lakes." In Cameroon in 1986 large volumes of carbon dioxide, which had formed from natural causes at the bottom of a volcanic crater lake, were suddenly and violently released from the water. The resulting cloud of carbon dioxide dispersed through a valley asphyxiating some 1,700 villagers.(3)

With the exception of this sort of rare phenomenon, natural disasters are fairly well understood and throughout history have had the potential for massive life loss. The following is a sampling of major natural disasters in the last fifty years, taken from the more comprehensive list you saw in Chapter 1.

- 1925—multiple tornadoes kill 689 in midwest United States
- 1931—flooding of the Huan He River kills almost 4 million in China
- 1969—flooding kills 189 in West Virginia
- 1970—a cyclone kills 300,000 in Bangladesh
- 1976—an earthquake kills 800,000 people in China
- 1979—Hurricane *David* kills 1,100 in the Caribbean and the United States
- 1986—an earthquake in Mexico City kills over 9,000
- 1990—a cyclone again strikes Bangladesh, killing 140,000 more

Chapter 9 contained a detailed discussion of the health effects and medical response problems posed by weather-related natural hazards prevalent in the United States—storms, hurricanes, tornadoes, and floods.

Chapter 10 contained information on terrain-based natural disasters—earthquakes, landslides, and volcanos. Most natural disasters pose the same general sorts of problems—interruption of utility services and communications, damage to key response resources and facilities, and regionwide effects that limit outside assistance. One all too familiar common denominator is the large number of persons affected.

Transportation Disasters

Chapter 11 covered transportation disasters. We saw how these can involve large numbers of persons, especially in aircraft and train accidents. Growing population density and technological advancements that increase the number of people who can be transported at one time raise the possibility for a very large mass casualty incident virtually anywhere in the country.

Most transportation accidents involve violent trauma either from impact or fire. Extended extrication efforts may be involved as they were in 1990 in the Amtrak derailment in Boston(4), and the Avianca air crash on Long Island.(5)

Major boat or ferry incidents are rare in this country but have occurred recently in Europe. Few can forget the ferry accident in Haifa, Israel, which killed 21 U.S. sailors during the Persian Gulf War.(6) Accidents can be caused by overturning in severe weather, collisions, and fires.

Highway transportation accidents normally result in fewer casualties

Mass transportation vehicles lead to mass casualties. Here, fire fighters work to extricate victims from an Amtrak train derailment. (Photo courtesy of the National Fire Academy.)

than other modes although chain reaction accidents in fog can require major extrication resources. We reviewed the tremendous challenge facing fire fighters from throughout McMinn County, Tennessee, during one I-75 pile up in 1991 that killed 13 and injured 50.(7)

HAZARDOUS MATERIALS INCIDENTS

As you learned in Chapter 12, hazardous materials incidents pose unique problems and compound mass-casualty response needs. Specific problems are listed here.

- The exact name of involved chemicals may not be known for quite some time if shipping papers or fixed facility operators are not available.
- The effects of chemical mixtures, even when the compounds are known, may not be predictable.
- Known chemicals can mix, creating new unknown hazards.
- The biological effects of chemicals are usually toxicological rather than traumatic. Thus, large-scale incidents result in numerous sick people rather than injured people. Resources may not be immediately available to deal with these patients.

- Patients and responders can become contaminated and thus continue to cause problems until they can be decontaminated. Decontamination can be difficult both in the field and in hospitals because of lack of special facilities.
- Expert technical advice on involved chemicals, including symptoms, personal protective measures, and medical treatment, may be difficult and time-consuming to locate.

Chief Warren Isman, who chaired the National Fire Protection Association (NFPA) Technical Committee on hazardous materials, pointed out the growing technical needs in the fire service in this field. The NFPA had just formed a committee to approve NFPA Standard 471, "Recommended Practice for Responding to Material Incidents"; 472, "Professional Competence of Responders to Hazardous Materials Incidents"; and 473, "Professional Competence of Emergency Medical Personnel at Hazardous Materials Incidents." They provide excellent guidelines for dividing personnel and resources among varying danger zones during a mass casualty or mass evacuation incident.(8)

War or Enemy Attacks

It was in Chapter 13 that we saw how the most deadly form of mass casualty could come if war were to precipitate an enemy attack. The principal distinguishing feature in an enemy attack is that the lead incident control agency will generally be law enforcement or the military. Providing medical care in the field could be difficult and involve substantial personal risk.

Two other types of "human-caused" incidents, civil disturbances and terrorist incidents, both of which were described in Chapter 8, should be specifically addressed because of added danger to fire and emergency medical service (EMS) personnel that complicate mass casualty incidents.

Civil Disturbances

Episodes of civil unrest occur periodically in the United States. In Chapter 8 we read about widespread violence that occurred in the late 1960s and 1970s which was associated with the anti-Vietnam War and Civil Rights movements. Since that time isolated events, such as those in Miami and various Persian Gulf War protests, erupted periodically, causing problems for fire and law enforcement forces in many cities throughout the world.

The factors that facilitate either isolated or widespread civil disturbances are numerous and interdependent, making it difficult to predict future episodes. However, incidents tend to occur in highly-populated urban areas that have heterogeneous populations in areas of relatively rapid social change in the distribution of power and wealth. They also occur among disadvantaged groups experiencing stress where large groups of persons are already congregated. In both situations the potential for a mass-casualty incident increases.

The specific causal factors for civil unrest may involve the response to a triggering event by government forces, particularly the police, but sometimes fire personnel as well. Policies on the role and responsibilities of the fire department in civil disturbances should be worked out in advance with police and senior government officials.

Medical care in mass casualty situations arising during civil disturbances will be similar to other incidents with the following exceptions.

- Field medical activities may be limited by difficulties in gaining access to victims until violent activity can be controlled.
- Hostile actions may be directed at fire and emergency medical services (EMS) personnel as part of the authority structure. This problem is worsened when fire department hose lines and personnel are used to control criminal activity.
- Fire stations and government facilities themselves may be the target of hostile activity.

Terrorism

Terrorism in this country has not been a major factor in mass-casualty incidents. However, experts warn that terrorism directed against U.S. citizens or factions will take place inside our borders with increasing frequency, especially if conditions in the Middle East further deteriorate.

Many acts of terrorism, such as hostage-taking, do not involve a large number of people. These incidents will not result in mass casualties, but may be highly charged because of diplomatic and political implications and intense news media coverage.

Aircraft hijackings and bombings are predicted to increase and may involve a much larger number of people. The large loss of life in the terrorist-activated passenger plane destruction over Lockerbie, Scotland, in 1989 could be repeated many times over if the utmost security measures are not maintained in every major airport.(9)

The possibility obviously exists for utilizing explosives or hazardous materials to take a very large number of people hostage, for example, at a sporting event. In the preceding chapter we noted real concern about the feasibility of fabricating a crude nuclear weapon. The detonation of such a device could be catastrophic even if an actual nuclear explosion did not occur because highly radioactive materials would be dispersed.

Medical response implications during terrorist incidents include the following.

- Federal agencies, such as the FBI, may be heavily involved in reacting to such incidents. Fire and EMS responders are not accustomed to coordinating their efforts with these agencies.
- Terrorist incidents may last for days and take unexpected turns.
- Medical responders or facilities may become targets if it is felt that such actions will place additional pressure on authorities to meet terrorist demands.
- Bombing incidents have been known to involve delayed secondary bombs designed to detonate after responders arrive. Responders

should closely coordinate with law enforcement commanders and watch for the presence of suspicious devices.

MASS CASUALTY PRIORITIES

Normally, two immediate needs face emergency responders at a mass-casualty incident:

- search and rescue
- incident stabilization.

If sufficient resources are available, both may be accomplished at the same time. Forces may be able to extinguish a fire or control a fuel spill, while conducting search and rescue (SAR) operations. Search and rescue is a multifaceted task at a mass-casualty incident. Alternatives not only involve whether to search and rescue but where to start.

After victims are rescued, many will need medical care. This is a high priority in all situations, but when large numbers are involved, a system should include the "Three T's"—triage, treatment, and transportation.

Doctors and nurses can be a valuable asset or a serious liability at the scene of a medical emergency. When they are properly identified and operating within the command system, they are extremely useful. When they do not understand their part in the plan and freelance, or try to take command, they become a serious liability.

Regardless of the planning effort, some emergency workers at a mass casualty incident will not be accustomed to working under a structured command system (doctors, nurses, private ambulance personnel), making it imperative that they be supervised by someone who understands the system.(10) We will first discuss the Three T's—triage, treatment, and transportation, and then discuss organizational considerations and incident command procedures that can facilitate these actions.

Triage

Triage is a French word meaning "to sort." Sorting victims as to need is an essential element of a multiple victim scene. Most fire fighters and EMS personnel are adept at performing this rapid evaluation on a small scale. At a multiple victim scene, it is not unusual to delay treatment of minor injuries, while turning all attention to a critically injured patient. At a scene with more victims than can be treated, triage sorts victims into the following categories:

1. those requiring immediate care (life-threatening injuries, but survival is probable with medical assistance)
2. severely injured victims who can withstand a short delay in medical care without immediate risk to life
3. victims with minor injuries (walking wounded)

4. dead or so severely injured that death is highly probable regardless of medical attention.

Triage tags identify victims as to the level of treatment required. See Figure 14-1 for a description of triage tags.

Three important points need to be made regarding triage teams.

1. They must be supervised so all victims are assessed and separated.
2. The most highly qualified medical personnel at the scene should perform triage.
3. Triage teams should not treat or transport victims until triage is complete. At the scene of a large disaster it may be best to have triage personnel reevaluate patients, rather than involving them in actual treatment.

The triage activity is the first medical priority in a medical disaster. The first arriving rescue unit should not leave the scene until relieved of triage responsibilities. Emergency medical teams are accustomed to treating one or two individuals, then transporting. Categorizing without treatment runs contrary to their normal role and must be overcome by rigid enforcement of the triage concept.

Another problem arises as EMS personnel are directed to pass by victims who are so severely injured that they are categorized with the known fatalities. When manpower permits, it is best to physically separate the dead and dying from other patients. The walking wounded should be directed to another location for future treatment.(11)

Treatment

Once the triage teams have sorted victims, treatment teams will follow. Treatment teams first treat those needing immediate care. Later, delayed treatment and minor injury treatment teams will be formed. The most highly qualified medical personnel are on the triage teams. The next highest level is the immediate treatment teams. The order of arrival also dictates who is on what team. Later arriving personnel may be assigned to delayed treatment teams, although they may be more qualified than others assigned to higher level activities.(12)

Transportation

Providing access to victims is an important role of the transportation officer. A "drive through" arrangement for ambulances will keep traffic lanes open. Helicopter landing zones need to be in safe areas, far enough away from treatment and triage areas so as to not interfere with those activities. Staging and categorizing transportation capabilities is essential in providing proper transportation.

It may be possible to transport several "delayed care" victims in the same vehicle. "Immediate care" victims often require advanced life support

Figure 14-1. Triage Tags.

Triage tags can take different forms depending on the jurisdiction. Below are both sides of a common tag used for quick triage (COURTESY OF JOURNAL OF CIVIL DEFENSE). *Right, a fairly common color-coding format used in many parts of the U.S. and Canada* (COURTESY OF THE PHILADELPHIA FIRE DEPARTMENT).

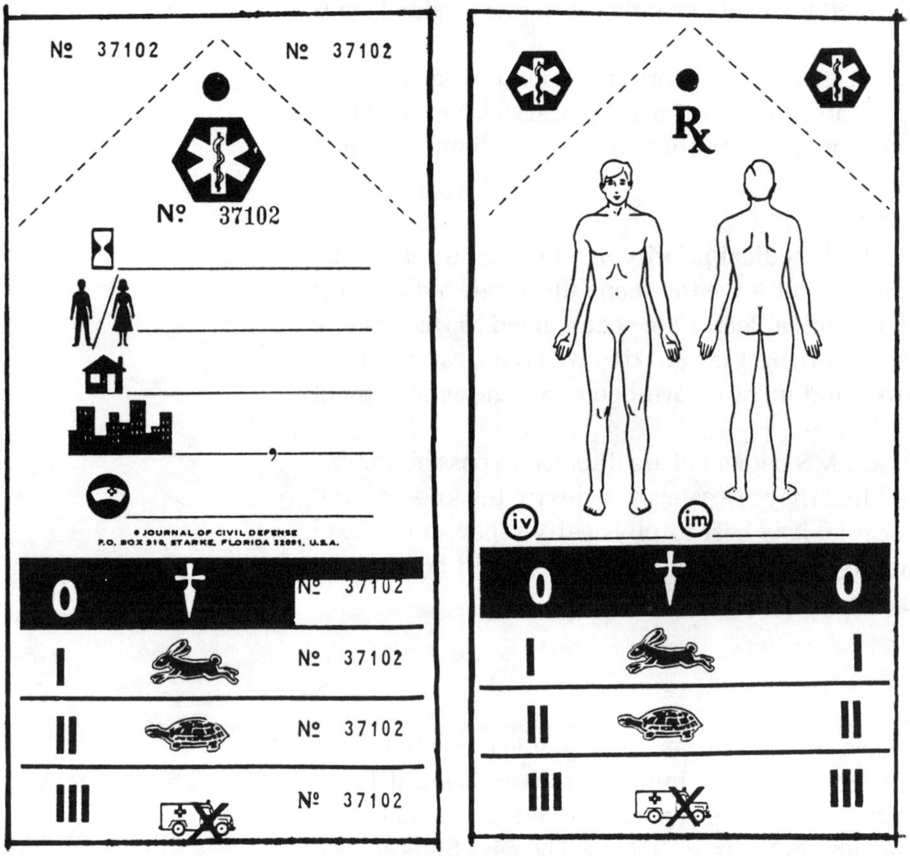

TRIAGE TAG CODING

Each tag contains name and age of the patient, injuries diagnosed, emergency treatment rendered, and a receipt tag with patient's name, ambulance description, and patient destination.

RED — Immediate Care, First Priority

This category must meet the following three criteria:

(a) Injuries are such that they have or will probably cause life-threatening shock or hypoxia;

(b) Casualty can be stabilized with intervention skills and equipment on hand and does not need constant care;

(c) Casualty has a high probability of survival if given level of care available and rapidly transported.

YELLOW — Secondary, Second Priority

(a) The casualty is not in life-threatening shock or hypoxia, but due to type of injuries, it may result;

(b) The casualty has a high probability of survival;

(c) The casualty can wait as long as 45 minutes before transporting.

YELLOW "PRIME"

Casualties with catastrophic injuries with a small chance of survival, e.g. flail chest.

GREEN — Delayed, Third Priority

(a) Casualties with injuries that do not have an immediate systemic implication;

(b) Casualties whose injuries, once stabilized, do not have an immediate systemic implication.

GRAY — Deceased, Fourth Priority

(a) Includes those in respiratory and/or circulatory arrest;

(b) Does not include those in a dying state.

(ALS) personnel on-board the transport vehicle, and only one can be transported.

Minor care victims usually are staged for later treatment near the site. If minor care victims need transportation, it can be provided using nonambulance vehicles. The only way to manage transportation vehicles is to utilize a staging area or areas. All incoming vehicles must be directed to a staging area.

Communications with hospital facilities to find out how many patients can be handled is necessary. Hospitals determine limits based on personnel and space. Limits established during normal conditions should not prevail during an emergency, but there is a limit. Patients should be directed to hospitals and trauma care centers based on facility capacities. Someone needs to document how many patients have been sent, or are enroute to all hospitals. The hospitals must be willing to exceed their norms, but also recognize when they have reached their expanded limits. The mass casualty incident may require that only "immediate care" patients be transported.

Figure 14-2. Mass Casualty Patient Flow Chart.

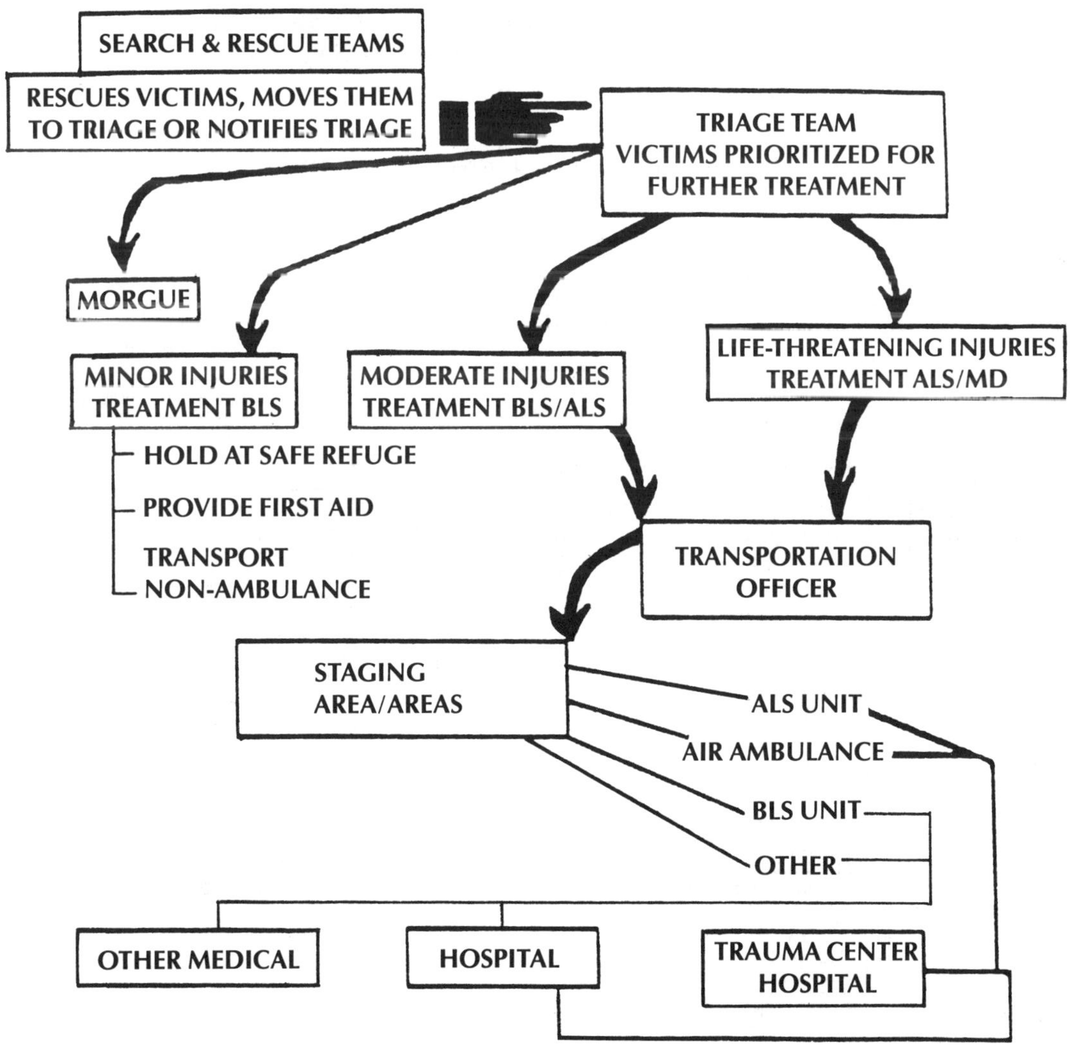

It may be necessary to set up field hospitals for "delayed care" patients. Even with transportation available, it is generally not a good idea to flood hospitals with the "walking wounded."(13)

Figure 14-2 shows a patient flow chart for a mass casualty incident.

ORGANIZING AND MANAGING MASS CASUALTIES

Operational procedures in mass casualty incidents differ considerably from routine EMS operations and require a special management structure that interfaces well with the incident command system (ICS). This section of the chapter discusses this management structure both in the field (on site) and off site settings.

Field Medical Operation

A number of organizational structures for managing multiple casualty situations have arisen around the country. One of the most comprehensive is the Medical Group structure, developed by the Los Angeles County Fire Chiefs Association, as a functional group in the ICS system. Figure 14-3 depicts this group.(14)

Figure 14-3. Medical Group Organizational Chart.

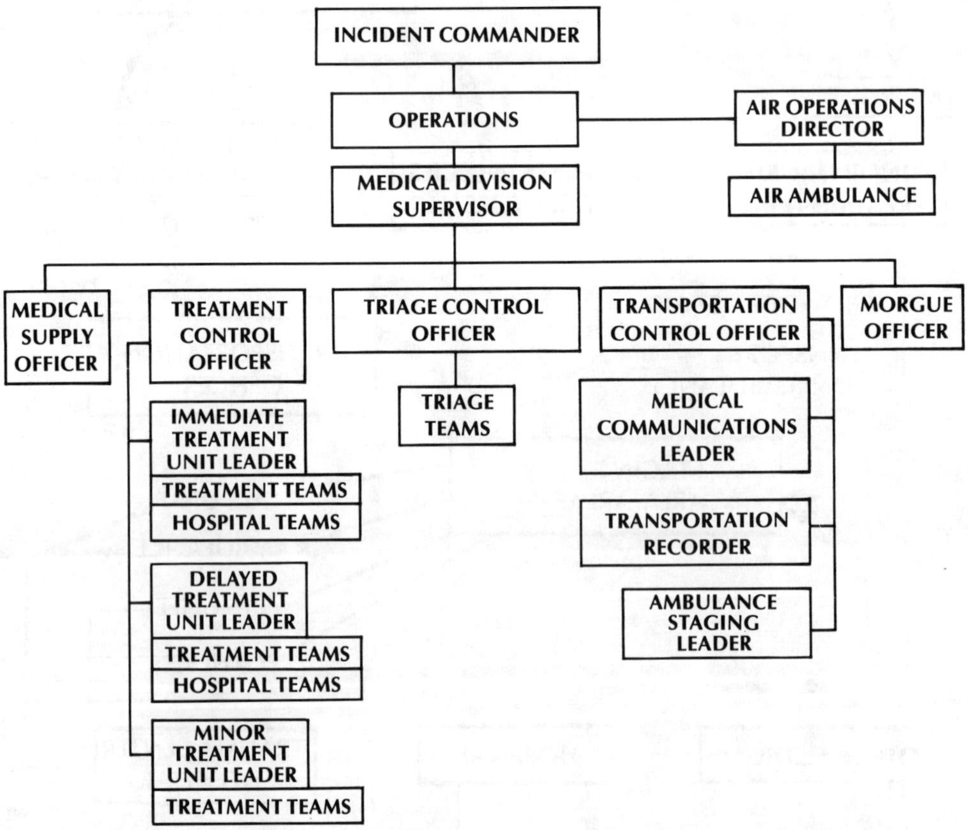

MEDICAL GROUP FUNCTION AND STRUCTURE

Field medical operations generally focus on the following sequence of activities:

1. identifying and triaging (sorting) casualties to establish priorities for treatment and transport
2. treatment of casualties in order of priority
3. loading of patients into paramedic or basic life support ambulances
4. distribution of patients to appropriate hospitals.

The Medical Group consists of a medical group supervisor and several groups that carry out these tasks.

MEDICAL GROUP SUPERVISOR

The medical group supervisor is responsible for deploying the medical group organization and for requesting medical resources from the incident commander. Group leaders such as triage and treatment report to the medical group supervisor who in turn reports to the operations chief. The medical group supervisor should usually be a field EMS supervisory person rather than a physician or EMS management.

TRIAGE CONTROL OFFICER

The triage control officer is responsible for identifying and deploying triage teams usually consisting of two EMT's (Emergency Medical Technician) or paramedics. Triage teams rapidly assess each victim and attach a tag or marker indicating a priority for treatment or transportation to the hospital based on the guidelines discussed previously in this chapter. The triage control officer coordinates with the treatment control officer in obtaining field medical care for high priority patients or moving them to one or more treatment areas. The triage control officer may be a physician with field medical experience or a paramedic or EMT.

TREATMENT CONTROL OFFICER

The treatment control officer is responsible for assembling treatment teams and establishing central treatment areas. In the treatment process victims receive a secondary triage to identify changes in their condition, appropriate field treatment, and preparation for loading into ambulances. In special cases, such as those involving extensive extrication, physicians or hospital medical teams may be utilized. The treatment control officer coordinates the activities of these teams.

TRANSPORT CONTROL OFFICER

The transport control officer has the following responsibilities:

Public officials can charter or commandeer public transportation vehicles such as buses to assist with mass evacuation. (Photo courtesy of the National Fire Academy.)

- requesting ambulances and other transport vehicles from staging
- establishing a loading zone at which victims are loaded onto stretchers and prepared for transport
- organizing vehicles adjacent to the loading zone
- coordinating with the command post or EMS coordination center to determine destination hospitals for each victim and keeping records of victim distribution.

MEDICAL SUPPLY OFFICER

The medical supply officer is responsible for requesting needed medical supplies from the logistics section or other available source and coordinating distribution to the various groups.

MEDICAL STAGING OFFICER

Medical staging may be part of a general fire or incident staging area or it may be a separate area. A general staging officer would report to the operations section chief while a medical staging officer would report to the medical group supervisor. The staging officer is responsible for keeping a designated number of ALS (advanced life support) and BLS (basic life support) ambulances and other medical equipment and transport vehicles on standby for deployment to the incident scene. Staging areas are generally several blocks from the incident site and in a location that permits uncongested ingress and egress.

MEDICAL COMMUNICATIONS

Medical operations may use one or both of two national radio communications bands.

The Hospital Emergency Administrative Radio (HEAR) network was established in the 1960s primarily to permit administrative coordination among medical facilities during large disasters. The system consists of two VHF simplex radio frequencies. The first, 155.34 MHz, was designated as a local frequency for communications among local hospitals, one of which was designated a regional coordinating facility. This frequency has also come to be used for communications between ambulances and hospitals. The second frequency, 155.28 MHz, was designated as a regional channel for communications among regional coordination hospitals.

The advent of Advanced Life Support programs vastly increased communications needs between paramedics in the field and hospital-based physicians. In the 1970s a set of ten frequency pairs in the 450 MHz UHF band were designated by the Federal Communications Commission (FCC) for use throughout the United States. Eight of these channels are for paramedic-physician communications and two are for ambulance dispatching. Certain local plans restrict the use of repeaters (devices that permit mobile-to-mobile communications) on some channels. However, the FCC does permit the use of repeaters on all ten channels. Agency planners should explore the possibility of obtaining local authorization to use repeaters on additional channels for major medical operations.

The widespread use of this ten-channel UHF system permits a variety of field and off-site communications schemes to be developed that can be used by the majority of EMS agencies. A communications officer should be established to coordinate on-site communications. This officer can report either to the medical group officer or the transport control officer.

Extrication Operations

Major extrication activities may be necessary in such incidents as transportation accidents, building collapses, earthquakes, and tornadoes. These activities are usually organized as a separate ICS group, but require very close coordination with the medical group. Physician or hospital-based medical teams may be very helpful in these situations and should be involved in key extrication decisions.

Hospital Coordination

The goal of mass casualty plans is to save the most lives by bringing appropriate field and definitive medical care to seriously injured victims in the shortest possible time. This does not necessarily occur by simply removing victims from the site as soon as possible. If activities such as distributing patients from the scene, transferring patients between facilities,

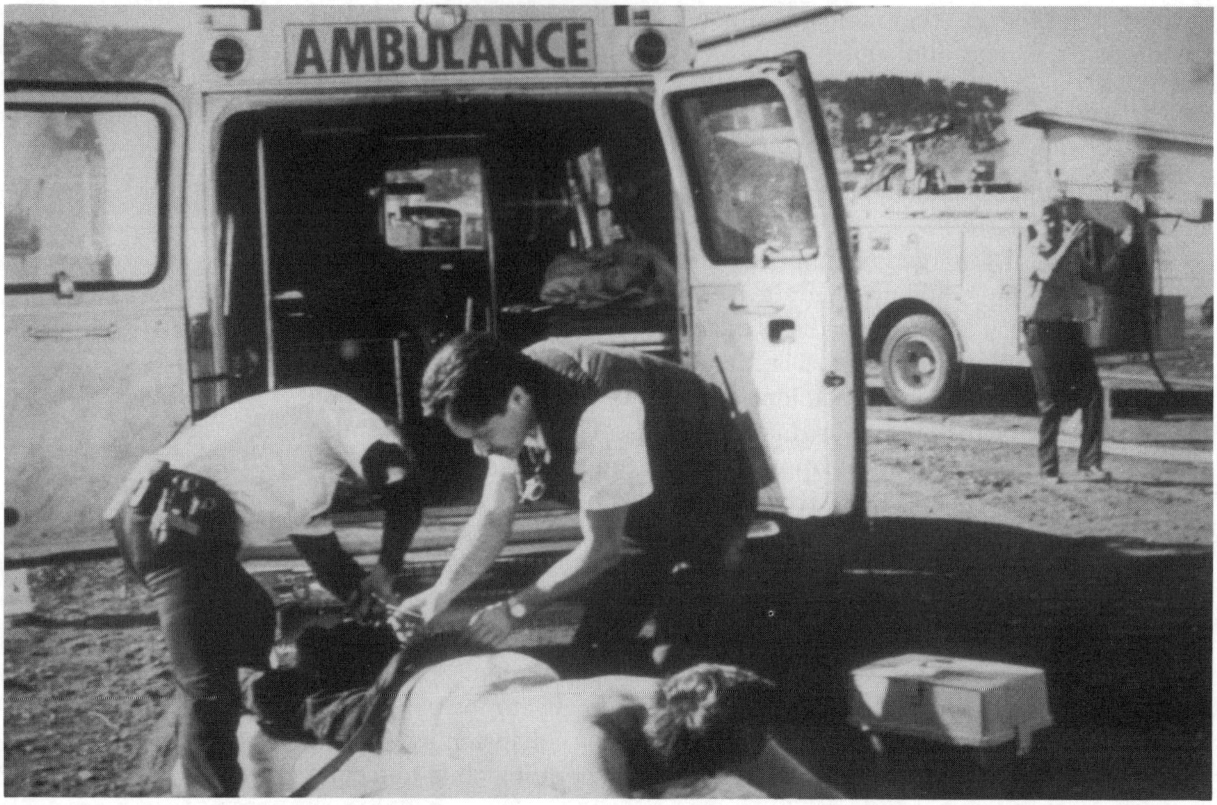

Pre-incident planning should arrange for abundant ambulances prior to a disaster. (Photo courtesy of the American Medical Association.)

and maximizing use of medical manpower are not coordinated, the disaster may simply be shifted from the site to hospital corridors.

Medical disaster plans should include identification of one individual, possibly a hospital administrator or physician, to coordinate with hospitals and other medical facilities concerning the following.

- Determination of the ability of surrounding hospitals to accept general trauma patients, chemical or radiological injuries, burns, spinal cord injuries, hand injuries, partial amputations, and other special cases.
- Coordination of hospital patient transfers.
- Coordination of medical manpower including physicians, nurses, technicians, paramedics and EMTs, and other allied health professionals.
- Coordination of medical supplies for use at hospitals.
- Establishment of auxiliary medical facilities in the event of very large numbers of casualties or damage to one or more primary medical facilities by fire, flooding, collapse, etc.

Appendix J of this book contains a Multiple Casualty Incident Plan submitted by Chief Gary Smith of the Watsonville, California, fire department.

Medical Assistance Teams

In some jurisdictions, the medical community is organized in advance to provide invaluable life-saving skills at a mass-casualty incident. One example is Hamilton County in southern Ohio. Here the medical assistance team (MATS Team) is comprised of physicians, nurses, and paramedics who have identification cards and will respond upon request.

The medical assistance team (MATS) is sponsored by the Cincinnati Academy of Medicine and the Hamilton County Disaster Council. All team members are linked by portable radio, which is integrated into the established disaster communication system for all the hospital emergency departments in the area. MATS in cooperation with the Greater Cincinnati Hospital Council has established seven sets of "Disaster Boxes." All of the disaster sets can be brought to the scene by emergency vehicle in event of a major mass casualty incident. This provides medical equipment for definitive critical care for 35 to 50 victims.

Physicians on the MATS team are identified by green helmets with MD in reflective letters. Registered nurses wear red jump suits, with red cross insignia, and a white helmet marked "nurse." Physicians and nurses have photo-identification cards. All MATS Team members wear patches and/or arm bands from the Cincinnati Academy of Medicine.

On notification of a possible disaster, the medical commanders of MATS are notified immediately through dispatch. The MATS commander then decides the extent of MATS response by evaluating the number of potential casualties, the type, size, and location of the disaster. A back-up notification system for MATS is conducted through the Hamilton County disaster net radios located in the emergency departments of hospitals. Hamilton County dispatch can also notify the MATS team.(15)

A publication entitled *Disaster Medical Assistance Team Organization Guide* is available from the Department of Health and Human Services.(16)

Evacuation by Helicopter

Occasionally evacuation by helicopter provides the difference between life and death to the most critically injured victims in a disaster. Even when the distance to a hospital is relatively short, a helicopter often provides a much smoother ride than one over roadways, thus reducing the amount of additional induced trauma and increasing the odds for survival. When a helicopter is to be employed, however, certain precautions are in order. The following is a list provided by the Training Bureau from the Cincinnati Fire Division.(17)

1. Landing site must be clear of overhead obstructions. Notify the pilot of the location and height of any overhead obstructions.
2. Clear the site of debris and loose objects, e.g., clothing, sheets, or blankets. You may expect a wind of 60 to 70 mph.
3. Wear eye protection.

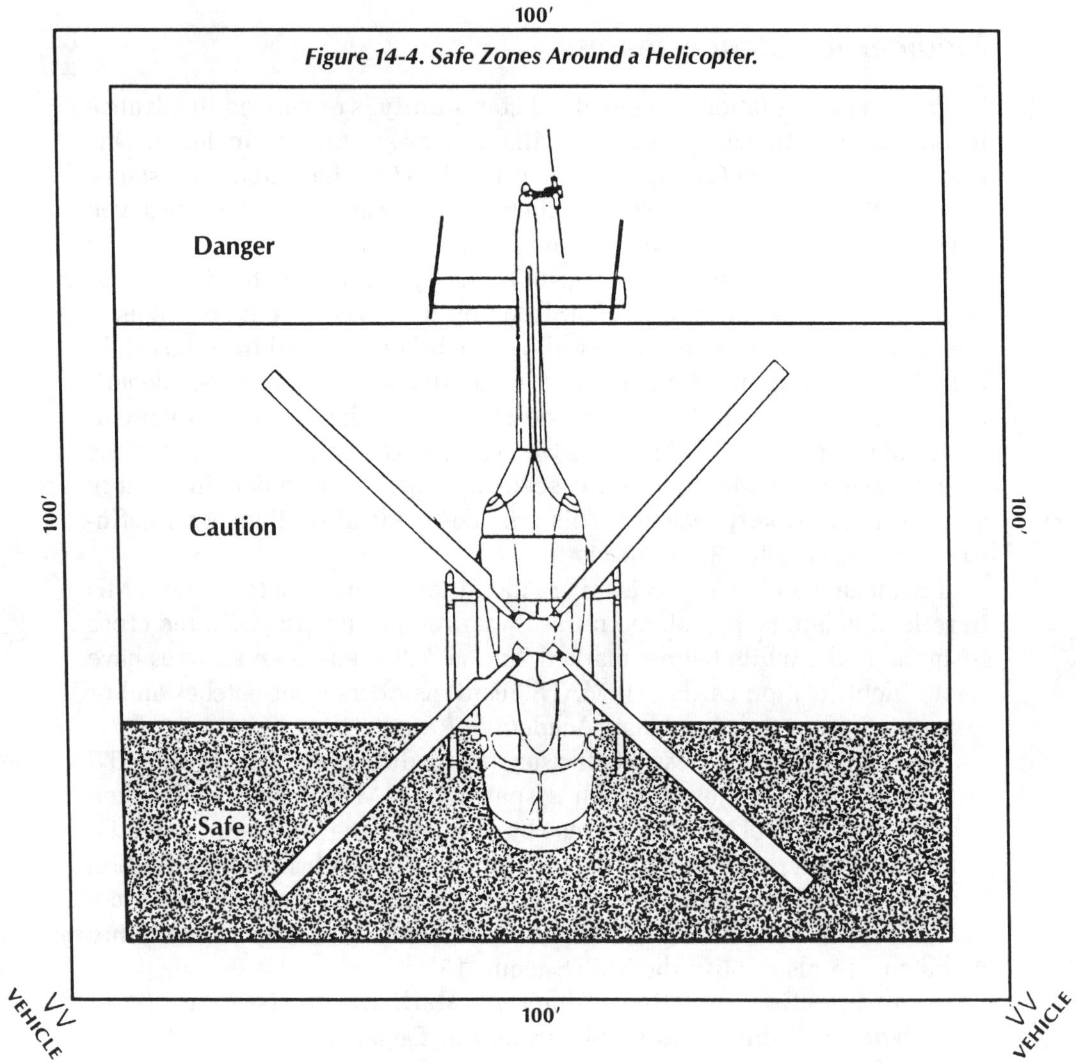

Figure 14-4. Safe Zones Around a Helicopter.

Figure 14-5. Night Landing Zone Markings.

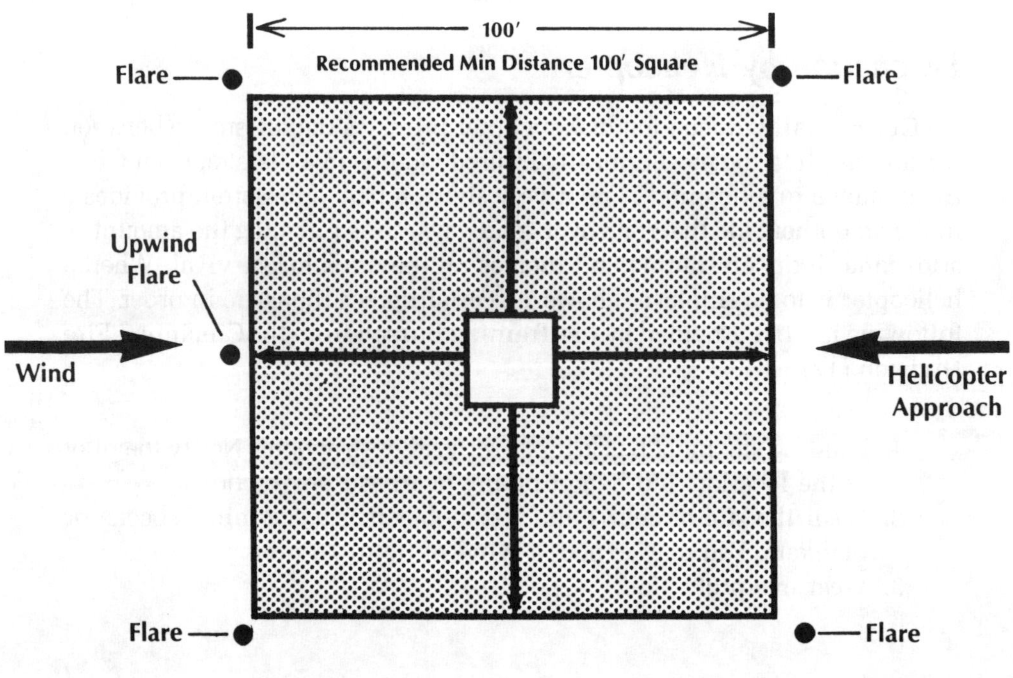

4. Never move toward the helicopter until signaled by the flight crew. Remember, if you can see the pilot's eyes, he can see you.
5. Only flight crew members may open and close aircraft doors.
6. Always approach and leave the helicopter from the front. (See Figure 14-4 for safe zones.)
7. When flammable or poisonous vapors are involved select a landing site at least one mile upwind.
8. At night:
   - ○ do not point lights at the helicopter while it is airborne.
   - ○ mark the corners of the landing site and the center of the windward edge with flares. (See Figure 14-5.)
9. Do not smoke near the helicopter.
10. Do not run near the helicopter.
11. Do not carry IV's or other objects above your head. Carry long objects parallel to the ground.
12. Be especially cautious in the tail rotor area.

Barrier tape should not be used to mark a landing zone because the heavy winds from the helicopter will cause it to be blown about severely in unpredictable ways perhaps adding to the calamity. See Figure 14-6 on how to be alert around helicopters.

Drills

The potential for mass casualty is present in every community, and many have conducted mock disaster drills to improve existing procedures so that the jurisdiction will be better prepared to deal with "the real thing." An excellent example was a mass casualty drill reported in the May 1989 issue of *Fire Engineering*. The drill was conducted by the Philadelphia Fire Department according to their "Philadelphia Regional Multiple/Mass Casualty Operation Plan" (PREMDOP RED), which utilizes full resources of the fire department, Delaware County, surrounding volunteer fire departments, helicopter support, and the Delaware County Hospital. The drill scenario involved a simulated Boeing 727 with 88 passengers and crew aboard. In the drill, the aircraft overshot the runaway and crashed at the far end of the field. A bus representing the main section of the plane with victims being volunteers from the U.S. Navy on the bus and scattered throughout a surrounding area simulated a mass casualty incident with a high degree of realism.(18)

In March 1991, a similar drill was conducted at Logan International Airport. Here, at the nation's tenth busiest airport, the Boston Fire Department, the Massachusetts Port Authority, and other emergency responders teamed up in a realistic drill that would test their abilities to handle mass casualties. Since the airport is located on a peninsula, there is a good chance that an aviation disaster could involve water. Hence this drill among other things tested the water rescue procedures at Logan. In this case, a barge simulated an aircraft section in the water, while off duty Logan airport fire fighters served as victims.(19)

Figure 14-6. Be Alert Around the Helicopter.

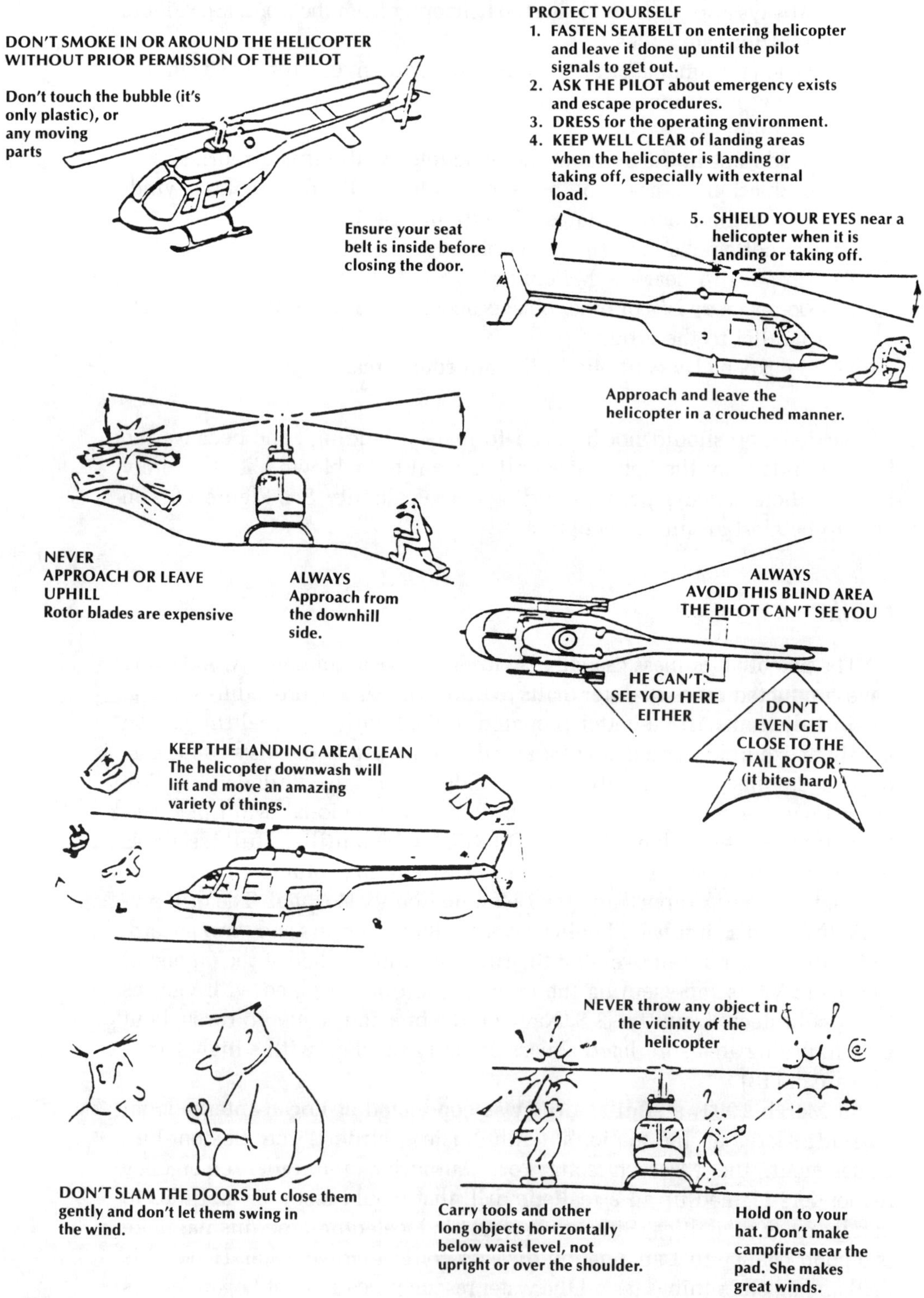

NATIONAL DISASTER MEDICAL SYSTEM

Most local disaster plans and resources are capable of managing casualties ranging in number from up to 100 in smaller communities to several hundred in larger cities. Even the largest municipalities would be severely stressed attempting to deal with 2,000 to 3,000 casualties. Yet a natural disaster, such as a predicted earthquake in California or a major toxic chemical release in a large metropolitan area, could generate tens of hundreds of thousands of casualties. Recognizing that such catastrophes are plausible, President Ronald Reagan established an Emergency Mobilization Preparedness Board in 1981 to develop a national medical response system.

The board is composed of representatives from four principal participating federal agencies:

- Federal Emergency Management Agency
- Department of Defense
- Veterans Administration
- Public Health Service of the Department of Health and Human Services.

This board developed the National Disaster Medical System (NDMS) and has identified 71 NDMS operational areas around the country. (See Figure 14-7.)

Elements of the National Disaster Medical System

MDMS consists of three principal elements:

- Disaster Medical Assistance Teams recruited from around the country that can be sent to an emergency site to assist local authorities in triaging, sorting, and clearing casualties.
- A large-scale airlift system, coordinated by the military and primarily utilizing Department of Defense aircraft, to transfer victims to any of the 71 NDMS operational areas around the country.
- A network of civilian hospitals, originally established under a Civilian-Military Contingency Hospital System, to receive these casualties.

Local Operational Areas

Each of the NDMS operational areas is in various stages of developing local operational plans, forming special teams, identifying resources, and identifying participating hospitals. Area planning is usually coordinated through a federal representative of the Veterans Administration or the Department of Defense working closely with local and state emergency management, EMS, and related emergency agencies.

Disaster Medical Assistance Teams

The Disaster Medical Assistance Teams (DMATs) are generally sponsored by local hospitals and consist of 23 people including physicians, nurses, technicians, and clerical staff.

A Casualty Clearing Unit is composed of three DMATs plus additional administrative personnel. These units are designed to function self-sufficiently to triage, definitively treat and prepare patients for transportation to distant medical facilities.

An important feature of NDMS is that DMATs and other local resources may be utilized in a local major emergency even when the NDMS is not activated federally.

Casualty Transportation and Distribution

Once casualties are cleared from the disaster locale, they are transferred to military aircraft and flown to various operational areas having available hospital beds. The U.S. Air Force is responsible for tracking all patients transported by NDMS. Upon arrival at local airports a DMAT will meet arriving patients and process them for ground transport to area hospitals.

In many cases operational areas may need to activate local medical disas-

Figure 14-7. Systems Areas of the National Disaster Medical System.

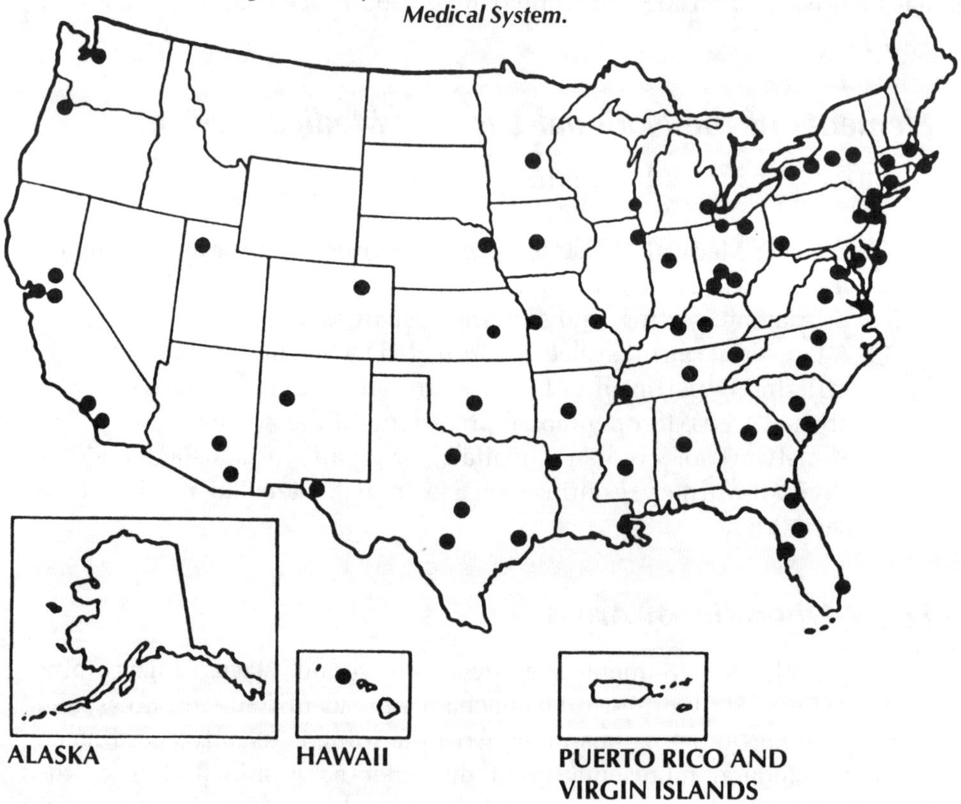

"National Disaster Medical System Implementation Task Force."

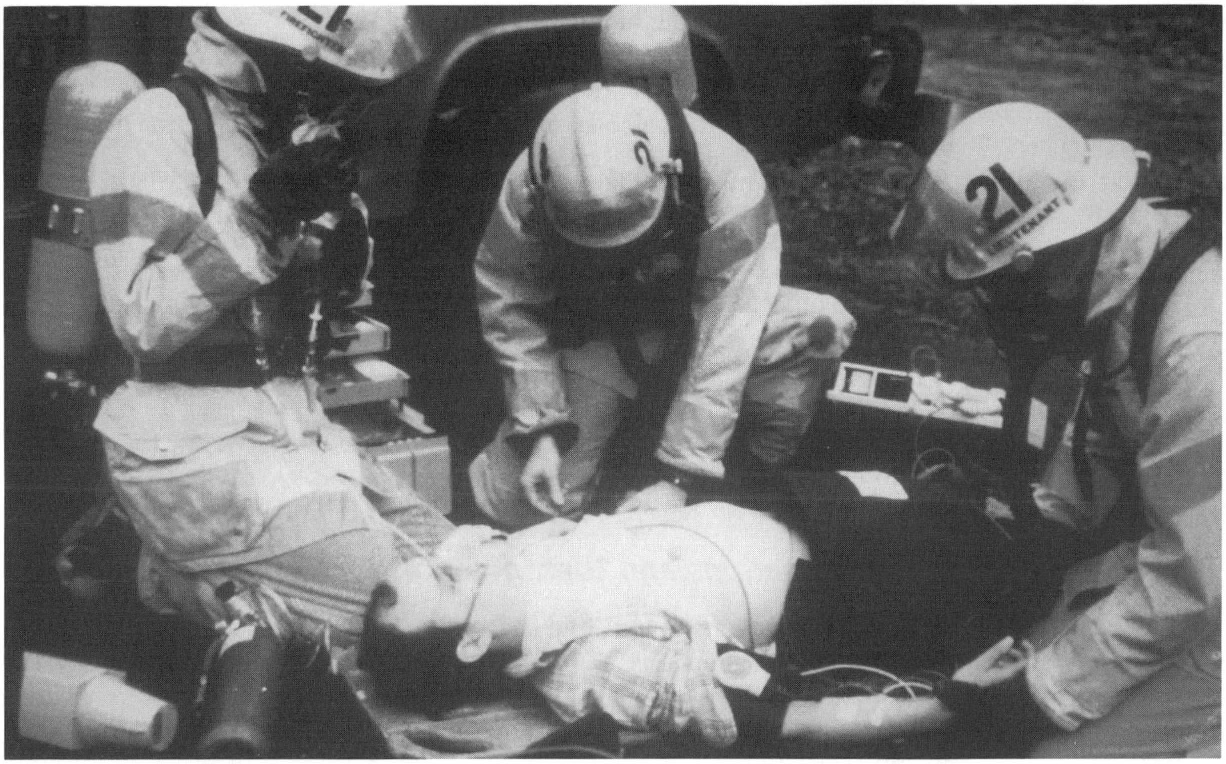

Cross-trained firefighter/paramedics sustain life in a victim following a transportation accident. (Photo courtesy of the National Fire Academy.)

ter plans to coordinate ground transportation from the airport to area hospitals, as well as hospital transfers and related activities.

Federal NDMS Activation

Federal NDMS activation may occur in one of three ways: upon declaration of a "major disaster" by the President at the request of a governor, by the Secretary of Defense in response to a military conflict, or by the Secretary of Health and Human Services in response to a peacetime disaster.

MULTIPLE FATALITY INCIDENTS

Incidents that result in multiple fatalities involve unique features that affect fire and medical operations.

Legal Authority in Multiple Fatality Incidents

In most states the coroner or medical examiner is considered a high-ranking law enforcement officer. The coroner or medical examiner may have the authority to do the following:

- determine the biological cause of death
- investigate deaths to determine the possibility of criminal acts
- identify fatal victims
- preserve the personal effects of victims at the scene of the death until investigations are complete.

The organization charge for the Medical Group under ICS includes a morgue officer (see Figure 14-3). This individual may come from the coroner's office, EMS, the fire department, or a police agency and is responsible for coordinating initial temporary storage of remains until longer-term arrangements can be made by the coroner. In most states the coroner or medical examiner has clearly specified legal authority and generally will not fall under the direct authority of the incident commander.

Investigations and Identification of Victims

Investigations of multiple fatality incidents can be extraordinarily difficult because remains are often badly burned or mutilated. Local laws or practices concerning movement or removal of remains by emergency personnel vary greatly, but the following general guidelines should apply in most multiple fatality incidents.

- The bodies of apparent fatalities should not be moved without approval of the coroner or medical examiner except to assess whether the victim is alive or viable for resuscitation, or to gain access to live victims.
- The coroner or medical examiner should be consulted, if possible, before moving remains that may be threatened by fire, rising water, or other conditions.
- The immediate personal effects found near victims may be valuable in identifying them. Items such as purses and briefcases should be packaged, labelled, and moved with the remains.
- Jewelry and clothing should not be removed.
- Surrounding items, such as furniture, should be moved as little as possible. If the remains must be moved, some sort of mark using chalk or masking tape should be made to indicate the location and position of the remains.
- If possible, EMS or fire personnel should keep careful records, including drawings or photographs, describing where victims were found and what articles were nearby.

In severe cases where numerous victims cannot be identified a special FBI victim identification team may be requested to assist.

Notification of next-of-kin is usually handled by law enforcement agencies although in some areas the Red Cross has that responsibility. In cases of multiple unidentified fatalities notifications may be delayed until all victims are positively identified. This will cause a great deal of emotional stress for relatives present at the scene. Fire or police chaplains may be helpful in such situations.

Managing the Remains

In large multiple fatality incidents local resources for storing and processing remains may be inadequate. The initial problem may be insufficient body bags for removing remains from the site. Local funeral directors or state or federal military departments may be able to assist.

Local hospital morgues may have insufficient storage space. Temporary facilities such as refrigerated semi-trailers or local ice skating rinks may be used.

Major incidents may involve multiple fatalities and multiple casualties or they may result almost entirely in fatalities. In either case, medical response personnel will be heavily involved. In dealing with remains, they must be sensitive to the emotional needs of relatives and bystanders by protecting bodies from view (especially by unauthorized photographers) and by handling them with respect. They must also be aware of the importance of preserving all items found both on the remains and in the immediate area, so that positive victim identification can be made as quickly as possible.

Coffins are stacked and staged, ready for use after the major earthquake in Mexico City. (Photo courtesy of Jim Lanagan.)

MASS EVACUATION

In order to prevent mass casualties, the fire service is often involved in mass evacuations of civilians from dangerous or potentially dangerous zones. The remainder of this chapter will deal with "potential victims"—the evacuees rather than the mass casualties. While police and occasionally military personnel are involved in a primary manner with evacuations, it is the fire department that usually provides expertise and direction.

Evacuation is an important protective action for hurricanes, floods, tsunamis, volcanoes, hazardous material accidents, nuclear power plant accidents, and crisis situations such as tornadoes or earthquakes. Another form of protective action is sheltering, and most forms of disasters will involve some combination of sheltering and evacuation.

Fires in occupied buildings frequently require removal of the occupants. Fire fighters have always been involved in such evacuation as they have been in removal of people from whole blocks.

For example, on January 12, 1990, in Menomonie, Wisconsin, a suspected arson fire which destroyed a two-story 27-unit apartment house required the evacuation of 44 persons.(20) In another example, on March 10, 1991, a fire that began in a second floor wooden locker forced 120 residents to evacuate a 16-story highrise YMCA in Ottawa, Ontario.(21) There is much less experience with large-scale, long-term dispersal of people due to natural or man-made disasters, such as that required in the Oakland, California, conflagration of 1991. But in recent years as technological mishaps have dramatically increased so has fire department involvement in mass evacuation.

Natural Events

Evacuation is an action taken as a protective measure necessitated by a dangerous situation. It sometimes stems from large, but fairly infrequent natural events such as hurricanes and floods. The impact of potential disasters is mitigated by removing the population from the dangerous area and the evacuation may be very large.

When the cyclone struck Bangladesh in May, 1991, killing 140,000, the world was struck with horror. Few people realized, however, that several million people had been evacuated to higher ground inland thus preventing an even more horrific casualty count.(22)

In 1961, Hurricane *Carla* caused the relocation of an estimated 750,000 people in the U.S., and in 1985 *Elena*, necessitation the relocation of 300,000 people. Many other evacuations affecting tens of thousands of people occur numerous times each year. Hurricanes and floods usually occur at certain times of the year, usually allow some warning time, and are mostly confined to certain areas of the country. (The Gulf Coast and eastern seaboard are hit with just about every hurricane.) Earthquakes and tornadoes usually give so little warning that evacuation is not a prime consideration. In these situations, subsequent sheltering is the dominant form of protection.

Technological Events

A second general type of occurrence that often results in evacuation is technological or human caused, principally hazardous material leaks, spills, or fires that more directly and immediately involve the fire department. These can be divided into two types:

- on-site as at chemical plants, factories, storage facilities
- transportation accidents.

On December 18, 1990, the discovery of 100 sticks of dynamite in a garage forced the evacuation of a Southgate, California, elementary school and hundreds of homes.(23)

Thousands of new chemicals are developed each year, and citizens and officials are concerned about accidents happening in their communities. Ninety-three percent of more than 3,100 localities completing a Federal Emergency Management Agency (FEMA) questionnaire in 1985 identified one or more hazardous materials risks as a significant threat to the community.

Table 14-1. Chemical Accident Evacuation by Cause and Year

Cause of Evacuation	*1980*	*1981*	*1982*	*1983*	*1984*
Train derailment	14	8	13	12	8
Train car spill/fire	3	6	5	4	5
Truck accident	9	9	6	6	5
Truck spill/fire	1	11	4	9	7
Chemical plant release	5	10	15	8	5
Industrial plant release	3	10	18	23	24
Pipeline	2	1	1	0	0
Ship incident	2	1	0	0	0
Waste site accident	0	1	2	3	1
Other	4	5	4	0	1
Actual Total (Chemical)	43	62	68	65	57
Est. Total (Overall)	86	90	100	110	125

	1985	*1986*	*1987*	*1988*	*1989*	*1990*	*1991*	*1992*	*1993*	*1994*	*1995*
Year Est. Totals	135	145	160	175	195	215	235	260	285	310	350

According to one study conducted in 1987 there were nearly 300 identified evacuations due to chemical releases alone, in the United States during a five-year period. They are listed in Table 14-1 by cause and year. On average, about 1,000 people were moved per event. The largest was estimated to include 30,000 people. In 1985 an estimated 125 evacuations took place because of the release of some form of hazardous materials.(24) Hence, total evacuations are approximately double those for chemical release. The figures for years 1985 to 1995 shown in Table 14-1 show increases estimated by the authors, reflecting both increases in the number of dangerous incidents and an increase in safety awareness. Accurate figures are difficult to obtain due to lack of centralized reporting of evacuations and lack of a concrete number beyond which the moving a few persons to safety is escalated to a *de facto* "evacuation."

Because of the risk of hazardous materials incidents and because local governments will be completely on their own in the first stages of almost any incident, communities need to maintain continuing preparedness. Specific evidence of being prepared is an emergency plan. A report prepared by FEMA on 242 evacuations of all types and sizes shows that communities rely upon evacuations as a primary means of protecting their citizens. However, the majority of evacuations cited were implemented on an ad hoc basis without a viable emergency operations plan (EOP).(25) Each jurisdiction should not only have an all-hazard EOP but also a complete and sound evacuation annex.

Joseph Weber, retired Deputy Chief for the City of Los Angeles and one of the most respected instructors at the National Fire Academy, has produced an excellent series of guidelines regarding mass evacuations. Joe has graciously agreed to allow the incorporation of these guidelines, with some modification by the authors, into this chapter on mass evacuations. Much of the information that follows originated with Joe and his research in producing the National Fire Academy's open learning course "Disaster and Fire Defense Planning."(26)

It is strongly recommended that a functional annex to a disaster plan be drafted. In response to Title III, the National Response Team (NRT) produced the *Hazardous Materials Emergency Planning Guide*, a product of the cooperative efforts of the 14 federal agencies that constitute the NRT. Below is the reproduced section of the NRT guide that corresponds to a functional annex on evacuation; it is not a model but a guide to the topics that should be covered.(27)

EVACUATION PROCEDURES

- ○ Title of person and alternate(s) who can order/recommend an evacuation
- ○ Vulnerable zones where evacuation could be necessary and a method for notifying persons in these places
- ○ Provisions for a precautionary evacuation
- ○ Methods for controlling traffic flow and providing alternate traffic routes
- ○ Shelter locations and other provisions for evacuations (e.g., special assistance for hospitals)
- ○ Agreements with nearby jurisdictions to receive evacuees
- ○ Agreements with hospitals outside the local jurisdictions
- ○ Protective shelter for relocated populations
- ○ Reception and care of evacuees
- ○ Re-entry procedures

Evacuation is the most sweeping response to an accidental release. The

Residents are being evacuated from a flooded area using rowboats. (Photo courtesy of the American Red Cross.)

plan should clearly identify under what circumstances evacuation would be appropriate and necessary. DOT's *Emergency Response Guidebook* provides suggested distances for evacuating unprotected people from the scene of an incident during the initial phase. It is important to distinguish between general evacuation of the entire area and selective evacuation of a part of the risk zone. In either case, the plan should identify how people will be moved (e.g., city buses, police cars, private vehicles). Provision must be made for quickly moving traffic out of the risk zone and also for preventing outside traffic from entering the risk zone. If schools are located in the risk zone, the plan must identify the location to which students will be moved in an evacuation and how parents will be notified of this location. Special attention must also be paid to evacuating hospitals, nursing homes, and homes for the physically or mentally disabled.

Maps (drawn to the same scale) with evacuation routes and alternatives clearly identified should be prepared for each risk zone in the area. Maps should indicate precise routes to another location where special populations (e.g., schools, hospitals, nursing homes, homes for the physically or mentally disabled) can be taken during an emergency evacuation. They should also show the methods of transportation during the evacuation.

Consideration of when and how evacuees will return to their homes should be part of this section. It should also include a description of how other agencies will coordinate with the medical community.

Copies of evacuation procedures should be provided to all appropriate agencies and organizations (e.g., Salvation Army, churches, schools, hospitals) and could periodically be published in the local newspapers.

KEY CONSIDERATIONS IN LARGE-SCALE EVACUATIONS

Items to be weighed when planning, or when actually evacuating, will surface by answering the questions, why, who, how, how long, where to, how much? Answers to these questions will determine what goes into the plan and what is important in the on-scene decision-making process.

Why Evacuate?

Listed below are factors to be weighed in making an evacuation decision.

- The accident expansion potential. How much material is left to explode, spread as a gas, etc?
- What are the odds for expansion?
- What are the possibilities of in-place shelter?
- How can conditions such as weather affect the outcome?
- What are the estimates of warning time if the condition worsens?
- What are the effects of a worst case situation compared with the adverse effects of an evacuation?
- What is the estimated time to evacuate compared with the time available in a worst case development?

Citizens evacuate when four conditions are met.

1. They have accounted for the safety of their immediate household.
2. They have been given by authorities, or have personally developed, a plan for protective action.
3. They believe that a threat does exist in the environment.
4. They perceive that upon impact this threat could result in some level of damage to their person or property.(28)

Who Is Evacuated?

Details of an evacuation are difficult to include in a hazardous material plan since the portion of the population at risk cannot be accurately predicted. In hazmat incidents the evacuation area depends upon the size and location of the incident, the material involved, geographical considerations, and weather conditions. The area to be evacuated may include the area of potential danger due to drifting gases, radioactivity, or subsequent explosions, as well as the area of actual damage. Large-scale hazmat evacuations have occurred because of gas plumes. In Chapter 6, we first saw a system called Computer Aided Management of Emergency Operations (CAMEO) for predicting the path and size of chemical gas dispersions.

If people can make their homes reasonably airtight, are they safer at home from poisonous gases or radiation than they are with the risks of evacuation?

If evacuation is decided upon, the area must be exactly defined. In a disaster exercise in Mississauga, near Toronto, Ontario, a grid was outlined by street names so that the public could easily understand the limits of evacuation. Additional zones for expanded evacuation were mapped out by street names and natural boundaries. This was in contrast to areas of 5, 10 and 20-mile radius at the Three Mile Island (TMI) nuclear power plant emergency in Pennsylvania in 1979.(29)

At TMI a selected portion of the population (pregnant women and children) were advised to evacuate, but this would probably never be the case in a chemical hazmat incident.

People will evacuate if they perceive the threat as real.(30) In March 1991, a train derailment in Cooper County, Missouri, involved a tank car of burning yellow phosphorus. Although the incident fortunately occurred in an area that was relatively sparsely populated, those residents who were in danger perceived a real threat and were easy to evacuate. When an incident gets everyone's attention, evacuation is easier.(31)

When an area is designated and evacuation is ordered, everyone must leave and security must be established. The knowledge that their property and pets are safe helps to persuade people to leave.

Who Orders Evacuation?

This is a matter of planning. In small-scale emergencies the first responder, first on scene fire chief, or whoever the Department SOP (Standard Operating Procedure) calls for, should have authority to evacuate exposed buildings or the immediate area. The plan must designate the person or persons with authority to order large-scale evacuations.

How?

Sometimes people leave by themselves without an order. Although only an evacuation advisory for pregnant women and young children was given at TMI, about 144,000 people, or 39 percent of the total population within 15 miles of the reactor, evacuated.(32) This may have been because the public perceived a unique danger that distance would minimize, but spontaneous evacuation is an uncertainty in emergency operations planning.

When an evacuation decision is made, the public must be informed. Public warning can be given by these means:

- structure-mounted sirens where the public is trained to turn on the radio upon hearing the proper warning signal
- radio and TV announcements
- phone, word of mouth
- mobile loudspeakers from vehicles traveling up and down blocks

- door to door.

Instructions to evacuees should be put on tape, if possible. If read by police over loudspeakers, each officer should be given the same printed copy. For the area surrounding a fixed facility such as a nuclear power plant, printed sheets can be, and in some areas have been, distributed to the public, even in advance of any incident. Some thought should be given to the instructions, which should include information regarding routes, means of transportation, articles to bring, destination, shutting off utilities. Hazardous materials accidents may preclude anything except an urgent directive to get out and head north as fast as possible. Language barriers are a possibility as in the Chelsea, Massachusetts, conflagration of 1973.

Especially in rural and suburban areas, almost everyone can travel in his own or in a friend's or relative's car. At Three Mile Island, 94 percent of households had a car and 84 percent of families left in their own cars. Many leave before an order is given, and many outside the order boundaries also leave.

By using public buses, school buses, or whatever resources are available in the community, transportation must be provided for those who lack it. Pick-up points can be identified in the evacuation instructions. The bedridden are an extra problem and provisions must be made for them. If at all possible, a door-to-door check should be made, and some standard mark or sign placed on the door of evacuated homes.

Large hazmat incidents rarely occur in inner cities but when they do they present different problems. In tenement areas, the majority of inhabitants may not own cars, door-to-door notification would most likely not be effective, securing the areas would be very difficult, and getting everyone out could become impossible. In Mississauga the evacuated population was upper middle class, relatively well off, in private homes with small families. That's a very different profile from that in the inner city.

Institutions can be more of a problem than the general public.

- Schools—Should we return students home and let them leave with their families or transfer the whole school population to another school outside the danger area? There are problems either way.
- Hospitals—Each should have its own plan for transfer to other hospitals. A hospital mutual aid plan with ambulances is a priority item.
- Nursing Homes—Transportation for almost all will be required.
- Jails—Difficulties will occur in absorbing large numbers into other jails. Special transportation will be required so advanced planning is a must. In one instance, state penitentiary inmates were transported to a football field.
- Animal Shelters—Tranfer of animals to a safer location may be feasible.

After evacuation, the area must be secured against re-entry.

Studies have not been done on topics such as how many police it takes to evacuate a given number of blocks, or how long it takes at different times of day or night. Such studies would probably only be valid if they were site

Evacuation routes may be blocked following a widespread disaster. Here, Hurricane Hugo has succeeded in knocking a bridge from its moorings. (Photo courtesy of Ron Buckingham.)

specific, but they would contain the kinds of information a planner needs in order to make reasonable estimates.

How Long?

If evacuation time is the total time from the onset of the incident until everyone has cleared the area, then it is the sum of the following:

(*Td*) time delay after occurrence until evacuation order
(*Tn*) time required to notify population
(*Tm*) time required for people to mobilize and get underway
(*Tt*) travel time required to leave the affected area(33)

- *Td* can be reduced by effective planning since it includes the notification of officials, interpretation of data, decision making, etc.
- *Tn* seems to be inversely proportional to the population density.(34) The closer people are the quicker they can be notified to evacuate. It is estimated under the best conditions of local planning to be from 15 minutes to one hour.
- *Tm* depends upon such things as families getting together, securing homes and farms, type of population, institutional preparation.
- *Tt* is a consequence of the number of people, highway capacity, distance to travel, highway congestion.

Since almost all evacuees travel by private car, highway capacity is most important. Surveys during test evacuations have shown 1,100 to 4,080 cars per lane per hour. Four persons per car at 2,600 cars per lane would move 10,000 per lane per hour.(35) Test evacuations under ideal conditions do not necessarily mirror real life.

Converging roads, traffic lights, barriers, and disabled vehicles will all cause long evacuation times, and all such factors must be overcome.

Gas station owners jump in their cars and evacuate too, so provisions must be made for fuel. Certain stations in selected locations can be kept open, not only for the public but also for emergency vehicles.

Re-entry

The longer the evacuation lasts, the more tension builds among the evacuees. Keeping the area sealed off becomes a growing police problem, especially if no danger is perceived. People have all kinds of urgent reasons for re-entry, from getting their medicine to feeding their pets who were left behind.

Where To?

In past events, people have generally shifted for themselves. At Mississauga fewer than 1 in 20 stayed in mass care centers. Eighty-four percent went with friends or relatives, and 12 percent arranged their own accommodations.(36)

Of 223,000 evacuees, between 30,000 and 40,000 registered at reception centers, but there were never more than 3,000 in all of them, and no more than 1,500 ever stayed overnight.

How are shelters selected? They should be far enough away to eliminate any chance of secondary evacuation. In hazmat incidents, wind direction is almost always a factor.

Schools are best for shelters, especially high schools. They have adequate bathrooms for a large daily population, a cafeteria, water supply, separate classrooms for more privacy, a large hall, and an equipped nurses room.

The Red Cross and Salvation Army have experience running shelters. People are registered, even if they don't remain.

Pets can become a problem at a shelter and at motels or hotels. Most people seem to take their dogs along. Instructions regarding pets should be given at time of evacuation. Humane societies should assist since in large-scale disasters the number of pets will reach the thousands.

How Much?

There is very little solid data on the detailed costs of past evacuations. Total ball park figures for an accident are sometimes stated, but a breakdown into component costs such as evacuation are not. Even if all the data could be collected on a specific past event, its applicability to a future event may be marginal.

Cost of evacuation is not considered where danger to life is obvious.

Where the effects of a hazmat accident on the populace may be minute, the cost of evacuation as opposed to seeking shelter may act as a constraint on ordering the evacuation.

Costs will involve the evacuees, the evacuators, financial loss in farm areas and in urban and industrial areas. The cost components are listed here:

- evacuee costs: food and housing, commercial or mass care; transportation—private, commercial, or government provided; loss of income; loss to unattended property
- evacuator costs: food, housing, transportation; salary; extra personnel or overtime; extra or special equipment
- farm loss: loss of finished product; harm to unattended animals; possible loss of crops
- urban and industrial loss: value of product; damage to unattended plant; closing down and re-opening costs; loss to suppliers or purchasers.

Using cost factors applicable to the area, a planner can make estimates of approximate monetary costs for evacuation of specific sites.

In summary, the cost of evacuation must be balanced against the potential cost of not evacuating.

CONCLUSION

When we deal with disasters, we invariably deal with large numbers of persons. These have the capability of overwhelming the fire service if planning and preparation are not made in advance to handle both the large number of casualties (dead and injured) known as mass casualties and in handling large numbers of persons who must be removed from a danger area, known as mass evacuees. The information and guidelines presented in this chapter can well mean the difference between order and chaos when a community is suddenly struck with a calamity affecting large numbers of its populace.

CHAPTER 14 REFERENCES

1. Uebelacker, Carl. "On the Job, Ohio" in *Firehouse*, Vol. 15, No. 10 (October 1990), p. 43.
2. Collier, W. Steve. "Mass Casualties," Unit 12, in *Disaster and Fire Defense Planning*, Open Learning Course, National Fire Academy, Emittsburg, MD, 1988.
3. Ibid. p. 12-2.
4. *Newsweek*, February 5, 1990, p. 22.
5. *Cincinnati Post*, December 13, 1990, p. 2A.
6. *Cincinnati Post*, January 3, 1991, p. 2A.

7. "Major Freeway Pileup," in *American Heat*, St. Louis, MO, Vol. 5, Prog. 8, February 1991.
8. Isman, Warren E. "Wanted: Comments on Response Standards" in *NFPA Journal*, March/April, 1991, pp. 110–111.
9. Brown, David A. "Bomb Destroys Pan Am 747 In Blast Over Scotland" in *Aviation Week & Space Technology*, January 2, 1989, p. 28.
10. Klaene, Bernard, Dan Kuhn, and William Brewster, "Operations Manual" *Cincinnati Fire Division* Section 204.02, p. 3.
11. Ibid. p. 7.
12. Ibid. p. 8.
13. Ibid.
14. Collier, W. Steve. *See note 2*. p. 12-9.
15. Klaene, Bernard, et al. See note 10, p. 4.
16. Mahoney, Emmet Louis. *Disaster Medical Assistance Team Organization Guide*. Report NMDS-86. Manpower Task Force, National Disaster Medical System, Dept. Health and Human Services, Public Health Service, Health Resources and Services Administration, Bureau of Resources Development, 25 July 1986.
17. Klaene, Bernard, et al. See note 10, section 204.04, pp. 1 to 5.
18. "Until the Real Thing Comes Along" in *Fire Engineering*, Vol. 142, No. 5 (May 1989), pp. 52–56.
19. Donahue, Robert. "Rescue by Land and by Sea" in *NFPA Journal*, March/April 1991, pp. 55–58.
20. "Fire Wire" in *Firehouse*, Vol. 15, No. 3 (March 1990) p. 67.
21. "Fire Wire" in *Firehouse*, Vol. 16, No. 5 (May 1991) p. 84.
22. Johnson, Sharon Shaw. "Cyclone Toll Could Hit 200,000" in *USA Today*, May 6, 1991, p. 5A.
23. "Fire Wire," in *Firehouse*, Vol. 16, No. 2 (February 1991) p. 69.
24. Sorensen, John, Barbara M. Vogt, and Dennis S. Mileti. *Evacuation: An Assessment of Planning and Research*, Oakridge National Laboratory, July 1987, (Washington, D.C.: Federal Emergency Management Agency, 1987, ORNL-6376), p. 44.
25. *Evacuate!* (Washington, D.C.: Federal Emergency Management Agency, HV 554.E92), p. i.
26. Weber, Joseph. "Mass Evacuation" in *Disaster and Fire Defense Planning*, National Fire Academy, Emmitsburg, MD, 1989, pp. 13-1 thru 13-18.
27. *Hazardous Materials Emergency Planning Guide* (Washington D.C., National Response Teams, 1987).
28. Perry, Ronald W. *Citizen Evacuation in Response to Nuclear and Nonnuclear Threats* (Washington, D.C.: Federal Emergency Management Agency, 1981), p. 77.
29. Ibid. p. 31.
30. Ibid. p. 52.
31. "Yellow Phosphorous Train Derailment" in *American Heat* video journal, St. Louis, MO, April 1991.
32. Perry, Ronald W. See note 28, p. 64.
33. *Manual of Protective Guides and Protective Actions for Nuclear Incidents* (Washington, D.C.: Environmental Protection Agency, 1980), p. 1.31.
34. Hans, Jr., Joseph M., and Thomas C. Sell. *Evacuation, Risks—An Evaluation*, Las Vegas, NV Environmental Protection Agency, 1974, p. 41.
35. Ibid. p. 42.
36. Sorenson, John H., Barbara M. Vogt, and Dennis S. Mileti. *Evacuation: An Assessment of Planning and Research*, Washington, D.C.: Environmental Protection Agency, 1987, ORNL-6376, pp. 210–215.

CHAPTER 15

AFTERMATH AND RECOVERY

As conditions are brought under control in a major emergency or disaster, the fire department and community enter the recovery phase. This period in which families, neighborhoods, and government agencies attempt to return to normal, can be emotionally intense as well as crucial to the future economic well-being of the community. Although community recovery efforts are not a primary responsibility of the fire service, fire departments are involved.

This chapter will examine the key features of the recovery phase:

- psychological and sociological effects in victims and responders
- fire department and community recovery process
- overview of the federal disaster relief process
- evaluation of emergency response operations.

The chapter will focus first on the individuals affected and then consider the community and the levels of government assistance.

PSYCHOLOGICAL AND SOCIOLOGICAL DIMENSIONS OF DISASTERS

Major emergencies are stressful experiences, not only for victims, but also for responders and many others in the immediate community. Many persons associated with such episodes will exhibit or experience some degree of psychological trauma.

As late as June of 1991, many residents along both the east and the west coasts were still experiencing certain degrees of psychological trauma.

Those on the east were recovering from Hurricane *Hugo*, which had struck in 1989, and those on the west were still recovering from the Bay Area earthquake, which had struck several months later.(1) The subsequent conflagration in Oakland and Berkeley in October 1991 has further augmented the anxiety of the Bay Area residents.

PSYCHOLOGICAL EFFECTS ON THE COMMUNITY: MYTHS AND REALITY

Victims of disasters tend to go through a series of phases in reacting to the incident. If there has been warning, pre-incident activities focus on attempts to validate the severity of the threat. As emergency conditions strike, immediate reactions turn to protecting self and family. An immediate inventory phase then follows in which people anxiously attempt to account for their homes and loved ones. Highly altruistic behavior is often seen as victims and unaffected volunteers work to assist those in need.

Once the emergency needs have been brought under control, those who are unhurt begin an energetic process of securing possessions, if their homes have been damaged or destroyed, and cleaning up the mess.

The aftermath of a disaster can be devastating as in this scene following the Loma Prieta earthquake. (Photo courtesy of Ron Buckingham.)

Psychological reactions in the early recovery phase focus very strongly on returning to some semblance of a normal routine.

In South Carolina, in the aftermath of Hurricane *Hugo*, hundreds of families went without heat or electricity for days, and yet they improvised in varying ways and were able to carry on a semblance of reasonable, normal living. This, by the way, prompted Congressman Curt Weldon to proclaim, "Can you imagine? A nation that can send 400,000 troops halfway around the world in four months cannot get generators to families that are without heat and electricity."(2)

DISASTER SYNDROME

Victims and survivors of the immediate impact area frequently exhibit a typical behavior pattern called disaster syndrome that psychologists have characterized as a defensive reaction. The behavior appears in several phases as described by Anthony Wallace, an early researcher in disaster behavior:

> During the first stage, the person displaying it appears to the observer to be "dazed," "stunned," "apathetic," "passive," "immobile," or "aimlessly puttering around." This stage apparently varies in duration from person to person, depending on circumstances and individual character, from a few minutes to hours; apparently severely injured people remain "dazed" longer than the uninjured although this emotionally dazed condition is no doubt often overlaid by wound shock.
>
> The second stage is one of extreme suggestibility, altruism, gratitude for help, and anxiousness to perceive that known persons and places have been preserved; personal loss is minimized, concern is for the welfare of family and community. This stage may last for days.
>
> In the third stage, there is a mildly euphoric identification with the damaged community, and enthusiastic participation in repair and rehabilitation enterprises; it sometimes appears to observers as if a revival of neighborhood spirit has occurred. In the final stage the euphoria wears off, and "normally" ambivalent attitudes return, with the expression of criticism and complaints, and awareness of the long-term effects of the disaster. The full course of the syndrome may take several weeks to run.(3)

More recent research suggests that disaster syndrome may not be a universal occurrence and that disaster behavior may vary considerably depending on the type of disaster and the social and cultural characteristics of the community.

THE MYTH OF PANIC

Several years ago, Steve Collier (previously introduced as co-author of the National Fire Academy's course, "Disaster and Fire Defense Planning") was reviewing the recruit training curriculum at a major law enforcement academy when he came across a section on disaster response. He noted that the section consisted of only a few paragraphs that described the principal law enforcement role at such incidents as the control of looting and public panic. Fortunately a substantial body of research indicates that actual occurrences of either are relatively rare and take place only under specific conditions. Panic, characterized as totally irrational fear, generally occurs only when individuals perceive themselves to be trapped or threatened with no option for escape, such as being trapped in a burning building. Panic, however, is clearly not a characteristic response to stress or loss of possessions.(4)

COUNTER-DISASTER SYNDROME

Another type of reaction is often seen in individuals who were outside of the impact area, but have strong emotional ties to the people and locale that were affected. The reaction is characterized by intense, almost frantic activity directed toward rescue, first aid, and providing immediate assistance. Psychologists generally attribute the reaction to feelings of guilt at having survived the incident when a great many others did not.(5)

CONVERGENCE

Convergence refers to the tremendous influx of people and equipment that typically occurs moments after a major incident and may continue for a period of hours or days. Convergence behavior can be divided into three sources; the general public, emergency response agencies, and the news media.

Convergence By the General Public

Disasters that are particularly violent or spectacular tend to draw tremendous numbers of people even in the first few moments. These people may be motivated by one or more of the following factors:

- ○ curiosity about the nature of the incident
- ○ desire to be part of or involved in the incident: "to have been there"
- ○ intent to volunteer assistance, particularly if they have medical or other emergency skills
- ○ desire to learn the status of friends or relatives who may be involved.

Barriers and signs must be adequate to warn an unsuspecting public. (Photo courtesy of the Federal Emergency Management Agency.)

The convergence of looters, as accepted notions hold, has not been extensively documented in the immediate aftermath period. Looting, of course, is a major factor in civil disturbances or social unrest since it is a tactic in itself. Souvenir hunting in incidents such as airplane crashes has been reported in a few cases.(6)

Convergence By Emergency Responders

Convergence is also a phenomenon among emergency responders. Virtually every major incident attracts law enforcement officers, fire fighters, and emergency medical personnel from miles away who can usually gain access to the site because of their vehicles and uniforms. In some cases this convergence is in response to general calls for assistance, but in many cases it is simply a desire to be part of the action. The convergence often creates serious problems for jurisdictional agencies since many of the outside responders are not familiar with local plans and procedures.

Convergence By the News Media

Any significant incident, particularly a spectacular one, will attract local news media from up to several hundred miles away as well as network broadcast and cable news crews. This type of convergence intensifies the feeling that the community is in a fish bowl being watched by millions. The frantic competition of outside, unfamiliar reporters and the influx of high tech broadcast equipment can be particularly unnerving to incident commanders and response agencies. At one point during the crash of an Air Force jet in Indianapolis in 1987, a row of thirteen mobile satellite dishes were in place within a block of the site.(7) In a very similar incident, a Navy A-7 Corsair fighter jet crashed into an apartment building in Smyrna, Georgia, an Atlanta suburb. Here, four fire departments responded and about a dozen different media organizations.(8)

Techniques for Dealing with Convergence

Fire, law enforcement, and other emergency response agencies can minimize the effects of convergence by doing the following.

- Anticipate convergence problems at major incidents and address them in plans so that field personnel will be prepared.
- Work together to develop major incident plans that prevent congestion, thus preventing the need to clear it out after it occurs.
- Provide a responsive media briefing center so that the news media can get valid information. Release technical and factual information only through designated public information officers. News media representatives should not be permitted to randomly interview agency officials.
- Try to release information as quickly as possible so that those converging on the scene simply to find out what happened will have valid information from an informed source. In the San Francisco Bay

The members of the press are restricted to a safe area during the Miamisburg, Ohio, train derailment and phosphorous fire in 1987. (Photo courtesy of the Miamisburg, Ohio, Fire Department.)

Area earthquake in 1989, area fire departments were successful in handling convergence due in large measure to earthquake preparedness training and education.(9)

CRISIS COUNSELING FOR VICTIMS

Various phases of the disaster syndrome may last for several days or weeks depending on the type of incident. In the airplane crash in Indianapolis, when the jet fighter struck the hotel, damage was so extensive that the hotel was closed for months. Most of the ten people killed were employees of the hotel as were many of the affected survivors. These survivors experienced severe difficulties because they had lost fellow employees and friends, had lost their jobs because the hotel was closed, and narrowly missed being killed themselves, and, to some extent, had experienced guilt feelings because they survived and their friends did not. Some of these employees experienced reactions so severe that they could not function well enough to take advantage of available help or seek employment.

Many of these victims eventually received counseling assistance from local mental health counselors and from an outside counseling team sent in by the National Organization for Victim Assistance. This team works with local mental health practitioners to counsel victims and to set up longer-term counseling programs.(10)

Most adverse psychological reactions can be dealt with effectively by early intervention that focuses on sharing feelings and receiving reassurance that these feelings and emotions are completely normal. Local mental health associations are routinely involved with crisis counseling, but the stigma of these agencies as dealing primarily with the mentally ill prevents many victims from seeking their assistance. Agency complaints and local or outside victim assistance groups can act as intermediaries in getting those who need help into the system.

CRITICAL INCIDENT STRESS IN RESPONDERS

Psychological reactions are also seen frequently in responders involved in major incidents. Certain types of incidents commonly result in problems. Such incidents include those causing significant loss of life, particularly when it is violent (such as an airplane crash), when it involves children (San Diego McDonald's sniping incident), when protracted rescue efforts are involved (Kansas City Hyatt Regency collapse), or when rescuers are exposed to sustained personal danger (e.g., pinned down by sniper fire). These so-called "critical incidents" can lead to "critical incident stress."

When Leo D. Stapleton, retired Fire Commissioner of Boston, launched a new monthly newsletter for fire chiefs and others in the fire fighting community, he considered critical incident stress a significant enough topic to be covered in the charter issue.(11)

Even when there is but one victim, the same psychological reactions are manifested in the emergency responders, and, for that matter, the public at large. Recall how the nation watched and waited for the rescue of little Jessica McClure. "The eyes of the nation were glued to the television as rescuers tunneled frantically for more than two days to save the life of a little girl trapped in an abandoned well."(12) Consider, then, how the emotional involvement of emergency responders can multiply with multiple similar victims.

The symptoms of critical incident stress include one or more of the following:

- compulsion to recount details of the incident for days, weeks, or months
- difficulty in concentrating
- difficulty in making decisions
- disciplinary problems at work
- sleep problems or recurring nightmares
- dependence on alcohol or drugs
- marital or sexual difficulties
- feelings of guilt ("Isn't there something more I could have done?")
- suicide (rare).

Although critical incident stress (CIS) has been around for a long time, the fire service until recently failed to recognize it or failed to provide a program to deal with it, primarily because there was little direct research or literature regarding the problem. As the 1990s began, research was conducted regarding the effects of CIS on emergency service providers. Literature also began to be published on the subject. One of the leading CIS authorities is Jeffrey T. Mitchell, Ph.D., an assistant professor with the Emergency Health Sciences Department at the University of Maryland, Baltimore County. Dr. Mitchell and other experts who have studied the effects of CIS on emergency service providers agree that when a person has been stressed beyond his or her stress threshold, normal stress coping mechanisms no longer are effective in reducing stress levels. At that point, other ways of dealing with this level of stress are needed.(13)

Most experts agree that an effective program for dealing with critical stress must have several component phases. There must be an education component that allows all persons in an emergency response unit to learn what critical stress is and how it could affect them. Fire chiefs and officers need to learn how to recognize the signs of stress in fire department personnel. There needs to be a peer support unit of individuals (co-workers) who have been professionally trained. These peers need to be available at all times and be well versed in the importance of confidentiality. Another part of an effective program is the availability of a Critical Incident Stress Debriefing (CISD) team, including professionals from outside the fire department. The CISD team performs two functions: (1) defuses individuals or small groups of individuals immediately after an incident; and (2) performs a full-scale debriefing about 48 hours after an incident. See the section below on debriefing.

Educational Programs

Information on CIS is increasingly available for fire departments. Educational programs and literature are available to educate fire fighters. Group lectures presented by mental health professionals, employee assistance counselors and chaplains can be effective. Many articles on the subject have been published in *The Journal of Emergency Medical Services* (JEMS), and books on the subject are available. Existing CISD teams, eager to reach out to new fire agencies, will usually be glad to make a presentation on the subject. The educational portion of the program is so important that it should be conducted by qualified professionals.

Peer Support

The next consideration is peer support. The people selected for the program should be volunteers who are carefully screened to ensure that they support the CIS concept, are compassionate, can be available, and most importantly will keep information they receive confidential. The confidentiality of the program must be emphasized to all members so that in time of need individuals will feel free to share feelings. Peer supporters should be trained in active listening skills as well as what to say and what not to say to victims of critical stress. They should be taught to read nonverbal signs and body language. Training is ongoing so that peer supports should meet on a regular basis to share information and techniques. They should develop a knowledge of the experiences of one another because victims may wish to talk to someone who has been through a similar traumatic experience.(14)

Debriefing

A necessary component of the CIS program is the Critical Incident Stress Debriefing (CISD) Team. "CISD teams are made up of dedicated and trained mental heath professionals who combine their expert knowledge and talents with specially trained peer support personnel drawn from the emergency service's ranks. The CISD team is essentially a partnership between the two groups with a common goal—the reduction of critical incident stress in emergency personnel."(15) There are three general areas in which CISD teams function: pre-incident, incident, and postincident.

CISD Functions

Pre-incident functions would include education of personnel, command staff, and the families of emergency personnel. They would also include development of stress management protocols to be used in the field. Functions of CISD teams during an incident would include on-the-scene support services (to provide assistance to obviously distressed personnel), defusing (a short, usually unstructured, debriefing that would encourage discussions of the events and reduce acute stress) and demobilizations (used on large-scale incidents in place of defusings). Demobilizations take place away from the scene and are held in a large meeting facility where mental-health professionals can address large numbers of emergency service providers about signs and symptoms of critical stress. All personnel being demobilized are given the opportunity to have something to eat and drink. They are also provided with the opportunity to talk to mental health professionals or chaplains before they return to duty or go home.

After demobilization or defusing, emergency personnel usually enter a phase lasting about 24 hours where they try to sort out events and compare their actions with accepted procedures and protocols. During this period a full-scale debriefing would probably not be very effective as fire service personnel are usually not ready to deal with whatever feelings have been generated during the incident. After about 48 hours, the third phase, the postincident function, takes place.

Postincident Functions

The postincident functions include the formal critical incident stress debriefing (CISD). Formal CISD is a carefully designed process that progresses through seven different phases and provides important stress-reduction information. "Events that require a CISD include the following: Any event that has significant emotional power to overwhelm usual coping mechanisms. Line-of-duty deaths, serious line-of-duty injuries, emergency worker suicide, disasters, unusually tragic deaths to children, significant events where the victims are relatives or friends of emergency personnel, events that attract excessive media attention, and events that seriously threaten the lives of the responders. Because overuse of CISDs dilutes their effectiveness, they are reserved for only those events that overwhelm the usual coping methods of emergency personnel."(16)

Eight basic steps are involved in a debriefing period. These are summarized in Table 15-1 and described briefly in the next several paragraphs.

Table 15-1. Eight Basic Steps in Debriefing

1. Selection of a place, notification of participants, review of videos and reports, and occasionally a site visit.
2. Collection of facts from personnel who are asked to describe what happened at the scene.
3. Thought phase in which people are asked their thoughts regarding the incident.
4. Reaction phase in which personnel are asked to describe the most mentally traumatic parts of the events including the venting of thoughts and feelings.
5. The symptom phase in which the group is asked to describe stress symptoms felt at three times:
   (a) during the incident
   (b) three to five days after the incident
   (c) those still remaining
6. Teaching phase in which Critical Incident Stress team members furnish stress reduction information to the group.
7. Reentry phase where the group is encouraged to ask questions and resolve doubts.
8. A summary and end to the Critical Incident Stress Debriefing.

As a first step the CISD team selects a place, notifies the personnel that are going to participate, and reviews the incident by viewing video tapes and reading reports of the incident. Many teams also visit the scene of the incident. Only the personnel involved in the incident should attend the debriefing and they should not be required to respond to calls during the debriefing. At this point the peer support people from the emergency response agency involved in the incident become important. Personnel will feel more comfortable with and are more likely to trust fellow workers. It should be made plain from the beginning that everything discussed is strictly confidential and it should be stressed that debriefing does not involve a critique of operations at the incident. At the outset the attendees need to be told what to expect in the debriefing, what the rules are, and that "the main objectives of CISD are to mitigate the impact of a critical incident and accelerate the return of personnel to routine functions after the incident."(17)

The second phase of the CISD is the fact phase where personnel are asked to describe what happened at the scene.

In the third phase, the team leader will then lead the discussion into the thought phase where people are asked what their thoughts were. In the fourth phase, the reaction phase, personnel are asked to describe the worst part of the events for them. This phase allows for venting of thoughts and feelings associated with the incident and prepares the participants for useful stress reduction information.

"The fifth phase of the CISD process is the symptom phase. At this point the group is asked to describe stress symptoms felt at three different times: first, those symptoms experienced during the incident; second, those that appeared three to five days after the incident; and third, those symptoms that might still remain at the time of debriefing."(18)

The sixth phase is the teaching phase wherein team members furnish stress-reduction information to the group. The last phase is called the re-entry phase where the group is encouraged to ask questions and resolve any doubts. There is a summary of the debriefing and the CISD is ended. After the debriefing, team members make themselves available to individuals who may require more assistance. If necessary, referrals are made for counseling. After all the emergency personnel have left the debriefing site the CISD team has a post-debriefing meeting to review the process and to discuss ways to improve future debriefings.

Follow-up Services

There should be follow-up services after all defusing, demobilizations, and debriefings. Some of the ways that follow-up can be accomplished include telephone calls to the individuals, discussions with fire company officers or commanders, visits to stations, through peer supporters, continuing education programs, support services for spouses, and counseling sessions with individuals. Follow-up services usually begin about 24 hours after the debriefing and are very important to the success of the program.

CISD Training

A needed consideration for a CISD team is the amount of training required for its members. There should be a minimum of two days training, in conjunction with continuing education on a regular basis. CISD teams should have good peer support personnel and mental health professionals in order to be effective. Without mental health professionals, CISD teams will not only be ineffective but also dangerous. Mental health professionals are needed for their skills in supervision and leadership, and for their diagnostic skills. Undetected symptoms of critical stress could contribute to the suicide of an emergency worker.

It may take six months to a year to organize a CISD team, and it is very important that all members of the team receive the same training. Operating procedures should be established in writing before the team starts to do formal debriefings. Competency should be the most important consideration when selecting team members.

"Critical incident stress debriefing teams" have experienced phenomenal growth in recent years, overcoming many problems and achieving many successes. They assume an important place within emergency services organizations and are likely to continue their support services into the future."(19)

OVERALL FIRE DEPARTMENT RECOVERY

In disastrous situations, the fire department may be severely affected, at times almost catastrophically. Damaged apparatus and equipment may need to be replaced without funds available; fire stations, maintenance facilities, and communications centers may require repair or reconstruction; and most seriously of all, fire fighters may have been killed or injured. When five fire fighters died in Hackensack, New Jersey, under a collapsing truss roof, fellow fire fighters were emotionally affected and sought critical incident stress counseling. This allowed them to function in their jobs and personal lives with a certain degree of normalcy.(20)

The fire department may well have difficulty solving its own problems because other agencies in the community may be in the same or worse predicaments, and adequate government funds may be unavailable. In fact, the local tax base may be severely affected by decreased revenues due to business closings, failures, or relocations.

The fire department may also be affected by damage to lifelines or infrastructure it depends on such as roadways, water supply systems, utilities, and telephone or communications services. If state or federal disaster funds do not become available, complete fire department recovery may be a long process during which operation at previous levels may not be possible

RESPONSE DOCUMENTATION

One very important activity during both the response and recovery phases of a disaster is that of response documentation. In applying for federal disaster funds, discussed later in this chapter, it is critical that all costs associated with fire department operations be documented. These include

- manpower overtime
- apparatus damages
- apparatus usage above normal
- fuel
- equipment and supplies used or damaged
- administrative costs associated with response such as those involved with purchasing, vendor payment, legal advice, etc.

Many fire departments are becoming quite skilled at accounting for their costs—even at relatively small-scale events, such as hazardous materials spills—and billing the responsible parties. Equipment, manpower, and materials expenses can be recovered. Volunteer fire departments can also bill for services, and according to a well-researched article by Michael Ward, can charge the same rates for labor as comparable career departments.(21)

COMMUNITY RECOVERY

Community recovery operations generally divide into short- and long-term activities that take place more or less concurrently. Short-term activities are those focusing on the return to normal life and may last from several days to several weeks. Examples of short- and long- term recovery activities are listed below.

Short-Term Recovery Activities include the following.

1. Development of short-term action plan
2. Preliminary damage assessment
3. Debris clearance
4. Temporary housing
5. Immediate financial assistance to individuals
6. Expedient repair of lifelines and utilities
7. Crisis counseling

Long Term recovery activities include the following.

1. Development of long term action plan
2. Acquisition of financial aid
3. Reconstruction of lifelines and utilities
4. Reconstruction of community business section
5. Preventive legislation, zoning, building codes

6. Economic development programs
7. Improvement of emergency response capabilities.

Combined short term and long term activities include the following.

1. Review recovery elements of emergency operations plans
2. Determine if an emergency has been declared
3. Review preliminary damage assessments and prepare comprehensive assessments
4. Identify and prioritize recovery needs
5. Identify sources of assistance
6. Identify specific funding sources
7. Develop and implement a detailed recovery action plan
8. Submit documentation/application for state and federal funding
9. Document recovery effort
10. Monitor recovery activities and fine tune
11. Evaluate entire emergency responses and recovery operation

The Federal Emergency Management Agency with guidance from Professor Jack Neal of the American University has published a booklet entitled *When Disaster Strikes*. It contains much valuable information for communities coping with disasters that have occurred. While the original booklet was prepared as a media guide, portions are reproduced in a modified form in the paragraphs that follow so as to be of value to fire officials and other public emergency officials.(22) The Federal Emergency Management Agency (FEMA) has enthusiastically encouraged us to share this information.

THE RESPONSE PROCESS

People expect their government to protect them, assist them, and inform them when a disaster threatens or occurs.

They have these expectations for good reasons. Thousands of hours and many dollars—local, state and federal—have gone into the development of comprehensive plans of action to be followed before, during, and after disaster strikes. Fire service personnel and police at the local level, and emergency management personnel at all levels of government have undergone extensive training. Communications equipment and procedures are in place; duties and responsibilities assigned. In late 1989, when hurricane *Hugo* and the San Francisco earthquake hit the United States in rapid succession many of the nation's lawmakers questioned the speed and thoroughness of the FEMA response, no doubt feeling pressure from their constituents. The agency was, however, extremely strapped as it was spread across two major disasters.(23)

Should the local government need outside assistance, the governor has authority to activate the state's emergency plan. The governor may find it appropriate to declare a state of emergency in the affected area, thus freeing state resources to provide the appropriate response and recovery activities.

Should combined local and state resources be inadequate, the governor may turn to the federal government with a request that the president declare a "major disaster" or, perhaps, an "emergency."

Thus, there is a clear-cut chain of events leading from the local level, through the state, and, if necessary, to the federal level. Each step to higher authority is predicated upon the inability of the preceding authority to adequately meet emergency needs.

Local Government

A local government's disaster authority and responsibility is centered in the mayor, city manager, or county executive's office. Most cities and populous counties direct disaster relief activities from emergency operating centers (EOC) discussed in Chapter 3. Fire department and local government EOC priorities in the early phases of a disaster are to warn and evacuate citizens and protect life and property.

Immediately following the emergency, local government must ensure public order and security. Vital services such as water, power, communications, transportation, shelter, and medical care must be continued uninterrupted if possible or be restored quickly if disrupted, and debris removal must begin as noted in Chapter 1.

Local officials rely on help from the private sector and individual volunteers and charitable organizations to cope with immediate post disaster tasks. Disaster plans and annexes should include liaison with such organizations as medical care facilities, gas and electric companies, the telephone company, transportation companies, chemical companies, and firms with heavy equipment such as cranes and bulldozers that can supplement city equipment.

In the wake of several disasters, including the Kansas City Country Club Plaza flood of 1977 which killed 18 people and the Hyatt Regency Hotel disaster in 1981 which killed 113 people, the city of Kansas City, Missouri, has put together an agreement with the Heavy Contractors Association of Greater Kansas City, Mid-American Regional Council, and various local governmental bodies. A published catalog is produced annually by the "Plan Bulldozer Committee of the Heavy Contractors Association" providing lists of hundreds of resources that can be used in the event of a disaster. Contracted names and key figures are provided. This could serve as an excellent model for other communities who have yet to establish an effective link to the private sector and its numerous invaluable resources.(24)

Invaluable assistance is provided by charitable institutions such as the American Red Cross, the Salvation Army, church groups, and others. These private organizations remove a great burden from local government during an emergency by being partly or totally responsible for distributing food, medicine and supplies, and providing emergency shelter.

Many agencies are busy in the aftermath of a disaster. Here, a civil defense unit assists with postincident recovery. (Photo courtesy of the Civil Defense Agency.)

State Government

If local officials need state help, state governors can provide several valuable services, such as state police to help ensure public safety and protection; state national guard troops to provide security, transportation, medical, food and temporary shelter services; state transportation services, including highway equipment and personnel; and state funds for disaster relief and recovery.

At any time in the disaster or emergency, local government or state officials may turn to a number of available federal agencies for assistance. Most can act quickly under their own statutory authorities without a presidential declaration. These agencies include the U.S. Coast Guard and armed forces for search and rescue operations; the U.S. Army Corps of Engineers for emergency flood protection; the Small Business Administration for homeowner and business loans; the Federal Highway Administration for road and bridge repairs; the U.S. Department of Agriculture for emergency loans or grants to farmers and ranchers; and the Department of Health and Human Services for assistance in public health and welfare measures.

Federal Government

When a disaster situation is beyond the capabilities of local and state forces, supplemented by private and volunteer organizations and limited assistance of federal agencies on the scene, the governor—and only a governor—may request that the president declare a "major disaster" or an "emergency."

A "major disaster" is defined in the Disaster Relief Act of 1974 (Public Law 93-288) as any "hurricane, tornado, storm, flood, high water, wind-driven water, tidal wave, tsunami, earthquake, volcanic eruption, landslide, mudslide, snowstorm, drought, fire, explosion, or other catastrophe in any part of the United States which, in the determination of the President, causes damage of sufficient severity and magnitude to warrant major disaster assistance above and beyond emergency services by the federal government to supplement the efforts and available resources of the state, local governments, and private relief organizations in alleviating the damage, loss, hardship, or suffering caused by a disaster."(25) Appendix K of this book contains a listing of the FEMA regional offices and state emergency offices divided by FEMA region.

FEDERAL RESPONSE

In Executive Order 12148, the president delegated the primary responsibility for administering the Disaster Relief Act (Public Law 93-288) to the Federal Emergency Management Agency (FEMA). Since 1979 this agency has served as the central point of contact within the federal government for a wide range of emergency response activities.

A governor's request for federal aid, addressed to the president, is forwarded to the appropriate FEMA regional director, who evaluates the on-scene damage assessments. The regional director's analysis and recommendations are forwarded to FEMA headquarters for review. The FEMA director's recommendation is then forwarded to the president.

It should be understood that the analysis and review process can take a few hours or a few days. At all levels of FEMA, responsible officials charged with accountability for federal disaster funds—taxpayers' dollars—must be certain that there truly is a need for supplemental federal aid.

For this reason, it should not be assumed that action on a request for federal aid will be either quick or affirmative. In fact, history shows that more requests for assistance are turned down by FEMA than are approved. The key question that must be answered in the affirmative is: Is the magnitude of the crisis beyond state and local capability?

As soon as possible following a presidential declaration, FEMA designates, on the basis of damage assessments, the counties or political subdivisions that are eligible for federal aid. The president's declaration also names a federal coordinating officer (FCO) who, as the president's representative, will direct federal recovery and coordination activities. FEMA then signs a disaster assistance agreement with the governor of the affected state outlining what the state's role will be and its commitment to the recovery operation.

FEMAs responsibility is to supplement disaster assistance available through state and local governments, not to supersede it. FEMA is also responsible for coordinating disaster assistance provided by all other federal

Federal programs assist disaster-stricken victims. Here, an impromptu office is set up in the field. (Photo courtesy of the Federal Emergency Management Agency.)

agencies, the military, state and local governments, and, with their consent, those of private relief agencies as well.

Basic disaster assistance from the federal government falls into two possible categories:

- Individual assistance, which can provide help to individuals and families, coordination of aid to businesses, including farmers; and
- Public assistance, which provides funding assistance and technical expertise to aid local and state governments and certain private, nonprofit organizations to repair essential facilities. (Reporters often call this assistance "civil" or "public works.")

Whether both categories of assistance are made available depends on the needs identified during damage assessments by teams usually made up of federal, state, and local officials.

FEMA's first concern is for families and individuals adversely affected by the disaster. The degree of suffering and deprivation plays a major role in determining the need for federal assistance.

There are no hard and fast rules within FEMA requiring that a specified number of people must be affected before a request can be considered. FEMA has recommended approval of federal assistance in situations involving only a few dozen disaster victims. The critical consideration is whether those impacted by the disaster have needs that state, local, and/or volunteer recovery efforts cannot meet for one reason or another.

While most major disaster declarations are aimed at providing assistance to individuals and families, there are occasions when the only needs determined to be beyond state and local capabilities are those addressed in the public assistance categories.

Public assistance is designed to provide federal aid to repair damaged or destroyed facilities essential to the overall integrity of the affected political jurisdiction and to protect public health, safety, and welfare. This can include repairs to roads, bridges, parks, city halls, public libraries, publicly held hospitals and utilities, and other facilities of this type. It can also be used to clear debris.

A governor's request may seek individual assistance, public assistance, or both. FEMA may recommend that the president authorize either or both, depending on the needs found to be beyond the state and local recovery efforts.

APPLYING FOR FEDERAL FUNDS

Application for federal disaster assistance involves specific damage assessment and application procedures. The official damage assessment is conducted by a team consisting of officials from the regional office of FEMA, the state emergency management agency, and representatives from local government. Assessments are detailed including specific numbers of damaged dwellings, business and public structures, assessed valuations, and percent of incurred damages. Funding applications normally must be submitted within 90 days of the presidential declaration.

The application is submitted through the state office of emergency services to the regional office of FEMA and then to the director of FEMA. The FEMA director will make a determination on whether the application meets federal guidelines. If the decision is affirmative, a recommendation will be made to the president to make the appropriate declaration. If the application is rejected the director will notify the state.

Figure 15-1 depicts this application and decision making process.

DISASTER APPLICATION CENTERS

To make it easier for disaster victims to obtain individual assistance, the FCO (federal coordination officer) establishes one or more Disaster Application Centers (DAC) where, under one roof, representatives of federal agencies, local and state governments, and voluntary relief agencies can counsel and accept applications for aid from individuals, families, and business firms.

These federal-state DACs are different in nature and purpose from assistance centers operated by the Red Cross and other volunteer groups. While

Figure 15-1. Sequence of Steps for Obtaining Federal Disaster Assistance.

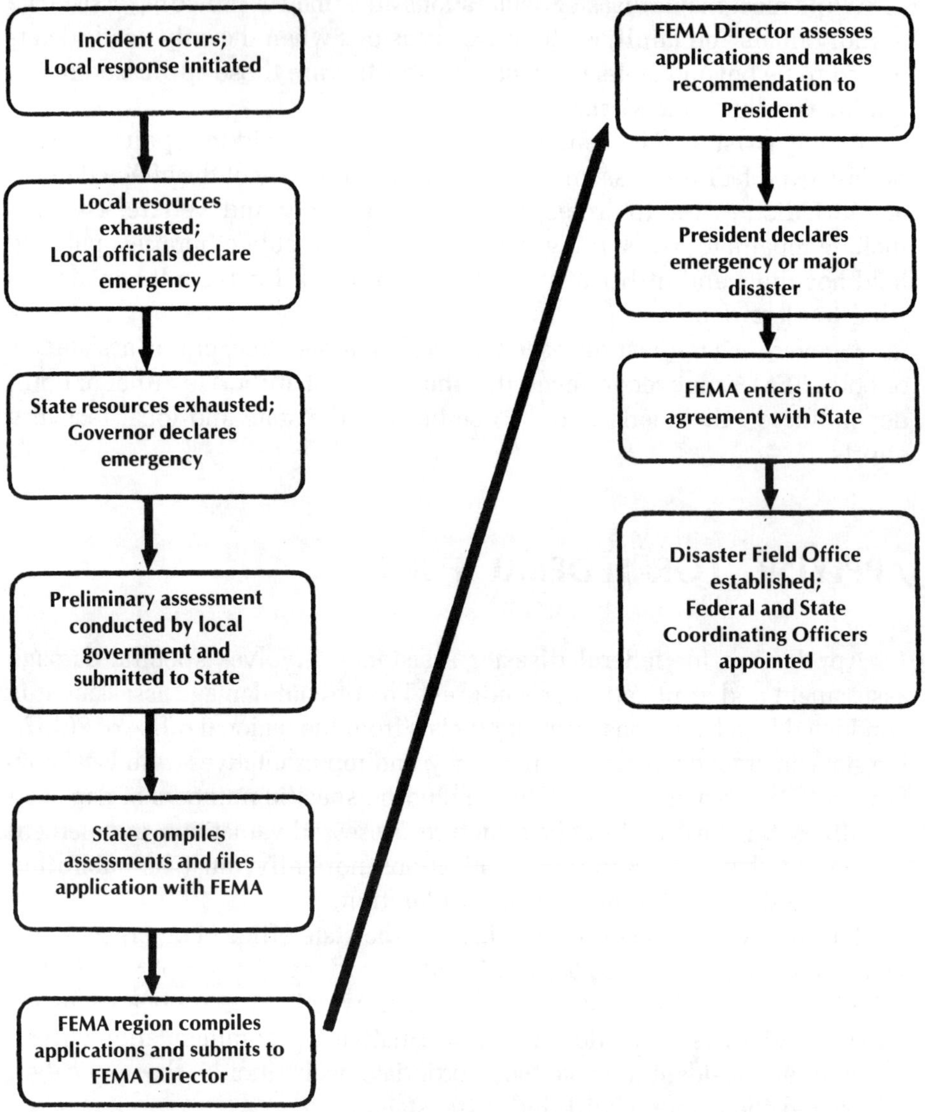

both usually are referred to as "center," the volunteer center is established to dispense emergency aid—food, clothing, medical supplies, shelter. A DAC, on the other hand, is a centralized location where the application process for long-term recovery aid—grants, loans, and so forth—is begun.

Disaster Application Centers are established close to the stricken area in large facilities, often in schools or armories, to accommodate many workers and the applicants they will serve. Following an initial interview, each applicant is directed to the desks of one or more of the government or private assistance programs most likely to meet the needs of that applicant. Since this is the beginning of the application process, no decision is made in the DAC whether the applicant is eligible for one type of aid or another. Such decisions are made by the various agencies on the basis of information provided by the applicant on forms obtained during this interview.

DACs remain open as long as the FCO and his or her counterpart, the state coordinating officer (SCO), deem necessary. New DACs may open in the disaster area and mobile DACs may be sent to remote areas if needed.

When DACs do close, the FCO makes available the forwarding addresses and phone numbers of the participating agencies where individuals may make applications or obtain needed assistance or information.

GETTING OUT THE WORD

Prior to the opening of the Disaster Application Center (DAC) the federal coordinating officers staff conducts a brief but intense public information campaign to advise disaster victims about the types of federal aid that will be available, eligibility requirements, and how to apply. The objective is to prepare people for the disaster relief application process, to keep expectations to realistic levels and to encourage only the most seriously affected individuals to visit DACs on opening day. This information effort, with appropriate media support, can save individuals much time, trouble, and disappointment at DACs, and may save them an unnecessary trip altogether.

The fire department can play a pivotal role in public information. A *Newsweek* article pointed out that the profession Americans trust most is that of a fire fighter.(26) They should be willing to cooperate with FEMA in providing public information.

The FEMA public information campaign makes these important points.

- A visit to a DAC is the first of several steps in the recovery process. Applicants will receive information and applications at these centers, not goods or cash.
- Long lines on opening day can be avoided by those not severely affected by the disaster by waiting a day or two before visiting a DAC.
- Basic information will be required from individuals at the DAC, such as the applicant's name, current and predisaster address (if appropriate), proof of residence, current phone number, insurance coverage, policy number(s) and agent's name. Applications will be taken without this information, but the process can be speeded up if it is readily available.
- Applicants have a responsibility to contact their insurance company or companies and file necessary claims. Government assistance does not cover damages or losses already covered by private insurances.
- Federal disaster aid does not guarantee total recovery from all disaster losses. Assistance is limited to that necessary to return damaged or lost property to predisaster condition.
- Federal and state assistance is long-range in nature. Therefore, it should be understood that recovery will not occur overnight.

DISASTER HOTLINE

Another effective tool the FCO may use to communicate with disaster victims is the Disaster Hotline, a telephone question and answer service which helps both fire victims and the federal response officials. In all disaster operations where such communications links can be established, FEMA sets up a special telephone number—usually toll free 1-800 lines—which victims may use to obtain basic information. Through this service, victims can obtain such information as where the DACs are, operating hours, what documents (if any) will be needed at the DAC, and so forth. Hotline operators, trained by FEMA, maintain a daily log of the types of questions asked. An unusually large number of calls relating to one program or another suggests possible confusion about the program and may indicate a need for greater emphasis in this area by public information officers and the media.

HOW MUCH WILL IT COST?

Monetary disaster losses must eventually be calculated but this process is not easy. It is virtually impossible to assign a dollar value to the amount of property lost or to the amount that will be needed to restore those losses. Even if it could be done, it is unlikely that the two figures—loss and cost—would balance neatly.

A governor's request for federal assistance almost always includes a dollar estimate of the damages. But, as the record shows, these estimates can be far too high or much too low. Such estimates are developed quickly, in most cases, and their only role in the decision process is to determine whether the recovery effort is beyond state and local response capability.

To illustrate the problem with assigning dollar values on loss and cost, consider this. Anybody with eyes can see that the city's $1 million city hall was destroyed. The mayor cites the $1 million loss and he is correct. But then it is determined that the city government carried $800,000 worth of insurance on the structure.

Thus the city lost its $1 million city hall, but insurance will reimburse the local government $800,000. Net loss to the city: $200,000.

But if there is a presidential declaration, the city would be eligible to seek federal reimbursement from FEMA to rebuild the structure. FEMA can reimburse 75 percent of the uninsured loss, or $150,000. The state and/or local government pays the remaining $50,000.

So what is the actual loss? Depends on whom you ask.

Mayor: $1 million

Insurance company: $800,000

FEMA: $150,000

State/Local government: $50,000

Factors such as these can make damage estimates and repair costs confusing and seemingly contradictory.

In most cases the actual dollar loss will not be known until the last damaged bridge is rebuilt, the final roofing shingle nailed in place, the final recovery loan disbursed, the last insurance claim paid.

Until then, there can be only estimates.

And it is extremely unlikely that the federal coordinating officer or the disaster public information officer on the scene will make estimates. Not unless there is solid information that such an estimate will be accurate.

There will be some figures available, however, that may be useful. For instance, there will be daily reports on the number of disaster victims who have visited a disaster assistance center and what kinds of assistance they sought. The American Red Cross and other volunteer organizations will be able to say how many victims they have helped and the kind of help provided. Local governments or the U.S. Army Corps of Engineers may let repair contracts and these details will be available.

Once the DACs have been operating for a few days, federal officials may be able to make projections. These can be developed on the basis of the number of aid applicants visiting a center, the needs they have, and the average cost of providing these needs. But there again, these will be only estimates.

One of the major programs usually involved in disaster recovery is the disaster loan program. Figures generated by applications to repair or replace homes, personal property, and businesses can make headlines, and they do represent losses. The record may show, for instance, that 500 homeowners and businesses applied for $60 million in disaster loans. This clearly represents at least part of the disaster loss, but is it a cost? Remember that these are loans, virtually all of which will be repaid to the federal government with interest.

Another important point to be remembered is that the federal funds needed for response and recovery in a presidentially declared major disaster are available no matter how much is required. This is because the President's Disaster Relief Fund, a line item in the federal budget, is replenished as needed.

None of this is intended to suggest that reporters will never get those all-important dollar figures that are so much a part of telling the story of the disaster. It is intended only to illustrate the problem facing the disaster information officer who tries to respond to the question, "How much will it cost?"

DISASTER RECOVERY PROGRAMS

Presidential approval of a governor's request for federal supplemental assistance activates a wide range of disaster recovery programs from a variety of federal agencies. Representatives of these agencies will move into the disaster area to provide the long-term recovery assistance not available through emergency response groups such as the American Red Cross, Salvation Army, and others.

While each federal agency is responsible for administering its programs, management and coordination of the overall recovery operation are the responsibilities of FEMA through the federal coordinating officer. In addition, FEMA, in cooperation with state officials, arranges for establishment of Disaster Application Centers, sets up appropriate communications and provides personnel to ensure the needs of the disaster victims are met in a timely and efficient manner.

While there is a wide range of programs that can aid disaster victims, it should not be assumed that all of them will be activated in a given situation. The determination of which programs will be used is made by the actual need found during the damage assessment and any subsequent information that may be developed.

Following is a brief explanation of the major provisions of the various disaster recovery programs. There is also an outline of the capabilities of the various volunteer agency programs.

These explanations are general because of space limitations. However, they are suitable for publication or broadcast should your area be declared by the president to be a major disaster area.

INDIVIDUAL AND BUSINESS ASSISTANCE

Temporary Housing

FEMA, or an agency designated by the state, can provide housing assistance to those whose homes are damaged, destroyed, or otherwise rendered uninhabitable by the disaster. The assistance may be provided for up to one year, but the applicant's need is reviewed at least quarterly.

Temporary accommodations may be uninhabited government-owned or subsidized housing units, commercial rental property, or, as a last resort, mobile homes or travel trailers. Mobile homes are seen as a last resort because of the time required to move them to the area of need, development of suitable sites, and construction of necessary utilities.

Funds may also be provided for use in making habitable those homes which sustained only minor damage.

Disaster Loans

The Small Business Administration (SBA) can make federally subsidized loans to repair or replace homes, personal property, or businesses that sustained damages not covered by insurance. Maximum loans are $100,000 for homes, $20,000 on personal property, and $500,000 on businesses. Over the years, the interest rates on disaster loans have varied in relation to the cost to the federal government to borrow money. Since interest rates can vary from time to time, it is recommended that media confer with the local SBA office or the FEMA regional public affairs officer to determine the applicable rates at the time of the declaration. Repayment of loans may go to 30 years, but actual terms are based upon the applicant's ability to repay. Application deadline is 60 days from the date of the president's declaration. (Note: The administrator of SBA is empowered to make the disaster loan program avail-

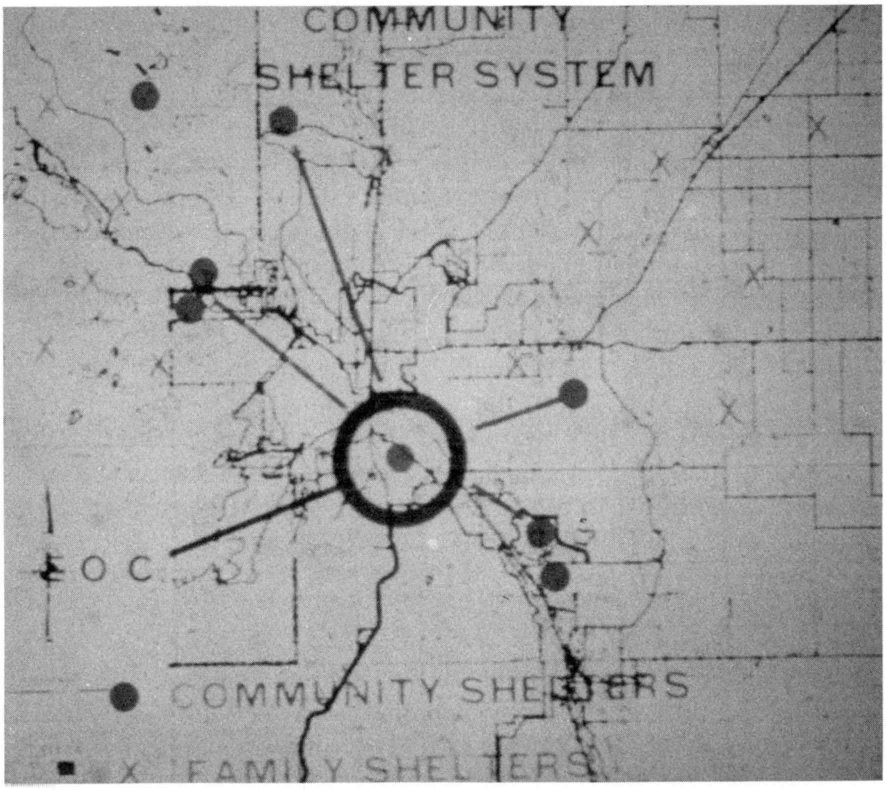

A series of community shelters, planned in advance, can provide an excellent relief valve in the wake of a disaster. (Photo courtesy of the Civil Defense Agency.)

able in a designated area without presidential declaration of a major disaster.)

Individual and Family Grants

Upon request of the governor, the individual and family grant program may be made available. This program, administered by the state but underwritten with 75 percent FEMA funding, can provide grants of up to $5,000 through an agency designated by the state to meet disaster-related serious needs and necessary expenses not provided before by other programs, insurance, or other means. One test for grant eligibility requires, among other things, a decision by the SBA that the applicant is ineligible for a disaster loan. Among needs that can be met with grant funds are housing repairs, home cleaning and sanitation, necessary household items, replacement of tools of the applicant's trade, clothing, transportation, and medical and dental expenses, among others. Application deadline is 60 days from the declaration date.

Agricultural Assistance

The Farmers Home Administration (FmHA) may make emergency loans to farmers and ranchers—owners or tenants—who were operating and managing a farm or ranch at the time of the disaster. These loans are limited to the amount necessary to compensate for actual losses to essential property and/or production capacity. Loans are limited to 80 percent of the actual production loss of up to $500,000. As with SBA, FmHA interest rates have varied through the years. It is recommended that media contact the Farmers Home Administration or the FEMA regional public affairs officer to determine the applicable rates at the time of the declaration. Disaster loans from FmHA also are available to agriculturists in counties that are contiguous to those counties designated as a result of a presidential declaration. This contiguous rule applies across state lines as well.

The U.S. Department of Agriculture has other programs that may come into play in varying degrees during recovery operations. Among these programs are those that include assistance in removing debris from agriculture lands, fence repair, crop insurance, and others. (Note: The Secretary of Agriculture is empowered to make disaster loans available without presidential declaration of a major disaster.)

Disaster Unemployment Assistance

This program, administered by the state's employment commission in cooperation with the U.S. Department of Labor, may provide weekly benefit payments to those out of work because of a major disaster, and who have no entitlement to regular state unemployment insurance programs. Examples include self-employed persons or farm and ranch workers, among others. The president must declare a major disaster for this program to be activated.

Food Stamp/Commodity Assistance

Food stamps and commodities are made available by the Food and Nutrition Service of the Department of Agriculture (USDA) after designation by the Secretary of Agriculture following a presidential declaration. Assistance includes:

- 30-day allotment of food stamps to low-income households in areas affected by the disaster
- food commodities to recognized disaster organizations for mass feeding
- food commodities for household distribution in areas served by the Food Stamp program whenever commercial food sources remain disrupted as a result of the disaster.

Legal Services

This program, usually supported by the Young Lawyers Division of the American Bar Association, can be activated by the appropriate FEMA regional director to provide free representation to low-income disaster victims requiring such assistance. Fee-generating cases—that is, cases which a local attorney likely would accept—are not accepted by these disaster counselors.

Crisis Counseling

This program, funded by FEMA through the National Institute of Mental Health, may be implemented if requested by the governor. Its purpose is to relieve those possible mental health problems caused or aggravated by the disaster or its aftermath. The assistance is provided by state and local health agencies.

Federal Tax Assistance

The Internal Revenue Service makes available advisors to assist disaster victims who wish to deduct uninsured casualty losses from the current tax year's return or the prior year's return. This action can result in an expedited tax refund useful to a disaster victim's recovery effort. The Bureau of Alcohol, Tobacco and Firearms helps businesses recover excise taxes and customs duties paid on distilled spirits, wines, beer, tobacco, etc.

Aid to Elderly

The Social Security Administration provides assistance to annuitants who require address changes and can expedite benefit check delivery. It also can assist disaster victims in applying for disability, death, and survivor benefits.

Assistance to Veterans

The Veterans Administration provides assistance in applying for VA death benefits, pensions, insurance settlements, and adjustments to VA-guaranteed home mortgages made appropriate because of financial burdens caused by disaster-related losses or damages.

Volunteer Assistance

The American Red Cross provides individual and mass shelters, feeding, vouchers, clothing or funds for clothing, cleaning supplies, supplemental medical care, blood and blood products. It also can provide funds for certain essential home furnishings, minor repairs to homes, rent, transportation, funeral expenses, personal occupational supplies and equipment, eyeglasses, prescription medicines, lost dentures, and prosthetic devices.

The Salvation Army provides emergency food, clothing, furniture, and financial assistance for immediate needs.

Other Emergency Aid

Emergency assistance programs also may be available through private relief organizations such as Church World Services, Seventh Day Adventists, Southern Baptist Disaster Services, local churches, private organizations, and others.

Mennonite Disaster Services may provide skilled and unskilled labor for debris removal, home repair and reconstruction, and home cleaning.

PUBLIC ASSISTANCE

At the direction of the president, FEMA can provide funding and technical assistance to the state and local government and certain private, nonprofit institutions for the repair or replacement of public properties damaged or destroyed by the disaster. Among these can be restoration of public-owned facilities (roads, bridges, buildings, parks, etc.) and other facilities considered necessary to re-establish the integrity of the disaster area's infrastructure.

FEMA also can provide assistance for emergency protective measures, debris removal, emergency communications, and emergency public transportation.

The agency may also provide assistance to state government for fire suppression for any fire that threatens such destruction as would constitute a major disaster. This can be accomplished by FEMA without a presidential declaration.

In addition, the agency may make limited loans to local governments that have suffered a substantial loss from their tax rolls and other revenues as a result of the disaster.

Federal Aid for Highways

The U.S. Department of Transportation's Federal Highway Administration can provide financial and technical assistance for repair or reconstruction of highways on the federal aid system, forest highways, park roads and trails, Indian reservation roads and other similar projects damaged by the disaster.

Restoration of Public Schools

The Office of Elementary and Secondary Education, U.S. Department of Education, can provide grants to restore public schools damaged or destroyed by disasters which are declared by the president. The agency also may provide grants to construct minimum public school facilities to take the place of a destroyed nonpublic school which will not be replaced.

FIRE SUPPRESSION ASSISTANCE

In Chapter 6, we read about the ever-present threat of wildland/urban interface fires. These occur often in many parts of the country, particularly during extended periods of drought. Careless campers, lightning, and arson are the most common causes of such fires.

Communities must learn to cope with missing bridges, missing highways, and other inconveniences following a disaster. (Photo courtesy of Ron Buckingham.)

In most cases, state and local resources—often augmented by trained fire fighters brought in by the state and local volunteers—are adequate to contain, control, and ultimately extinguish these fires. But there are times when a governor foresees a need of federal assistance from the Federal Emergency Management Agency (FEMA).

The Disaster Relief Act in effect since 1974 contains a provision that authorizes the agency to provide grants, equipment, supplies, and personnel to state government, but not specifically to local governments. Though the agency can provide any or all of these forms of assistance to a state, the most common is financial reimbursement for a portion of its actual cost to battle the fire.

Financial reimbursement is determined by a formula based on the state's average annual expenditure to fight these types of fires. A "floor cost," based on a five-year average of state expenditures is used to determine the amount of extraordinary costs incurred by the state to fight a specific fire.

If the request for federal aid is approved and there are found to be extraordinary costs beyond the state's floor cost for the year, FEMA can provide reimbursement of 70 percent. It should be noted that fires on federal national forests and Bureau of Land Management lands are the responsibility of the federal agency concerned.

Whether assistance can be provided by FEMA depends upon a key phrase in the enabling legislation. According to the law, the specific fire—on nonfederal public or private lands—must "threaten destruction as would constitute a major disaster." In other words, the fire must pose a present or continuing threat to public safety or to property necessary to the public welfare.

The authority to grant fire suppression assistance has been delegated to FEMA. That means no specific action by the president is required.

EVALUATION OF DISASTER RESPONSE OPERATIONS: IMPROVING THE EMERGENCY MANAGEMENT PROGRAM

An important step in the recovery process that normally takes place near the end of the short-term recovery phase is an evaluation of the emergency response effort. This evaluation is a systematic comparison of the outcome of emergency response efforts to the response goals, objectives, and priorities established in the emergency operations plan. It is intended to identify specific changes indicated in the preparedness program including the hazard identification and emergency operations plans, and in capabilities for carrying out those plans.

Evaluation Methods

There are a number of methods for evaluating response plans and operations which are listed below.

- Review the hazard analysis. (Was this emergency anticipated?)
- Review documentation compiled during the response phase.
- Obtain outside input (local, state, federal officials, researchers).
- Evaluate public reaction or criticism.
- Hold critique sessions among emergency responders.
- Interview involved agency heads.

Evaluation Criteria

The major areas to be evaluated should follow the major elements in the emergency operations plans. Those categories and examples of specific aspects to evaluate are listed below.

- Direction, Control, Coordination
  - Were strategy and tactics effective?
  - Was the command organization effective?
  - Were control preparations effective?
  - Were resources adequate?
  - Were mutual aid procedures adequate?
  - Was coordination among external agencies adequate?
- Alerting and Warning

Was warning information heeded?
Were there preventable injuries or fatalities?

- Communications
  - Were there major hardware failures?
  - Were emergency notifications made?
  - Were field tactical communications adequate?
  - Were EOC/field communications adequate?
  - Were there problems in receiving requests for emergency services from the public?
- Emergency Public Information
  - Were there complaints from the public?
  - Were there complaints from the news media?
  - Were there excessive rumors or misinformation?
  - Was conflicting information distributed by government agencies?
  - Were emergency information bulletins/recommendations heeded?
  - Was the public response appropriate?
  - Did the public receive sufficient information.
- Assessment of Hazards to Responders
  - Were personnel injured or resources lost unnecessarily?
- Size-up and Situation Status
  - Was available information sufficient for action planning?
- Evacuation and Sheltering
  - Was evacuation conducted in time?
  - Were transportation resources adequate?
  - Were shelter facilities adequate?
  - Was temporary housing (if long term) adequate?
  - Were food and support adequate?
- Law Enforcement
  - Was perimeter/traffic control adequate?
  - Was control of criminal activity adequate?
  - Were evacuation efforts of law enforcement agencies adequate?
- Fire Suppression
  - Were fires contained and controlled?
  - Were there adequate fire fighting sources (including water)?
  - Were search and rescue efforts effective?
  - Were hazardous material leaks controlled?
  - Were mutual aid plans effective?
- Medical/Mass Casualty
  - Was patient triage adequate?
  - Was field treatment appropriate and adequate?
  - Was distribution of casualties to hospitals adequate?
  - Were there sufficient resources including manpower, ambulances, equipment, and supplies?
  - Was management of fatalities appropriate and adequate?
  - Were casualty lists prepared and released appropriately?
- Public Works
  - Was debris cleared adequately?
  - Were public facilities protected (including sewers, waste treatment, dams, flood control facilities, reservoirs, aqueducts, cisterns)?

Was support in assessing damages adequate?

- ○ Transportation
  - Were transportation resources adequate?
  - Was maintenance/repair of transportation facilities (airports, roads, bridges, railways, mass transit systems, pipelines) adequate?
- ○ Human Services and Distance Relief
  - Were the following items available from relief agencies?
    - food
    - counseling
    - clothing
    - sheltering/temporary housing
    - nonemergency medical needs (medicine, oxygen, appliances)
    - money (for rent, food, etc.)
- ○ Utilities
  - Was service returned in a reasonable time?
  - Were resources (regular, standby, emergency alternate) adequate?
- ○ Administration
  - Were purchasing procedures adequate?
  - Was legal advice timely and adequate?
  - Were critical government records preserved?
  - Were overall emergency response costs adequately documented?
- ○ Recovery
  - Was there adequate funding available for necessary recovery activities?
  - Did recovery include provisions to prevent or minimize future emergencies?
  - Was there prompt response to requests for funds?
- ○ Hazard Identification
  - Was this hazard identified as a serious threat before the emergency?
  - Was the analysis of primary and secondary effects accurate?
  - Was the hazard corrected to prevent recurrence.

Incident Case Study

As a final step in the evaluation process it is often useful to summarize and document the response to the incident in a case study. A great deal will be learned about what actually happened as participants are interviewed and reports, documents, and photographs are collected.

Such a report is also helpful in sharing experiences and lessons learned with other agencies. The Learning Resource Center at the National Emergency Training Center maintains a comprehensive library of case studies of major incidents from around the world.

In compiling a case study it is important to identify problems and potential solutions. Generally it is more productive to focus on improving plans and response systems rather than to highlight poor individual decisions.

Improving the Emergency Management Program

After the conclusion of the evaluation process, efforts turn to preventing or mitigating future emergencies and to improving response. The hazard analysis should be reviewed and updated if the incident was unexpected, or if it caused effects that were different from those expected.

Plans at the departmental and community level should be reviewed and revised as necessary. Resources that were needed but unavailable should be identified and arrangements made for acquiring them in the future.

Finally, responder training programs and exercises should be developed to address the problems that invariably occur in large-scale incidents. The response will never be as the written plans describe, but substantial improvements in both the plans and in capabilities for carrying them out can be made.

CONCLUSION

Throughout this book, the seemingly endless complexities of disaster planning and control were examined. Regardless of where a disaster strikes, and regardless of the type of disaster, one fact remains clear. The greater the degree of advanced planning, the less serious will be the negative impact on the community. Those fire departments that have adopted a vigorous forward approach to disaster planning will be those most capable of serving their communities when disaster strikes. But keep in mind that despite every effort to anticipate all possible contingencies that could arise in every conceivable form of disaster, "Murphy's Law" may affect your best laid plans and unforeseen problems may arise. If this happens, you can console yourself with the old adage "and this, too, shall pass!"

CHAPTER 15 REFERENCES

1. Post-Disaster Stress in both South Carolina and California in Summer 1991 documented by William Kramer while researching information for Chapters 9 and 10 of this text.
2. Weldon, Curt. "Weldon's Wire—Finding Common Ground With a New Friend" in *Firehouse*, Vol. 16, No. 5. (May 1991), p. 21.
3. Wallace, Anthony F.C. *Tornado in Worcester: An Exploratory Study of Individual and Community Behavior in an Extreme Situation*, Washington, D.C.: National Academy of Sciences, National Research Council, 1956.
4. Collier, W. Steve. "Disaster Recovery" in *Disaster and Fire Defense Training*, Open Learning Course, Emmitsburg, MD: National Fire Academy, 1989.
5. Quarantelli, E.L., and Russell R. Dynes. "Community Response to Disasters," in *Disaster and Mental Health Selected Contemporary Perspectives*, ed. Barbara Sowder. Washington, D.C.: U.S. Government Printing Office, 1985, pp. 97-104.
6. A typical plane crash that would illustrate these characteristics was the USAir crash in Colorado Springs, Colorado; in *USA Today*, March 4, 1991, p. 1-A.

7. Collier, W. Steve See note 4, Unit 15.
8. "Navy Jet Crash" in *American Heat* video journal, St. Louis, MO, Vol. 3, Prog. 7 (January 1990).
9. "Northern California Earthquake" in *American Heat* video journal, St. Louis, MO, Vol. 4, Prog. 6 (December 1989).
10. Quarantelli, E.L., and Barbara Baisden. "The Recovery Period in U.S. Disasters: Problems and Opportunities" in *Proceedings: National Public Policy Forum for Disaster Relief*, Racine, WI: National Voluntary Organizations Active in Disasters, 1979.
11. Stapleton, Leo D. "Critical Incident Stress Debriefing Procedures" in *Commish*, Boston, MA, June 1991.
12. Moore, Jeffrey. "Self-Evaluation" in *Firehouse*, Vol. 16, No. 3 (March 1991), p. 79.
13. Mitchell, Ph.D., Jeffrey T. "Stress—Using the REAPER Model To Learn To Cope" in *Journal of Emergency Medical Services* (JEMS), September 1988.
14. Mitchell, Ph.D., Jeffrey T. "The History, Status and the Future of Critical Incident Stress Debriefings" in *Journal of Emergency Medical Services* (JEMS), November 1988.
15. Mitchell, Ph.D., Jeffrey T. "Stress—Development and Functions of a Critical Incident Stress Debriefing Team" in *Journal of Emergency Medical Services* (JEMS), December 1988.
16. Ibid.
17. Ibid.
18. Ibid.
19. Ibid.
20. "Firefighter's Deaths/Hackensack, N.J.—Firefighter's Stress" in *American Heat* video journal, St. Louis, MO, Vol. 3, Prog. 5 (November 1988).
21. Ward, Michael. "Recovering Costs for Better Service" in *Chief Fire Executive*, Vol. 2, No. 1 May/June 1987, pp. 41–43.
22. *When Disaster Strikes*, Federal Emergency Management Agency, U.S. Government Printing Office.
23. Weldon, Curt. See note 2.
24. Keith, William. "Plan Bulldozer" in *Firehouse*, Vol. 15, No. 10 (October 1990), p. 42.
25. "Major Disaster" is defined in the Disaster Relief Act of 1974. (Public Law 93-288).
26. "The Moral Minority" in *Newsweek*, May 6, 1991, p. 63.

GLOSSARY

ALS: Advanced Life Support.

Base: That location at which the primary logistics functions are coordinated and administered. (Incident name or other designator will be added to the term "Base.") The Incident Command Post may be co-located with the base. There is only one base per incident.

BLEVE: Boiling Liquid Expanding Vapor Explosion.

BLS: Basic Life Support.

Branch: That organizational level having functional/geographic responsibility for major segments of incident operations. The branch level is organizationally between section and division group.

CAD: See Computer-Aided Dispatch.

CAMEO: See Computer-Aided Management of Emergency Operations.

Camp: A geographical site, within the general incident area, separate from the base, equipped and staffed to provide food, water, and sanitary services to incident personnel.

CDC: Civil Defense Corps.

Cellular Mobile Telephone: A recently implemented mobile telephone system which utilizes trunked radio technology to virtually eliminate waiting time by efficiently utilizing a limited number of radio frequencies.

CEPP: Chemical Emergency Preparedness Program.

CERCLA: Comprehensive Environmental Response and Liability Act.

Chemical Transportation Emergency Center: (CHEMTREC) Offers guidance of the dangers and precautions associated with the uncontrolled release of hazardous materials.

CHEMTREC: See Chemical Transportation Emergency Center.

CHRIS: Chemical Hazards Response Information System.

Civil Disturbance: A demonstration of popular unrest that may manifest itself in acts of violence and destruction against human life and property.

Clear Text: The use of plain English in radio communications transmissions. No ten codes or specific codes are used when using clear text.

Command Staff: The command staff consists of the information officer, the safety officer, and the liaison officer, and reports directly to the incident commander.

Company: Any piece of equipment having a full complement of personnel.

Computer-Aided Dispatch: (CAD) Computer systems designed to enhance the speed and accuracy of dispatching equipment to emergencies.

Computer-Aided Management of Emergency Operations: (CAMEO) A computer program, developed by the federal government, which can be used in the field to provide real-time emergency information for managing chemical incidents.

Conflagration: Large continuous fire involving buildings on more than one city block.

COPE: Communities Organized for Potential Emergencies.

Disaster Medical Assistance Team: (DMAT) An organizational unit developed under the National Disaster Medical System.

Division: That organizational level having responsibility for operations within a defined geographic area or with functional responsibility. The division level is organizationally between the strike team and the branch. See also Group.

DMAT: See Disaster Medical Assistance Team.

DOD: Department of Defense.

EEM: Earthquake Emergency Mode.

EIS: Emergency Information System.

EMA: Emergency Management Assistance.

Emergency: As it relates to the federal disaster assistance program, a declaration made by the president of the United States authorizing certain specific assistance from one or more federal agencies. See Major Disaster.

EMS: Emergency Medical Service.

EOC: See Emergency Operations Center.

Emergency Operations Center: (EOC) A facility that serves as a multiagency focal point for coordination activities during disasters.

Engine Company: Any ground vehicle providing specified levels of pumping, water, hose capacity, and personnel.

EOP: Emergency Operations Plan.

EPA: Environmental Protection Agency.

Facsimile: (FAX) Electronic devices which can transmit copies or facsimiles of written documents via ordinary telephone lines.

FAX: See Facsimile.

Federal Coordinating Officer: A federal officer appointed to coordinate with state agencies following the establishment of a FEMA/state agreement for disaster assistance to communities in that state.

FEMA: Federal Emergency Management Agency.

FHA: Federal Housing Administration.

FIRESCOPE: FIre REsources of Southern California Organized for Potential Emergencies. (More recently—FIre RESources of California Organized for Potential Emergencies.)

General Staff: The group of incident management personnel comprised of the incident commander, the operations chief, the planning chief, the logistics chief, and the finance chief.

GHQ: General Headquarters.

Group: A functional division (e.g., air support, medical, salvage, etc.), having no geographical boundaries.

Group fire: Large fires involving several structures on the same city block.

HA MANEUVER: Sudden flight of personnel (Haul ass!).

Hazardous Material: Any substance that poses an unreasonable risk to health, safety, property, and the environment.

Hazardous Materials Response Team: A group of individuals whose abilities, training, and equipment inventory represent an asset in the handling of a hazardous materials incident.

HAZMAT: Hazardous Materials.

HEAR: See Hospital Emergency Administrative Radio.

Hertz: A unit of frequency of a periodic process equal to one cycle per second.

HICA/MYDP: Hazardous Identification, Capability Assessment, Multi-Year Development Plan.

Hospital Emergency Administrative Radio: (HEAR) A radio system developed to permit coordination among hospitals during disasters.

HWOs: Hurricane Warning Offices.

IAFC: International Association of Fire Chiefs.

IC: See Incident Commander.

Icon: A picture symbol utilized by certain computer programs to simplify use of the program by reducing the number of written terms which must be remembered.

ICP: See Incident Command Post.

ICS: See Incident Command System.

IEMS: Integrated Emergency Management System.

Incident Action Plan: The basic operational plan for the incident. The plan which is initially prepared at the first planning meeting containing general control objectives reflecting the overall incident strategy and specific suppression and rescue action plans for the next operational period.

Incident Command Post: (ICP) That location at which the primary command functions are executed.

Incident Command System: (ICS) The combination of facilities, equipment, personnel, procedures, and communications operating within a common organizational structure with responsibility for the management of assigned resources to effectively accomplish stated objectives pertaining to the incident.

Incident Commander: (IC) The individual responsible for the management of all incident operations.

MACS: Multiagency Coordination System.

MAJOR Disaster: As it relates to the federal disaster assistance program, a declaration made by the president of the United States authorizing broad federal assistance and funds as delineated by the Federal Disaster Relief Act of 1974. See Emergency.

MHz: Megahertz: a unit of frequency equal to one million hertz.

Multi-Hazard Approach: An approach to emergency planning in which one response plan is adapted for several types of emergencies.

National Disaster Medical System: (NDMS) A system being developed by the federal government to deal with very large mass casualty disasters.

NCP: National Contingency Plan.

NDMS: See National Disaster Medical System.

NHC: National Hurricane Center.

NIMS: National Interagency Management System.

NOAA: National Oceanic and Atmospheric Administration.

NOVA: National Organization for Victims Assistance.

NRT: National Response Team.

OCC: See Operations Control Center.

OES: Office of Emergency Services.

On Scene Coordinator: Federal official predesignated to coordinate and direct federal response to hazardous materials incidents.

Operations Control Center: A term used by some police and fire departments to designate the alarm office or dispatch center.

RACES: See Radio Amateur Civil Emergency Service.

Radio Amateur Civil Emergency Service: (RACES) Groups of amateur (HAM) radio operators formally affiliated with local and state emergency management agencies.

Repeater: An electronic device that allows 3-way communication on 2-frequency radio channels.

RESTAT: Resource Status.

Riot: The most serious variant of crowd behavior, characterized by the threat of destruction of lives and property.

SARA: Superfund Amendment and Reauthorization Act.

SBA: Small Business Administration.

Sector: Division under Incident Command.

Sectorization: Dividing the incident scene into more manageable command areas called sectors.

Single Hazard Approach: An approach to emergency planning in which each type of emergency has its own response plan.

SITSTAT: Situation Status.

Staging Area: That location where incident personnel and equipment are assigned on a three-minute available status.

State Coordinating Officer: An officer normally appointed by the state emergency management agency to coordinate federal disaster assistance with the federal coordinating officer and local officials.

Strike Team: Specified combinations of the same kind and type of resources, with common communications and a leader.

Task Force: A group of resources with common communications and a leader, temporarily assembled for a specific mission.

Terminal Node Controller: (TNC) An electronic device which divides and encodes computer data into "packets" and transmits it via regular two-way radio. The data are received and decoded by another TNC. The technology permits error-free transmission of data with minimal interruption of voice communications.

Terrorism: The use of nonmilitary violence by individuals and organizations for the attainment of political ends.

TMI: Three Mile Island.

TNC: See Terminal Node Controller.

Trunked Radio: A new technology designed to maximize the use of a limited number of frequencies by allowing them to be efficiently shared. Users of the system are linked on channels controlled by a computer rather than communication on dedicated radio frequencies.

UHF: See Ultra High Frequency.

Ultra High Frequency: (UHF) A segment of the radio spectrum at approximately 450 MHz.

USDA: United States Department of Agriculture.

USFS: United States Forest Service.

VA: Veterans Administration.

VHF: Very High Frequency; a segment of the radio spectrum at approximately 155 MHz.

Warning: A term developed by the National Weather Service meaning that a certain type of severe weather condition has actually been detected or spotted locally. Often follows the issuance of a severe weather "watch." Used in conjunction with specific weather types such as tornado, severe storm, blizzard, flash flood.

Watch: A term developed by the National Weather Service meaning that conditions are right for the formation of severe weather, but that conditions have not yet developed. Also used in conjunction with specific weather types.

Wildland Fire: Large unfriendly fires in which the natural cover of the fire area constitutes the primary fuel supply.

APPENDIX A

HAZARD IDENTIFICATION EXERCISE

The Federal Emergency Management Agency Integrated Emergency Management System (FEMA–IEMS) provides the following hazard identification exercise.

a. General. The Hazard Identification, section 200, contains questions on 22 hazards. The hazards included are those believed most likely to have a major impact on local jurisdictions throughout the United States. Space is provided at the end of the Hazard Identification format, however, to add other hazards not covered that are believed to pose a significant threat to the jurisdiction.

Answering the questions presented will require thought and judgment. The answers received will be viewed as the best estimates available from the local level. The answers are important since they reflect local knowledge and experience for which federal assumptions and statistics cannot be substituted.

b. Hazard Identification Questions. There are four questions for each hazard. The questions are listed along the top of a table format with the hazards and brief descriptions listed down the left column. Answer the questions for each listed hazard by marking "X" over the appropriate response (questions 1–3) or filling in the blank (question 4).

(1) *Question 1.* This asks if the hazard could potentially affect the jurisdiction. If the answer to this question is NO, skip the remaining questions for that hazard and go to the next hazard. If the answer is YES, go to question 2.

(2) *Question 2.* This asks if the hazard is a significant threat. The following four general factors should be considered in determining if hazard is significant:

(a) Historically, the hazard has affected the jurisdiction;

(b) Loss of life or property could result;

(c) The local emergency management organization would be involved if the event occurred; or

(d) Specific plans exist or are needed to respond to the hazard.

If the answer to question 2 is NO, skip the remaining questions on that hazard and go to the next hazard. If the answer is YES, also answer questions 3 and 4.

(3) *Question 3.* This asks how frequently the hazard occurs in the jurisdiction. Seven choices are provided.

(4) *Question 4.* This asks for an estimate of the population that could be affected by the hazard. When answering this question, consider the maximum number which could be affected by a worst case single event. For the purpose of this Hazard Identification, seriously affected means killed, injured, made ill, or requiring shelter, evacuation, or other measures to protect that portion of the population. In addition, any peak population that could be affected should be included in the population at risk estimate.

c. *Other Hazards.* Space is included at the end of the table to include hazards not listed that pose a significant threat to the jurisdiction. Please provide the hazard name in five words or less. Questions 3 and 4 should be answered for each of the added hazards.

d. *CPG 1-35a, Response Book.* Transfer the responses in the Hazard Identification section of CPG 1-35, Workbook to CPG 1-35a, Response Book. The Hazard Identification is presented in an abbreviated format in CPG 1-35a. It is important that all information requested be transferred to CPG 1-35a since this document will be transmitted to the State and, eventually FEMA. A black ball-point pen should be used for all entries, and please make the responses legible to facilitate data entry. When information is being transferred, care should be taken to assure that entries are copied correctly, and all appropriate questions have been answered.

NATURAL HAZARDS	1. Could this hazard affect your jurisdiction? If Yes, answer question 2. If No, go to next hazard. (Mark (X) the appropriate answer).	2. Is this hazard a significant threat to your jurisdiction? If Yes, answer questions 3 and 4. If No. go to next hazard. (Mark (X) the appropriate answer).	3. Approximately how often does this hazard occur in your jurisdiction? (Mark (X) the appropriate answer).	4. What is your best estimate of the total population that could be seriously affected by this hazard? Consider peak population if appropriate.
201. AVALANCHE Mass of sliding snow, occurs in mountainous terrain where snow is deposited on slopes of 20 degrees or more.	201.1 Ⓨ [N]	201.2 Ⓨ [N]	201.3 □ a. Once or more a year. □ b. Once every 5 years. □ c. Once every 10 years. □ d. Once every 50 years. □ e. Once every 100 years. □ f. Less than once in 100 years. □ g. Has not occurred.	201.4 __________ Enter a number
202. DROUGHT Prolonged period without rain.	202.1 Ⓨ [N]	202.2 Ⓨ [N]	202.3 □ a. Once or more a year. □ b. Once every 5 years. □ c. Once every 10 years. □ d. Once every 50 years. □ e. Once every 100 years. □ f. Less than once in 100 years. □ g. Has not occurred.	202.4 __________ Enter a number
203. EARTHQUAKE Sudden motion of the ground which may result in surface faulting (ground rupture), ground shaking, and ground failure.	203.1 Ⓨ [N]	203.2 Ⓨ [N]	203.3 □ a. Once or more a year. □ b. Once every 5 years. □ c. Once every 10 years. □ d. Once every 50 years. □ e. Once every 100 years. □ f. Less than once in 100 years. □ g. Has not occurred.	203.4 __________ Enter a number
204. FLOOD *Rivering:* Periodic overbank flow of rivers and streams. *Flash:* Quickly rising small streams after heavy rain or rapid snowmelt. *Urban:* overflow of storm sewer systems usually due to poor drainage, following heavy rain or rapid snowmelt.	204.1 Ⓨ [N]	204.2 Ⓨ [N]	204.3 □ a. Once or more a year. □ b. Once every 5 years. □ c. Once every 10 years. □ d. Once every 50 years. □ e. Once every 100 years. □ f. Less than once in 100 years. □ g. Has not occurred.	204.4 __________ Enter a number

NATURAL HAZARDS	1. Could this hazard affect your jurisdiction? If Yes, answer question 2. If No, go to next hazard. (Mark (X) the appropriate answer).	2. Is this hazard a significant threat to your jurisdiction? If Yes, answer questions 3 and 4. If No, go to next hazard. (Mark (X) the appropriate answer).	3. Approximately how often does this hazard occur in your jurisdiction? (Mark (X) the appropriate answer).	4. What is your best estimate of the total population that could be seriously affected by this hazard? Consider peak population if appropriate.
205. HURRICANE/TROPICAL STORM Large cyclonic storm accompanied by high winds, extreme rainfall, and storm surge.	205.1 Ⓨ N	205.2 Ⓨ N	205.3 □ a. Once or more a year. □ b. Once every 5 years. □ c. Once every 10 years. □ d. Once every 50 years. □ e. Once every 100 years. □ f. Less than once in 100 years. □ g. Has not occurred.	205.4 ________ Enter a number
206. LANDSLIDE A mass of sliding mud or rock.	206.1 Ⓨ N	206.2 Ⓨ N	206.3 □ a. Once or more a year. □ b. Once every 5 years. □ c. Once every 10 years. □ d. Once every 50 years. □ e. Once every 100 years. □ f. Less than once in 100 years. □ g. Has not occurred.	206.4 ________ Enter a number
207. TORNADO A small radius cyclonic windstorm.	207.1 Ⓨ N	207.2 Ⓨ N	207.3 □ a. Once or more a year. □ b. Once every 5 years. □ c. Once every 10 years. □ d. Once every 50 years. □ e. Once every 100 years. □ f. Less than once in 100 years. □ g. Has not occurred.	207.4 ________ Enter a number
208. TSUNAMI Seismic sea wave usually generated by submarine geophysical displacement.	208.1 Ⓨ N	208.2 Ⓨ N	208.3 □ a. Once or more a year. □ b. Once every 5 years. □ c. Once every 10 years. □ d. Once every 50 years. □ e. Once every 100 years. □ f. Less than once in 100 years. □ g. Has not occurred.	208.4 ________ Enter a number

209. VOLCANO Eruption from the earth's interior producing lava flows or violent explosions issuing rock, gases, and debris.	209.1 Ⓨ [N]	209.2 Ⓨ [N]	209.3 □ a. Once or more a year. □ b. Once every 5 years. □ c. Once every 10 years. □ d. Once every 50 years. □ e. Once every 100 years. □ f. Less than once in 100 years. □ g. Has not occurred.	209.4 ________ Enter a number
210. WILDFIRE Any instance of uncontrolled burning in grassland, brush, or woodlands.	210.1 Ⓨ [N]	210.2 Ⓨ [N]	210.3 □ a. Once or more a year. □ b. Once every 5 years. □ c. Once every 10 years. □ d. Once every 50 years. □ e. Once every 100 years. □ f. Less than once in 100 years. □ g. Has not occurred.	210.4 ________ Enter a number
211. WINTER STORM (SEVERE) Includes icestorm, blizzard, and extreme cold. Vulnerable areas would be subject to heavy snowfall, combined snow and high winds, or ice storms.	211.1 Ⓨ [N]	211.2 Ⓨ [N]	211.3 □ a. Once or more a year. □ b. Once every 5 years. □ c. Once every 10 years. □ d. Once every 50 years. □ e. Once every 100 years. □ f. Less than once in 100 years. □ g. Has not occurred.	211.4 ________ Enter a number

TECHNOLOGICAL/
MANMADE HAZARDS

212. CIVIL DISORDER Terrorist attack, riot, violent protest, demonstration, illegal assembly. Certain types of facilities are more vulnerable than others such as government buildings, universities, military bases, nuclear power facilities, correctional facilities.	212.1 Ⓨ [N]	212.2 Ⓨ [N]	212.3 □ a. Once or more a year. □ b. Once every 5 years. □ c. Once every 10 years. □ d. Once every 50 years. □ e. Once every 100 years. □ f. Less than once in 100 years. □ g. Has not occurred.	212.4 ________ Enter a number

TECHNOLOGICAL/ MANMADE HAZARDS	1. Could this hazard affect your jurisdiction? If Yes, answer question 2. If No, go to next hazard. (Mark (X) the appropriate answer).	2. Is this hazard a significant threat to your jurisdiction? If Yes, answer questions 3 and 4. If No, go to next hazard. (Mark (X) the appropriate answer).	3. Approximately how often does this hazard occur in your jurisdiction? (Mark (X) the appropriate answer).	4. What is your best estimate of the total population that could be seriously affected by this hazard? Consider peak population if appropriate.
213. DAM FAILURE Downstream flooding due to the collapse or failure of an impoundment. High area to the downstream inundation area as mapped by the Corps of Engineers or State/local agencies.	213.1 Ⓨ 🄽	213.2 Ⓨ 🄽	213.3 □ a. Once or more a year. □ b. Once every 5 years. □ c. Once every 10 years. □ d. Once every 50 years. □ e. Once every 100 years. □ f. Less than once in 100 years. □ g. Has not occurred.	213.4 ________ Enter a number
214. HAZARDOUS MATERIALS INCIDENT—FIXED FACILITY Uncontrolled release of hazardous materials from a fixed site. Jurisdictions with hazardous materials fabrication, processing, or storage site or hazardous waste treatment, storage, or disposal site are at risk.	214.1 Ⓨ 🄽	214.2 Ⓨ 🄽	214.3 □ a. Once or more a year. □ b. Once every 5 years. □ c. Once every 10 years. □ d. Once every 50 years. □ e. Once every 100 years. □ f. Less than once in 100 years. □ g. Has not occurred.	214.4* ________ Enter a number
215. HAZARDOUS MATERIALS INCIDENT—TRANSPORTATION Uncontrolled release of hazardous materials during transport. Areas at risk would be along highways, rail lines, pipelines, rivers, and port areas.	215.11 Highway: Ⓨ 🄽	215.21 Highway: Ⓨ 🄽	215.31 Highway: □ a. Once or more a year. □ b. Once every 5 years. □ c. Once every 10 years. □ d. Once every 50 years. □ e. Once every 100 years. □ f. Less than once in 100 years. □ g. Has not occurred.	215.41 Highway:† ________ Enter a number
	215.12 Rail: Ⓨ 🄽	215.22 Rail: Ⓨ 🄽	215.32 Rail: A B C D E F G □ □ □ □ □ □ □	215.42 Rail:† ________ Enter a number

*What is your best estimate of the total population within a 5 mile radius of all hazardous materials and/or waste sites in your jurisdiction?

†If an accident occurred in the most heavily populated area of your jurisdiction through which hazardous materials were transported, what is your best estimate of the total population within 5 miles of the accident site? (Answer for each of the transportation modes as applicable.)

215. *(cont.)*	215.13 Pipeline: Ⓨ N	215.23 Pipeline: Ⓨ N	215.33 Pipelines: A B C D E F G □ □ □ □ □ □ □	215.43 Pipelines:† ________ Enter a number
	215.14 River: Ⓨ N	215.24 River: Ⓨ N	215.34 River: A B C D E F G □ □ □ □ □ □ □	215.44 River:† ________ Enter a number
216. NUCLEAR ATTACK Blast and fallout effects assumed for a large-scale nuclear attack.	216.11 Ⓨ N Both Blast (2 psi or greater) and fallout.	NOT APPLICABLE	NOT APPLICABLE	216.41 Population in area assumed to receive 2 psi or greater ________ Enter a number
	216.12 All jurisdictions are assumed to be at risk from fallout.	NOT APPLICABLE	NOT APPLICABLE	216.42 (Number of people for whom fallout shelter planned) ________ Enter a number
217. POWER FAILURE Interruption or loss of electrical service for an extended period of time. (An extended period of time would be long enough to require emergency management organization response to needs for food, water, heating, etc., caused by loss of power.)	217.1 Ⓨ N	217.2 Ⓨ N	217.3 □ a. Once or more a year. □ b. Once every 5 years. □ c. Once every 10 years. □ d. Once every 50 years. □ e. Once every 100 years. □ f. Less than once in 100 years. □ g. Has not occurred.	217.4 ________ Enter a number
218. RADIOLOGICAL INCIDENT—FIXED FACILITY Uncontrolled release of radioactive material at a commercial power plant or other reactor facility. Areas at risk are designated as within the emergency planning zone of such facilities.	218.1 Ⓨ N	218.2 Ⓨ N	218.3 □ a. Once or more a year. □ b. Once every 5 years. □ c. Once every 10 years. □ d. Once every 50 years. □ e. Once every 100 years. □ f. Less than once in 100 years. □ g. Has not occurred.	218.4 ________ Enter a number

TECHNOLOGICAL/ MANMADE HAZARDS	1. Could this hazard affect your jurisdiction? If Yes, answer question 2. If No, go to next hazard. (Mark (X) the appropriate answer).	2. Is this hazard a significant threat to your jurisdiction? If Yes, answer questions 3 and 4. If No, go to next hazard. (Mark (X) the appropriate answer).	3. Approximately how often does this hazard occur in your jurisdiction? (Mark (X) the appropriate answer).	4. What is your best estimate of the total population that could be seriously affected by this hazard? Consider peak population if appropriate.
219. RADIOLOGICAL INCIDENT—TRANSPORTATION Any incident involving the shipment of radiological materials.	219.1 Ⓨ N	219.2 Ⓨ N	219.3 □ a. Once or more a year. □ b. Once every 5 years. □ c. Once every 10 years. □ d. Once every 50 years. □ e. Once every 100 years. □ f. Less than once in 100 years. □ g. Has not occurred.	219.4 ________ Enter a number
220. SUBSIDENCE Depressions, cracks, and sinkholes in the ground surface. Areas of high vulnerability are active or abandoned underground mining sites, areas subject to other hazards which could trigger subsidence (i.e., earthquake), or areas of extensive groundwater withdrawal.	220.1 Ⓨ N	220.2 Ⓨ N	220.3 □ a. Once or more a year. □ b. Once every 5 years. □ c. Once every 10 years. □ d. Once every 50 years. □ e. Once every 100 years. □ f. Less than once in 100 years. □ g. Has not occurred.	220.4 ________ Enter a number
221. TRANSPORTATION An incident involving passenger air or rail travel resulting in death or injury. Risk areas would be around airports with FAA control towers or with traffic flow heavy enough to pose a hazard, and passenger rail lines.	221.11 Air: Ⓨ N	221.21 Air: Ⓨ N	221.31 □ a. Once or more a year. □ b. Once every 5 years. □ c. Once every 10 years. □ d. Once every 50 years. □ e. Once every 100 years. □ f. Less than once in 100 years. □ g. Has not occurred.	221.41* ________ Enter a number
	221.12 Rail: Ⓨ N	221.22 Rail: Ⓨ N	221.32 Rail: A B C D E F G □ □ □ □ □ □ □	(Not applicable to rail)

*Total population in a 5 mile radius of all airports where accidents are considered a significant hazard.

	Enter names of other significant hazards (using five words or less). Answer questions 3 and 4.		3. Approximately how often does this hazard occur in your jurisdiction? (Mark (X) the appropriate answer).	4. What is your best estimate of the total population that could be seriously affected by this hazard? Consider peak population if appropriate.
222. URBAN FIRE Uncontrolled burning in residential, commercial, industrial, or other properties in developed areas. All urban areas are at risk to personal injury or property damage due to fire.	222.1 Ⓨ [N]	222.2 Ⓨ [N]	222.3 □ a. Once or more a year. □ b. Once every 5 years. □ c. Once every 10 years. □ d. Once every 50 years. □ e. Once every 100 years. □ f. Less than once in 100 years. □ g. Has not occurred.	222.4 __________ Enter a number
223. OTHER HAZARD			223.3 □ a. Once or more a year. □ b. Once every 5 years. □ c. Once every 10 years. □ d. Once every 50 years. □ e. Once every 100 years. □ f. Less than once in 100 years. □ g. Has not occurred.	223.4 __________ Enter a number
224. OTHER HAZARD			224.3 □ a. Once or more a year. □ b. Once every 5 years. □ c. Once every 10 years. □ d. Once every 50 years. □ e. Once every 100 years. □ f. Less than once in 100 years. □ g. Has not occurred.	224.4 __________ Enter a number

APPENDIX B

EMERGENCY OPERATIONS PLAN, MAPLEWOOD, MINNESOTA

Following is the basic Emergency Operations Plan of the City of Maplewood, Minnesota, revision 1. The plan is updated yearly.

I. Reason for Plan

Tornadoes, floods, blizzards, and other natural disasters can affect the City of Maplewood. In addition, major disasters such as train wrecks, plane crashes, explosions, accidental releases of hazardous materials, and enemy attack pose a potential threat to public health and safety in Maplewood. An emergency plan is needed to ensure the protection of the public from the effects of these hazards.

II. Purpose of Plan

The City of Maplewood has many capabilities and resources which could be used in the response to any major disaster. These include the facilities, equipment, manpower, and skills of both government and non-government professions and groups.

The purpose of this plan is to ensure the effective, coordinated use of these resources so as to:

A. Maximize the protection of life and property
B. Ensure the continuity of government
C. Sustain survivors
D. Repair essential facilities and utilities

III. Legal Basis and References

The following references establish the legal authority and responsibility for putting this plan into action:

A. Public Law 920, as amended
B. Public Law 99-499, (Superfund Amendments and Reauthorization Act (SARA) of 1986)
C. Minnesota Statutes, Chapter 12, as amended
D. Pipeline Safety Law, Chapter 353, State Statute
E. City of Maplewood Ordinance No. 378, dated Oct. 17, 1974
F. Minnesota Division of Emergency Management BULLETIN No. 87-1
G. Ramsey County-county wide Emergency Operations Plan
H. Mutual aid agreements
I. Hospital plan(s)
J. Facility hazardous materials plans

IV. City Organization

Existing government is the basis for emergency operations. That is, government agencies will perform emergency activities related to those they perform on a day-to-day basis. City organization and interrelationships are shown on chart A of this basic plan.

*Chart A: City Organization and Interrelationships**

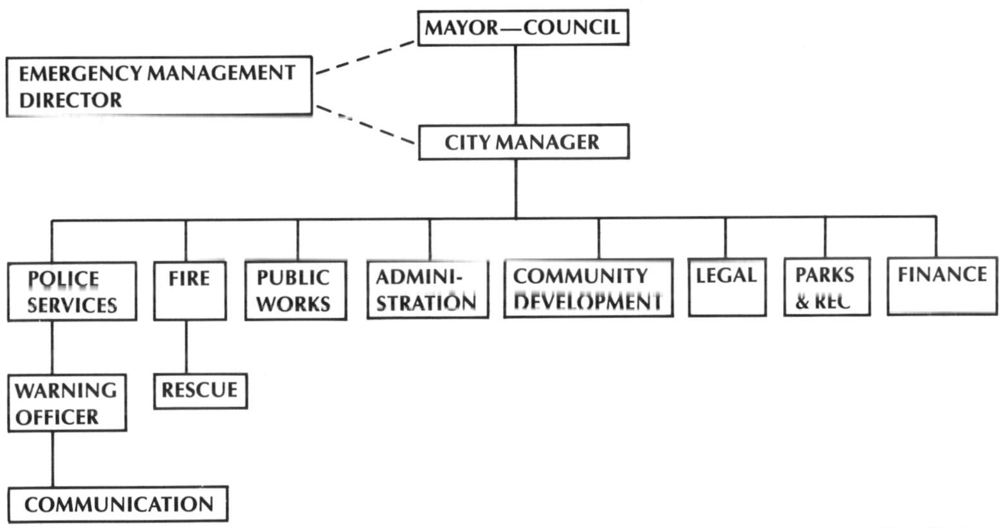

*Chart includes primary operational positions/departments only.

V. Direction and Control

The direction and control of government operations from a central, protected facility with adequate communications and key personnel is essential to the conduct of emergency operations.

In the City of Maplewood, the Mayor is responsible for providing overall direction and control of city government resources involved in the response to a disaster. The City Emergency Management Director will serve in a staff capacity to the mayor and city manager, and will coordinate all aspects of this plan. The

City Emergency Management Director will also serve as a liaison with the Ramsey County Emergency Management Director.

The City Emergency Management Director will serve as the Community Emergency Coordinator. The Hazardous Materials Facility Emergency Coordinator for the St. Paul Water Purification Plan is the plant operations manager. (See SOP R/HMP page 30 for additional information.)

Direction and control of the City's response to a major disaster will take place from the Maplewood Emergency Operating Center (EOC). The City EOC is located at 1902 East County Road B.

(For additional EOC information such as staffing, communications, etc., refer to the Direction and Control Annex to this Plan.)

VI. Emergency Responsibility Assignments

A. A summary of the City of Maplewood emergency responsibility assignments, by functions, is shown on chart B. Heads of the various city government departments and agencies will be responsible for carrying out the assignments shown on this chart.

B. Responsibilities have been assigned by a code letter: "P," "S," or "C."

1. "P" indicates primary operational responsibility, which means the official or agency is in charge of, and responsible to make provision for, that function.
2. "S" indicates support responsibility, which means the agency so assigned will, if possible, support and assist the official or agency designated primarily responsible.
3. "C" indicates coordination responsibility, and is assigned when several agencies have support capability but no specific official or agency has obvious primary responsibility. This will be especially true when non-government agencies are involved.

VII. Operations Policies

A. Protection of life and property during an emergency is the primary responsibility of government at all levels.
B. In an emergency affecting more than one political jurisdiction, officials of all jurisdictions involved will coordinate their services to the maximum extent possible.
C. The Maplewood Emergency Management Director will assist in providing resource coordination between government agencies and the private sector.
D. City government resources must be utilized to the maximum before state or federal assistance will be made available.
E. Each agency, department, or service of city government will provide for the maintenance of records during an emergency. These records should include work hours, equipment hours, supplies and materials consumed, injuries to personnel, and damage to public facilities and equipment.

Chart B: Emergency Responsibility Assignments

Function	*Responsible Agencies*	*Remarks*
1. Warning and Notification	Public Safety-P Emergency Mgmt-C	Warning sirens activated by Ramsey County Sheriff's Office
2. Direction and Control	City Elected Admin-P All City Depts-S Emergency Mgmt-C	
3. Emergency Public Information	Public Info. Officer-P Mayor-P Emergency Mgmt-C	Public Info dissemination will be by joint agreement between Mayor and Emerg. Mgmt. via the P.I.O.
4. Search and Rescue	Fire Departments-P Public Safety-S Emergency Mgmt-C	Additional volunteer units available for search
5. Health/ Medical	Police Paramedics-P Fire-P County Health Officer-C	
6. Evacuation, Traffic Control and Security	Police Department-P Police Reserves-S Emergency Mgmt-C	
7. Fire Protection	Fire Department-P Mutual Aid-S Fire Marshal-C	
8. Damage Assessment	Building Inspector(s)-P Red Cross-S Emergency Mgmt-C	
9. Congregate Care	Red Cross-P Salvation Army-S Emergency Mgmt-C	
10. Debris Clearance	Public Works-P Parks and Rec-S Emergency Mgmt-C	
11. Utilities Restoration	No. States Power-P St. Paul Water Dept-P No. States Gas-P No. St. Paul Util-P Maplewood Sewer Dept-S Public Works-S Emergency Mgmt-C	
12. Radiological	Emergency Mgmt-P Fire Departments-S	
13. Hazardous Materials Protection	Fire Departments-P Emergency Mgmt-S	

Code: P = Primary, S = Support, C = Coordination

VIII. Support

A. National Guard

1. Overview

When a natural disaster or other major emergency is beyond the capability of local government, support from National Guard units may be available. Only the governor, commander-in-chief of the Minnesota National Guard, has the authority to activate the Guard. The purpose of the activation is to ensure the preservation of life and property and to support civil law enforcement agencies:

a. National Guard assistance will complement, and not be a substitute for city and/or county participation in emergency operations.
b. If made available, National Guard personnel remain under military command at all times, but will support and assist county and/or city forces in the accomplishment of a specific task or tasks.

2. Request Procedure

In the case of the county and all cities that are not of the first class, the sheriff must submit the request for assistance to the governor's office. (In the case of cities of the first class, the mayor, after consulting with the county sheriff, will submit the request to the governor.) Requests for activation of the National Guard should be submitted to the State DEM duty officers by calling 1-800-422-0798 (outside the Minneapolis/St. Paul metro area), or 649-5451 (within the Minneapolis/St. Paul metro area).

B. State and Federal Agencies

1. Information and assistance in securing state or federal support may be obtained by contacting the County Emergency Management Director.
2. A summary of state and federal programs available to local governments which are experiencing or have experienced a disaster is contained in Disaster Response and Recovery: A Handbook For Local Government. This document was developed by the Minnesota Division of Emergency Management (DEM), and is on file with the Maplewood Emergency Management Director.
3. Federal Assistance—Hazardous Materials Incidents

The following types of assistance can be accessed through the State DEM duty officer.

The On-Scene Coordinator (OSC) is the federal official predesignated by the Environmental Protection Agency or the U.S. Coast Guard to coordinate and direct federal responses and removals under the National Contingency Plan; or the Department of Defense (DOD) official designated to coordinate and direct the removal actions from releases of hazardous substances, pollutants, or contaminants from DOD vessels and facilities. When the National Response Center (NRC) receives notification of a pollution incident, the NRC duty officer notifies the appropriate OSC, depending on the location of an incident. Based on this initial report and any other information that can be obtained, the OSC will go to the scene and monitor the response of the responsible party of state or local government. If the responsible party is unknown or not taking appropriate action, and the response is beyond the capability of state and local governments, the OSC may initiate federal actions, using funding from the Federal Water Pollution Control Act Pollution Fund for oils discharges and the CERCLA Trust Fund (Superfund) for hazardous substance releases.

The Regional Response Teams (RRT) are composed of representatives of federal agencies and a representative from each state in the federal region. During a response to a major hazardous materials incident involving transportation or a fixed facility, the OSC may request that the RRT be convened to provide advice or recommendations in specific issues, requiring resolution. Under the national Contingency Plan, RRTs may be convened by the chairman when a hazardous materials discharge or release exceeds the response capability available to the OSC in the place where it occurs; crosses regional boundaries; or may pose a substantial threat to the public health, welfare, or environment, or to regionally significant amounts of property. Regional contingency plans specify detailed criteria for activation of RRTs. RRTs may review plans developed in compliance with Title III, if the local emergency planning committee so requests.

IX. Plan Updating and Distribution

A. For purposes of this Plan, the City of Maplewood Emergency Management Director shall serve as the planning coordinator. As such, the director will have overall authority and responsibility for the maintenance of the plan.

B. This plan will be reviewed and updated as necessary, but at least once annually. The Maplewood Emergency Management Director will be responsible for ensuring that this updating occurs, and that it is in accord with the schedule and procedures established by the Minnesota Division of Emergency Management. In order to carry out this task, the Director may request assistance from the Ramsey County Emergency Management Director.

C. The City of Maplewood is in compliance with the Division of Emergency Management and Federal Emergency Management exercise requirements, as published.

D. This plan will be distributed to all city government departments and agencies which have emergency assignments in the event of a major disaster in the City of Maplewood. A plan distribution list will be maintained by the Maplewood Emergency Management Director.

A P P E N D I X C

MUTUAL AID PLAN: CITY OF DAVIS AND REGENTS OF THE UNIVERSITY OF CALIFORNIA

Agreement for Automatic Mutual Fire Protection Aid Between the City of Davis and the Regents of the University of California

THIS AGREEMENT, made and entered into this .... day of ...., 1976, by and between the CITY OF DAVIS, a municipal corporation hereinafter referred to as CITY, and THE REGENTS OF THE UNIVERSITY OF CALIFORNIA, hereinafter referred to as UNIVERSITY, both organized and existing under and by virtue of the laws of the State of California.

WITNESSETH:

WHEREAS, the parties hereto are geographically located in proximity to each other in the County of Yolo; and

WHEREAS, it is to the mutual advantage and benefit of the parties hereto that each of the other parties effectively assist the other in the event of a fire or other emergency that endangers life or property;

NOW, THEREFORE, in consideration of their mutual covenants, the parties hereto agree as follows:

1. That it is hereby set forth to be a policy of the City, and the Chief of the Davis Fire Department is hereby directed to respond, pursuant to the running card assignment system, with appropriate equipment and normal complement of personnel within those areas of the Davis Campus of the University designated by the running card assignment system. The City companies shall, under such circumstances, come under the authority of the senior officer of the University Fire Department at the scene.

2. That it is hereby set forth to be the policy of the University, and the Chief of the Fire Department on the Davis Campus of the University is hereby directed to respond, pursuant to the running card assignment system, with appropriate equipment and normal complement of personnel within those areas of the City designated by the running card assignment system. The University companies shall, under such circumstances, come under the authority of the senior officer of the Davis Fire Department at the scene.
3. The Chief of the Davis Fire Department and the Chief of the University Fire Department on the Davis Campus shall have the responsibility for coordinating running card assignments and response equipment and personnel in such a manner as to insure that an adequate level of fire protection service remains available to each jurisdiction.
4. No party to this agreement shall be required to pay any compensation to any other party to this agreement for services rendered hereunder, the mutual advantages and protection afforded by this agreement being considered adequate compensation to all of the parties.
5. This agreement is intended to cover day-to-day mutual aid only and shall be of no force and effect in cases in which a state of extreme emergency has been proclaimed and when the California Disaster and Civil Defense Master Mutual Aid Agreements become operative.
6. It is mutually understood and agreed that this agreement does not relieve any of the parties hereto from the necessity and obligation of providing adequate fire protection within its own areas, and each party hereto agrees that it shall use reasonable diligence in keeping the fire fighting equipment in its possession up to the minimum standards recommended by the Insurance Services Office for cities of comparable size.
7. This agreement shall not be construed as or deemed to be an agreement for the benefit of any third party or parties, and no third party or parties shall have any right of action hereunder for any cause whatsoever. Any services performed or expenditures made in connection with furnishing mutual aid under this agreement by either party hereto shall be deemed conclusively to be for the direct protection and benefit of the inhabitants and property of such party.
8. Any party to this agreement may withdraw at any time, upon thirty days' written notice to each of the other parties, and thereafter, such withdrawing party shall no longer be a party to this agreement.

IN WITNESS WHEREOF, this agreement has been executed the day and year first above written by the parties hereto through the mayor of the City of Davis, first thereunto duly authorized by the city council, and by The Regents of the University of California.

CITY OF DAVIS	THE REGENTS OF THE UNIVERSITY OF CALIFORNIA
By .... Mayor	By .... Vice President-Business & Finance
By .... Fire Chief	By .... Fire Chief
Attest:	Attest:
By .... City Clerk	By ....
Approved As To Form	Approved As To Form
By .... City Attorney	By .... Assistant Counsel of the Regents of the University of California

APPENDIX D

SANTA CRUZ COUNTY FIRE SERVICES MUTUAL AID PLAN

Mutual Aid/Disaster Preparedness Committee: Dave Mack—Chairman, Jerry Bowles, Ed Ekers, Don French, Steve Negro, Bud Tomlin, Dan Turner, Mel Angel—Advisory Member.

I. SANTA CRUZ COUNTY FIRE MUTUAL AID PLAN, JUNE 1989

A. PURPOSE

To provide a uniform mutual aid plan for the most effective mobilization and utilization of fire fighting resources within Santa Cruz County.

B. CONCEPTS

1. No department has the resources to effectively handle all potential emergencies. Fire officials must pre-plan for emergency operations to insure the efficient utilization of all available resources.
2. The foundation of California's fire service emergency planning is a statewide system of mutual aid. The County Plan is compatible with the California Fire and Rescue Emergency Plan thus insuring effective mutual aid operations.
3. The plan *does not* preclude neighbor-to-neighbor arrangements for mutual aid on a day-to-day basis. The County Plan is intended to provide mutual aid coordination on those major emergencies which exceed the day-to-day capabilities and require multi-zone response.
4. The Local Fire Official/Incident Commander (IC) requesting mutual aid assistance shall be in charge of all personnel and apparatus assigned to the incident.

C. BASIC ORGANIZATION

1. *Area Fire Coordinator (AFCO)*—A chief officer elected by the County Fire Chiefs, responsible for receiving and accounting for requests for mutual aid assistance from fire agencies within Santa Cruz County. The AFCO shall maintain a current inventory of the fire service resources within the County.

   In the event of a major County-wide emergency as many as fourteen fire agencies may process resource requests through the Area Fire Coordinator. To maintain span of control, the County is divided into three zones, each with a Zone Coordinator.

   The three zones are:

   North Zone: Ben Lomond, Boulder Creek, Felton, Zayante, and CDF/County Fire*

   Mid Zone: Aptos/La Selva, Branciforte, Central, Santa Cruz, Scotts Valley, and CDF/County Fire*

   South Zone: Aromas, Freedom, Salsipuedes, Watsonville, and CDF/County Fire*

2. *Zone Coordinators*—A fire agency representative and alternate from North, Mid, and South County will be selected to act as a Zone Coordinator for their area. During countywide or major emergencies, each Zone Coordinator will coordinate resources in the local area and keep the Area Fire Coordinator informed as to resource status and needs.
3. *Local Fire Official*—The fire official responsible for preplanning and emergency response activities within a jurisdiction. Provides the Area Fire Coordinator inventory information on resources within that department. In the event an emergency occurs which requires mutual aid assistance, the Local Fire Official/IC shall be responsible for initiating requests for assistance and shall be in charge of all assistance provided.
4. *County Communications (Fire Control)*—A service organization with no command function. Since most departments contract for dispatching service, this organization is a vital link to daily mutual aid in Santa Cruz County.
5. *Incident Command System*—The Santa Cruz County Fire Chiefs Association adopted the Incident Command System in an effort to standardize operational procedures and to put common terminology into effect when departments work together.

*CDF/County Fire Assignments:
North: Stations 21, 23, 31, 32, 34, and 36
Mid: Stations 33, 37, 43, 44, 45, and 47
South: Stations 41, 42, and 46

D. OPERATION AND IMPLEMENTATION OF LOCAL AND STATEWIDE MUTUAL AID

1. When an emergency exceeds a local jurisdiction's capabilities that jurisdiction should request mutual aid assistance utilizing this plan.
2. The AFCO will be notified when an incident goes to a third alarm or strike team assignment.
3. Upon the request of the Incident Commander (IC), the AFCO will place a Logistics Section Chief, with communications, at the scene to assist the IC.
4. The AFCO will activate the Zone offices when an incident(s) grows to the magnitude that delegation of the workload is necessary.
5. If the request for resources exceeds the capabilities of the County's fire service, the AFCO shall request mutual aid from the Region II Fire Coordinator or as stipulated in the State plan.

Exhibit I.C.

CHANNELS FOR FIRE SERVICE MUTUAL AID

STATE
FIRE COORDINATOR

REGION II
FIRE COORDINATOR

SANTA CRUZ COUNTY
AREA FIRE COORDINATOR

SANTA CRUZ COUNTY
ZONE COORDINATORS

LOCAL
FIRE AGENCIES

DAY-TO-DAY
MUTUAL AID

II. 2nd AND 3rd ALARM ASSIGNMENTS

A. PURPOSE

The 2nd and 3rd alarm assignment concept designates a predetermined response to incidents not requiring the need of a full strike team response.

B. GOAL

The goal of 2nd and 3rd alarm assignments is to provide a systematic fire response from fire agencies.

C. OPERATIONAL OVERVIEW

A 2nd and/or 3rd alarm assignment can be requested by any Incident Commander.

The typical 2nd and 3rd alarm assignment card will include the type of emergency (i.e., urban structure, rural structure, wildland, and medical), the responding agency, and the kind of equipment to be dispatched.

D. GENERAL INFORMATION

1. *Apparatus*
   a. Minimum staffing of an engine company is three people.
   b. All dispatched units will respond to the reporting location.

E. PROCEDURE

1. When the 2nd or 3rd alarm assignment is requested, the IC will specify the type of response, i.e., urban structure, rural structure, wildland, or medical.
2. The IC will designate the location of the staging area or other reporting location(s).
3. The IC will specify the code for response: Code 2 or Code 3.

III. THE STRIKE TEAM PLAN

A. PURPOSE

The Strike Team Plan recognizes the need for standardization and uniformity of response in coping with fires and other emergencies.

The following sections of this plan provide specific guidelines to assist in a smooth transition to a larger effort at the local level.

B. GOALS

The goals of the Strike Team Plan are:

1. To provide a systematic allocation of predetermined fire protection resources.
2. To provide back-up responses to those agencies responding or involved with major emergencies.

C. OPERATIONAL OVERVIEW

The Strike Team Plan can be activated by an Incident Commander. In addition, Santa Cruz County fire agencies will send strike teams to adjoining State Office of Emergency Service (OES) Operational Areas and throughout California as per the State OES Plan.

The Strike Team Plan divides the County into the zones used for AREA FIRE COORDINATION. Assigned to each zone are fire protection agencies having the same geographical proximity. (See Exhibit III.C.)

Exhibit III.C. Santa Cruz County Engine Strike Team Configurations

NORTH ZONE STRIKE TEAM AGENCIES

CDF/County Fire Boulder Creek Ben Lomond Felton Zayante ——— Scotts Valley (Alt.) Santa Cruz (Alt.)	For incidents occurring in the North Zone area, the 2nd strike team will come from the Mid Zone area and the 3rd strike team will come from the South Zone area.

MID ZONE STRIKE TEAM AGENCIES

Scotts Valley Branciforte Santa Cruz Central Aptos/La Selva ——— Central (Alt.) CDF/County Fire (Alt.)	For incidents occurring in the Mid Zone area, the 2nd strike team will come from the South Zone area and the 3rd strike team will come from the North Zone area.

SOUTH ZONE STRIKE TEAM AGENCIES

Freedom CDF/County Fire Watsonville Salsipuedes Aromas ——— Aptos/La Selva (Alt.) Central (Alt.)	For incidents occurring in the South Zone area, the 2nd strike team will come from the Mid Zone area and the 3rd strike team will come from the North Zone area.

Each agency requesting an Engine strike team will receive five (5) engines and a Strike Team Leader.

D. GENERAL INFORMATION

1. *Apparatus*
   a. The minimum staffing for each engine company is three people; four people is preferred.

b. All dispatched units will respond to the reporting location or staging area.

2. *Operational*
   a. Specialized equipment will be dispatched upon request of the Incident Commander (truck, rescue companies, light towers, foam, etc.).
   b. If initial commitments to apparatus have been made by neighboring departments prior to the activation of the County Strike Team Plan, those participating departments are not obligated nor are they expected to send additional engine companies when notified.

3. *Response Times*

In-County and Adjoining Operational Area Strike Teams are normally an "immediate need" type of request. To insure the most effective response, the following getaway times are established.

   a. IMMEDIATE NEED RESPONSES: individual units shall respond within 5 minutes of dispatch.
   b. PLANNED NEED RESPONSES: individual units shall respond within 10 minutes of dispatch.
   c. LONG-TERM/STATEWIDE RESPONSES: individual units shall be enroute to the staging area within 15 minutes of dispatch.

Note: Participating agencies, including alternates, will be given as much notice as possible of pending dispatch involving Long-Term assignments.

4. *Communications*
   a. All dispatched Fire Agencies will *immediately* confirm their response status with County Communications.
   b. Engines going in service as a part of a strike team assignment will be assigned a tactical frequency.
   c. When calling for an In-Zone Strike Team, it must be remembered that part of the assigned strike team may be on the scene already through normal automatic or mutual aid. In those instances, the Second Out Strike Team will be required.
   d. Engine companies dispatched with a strike team will be directed to a reporting location or staging area. If directed to the emergency, the company officer in charge of the engine shall report to the Strike Team Leader or the Incident Commander at the Scene. [See Exhibit III.E.-1, Predetermined Staging Areas (In-County Response).]
   e. Strike Teams and Leaders will be identified by using the State OES Resource Designation System. Any need to contact the Strike Team Leader by radio will be stated as in this example: "Strike Team 2325 B(aker), Fire Control."
   f. The IC shall make every effort to release responding strike team resources as soon as possible.

Exhibit III.E.–1: Predetermined Staging Areas (In-County Response)

Staging Area Number and Location
1. Highway 9 and Highway 236.
2. Big Basin State Park Headquarters.
3. Railroad Avenue, off Middleton, Boulder Creek.
4. Felton Faire Shopping Center, Mt. Hermon and Graham Hill.
5. Scotts Valley Fire Station #1, 7 Erba Lane, Scotts Valley.
6. Highway 1 in Davenport, across from Cement Plant.
7. Entrance to UCSC, High Street and Glen Coolidge Drive.
8. Ocean Street Extension, adjacent to Cemetary.
9. De Laveaga Park parking lot, off Branciforte Drive.
10. Commercial Way, next to Standard Station, Live Oak.
11. 41st Avenue, K-Mart Parking Lot, Soquel.
12. Resurrection Church parking lot, State Park Drive, Aptos.
13. San Andreas Road, just south of Highway 1, Larkin Valley.
14. Highway 129 (Riverside Rd.), just east of Highway 1.
15. Rear of Freedom Fire Station, Freedom Blvd. and Airport Blvd.
16. Salsipuedes Fire Station, Casserly Road.
17. Corralitos Fire Station, Eureka Canyon Road.
18. Eureka Canyon Road and Buzzard Lagoon Road.
19. Summit Road and Loma Prieta Elementary School.
20. Skyline Blvd. at Castle Rock State Park parking lot.
21. Highway 9 entrance to Highlands Park.

g. The IC will have command authority of all personnel and apparatus assigned to the incident from other jurisdictions.

h. The receiving department should make every effort to place someone knowledgeable about the area with the incoming Strike Team Leader(s).

5. *Strike Team Leader Responsibilities (on Out-of-County Assignments)*

   a. Will call the AFCO daily to provide status and receive any instructions from sending agencies.

   b. Is responsible for the direct supervision of all Engine Company Officers on the strike team and compliance with local agency policies.

Exhibit III.E.–2

STRIKE TEAM LEADER ASSIGNMENTS BY DEPARTMENT

NOTE: Department in top position each month has primary responsibility to provide the Strike Team Leader for that Zone. Department in bottom position each month is the alternate.

Zone						
North Zone	*January*	*February*	*March*	*April*	*May*	*June*
	Boulder Creek Page 286	Scotts Valley Page 558	CDF/CO. Fire Hotline CDF	Boulder Creek Page 286	Scotts Valley Page 558	CDF/CO. Fire Hotline CDF
	CDF/CO. Fire Hotline CDF	Boulder Creek Page 286	Scotts Valley Page 558	CDF/CO. Fire Hotline CDF	Boulder Creek Page 286	Scotts Valley Page 558
	July	*August*	*September*	*October*	*November*	*December*
	Boulder Creek Page 286	Scotts Valley Page 558	CDF/CO. Fire Hotline CDF	Boulder Creek Page 286	Scotts Valley Page 558	CDF/CO. Fire Hotline CDF
	CDF/CO. Fire Hotline CDF	Boulder Creek Page 286	Scotts Valley Page 558	CDF/CO. Fire Hotline CDF	Boulder Creek Page 286	Scotts Valley Page 558
Mid Zone	*January*	*February*	*March*	*April*	*May*	*June*
	Central Page 509	Santa Cruz Page 288	Central Page 509	Santa Cruz Page 288	Central Page 509	Santa Cruz Page 288
	Santa Cruz Page 288	Central Page 509	Santa Cruz Page 288	Central Page 509	Santa Cruz Page 288	Central Page 509
	July	*August*	*September*	*October*	*November*	*December*
	Central Page 509	Santa Cruz Page 288	Central Page 509	Santa Cruz Page 288	Central Page 509	Santa Cruz Page 288
	Santa Cruz Page 288	Central Page 509	Santa Cruz Page 288	Central Page 509	Santa Cruz Page 288	Central Page 509
South Zone	*January*	*February*	*March*	*April*	*May*	*June*
	Freedom Page 566	Salsipuedes Page 805	Watsonville Hotline WATS	Aromas Hotline Mtry Cnty	CDF/CO Fire Hotline CDF	Freedom Page 566
	CDF/CO Fire Hotline CDF	Freedom Page 566	Salsipuedes Page 805	Watsonville Hotline WATS	Aromas Hotline Mtry Cnty	CDF/CO Fire Hotline CDF
	July	*August*	*September*	*October*	*November*	*December*
	Salsipuedes Page 805	Watsonville Hotline WATS	Aromas Hotline Mtry Cnty	CDF/CO Fire Hotline CDF	Freedom Page 566	Salsipuedes Page 805
	Freedom Page 566	Salsipuedes Page 805	Watsonville Hotline WATS	Aromas Hotline Mtry Cnty	CDF/CO Fire CDF/CO Fire	Freedom Page 566

E. PROCEDURE: IN-COUNTY RESPONSES

1. The department requesting assistance will notify County Communications who will then coordinate dispatch of all equipment.
2. The Incident Commander will provide the following information to County Communications.
   b. Nature and location of emergency
   b. Amount of assistance being requested: one, two or three strike teams
   c. Anticipated strike team assignment
   d. Reporting location for responding units:
      1) To the scene
      2) To the pre-determined staging area (See Exhibit III.E.-1)
      3) Cover in stations
   e. Incident Commander and Command Post location
   f. Incident Commander to designate whether Code 3 or Code 2 response is desired
3. Strike Team Leaders for each Zone will be pre-scheduled each December and identified by Department for monthly assignment. (See Exhibit III.E.-2)
   a. County Communications will announce agencies designated for Strike Team Leader during dispatch of strike team. That agency will then identify the Strike Team Leader by radio identifier.

F. PROCEDURE: ADJOINING OPERATIONAL AREA

1. The State Mutual Aid Plan allows Adjoining Operational Area Fire Coordinators to dispatch five (5) units across county lines when requested by the adjoining AFCO. After dispatch, this request is then processed with the OES Regional Coordinator.

   Santa Cruz County fire agencies in accordance with the State Plan will interact with our counterparts in San Mateo, Santa Clara, San Benito, and Monterey Counties in sending fire apparatus across county lines.

   The request will normally be for a short term, to fill the gap between day-to-day mutual aid and the arrival of other OES dispatched equipment.
2. Direct dispatch with adjoining counties will be made utilizing the strike teams made up within the three zones of Santa Cruz County. Strike Team Leaders will be assigned using the same system of assignments for an In-County Response. (See Exhibit III.F.)
3. Strike teams will assemble at a pre-determined Initial Staging Area prior to leaving the County, unless otherwise directed.

Exhibit III.F. Direct Dispatch to Adjoining OES Operational Area

Strike Team Assignments

ICS Identifier XCZ 2325 (North Zone Strike Team)	*ICS Identifier XCZ 2326 (Mid Zone Strike Team)*	*ICS Identifier XCZ 2327 (South Zone Strike Team)*
1. Boulder Creek	1. Scotts Valley	1. Watsonville
2. Ben Lomond	2. Santa Cruz	2. Freedom
3. Zayante	3. Central	3. Salsipuedes
4. Felton	4. Aptos/La Salva	4. Aromas
5. CDF/County Fire	5. Branciforte	5. CDF/County Fire
Alternates	Alternates	Alternates
1. Scotts Valley	1. Central	1. Aptos/La Selva
2. Santa Cruz	2. CDF/County Fire	2. Central

Assignments by County

Requesting County	*Travel Route*	*Assignment Sequence*	*Initial Staging Area*
Monterey	HWY 1, 101	1. South (XCZ 2327) 2. Mid Zone (XCZ 2326) 3. North Zone (XCZ 2325)	Riverside Rd @ HWY 1
San Benito	HWY 129, 101	1. South Zone (XCZ 2327) 2. Mid Zone (XCZ 2326) 3. North Zone (XCZ 2325)	HWY 129 @ Rogge Lane
South Santa Clara	HWY 129, 152, HWY 101	1. South Zone (XCZ 2327) 2. Mid Zone (XCZ 2326) 3. North Zone (XCZ 2325)	HWY 152 @ Fairground
North Santa Clara	HWY 17	1. Mid Zone (XCZ 2326) 2. North Zone (XCZ 2325) 3. South Zone (XCZ 2327)	HWY 17 @ Cloud 9
San Mateo, Coastal	HWY 1	1. Mid Zone (XCZ 2326) 2. North Zone (XCZ 2325) 3. South Zone (XCZ 2327)	HWY 1 @ Cement Plant
San Mateo, Inland	HWY 9, 35 (Skyline Blvd)	1. North Zone (XCZ 2325) 2. North Zone (XCZ 2326) 3. South Zone (XCZ 2327)	HWY 9 @ HWY 35

G. PROCEDURE: LONG-TERM/STATEWIDE USE

1. Long-Term/Statewide responses will be coordinated by the Area Fire Coordinator.
2. Each December, two strike teams will be identified by the Fire Chiefs Association. Each strike team will have at least two alternate units identified. (See Exhibit III.G.)
3. Each December, the list of Statewide Strike Team Leaders will be reviewed and updated, as needed.
4. When the OES Regional Coordinator makes a request for one or two strike teams, the AFCO will notify the appropriate departments/districts providing the necessary information. County Communications may assist in the actual dispatch.
5. The AFCO will select a Strike Team Leader, make personal contact, and provide all necessary information.

EXHIBIT III.G. Long-Term/Statewide Strike Team Assignments

FIRST OUT	ICS IDENTIFIER	XCZ 2328 A (4 people/engine) or XCZ 2328 B (3 people/engine)
	1. Freedom 2. Central 3. CDF/County Fire 4. Watsonville 5. Scotts Valley	
	Alternates	
	1. Felton 2. Zayante	
SECOND OUT	ICS IDENTIFIER	XCZ 2329 A (4 people/engine) or XCZ 2329 B (3 people/engine)
	1. Aptos/La Selva 2. Santa Cruz 3. Aromas 4. Salsipuedes 5. Boulder Creek	
	Alternates	
	1. Ben Lomond 2. CDF/County Fire	

NOTE: This form will be filled out each December and filed with the AFCO.

APP. A

Incident Command System Santa Cruz County Fire Services

INTRODUCTION

The Incident Command System is a management system consisting of personnel, facilities, equipment, communications and procedures operating within a common organizational structure to accomplish stated objectives pertaining to an incident.

The Incident Command System constitutes a somewhat different concept of organization and operations management from that which may exist within many of the participating organizations. The principal difference is in the method of organization.

The Incident Command System is designed to become the *basic operating* system for every incident within each agency on a day-to-day basis. The transition from normal day-to-day operations to that of multi-agency involvement should require a minimum of transition for any agency. There are advantages to be gained by the timely combining of resources under a commonly understood organizational structure to combat a serious incident. Combining forces requires a need for mutual understanding and agreement on organizational structure, common terminology, operating procedures and acceptance of the qualifications of all personnel who may be assigned.

The Incident Command System begins to develop from the time an incident occurs until the requirement for management and operations no longer exists. The Incident Commander is a title that can apply equally to any officer of a fire or law enforcement agency or to the chief of an agency. The title only implies the *function* which is to be carried out. The Incident Command System is intended to be staffed and operated by qualified personnel from any fire suppression or law enforcement agency and could involve personnel from a variety of agencies.

It is important that there be only one Incident Commander at any given time on a single agency incident. There may be Deputy Incident Commanders appointed depending on the complexity of the emergency and the need to encompass primary responsibilities. For example, during and immediately after an earthquake, first consideration would be given to life rescue and life preservation, containment of fires, etc., then later primary concern might shift to prevention of looting and willful damage to private property. At that time the Incident Command might shift from a fire officer to a law enforcement officer, or if the emergency moves, from one jurisdiction to another.

One of the more complex problems facing the development of the Incident Command System is command of a multi-jurisdictional, multi-agency incident.

This is based on the premise that jurisdictional authority of the participating agencies will not be compromised—only united. Each agency has legal responsibility with the Incident Command function. Agencies which assist in support of a jurisdiction on a multi-agency incident will function under the direction of the jurisdiction's designated Incident Commander.

In case of a multi-jurisdictional incident, the respective jurisdiction Incident Commanders will establish unified command and planning capability, effectively pool resources, and develop strategies for combating the overall incident situation.

On large-scale multi-jurisdictional incidents, a single location for a multi-agency incident command post must be established. This can include but is not limited to fire, police, medical, public works, and others as necessary to provide an integrated system.

The total Incident Command System is very complex. The use of the management portion of this system will not require extensive modification of the Santa Cruz County Mutual Aid Plan.

Each fire agency is responsible for training its personnel in the Incident Command System.

The following procedures are important to the success of the Incident Command System in Santa Cruz County.

A. COMMAND POST—INCIDENT COMMAND SYSTEM

1. When the need dictates, the Incident Commander will establish a Command Post and announce to those concerned the location of the Command Post.
2. The Command Post should be identified by a clear or green flashing light mounted on top of the vehicle and a 16″ × 17″ flag with the words "COMMAND POST" written on the sides. The flag will be international safety orange in color, and will be mounted on the vehicle in such a way that the bottom of the flag clears by at least 6″ the top of the vehicle or light bar.

B. INCIDENT NAMING AND POSITION TITLES

1. Working incidents shall be identified by name, e.g., Boulder Creek, Smith Road, Big Basin.
2. From this point the Incident Commander's radio identifier will be "*NAME OF INCIDENT* IC."
   a. Example: "Fair Grounds IC."
   b. Should the incident grow in complexity, the Incident Commander may change, but the radio identifier will not.
3. Operations staff radio identifiers will be their incident title. For example:

Working Title	*Radio*
Operations Chief	Operations
Division Supervisor	Division A etc.
North Zone Strike Team	North Zone Strike Team

C. COMMAND IDENTIFICATION AT SCENE

The Incident Commander and staff down to Strike Team Leader may be identified by wearing orange colored traffic vests with their position for the incident written on the front and back of the vest.

Each department may assign one traffic vest to each chief officer; by using masking tape, the vest can be marked to identify a particular position.

D. TRANSFER OF COMMAND FROM ONE AGENCY TO ANOTHER

PROCEDURE:

1. When a mutual aid unit is first on the scene, command will be established by that unit. Upon arrival of the first unit from the responsible department, face to face contact will be made with the initial Incident Commander to get details of attack plans and make an orderly transfer of command or set up a unified command.
2. Each fire department should have transfer of command worked out with their mutual aid partners in advance to avoid confusion at the incident scene.

E. MISCELLANEOUS

Details on the Incident Command System are found in the Incident Command System Operational System description and the Incident Command System Field Operations Guide.

APP. B

Kind 1	Strike Team Types	Number/ Type	Minimum Equipment Standards							Minimum Staffing		
			Pump Capac.	*Water Capac.*	*2½" Hose*	*1½" Hose*	*1" Hose*	*Ladder*	*Heavy Stream*	*S/T Leader*	*Per Increment*	*Total Personnel*
ENGINES	1 (A)	5—Type 1	1000 GPM	400 Gal.	1200 Ft.	40 Ft.	200 Ft.	20 Ft. Ext.	500 GPM	1	4	21
	2 (B)	5—Type 2	500 GPM	400 Gal.	1000 Ft.	500 Ft.	300 Ft.	20 Ft. Ext.	N/A	1	3	16
	3 (C)	5—Type 3	120 GPM	300 Gal.	N/A	1000 Ft.	800 Ft.	N/A	N/A	1	3	16
	4 (D)	5—Type 4	50 GPM	200 Gal.	N/A	300 Ft.	800 Ft.	N/A	N/A	1	3	16

APP. C

Company Officer's Checklist

The purpose of this checklist is to provide the Fire Company Officer with a step by step procedure to follow when initiating a response as part of the County Strike Team/Mutual Aid Program.

I. PREPLAN: BEFORE LEAVING YOUR STATION

A. Identify your engines by TYPE 1, 2, 3, or 4 (See App. B above)
B. Preferred options
   1. Closed cab unit
   2. Pump and roll capabilities
   3. Hard suction
   4. Red, White and assigned radio frequencies with scanner
   5. 400 gallon (or larger) booster tank
   6. Hard lines
   7. Crew of three (minimum) with proper footwear if a wildland response
   8. Portable pump capability
   9. Spare tire/tools

II. INITIATING RESPONSE: ANSWERING THE REQUEST FOR STRIKE TEAM RESPONSE

A. Implement your department's strike team response policy.
B. Understand dispatch instructions.
   1. Location of incident/staging area
   2. Type of incident
   3. Nature of assignment
C. Report your response:
   1. Contact Fire Control on Red Frequency
   2. Give unit number
   3. Remain on Red Frequency unless otherwise directed
   4. Keep radio traffic to a minimum
D. Normally you will be advised of the identity of your Strike Team Leader while you are enroute from your station. If not, this information should be secured at the location of the incident/staging area.
E. Monitor radio transmissions closely for additional assignments, changes, etc.
F. Fill out the apparatus card from the strike team material while enroute. Present this information to the Strike Team Leader upon arrival at the incident/staging area.

III. STAGING AREA/INCIDENT LOCATION

A. Company Officer (only) report to the Strike Team Leader:
   1. Get all relevant information
   2. Hand in apparatus card
   3. Collect Strike Team I.D. package
      a. Ribbon
      b. Strike Team I.D. cards
      c. Additional apparatus card
B. If you have not completed the apparatus card, do it now and turn it in.
C. As time permits, place I.D. package as follows:
   1. Tape cards on driver side of apparatus, front and back.
   2. Secure ribbon to radio antenna.
D. Stand by for assignment
   1. Keep your personnel together.
   2. Stay near your apparatus.
E. All communications are directed through the Strike Team Leader.

IV. TACTICS AND SAFETY FOR WILDLAND OPERATIONS

A. The strike team stays together. It is set up as a team and should work as such. Crews park their vehicles together, eat together and sleep together as required.
B. When water supply is a problem, as most often it is, place two engines on each involved structure; one of the two can shuttle water. If structures are directly threatened, place one engine for every one or two structures. If a structure is totally involved, let it go and move on.

C. Stay mobile and move with the fire. Long supply line hose lays greatly decrease mobility. Make only short lays.
D. If the fire head is too hot, work the flanks. Fire engine stands on fuel breaks and roads can be effective but are also very dangerous. Don't waste time and water wetting down brush ahead of the fire. Use your resources to knock the fire down at the break or road and look for spots across your control line.
E. Water Sources:
   1. Swimming pools
   2. Garden hoses
   3. Ponds
   4. Lakes
   5. Irrigation lines
   6. Wells
   7. Hydrants
F. Use water wisely; make every drop count. Keep a reserve in your booster tank for the unknown emergency and your own protection.
G. In hot fire situations, have your crew coil a charged 1½" line on top of your hose bed. This will give your crews instant protection with a good volume.
H. Don't leave paper, sleeping bags and lunch sacks laying on top of your units. They catch fire quickly. Also, a clean undercarriage will not catch fire as quickly as a dirty one.
I. In open-cab engines, have a salvage cover and fire shelters in the front seat for crew personnel protection.
J. The Strike Team Leader should reconnoiter narrow roads and weak back-country bridges. *When possible have units back into an area.* This prevents turning around and confusion when you have to pull out in a hurry.

APP. D

Strike Team Leader Qualifications

Recommended Minimum Qualifications for In-County and Adjoining Operational Area Strike Team Leaders

a. Minimum Qualifications:
   1) Fire Captain or above
   2) Vehicle available with mobile and portable communications
   3) ICS 220—Basic ICS
   4) ICS 334—Strike Team Leader (Engine)
   5) Recommended by the agency Chief
   6) Demonstrated willingness and interest in the program
b. Desirable Qualifications:
   1) ICS 339-Division/Group Supervisor
   2) Command 1A
   3) Command 2E or Intermediate Fire Behavior

Recommended Minimum and Desirable Qualifications for Statewide Strike Team Leaders

a. Minimum Qualifications
   1) Approved by Fire Chiefs Association
   2) Chief Officer
   3) ICS 220—Basic ICS
   4) ICS 334—Strike Team Leader (Engine)
   5) Ability to leave for a minimum of five days
   6) Participated as a Strike Team Leader Trainee under supervision of a qualified Strike Team Leader

b. Desirable Qualifications
   1) Command 2E or CDF Fire Management I
   2) Intermediate Fire Behavior
   3) ICS 339—Division/Group Supervisor

c. Trainee Positions
   Individuals who have received the ICS STL training and wish to become Statewide Strike Team Leaders will be placed on a list and may be assigned with a CDF or local government strike team in order to obtain the experience required.

APP. E
Qualified Strike Team Leaders 1989

Agencies Providing In-County and Adjoining Operational Area Strike Team Leaders

NORTH ZONE:	MID ZONE:	SOUTH ZONE:
Boulder Creek	Central	Aromas
CDF/County Fire	Santa Cruz	CDF/County Fire
Scotts Valley		Freedom
		Salsipuedes
		Watsonville

Approved Long-Term/Statewide Strike Team Leaders

Aromas:	Ken French		
Boulder Creek:	Bud Tomlin		
	Sam Robustelli		
CDF/County Fire:	Jim Asche	Steve Miller	Dan Turner
	Russ Borelli	Les Parr	
	Craig Herrett	Bill Ruskin	
	Frank Lewis	Rich Seiple	
	Jeff Malmin	Shep Shepherd	
Central:	Steve Negro		
	Steve Van Den Heuvel		
Freedom:	Ben Evans		
Salsipuedes:	Ron Hart		
Scotts Valley:	Eldon Nagel		
	Mike Phinn		
	Mark Trancinna		
Watsonville:	Don French		
	Gary Smith		

FOR: CENTRAL

OUT OUT

STAGING AREA # 11

K-Mart Parking Lot, 2600 41st Avenue

NOTIFY FORESTRY

First Strike Team Mid County Zone			Second Strike Team South County Zone			Third Strike Team North County Zone		
Department		*Page/Phone*	*Department*		*Page/Phone*	*Department*		*Page/Phone*
APTOS/LSB	(2)	354	Salsipuedes	(2)	808 Hotline	Ben Lomond	(3)	648
Branciforte	(2)	359	Aromas		Monterey Co.	Boulder Creek	(3)	286
Santa Cruz	(2)	257, 288	Watsonville		Hotline	Scotts Valley	(3)	558
Freedom	(2)	536	CDF/County Fire		Hotline	Zayante	(3)	644
Scotts Valley	(3)	558	Felton	(3)	647	CDF/County Fire		Hotline
Additional			*Additional*			*Additional*		
CDF/County Fire		Hotline	CDF/County Fire		Hotline	CDF/County Fire		Hotline
Felton	(3)	647	Ben Lomond	(3)	648			
Watsonville		Hotline	Zayante	(3)	644			

LEGEND: ( ) repeater

APPENDIX E

MUTUAL AID RESPONSE PLAN

GDAFD MUTUAL AID AGREEMENT

WHEREAS, the political subdivisions in Montgomery County and counties adjoining said Montgomery County, to-wit: Preble County, Butler County, Warren County, Greene County, Clark County, Miami County and Darke County, collectively referred to as the Greater Dayton Area Fire Departments (GDAFD), and Wright-Patterson Air Force Base Fire Department (WPAFB Fire Department), which are parties to this Agreement, are desirous of obtaining additional fire protection for the citizens of the various political subdivisions and WPAFB Fire Department in time of emergency, pursuant to Sections 505.44 and 9.60, Ohio Revised Code.

It is hereby mutually agreed by the parties hereto as follows:

(1) Definitions:

(a) Fire service agencies shall include, in addition to general fire departments, emergency operations, hazardous materials units, environmental protection units and other related organizations dedicated to protecting life, property and the environment (excluding police services).

(b) The term "fire protection" includes personal services and equipment required for the protection of life and property from fire, fire-fighting and/or emergency medical services.

(c) The term "emergency response" includes necessary services to respond to a fire, health, safety (not including law enforcement) or environmental incident that threatens life, property, or environment.

(2) All Mutual Aid Agreements previously entered into for the purpose of obtaining additional fire protection are hereby superseded by this Agreement.

(3) In the event of emergency, and upon request of another fire department by the highest ranking officer of that department on duty at

the time of the emergency, each party which is a signatory to this Agreement will furnish fire department personnel with equipment, if the highest ranking officer on duty of the requested fire service agency is of the opinion that such personnel and equipment are available. Such personnel and equipment may be recalled at the sole discretion of the highest ranking officer on duty of the fire service agency furnishing such personnel and equipment.

(4) The execution of this Agreement shall not give rise to any liability or responsibility, including but not limited to failure to respond to any request for assistance, lack of speed in answering such a request, inadequacy of equipment, negligent operation of equipment, failure to extinguish any fire, or any cause whatsoever growing out of such use of fire or other equipment and personnel, nor shall the party which issued the call be liable in any manner for damages or loss of equipment or personnel suffered by the party answering the call. This Agreement shall not be construed as or deemed to be an Agreement for the benefit of any third party or parties, and no third party or parties shall have any right of action whatsoever hereunder for any cause whatsoever.

(5) No charge shall be made to any party (parties) to this Agreement for services rendered by any other contracting party (parties) under the provision of this Agreement.

(6) There shall be no reimbursement for loss or damage to equipment while engaged in an activity in accordance with this Agreement, nor shall there be any reimbursement for any indemnity award or premium contribution assessed against the employing party for workmen's compensation benefits arising by reason of injury or death to a member of the defined agency of said party while engaged in rendering services under this Agreement.

(7) This Agreement shall become effective on January 1, 1989, and shall continue as a non-expiring Agreement; provided, however, that any party to this Agreement may withdraw at any time upon thirty (30) days written notice, addressed to the Chief of the defined agency or other officer in charge of each of the other parties to this Agreement, and thereafter such withdrawing party shall no longer be a party to this Agreement, but this Agreement shall continue to exist among or between the remaining parties.

(8) This Agreement may be executed in any number of counterparts, all of which together shall be a single instrument. It shall not be necessary for any counterparts to be signed by more than one party. All counterparts shall be filed with the office of the County Prosecutor of Montgomery County, Ohio, which shall be the official depository for this Agreement. The County Prosecutor of Montgomery County, Ohio, shall send to each party to this Agreement a certificate showing the names of the signatories which have executed this Agreement, and any additions or deletions of participating parties as they occur.

(9) This is a non-expiring agreement.

IN WITNESS WHEREOF, the ______________________________

located in ________________County, ________________has accepted

this Agreement and authorized its execution by its ______________

____________________, and passed the ____________________day of

__________, 1989, in accordance with Section 9.60, Ohio Revised Code.

____

By: ____________________________

Title: __________________________

AUTOMATIC MUTUAL AID RESPONSE SYSTEM GUIDELINES

1. The program is not a substitute for any phase of fire department operations, but a supplement to the forces of each community to expand and refine existing fire suppression activities.
2. Use of current Montgomery County Mutual Aid Agreement is the basic working document. To activate an automatic response, consenting parties need only to addend the Mutual Aid Agreement with a Letter of Understanding for the Automatic Mutual Aid Response System (AMARS).
3. Agreeing parties will automatically request nearest equipment from each other with the following guidelines:
   a) Response areas for neighboring apparatus will be determined by average response times.
   b) For a maximum period of six (6) months, Fire Departments participating in the AMARS Program will maintain their normal response pattern, in addition to the automatic assistance.
   c) Agreeing parties will notify affected fire departments of any situation that will take an AMAR station out of service. If a fire company assigned to AMAR responsibilities is unavailable, the community(s) affected must be made aware immediately. Such notification can be made by the mutual aid radio frequency or telephone.
   d) Agreeing parties will automatically take an apparatus out of service when there are less than three (3) people available for response. Preferred crew staffing is four (4) people per crew. Notification will follow procedure (c) listed above.
4. There will be frequent joint training exercises between agreeing parties. Such exercises will be conducted at least four (4) times per year and a minimum of two (2) hours in length.
5. Agreeing parties may use the frequencies of the Dayton Fire Department, but must be able to communicate on the fire mutual aid frequency.
6. As prescribed by the State of Ohio Statutes and the Basic Mutual Aid Agreement, the highest ranking fire official in whose jurisdiction an incident occurs, is the final authority.

7. AMARS Agreements will be signed by the Fire Chiefs of cooperating jurisdictions and will be reviewed annually for possible revisions.
8. AMARS Agreements can be cancelled by either party upon presentation of written notice to the affected community thirty (30) days in advance of the cancellation date. Cancellation of an AMARS Agreement does not affect the Basic Montgomery County Mutual Aid Agreement.

The Agreement will become effective on Monday, December 17, 1984.

Approved by: ________________ on December 17, 1984
Chief of Kettering Fire Department — Date

Approved by: ________________ on December 17, 1984
Director & Chief Dayton Fire Department — Date

APPENDIX F

CAMEO II SOFTWARE

ABOUT CAMEO II™—AN OVERVIEW

The Computer-Aided Management of Emergency Operations (CAMEO™ II) program is designed to help emergency planners and first responders both plan for, and safely handle, chemical accidents. CAMEO II contains response information and recommendations for 2,629 commonly transported chemicals; an air dispersion model to assist in evaluating release scenarios and evacuation options; and several easily adaptable databases and computational programs that address the emergency planning provisions of Title III, the Emergency Planning and Community Right-to-Know Act of 1986.

Figure 1. The Navigator, CAMEO II's main screen.

CAMEO can include such diverse information as facility floor plans with chemical storage locations; contacts lists; locations of schools, hospitals, and other population concentrations; response resources; Material Safety Data Sheets; and digitized maps of the planning area, overlaid with plumes calculated by the air model.

CAMEO II requires a Macintosh Plus, SE, or II with a hard disk; Apple Computer's HyperCard 1.1; and drawing and communications software applications.

The CAMEO chemical database of 2,629 chemicals is also available from NOAA in MS-DOS format that may be used in a stand-alone application that does not include graphics, the CAMEO air model, or Title III functions.

CAMEO Maps

You can enter CAMEO maps using one of several conventional Macintosh drawing programs.

The sample map section below was digitized using a stylus and digitizing tablet. The map contains city boundaries, street names, parks, and major landmarks (Figure 2).

Drawings of chemical facilities provide a close-up view of buildings, street access, and information on chemical inventories for selected locations. You may also show the interior floor plans of selected buildings (Figure 3).

There are several primary methods of entering a map into CAMEO, including sketching the map with the mouse, tracing it from an original drawing using a digitizing tablet, scanning the image with a desktop scanner, or transferring an existing electronic file from another Macintosh or a mainframe computer.

All maps may be linked within the computer's memory so that a first responder may quickly "zoom in" on a location of interest. Symbols on the maps provide additional detail on the facility or chemical being represented. For example, clicking on a chemical symbol from a facility floor plan shows you a card containing specific information on the facility's chemical inventory (Figure 4).

CAMEO CODEBREAKER

Codebreaker (Figure 5) is a chemical identification database containing 50,000 synonyms, identification numbers, and labeling conventions for 2,629 chemicals, each cross-referenced with standard names and numbers used in the transportation industry. The result is a simple and rapid way to converge on the identity of a specific chemical.

Codebreaker addresses several problems that are well-known to the first responder. Using several different references on-scene is time-consuming and awkward; confusion about the true identity of a product can lead to the use of improper guidelines and protocols. CAMEO helps to solve this problem through its compilation of synonyms and other identification data, minimizing reliance upon library references at the scene of an emergency.

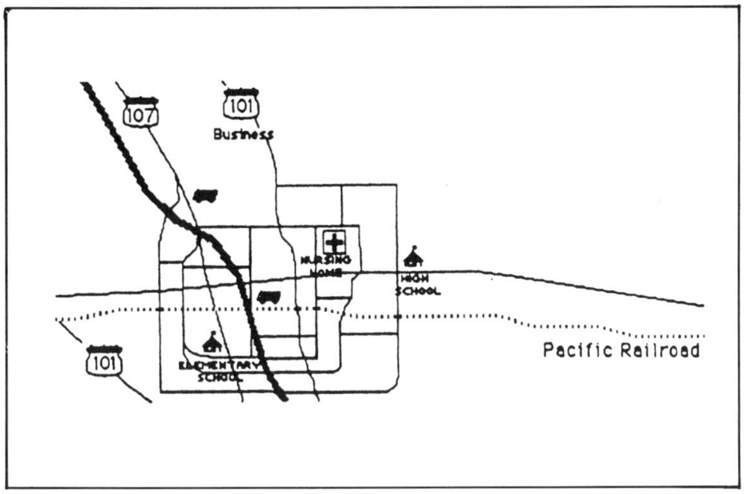

Figure 2. Example of map detail.

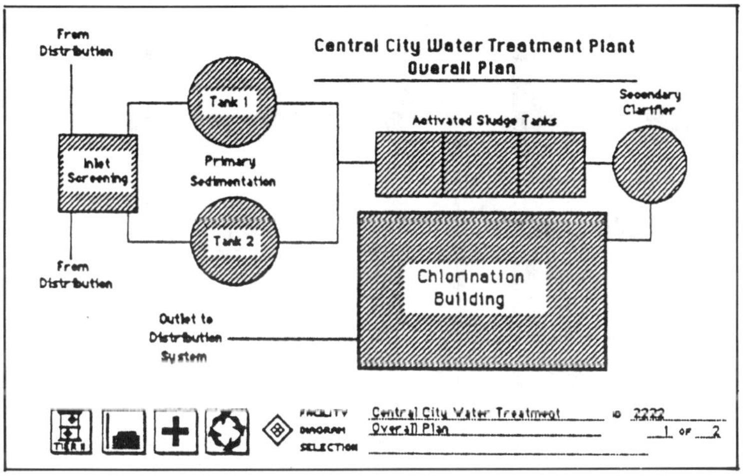

Figure 3. Sample building drawing.

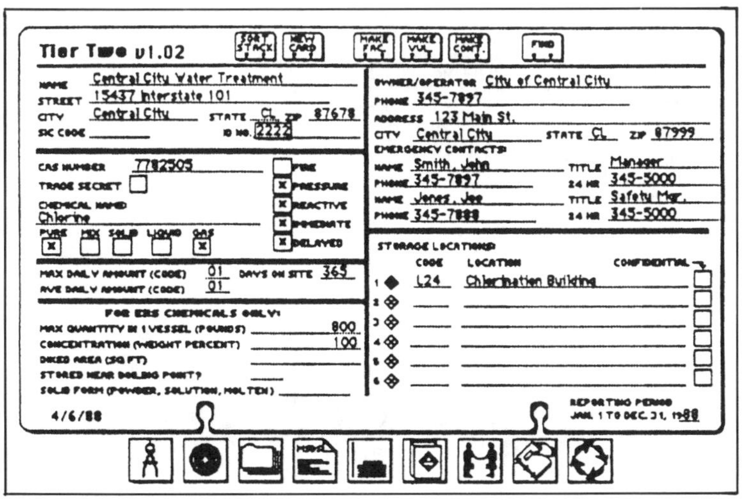

Figure 4. Sample chemical facility information screen.

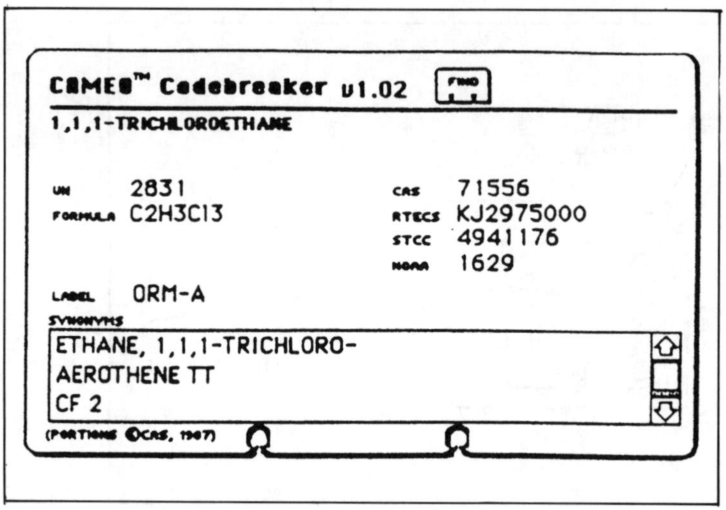

Figure 5. Sample Codebreaker record.

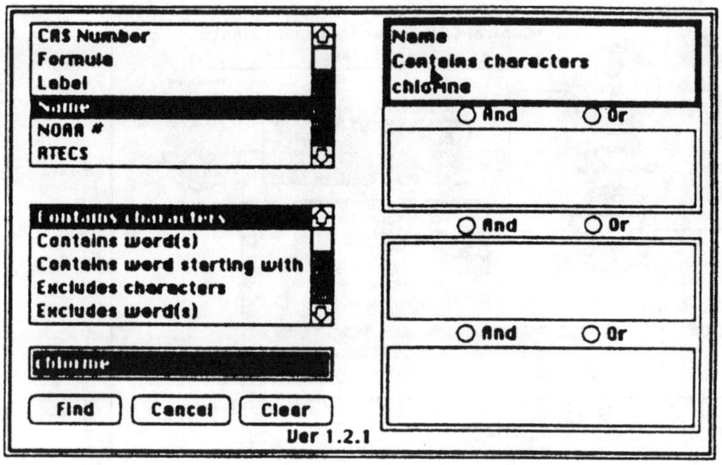

Figure 6. Sample chemical search.

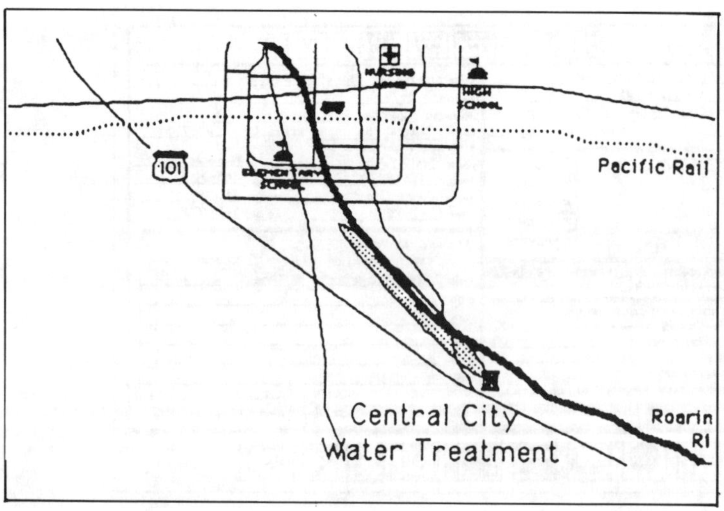

Figure 7. Sample chemical plume "footprint."

Codebreaker may be activated by having the computer search its identification files for one or more characteristics, such as a chemical's label or trade name. Codebreaker can not only conduct searches based on fragments of information, such as the first four digits of a CAS number, but also can conduct multiple searches on four search criteria simultaneously (Figure 6).

CAMEO CHEMICAL DATABASE

The chemical database contained in CAMEO streamlines the transfer of information from expert sources to the on-scene responder. The CAMEO chemical database contains text from the most commonly used reference sources. Subjects most critical to first responders are emphasized, including personal protective measures, fire and explosion hazards, firefighting techniques, human health threats, and spill cleanup procedures.

References familiar to first responders were selected; these include the EPA *Chemical Profiles*, the U.S. Department of Transportation's *Emergency Response Guidebook*, the Association of American Railroads' *Emergency Handling of Hazardous Materials in Surface Transportation*, the U.S. Coast Guard's *Chemical Hazards Response Information System (CHRIS)*, as well as a number of other chemical references. Supplementary information, where necessary, was devoloped by NOAA.

CAMEO AIR MODEL

CAMEO can also help estimate downwind chemical concentrations resulting from an accident, drawing on information from a radio-controlled atmospheric station linked to the computer and its chemical database. You may also enter atmospheric information directly from the keyboard to create accident scenarios. The system uses a dispersion model based on the U.S. Environmental Protection Agency's *Workbook of Atmospheric Dispersion Estimates*.

CAMEO allows a first responder to quickly estimate the extent of a plume downwind from a chemical spill. The first responder sets several "switches" on the computer screen to indicate the name of the chemical involved, the roughness of the terrain, the size or nature of the spill, etc. The "footprint" of the chemical plume is then drawn by the computer from the defined location on a CAMEO map (Figure 7).

USING CAMEO TO COMPLY WITH SARA TITLE III

Recent additions to the original CAMEO program allow it to be used for contingency planning and to facilitate the hazards analysis process outlined in EPA's technical guidance document for SARA Title III

(Figure 8). CAMEO can be used to help define high risk areas through identification of possible accident locations and their proximity to highly sensitive areas (schools, hospitals, areas of high population density, etc.). Using data specific to a community, contingency planning efforts can focus on realistic accident simulations.

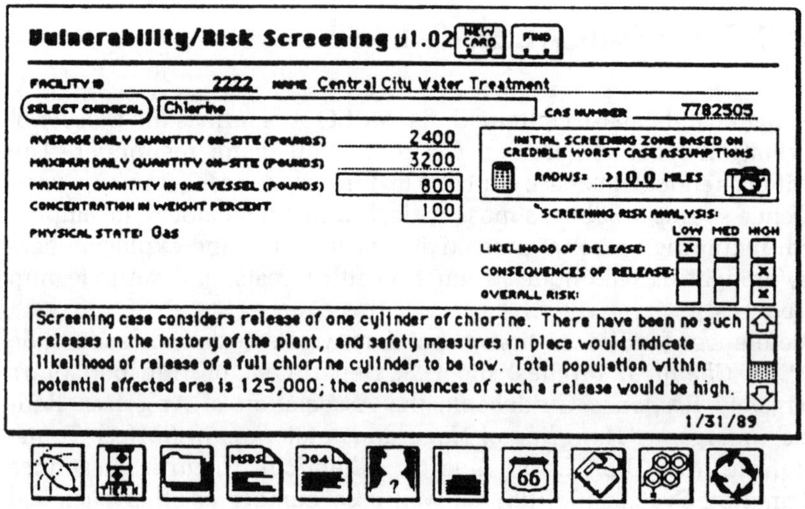

Figure 8. Sample hazards analysis card.

Trademark Protection

All of the products mentioned here and elsewhere in *The CAMEO™ II Manual* are registered trademarks of their holders. Use of a product by an agency of the U.S. Government does not constitute endorsement of that product.

For further information on CAMEO II, call (206) 526-6317 or write the CAMEO Database Manager at

NOAA/Hazardous Materials Response Branch
7600 Sand Point Way N.E.
Seattle, Washington 98115

APPENDIX G

TORNADO FACTS

The following text appears in "Funnel Facts," a brochure prepared in the interest of public safety by State Farm Fire and Casualty Company and the National Coordinating Council on Emergency Management.

Tornadoes are frightening realities. We can't prevent them or even predict where or when they will strike. Many of us aren't sure what we should do during a tornado. Some people aren't even clear about the difference between a tornado "watch" and a tornado "warning."

But there are some precautions we can take to minimize danger—to people and property.

When tornado season approaches, the watchwords are: Be informed and be prepared.

FUNNEL FACTS

Knowing the difference between a watch and a warning is a good first step:

Tornado Watch

A tornado watch simply means that conditions are favorable for tornadoes to develop. In this case you should take precautions to protect you and your property, and listen to the radio to keep informed.

Tornadoes are most likely to occur in the late afternoon on a hot spring day. However, tornadoes have occurred in every month at all times of the day or night. When a tornado *watch* is issued, be alert for changes in the weather. And be prepared to act quickly.

Tornado Warning

A tornado warning means that a tornado has actually been sighted. If one is issued for your area, you should seek shelter immediately!

There is little time for closing windows or hunting for flashlights. It's a good idea to know where things are, and to have an emergency storm kit already prepared.

Before the Storm Hits

A little planning can prevent unnecessary panic and confusion if a tornado does strike.

- Learn the warning signals used in your community. If a siren sounds, that means **stay inside** and take cover.
- Consider setting up a neighborhood information program through a club, church group or community group. Hold briefings on safety procedures as tornado season approaches. Set up a system to make sure senior citizens and shut-ins are alerted if there is a tornado warning.
- Put together an emergency storm kit including candles, matches, transistor radio, flashlight, batteries, and simple first aid items in a waterproof container.
- Make a complete inventory of your possessions for insurance purposes. Keep that list in a bank safe deposit box or other safe place away from your home.
- Conduct drills with your family in the home; make sure each member knows the correct procedures if they are at work or school when a tornado hits.

When a Tornado WATCH is in Effect

You can take certain precautions to lessen danger.

- Move cars inside a garage or carport, if possible, to avoid damage from hail that often accompanies severe storms. Keep your car keys and house keys with you.
- Move lawn furniture and yard equipment such as lawnmowers inside *if time permits.* Otherwise they could become damaged or act as dangerous projectiles, causing serious injury or damage.
- Account for family members at home. Have your emergency kit ready.
- Keep your radio or TV tuned in to the weather reports.

When a Tornado WARNING Has Been Issued on the Radio or by Siren

Seek shelter immediately.

AT HOME: In the basement under something sturdy, like a bench. If there is no basement, a small room in the middle of the house (a closet or bathroom) is best. Always stay away from outside walls and windows.

AT WORK OR SCHOOL: Designated shelter areas are best. Stay away from large open rooms like auditoriums and gymnasiums, and rooms with windows. Lie low with hands covering the back of your head to reduce neck injury.

IN SHOPPING MALLS: Go to a designated shelter area or to the center of the building on a low level. Stay away from large, open rooms and windows. Never seek shelter in cars in the parking lot.

IN MOBILE HOMES OR CARS: Leave the vehicle. Seek a safe structure or lie down in a low area with your hands covering the back of your head and neck. Keep alert for flash floods that often accompany such storms.

After a Tornado

Keep calm. Stay in your shelter until after the storm is over.

- Check people around you for injuries. Begin first aid or seek help if necessary. Always cooperate with local officials.
- Check utility lines and appliances for damage. If you smell gas, open the windows and turn off the main valve. Don't turn on lights or appliances until the gas has dissipated. If electric wires are shorting out, turn off the power.
- When you go outside, watch out for downed power lines.
- Notify your insurance agent and provide as much detail as possible about damage to your property. Follow the agent's directions on filing your claim.
- Take steps to protect your home and furniture from further damage:
  - Clean and dry your furniture, bedding, rugs and carpeting as soon as possible.
  - Board up windows and holes in the walls or roof.
  - Don't be rushed into signing repair contracts. Deal with reputable contractors. If you're unsure about a contractor's credentials perhaps your agent, claim adjuster, Better Business Bureau or Chamber of Commerce can help. Make sure the contractor you hire is experienced in repair work—not just new construction. Be sure of payment terms and consult your agent or adjuster before you sign any contracts.
  - Keep receipts for living expenses beyond your normal ones (such as temporary quarters) and for temporary repair costs so you can seek insurance reimbursement.

FACT OR MYTH

Separating fact from myth can mean the difference between injury and safety in case of a tornado. Here are four of the most common and dangerous tornado myths and the truth about them.

Myth: The best place to be during a tornado is in the southwest corner of a building
Fact: The southwest corner is no safer than any other part of the building. The safest place to be is **in a basement** under something sturdy, like a workbench. If there is no basement, seek shelter in a small interior room in the middle of the building, like a closet or a bathroom. *Always stay away from outside walls and windows.*

Myth: Windows should always be opened to equalize pressure
Fact: Opening windows to equalize pressure during a tornado is ineffective in reducing damage. Don't worry about the windows; worry about protecting yourself. Also, flying glass is a real hazard.

Myth: You can outrun a tornado in a car
Fact: Don't bet your life on it. A tornado is unpredictable; you can't

know which way it's going to go, or how fast. If you're in a car and a tornado is near, get away from the car and lie in a ditch or low area, protecting your head with your hands.

Myth: Mobile homes are safe if they're tired down
Fact: A mobile home is *never* safe in a violent windstorm such as a tornado. If you're in a mobile home when a tornado **watch** is announced, leave and go immediately to a safe structure. Or be prepared to take cover in a low area, covering your head and the back of your neck. **Remember:** If you're in a ditch or ravine, be alert for flash floods that often accompany tornadoes.

APPENDIX H

BIBLIOGRAPHY ON UNIDENTIFIED FLYING OBJECTS (UFOs)

Above Top Secret, Timothy Good, William Morrow, NY, 1988.

Aids to Identification of Flying Objects, Air Technical Intelligence Center, Gov't Printing Office, Supt. Doc., 1966.

Aliens Among Us, Ruth Montgomery, Fawcett Crest, NY, 1985.

Aliens from Space—The Real Story of UFOs, Donald E. Keyhoe, Doubleday, NY, 1972.

Beyond Earth: Man's Contact with UFOs, Ralph and Judy Blum, Bantam Books, NY, 1974.

Breakthrough to Creativity, Shafica Karakulla, M.D., De Vorss and Co., Marina Del Rey, CA, 1967.

Chariots of the Gods, Erich Von Daniken, Putnam, NY, 1970.

Clear Intent, Barry Greenfield, Prentice Hall, Englewood Cliffs, NJ, 1987.

Communion, Whitley Streiber, William Morrow, NY, 1987.

Extraterrestrial Visitations from Prehistoric Times to the Present, Henry Regnery Co., Chicago, 1970.

Flying Saucers—Letters to the Air Force on UFOs, Bill Adler, Dell Books, NY, 1967.

Flying Saucers—Serious Business, Frank Edwards, Lyle Stuart Inc., NY, 1966.

Identified Flying Saucers, Robert Lofton, David McKay Co., NY, 1968.

In Search of Extra Terrestrials, Alan Landsburg, Bantam Books, NY, 1967.

Inside the Flying Saucers, George Adamski, Paperback Library, NY, 1967.

Insights for the Age of Aquarius, Gina Cerminara, Theosophical Pub. House, Wheaton, IL, 1973.

Intruders, Budd Hopkins, Random House, NY, 1987.

Mysteries of the Unexplained, Readers Digest, Readers Digest Assn., Pleasantville, NY, 1982.

New Psychic Frontiers, Walter and Mary Jo Uphoff, Colin Smythe Ltd., and Bolger Pubs., Minneapolis, MN, 1975.

Project Blue Book, Brad Steiger, Editor, Ballantine, NY, 1976.
Strange World, Frank Edwards, Lyle Stuart, NY, 1965.
Strangers Among Us, Ruth Montgomery, Fawcett Crest, NY, 1979.
The Intruders, Budd Hopkins, Random House, NY, 1987.
The Invisible College, Jacques Vallee, E.P. Dutton, NY, 1975.
The Possibility of Intelligent Life Elsewhere in the Universe, U.S. House Committee on Science and Technology, Govt. Printing Office, 1975.
The Roswell Incident, William L. Moore, Grosset and Dunlap, 1980.
The UFO Conspiracy—The First Forty Years, Jenny Randles, Sterling Pub. Co., 1989.
The Unexplained, Allen Spraggett, Signet, NY, 1967.
The World's Greatest UFO Mysteries, Nigel Blundell and Roger Boar, Berkeley Book, 1990.
The World's Last Mysteries, Readers Digest, Pleasantville, NY, 1978.
UFO Abductions, Philip Klass, Prometheus Books, Buffalo, NY, 1989.
UFO . . . Contact from the Pleiades, Lee and Brit Elders, Genesis III Pub. Co., Phoenix AZ, 1984.
UFOs From Behind the Iron Curtain, Ian Hobana and Julien Weverbergh, Bantam Books, NY, 1975.
UFOs Over the Americas, Jim and Coral Lorenzen, Signet Books, NY, 1968.
UFOs—A Scientific Debate, Carl Sagan and Thornton Page, Cornell University, NY, 1972.
We Are Not The First, Andrew Thomas, Putnam & Sons, NY, 1971.

APPENDIX I

EARTHQUAKE SCALES: MODIFIED MERCALLI AND ROSSI-FOREL

THE MODIFIED MERCALLI SCALE OF EARTHQUAKE INTENSITY

I. Not felt except by a very few who might live in an extremely undisturbed environment—no vibrations from passing cars, refrigerators, or other machinery.

II. Felt by only a few persons who are resting, especially on upper floors of buildings. Delicately suspended objects may swing.

III. Felt quite noticeably indoors, especially on upper floors of buildings, but many people do not recognize it as an earthquake. Standing automobiles may rock slightly. Vibrations are similar to those of a passing truck. It is possible to estimate the duration of the tremor.

IV. During the day felt indoors by many, outdoors by a few. At night some are awakened. Dishes, windows, and doors are disturbed. Walls make cracking sounds. Sensation similar to a heavy truck striking a building. Standing automobiles rock noticeably.

V. Felt by nearly everyone. Many are awakened at night. Some dishes, windows, and other glass objects are broken. In some places, there is broken plaster. Unstable objects are turned over. Trees, poles, and other tall objects are visibly disturbed. Pendulum clocks may stop.

VI. Felt by all. Many frightened and run outdoors. Some heavy furniture moves. Some instances of falling plaster and damaged chimneys. Other damage slight.

VII. Everyone runs outdoors. Damage is negligible in buildings of good design and construction, slight to moderate in well-constructed buildings, considerable in poorly built or badly designed structures. Some chimneys broken. People driving cars notice shaking.

VIII. Damage slight in specially designed structures; considerable in ordinary substantial buildings, with some suffering partial collapse; great in poorly built structures. Panel walls torn out of frame structures. Chimneys, factory stacks, columns, monuments, and walls fall. Heavy furniture is overturned. Underground sand and mud are ejected to the surface in small amounts. There are changes, in level and clarity, in well water. People driving automobiles are perturbed.

IX. Damage considerable in specially designed structures; well-designed frame structures thrown out of plumb; damage great in substantial buildings, with some suffering partial collapse. Buildings shifted off foundations. Ground is conspicuously cracked. Underground pipes are broken.

X. Some well-built wooden structures are destroyed. Most masonry and frame structures destroyed along with their foundations; ground badly cracked. Rails are bent. Numerous landslides around river banks and steep slopes. Sand and mud are shifted. Water is splashed over banks.

XI. Few, if any, masonry structures remain standing. Bridges are destroyed. Broad fissures open in the ground. Underground pipelines are put completely out of service. Earth slumps and land slips in soft ground. Rails are greatly bent.

XII. Damage total. Waves seen are reportedly moving across solid surfaces. Lines of sight and level are distorted. Objects are thrown upward in the air.

ROSSI-FOREL SCALE FOR EARTHQUAKE CLASSIFICATION

1. Microseismic shock—recorded by a single seismograph.
2. Extremely feeble—felt by a number of persons at rest.
3. Very feeble shock—strong enough for the duration and direction to be noticed.
4. Feeble shock—felt by persons in motion; disturbance of moveable objects, doors, windows, cracking of ceilings.
5. Shock of moderate intensity; felt generally by everyone; disturbance of furniture, beds, ringing of bells.
6. Fairly strong shock; general awakening of those asleep; ringing of bells, oscillation of chandeliers, stopping of pendulum clocks, people leave their dwellings.
7. Strong shock; overthrow of moveable objects; falling plaster; ringing of church bells, general panic, but no damage to buildings.
8. Very strong shock; fall of chimneys, cracks in walls of buildings.
9. Extremely strong shock; partial or total destruction of some buildings.
10. Shock of extreme intensity; great disaster; ruins, disturbance of strata, fissures in ground, rock falls from mountains.

APPENDIX J

MULTIPLE CASUALTY INCIDENT PLAN

The following is from the Santa Cruz County Emergency Plan, Medical and Health Annex, Attachment No. 8.

MULTIPLE CASUALTY INCIDENT PLAN

Part One, Administrative Plan

I. PURPOSE

The purpose of this MULTIPLE CASUALTY INCIDENT PLAN is to define the authority, responsibility, and function of the various emergency responders who will be called upon to coordinate their emergency response activities at the scene of a major accident.

II. GENERAL

Authority

1. Santa Cruz County Emergency Plan, dated January 1984

References

1. Operational System Description ICS 120-1, Incident Command System Publication, December 12, 1981.

Applicability

1. This Multiple Casualty Incident Plan is applicable to *ALL* emergency responders and to *ALL* governmental and non-governmental medical support services in Santa Cruz County.

III. SCOPE

This Multiple Casualty Incident Plan is intended to coordinate multi-agency response to the *single site* disaster which could overwhelm the day-to-day emergency medical response system. The Multiple Casualty Incident Plan is *NOT* designed to accommodate a County-wide ongoing disaster.

Few, if any government agencies have the resources to properly respond to multiple casualty incidents involving fifteen or more casualties. Therefore, multiple casualty emergencies become multi-agency incidents. Law Enforcement, fire, ambulance, hospitals, Cal Trans, Public Works, American National Red Cross, and Amateur Radio personnel are among the public and private agencies that become involved upon activation of this plan. Each has a specific role defined in this plan.

IV. OBJECTIVES

The objectives of this Multiple Casualty Incident Plan are to ensure adequate and coordinated efforts to minimize loss of life, disabling injuries, and human suffering by providing effective emergency medical assistance through efficient utilization of medical and other resources in the event of an incident which results in a number of injured persons.

Further, the objectives of this plan are to identify and provide the resources necessary for mobilizing teams to effectively deal with the victims resulting from an accident or incident while simultaneously providing teams to deal with the other emergency situations associated with the accident or incident.

V. ORGANIZATION

A. Command Authority

Each agency shall retain full command authority within its jurisdiction at all times. Agencies which are assisting in support of a single jurisdiction will function under the direction of that jurisdiction's designated incident commander.

In multi-jurisdictional incidents, Incident Commanders *may establish unified command*, planning, and coordinated strategies for controlling the overall incident at a *single location command post*.

B. Incident Command System

The incident scene is organized using the Incident Command System (ICS) which provides a common organizational structure to accomplish set incident objectives and provides a means to interface all agencies at any type of major emergency. The Incident Commander only needs to designate persons to fill positions needed for the size of the incident. Refer to organizational chart on facing page.

C. General Duty Statements

The general duty statements of the Command Staff and the specialized multiple casualty positions are contained below:

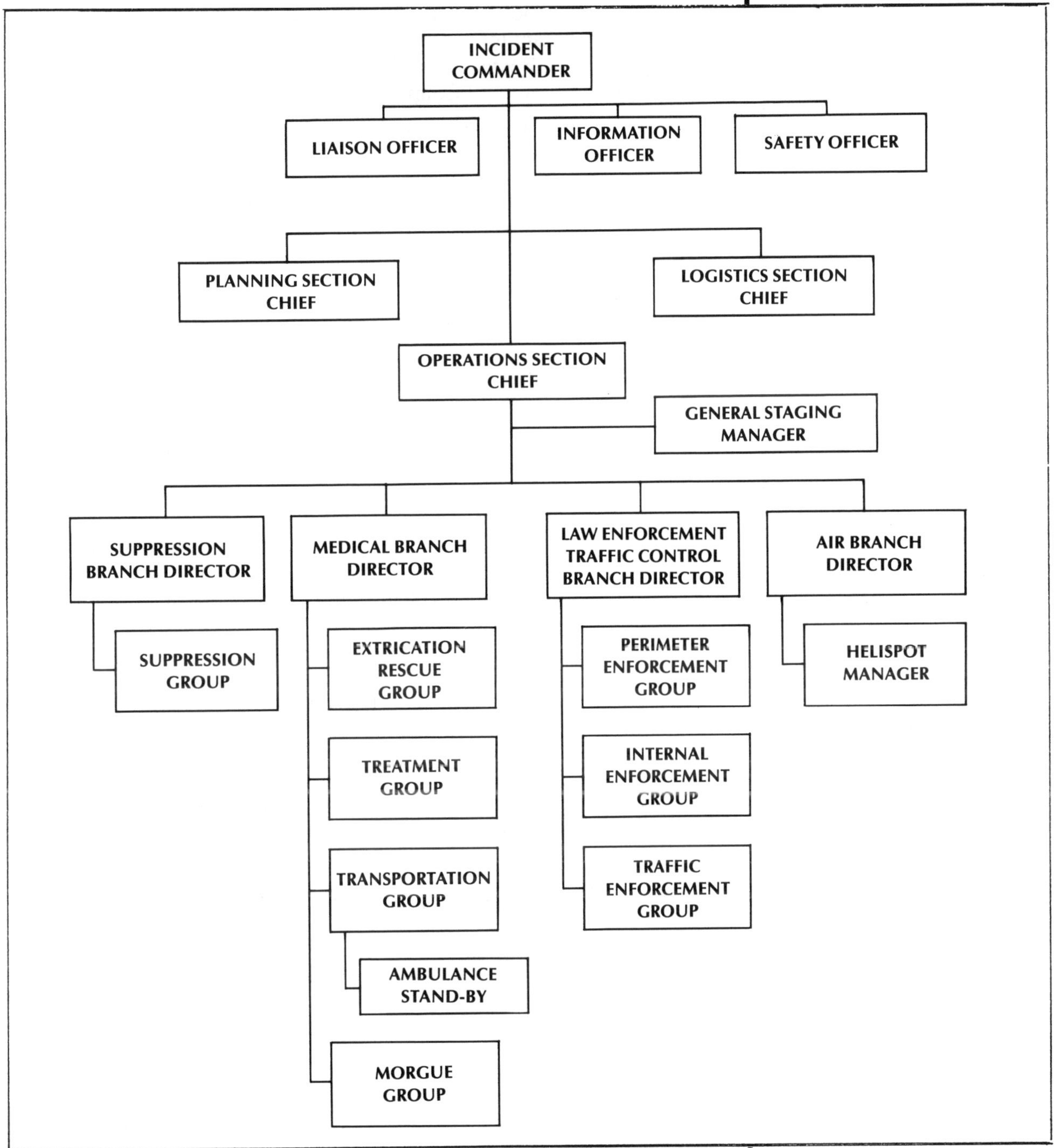

1. *Incident Commander* ("IC")

   The Incident Commander is responsible for overall scene incident activities. The IC may be a law enforcement, fire, or health person. The ultimate IC may be a senior law enforcement, fire or health person based upon the rules, policies or regulations established by the jurisdiction in which the incident has occurred.

2. *Liaison Officer* ("Liaison")

   The Liaison Officer's function is to be a point of contact for representatives from other agencies which are directed to send their respective liaison officers to the Incident Command Post.

3. *Information Officer* ("P.I.O.")

   The Information Officer is the point of contact for the media and other governmental agencies which desire information *directly from the incident.*

4. *Safety Officer* ("Safety")

   The Safety Officer's function is to assess hazardous and unsafe situations and to develop measures for assuring personnel safety.

5. *Planning Section Chief* ("Plans")

   The Planning Section Chief will normally be from the jurisdiction which has incident responsibility. The Planning Section Chief is responsible for the gathering and analysis of all data with respect to incident operations and assigned resources, developing alternatives for tactical operations, conducting the planning meetings and preparing the action plan for each operational period.

6. *Logistics Section Chief* ("Logistics")

   The Logistics Section Chief is responsible for providing all support needs, with the exception of air support, to the incident. Responsibilities include ordering all resources from off-incident locations; providing facilities, transportation, supplies, equipment maintenance and fueling, feeding, communications and medical services for emergency responders.

7. *Operations Section Chief* ("Operations")

   The Incident Operations Section Chief is responsible for the direct management of all incident tactical activities.

8. *Staging* (Staging Manager)

   The Staging Manager answers directly to the Operations Chief and is responsible for the temporary location of resources (except ambulances) which are available for immediate assignment. All resources within the designated staging area are under the direct control of the Operations Chief and are on a three minute availability status.

9. *Medical Branch Director* ("Medical")

   The Medical Branch Supervisor manages the medical tactical elements within the area of the incident. The Medical Group Supervisor supervises the units who triage, extricate, treat, and transport accident victims;

and establishes and maintains liaison with other participating medical service providers.

10. *Extrication/Rescue Group Supervisor* ("Extrication")

    The Extrication/Rescue Group Supervisor is a firefighter who is responsible for site safety, initial point triage, disentanglement of victims and movement of victims/patients to the treatment/tagging area.

11. *Treatment Group Supervisor* ("Treatment")

    The Treatment Group Supervisor (a firefighter, Emergency Medical Technician I) is responsible for establishing and operating a treatment area at a suitable location; and is also responsible for establishing and operating a tagging area.

12. *Transportation Group Supervisor* ("Transportation")

    The Transportation Group Supervisor is responsible for providing and coordinating patient transportation. The Transportation Group Supervisor, equipped with two radio frequencies (fire tactical channel and medical channel) must have in-depth field medical knowledge (Advanced Life Support Provider) and be familiar with local hospital capabilities.

    a. *Ambulance Stand-By Manager* (Ambulance Stand-By)

       The Ambulance Stand-By Manager must maintain close contact, visually or by radio, with the Transportation Team Leader. The Ambulance Stand-By Manager is responsible for advising incoming ambulances to unload appropriate equipment in/for the treatment area.

13. *Medical Control* ("Medical Control")

    Medical Control will be the base station hospital that the first arriving ambulance has established or would have established a working field relationship with. All *OTHER* hospitals, regardless of their normal everyday designation, will be referred to as "receiving hospitals" in this plan.

14. *Morgue Group Supervisor* ("Morgue")

    A Deputy Sheriff-Coroner will be assigned to establish and maintain a temporary morgue, and to carry out the necessary coroner's investigation.

15. *Suppression Branch Director*

    The Suppression Branch Director is responsible for the overall tactical management of fire suppression.

16. *Suppression Group Supervisor* ("Suppression")

    The Suppression Group Supervisor coordinates fire suppression activities.

17. *Law Enforcement/Traffic Control Branch Director* ("Law")

    The Law Enforcement Traffic Control Branch Director works closely with the Incident Commander or Incident Liaison Officer (#2 Liaison) and directs the overall tactical management of the Law Enforcement Perimeter, Internal Enforcement and Traffic Control Groups.

18. *Perimeter Enforcement Group Supervisor* ("Perimeter Enforcement")

    The Perimeter Enforcement Group Supervisor is responsible for sealing off the perimeter, as necessary, and for directing emergency responders to their respective staging areas. All non-essential traffic will be directed away from the incident site.

19. *Internal Enforcement Group Supervisor* ("Internal Enforcement")

    The Internal Enforcement Group Supervisor is responsible for security of persons, property, evidence preservation, and general enforcement/exclusion of nonessential personnel in the internal perimeter of the incident site.

20. *Traffic Enforcement Group Supervisor* ("Traffic")

    At the site of those incidents which occur in any unincorporated area of the County (California Highway Patrol traffic jurisdiction) a California Highway Patrol person will assume duties of the Traffic Enforcement Group Supervisor. The Traffic Enforcement Group Supervisor will report to the Incident Commander and be responsible for the direction of all non-essential traffic away from the incident site.

21. *Air Branch Director* ("Air Group")

    The Air Branch Director, working closely with the Medical Branch Director, will arrange for air transportation of patients, supplies, or material to the incident scene.

    a. *Helispot Manager*

       The Helispot Manager is responsible for arranging a helicopter landing area and for coordination of activities with the Medical Group, for off-loading of supplies and material and for loading of evacuees.

VI. OPERATIONS

This Multiple Casualty Incident Plan will address specific problem areas representative emergency responders identified in preplanning agreements. The specific areas to be addressed are those where cooperative effort, mutual understanding, and a coordination of activities are the essential ingredients for success.

All aspects of on-scene operation at a Multiple Casualty Incident Plan will not be addressed because: 1) emergency responder's standard operating procedures for particular situations vary; 2) variety and complexity of the varied tasks would produce an unwieldy document. This plan is not intended to be a review of the Incident Command System; but rather an explanation of how the various emergency responders in Santa Cruz County will coordinate their respective responses and cooperate at the site of a multiple casualty incident.

It should be noted, particularly by those who would be Incident Commanders, that not all of the positions outlined in this plan—and organizational chart (page 57)—need be filled on any given incident. However, the Span-of-Control recommended by the Incident Command System should be observed.

A. *Plan Activation*

1. WHEN TO ACTIVATE THIS PLAN:

   A single site incident which overwhelms the initial responder's human resources and/or equipment.

2. WHO MAY ACTIVATE THIS PLAN:

   Any fire, law enforcement or Medical-Health Services personnel

3. HOW TO ACTIVATE THIS PLAN:

   Report to the EMERGENCY CENTER that the *Multiple Casualty Incident Plan* should be activated.

4. PROVIDE THE FOLLOWING INFORMATION AT TIME OF ACTIVATION:

   —Type of Incident
   —Location of Incident
   —Best Access Route to Incident Site
   —Number of Injured (Approximate)
   —Types of Injuries (Burns, Severe Trauma, Respiratory, etc.)
   —Need for More Than Initial First Response

5. REQUEST SPECIAL RESOURCES: (such as)

   —Special Rescue Equipment/Additional Fire Personnel
   —Additional Law Enforcement (Perimeter Control/Traffic Enforcement)
   —Cal Trans/Public Works
   —Coroner
   —Lighting, Other Equipment
   —Additional Ambulance Transportation Units

B. *Incident Commander*

   The Incident Commander may be the first supervisory fire, law enforcement, emergency medical care provider or health officer at an incident site. The Incident Commander will identify himself by using the name of the facility, street or road where the incident has occurred.

For example, the Incident Commander may identify himself as "Rodeo Gulch Command"; Riverside Road Command"; or by some such other common landmark or area. In any case, once the Incident Commander chooses a name and identifies him or herself by that particular name, that name will be used throughout that particular incident.

C. *Command Post*

A single site command post shall be established and made immediately identifiable by the Incident Commander, pursuant to Incident Command System directions.

1. Command Post Location

The Command Post shall be established keeping in mind:

a. Ease of access
b. Perspective of operations which does not conflict with emergency operations and provides an adequate margin of safety for all command and liaison personnel.

2. Command Post Functions

The Command Post is the location from which all incident operations are directed. Law enforcement, Liaison, Public Information, and Safety will co-locate at the Command Post. All other agencies or responders will report to the Staging Area, when a staging area is set up.

D. *Staging*

This plan assumes incidents may dictate establishing two staging areas, a General Staging Area and an Ambulance Stand-by Area.

1. General Staging Area

A General Staging Area may be established for temporary location of all arriving resources. For this plan, all medical supplies will be *immediately* relocated to the Treatment Area.

2. Ambulance Stand-by Area

The Ambulance Stand-by Area will be established for the location of all ground medical transportation units. Ambulances will be directed to off-load certain supplies for the Treatment Area. Ambulance Stand-by will be located so that move-up for loading can be signalled visually or by radio.

E. *Ambulance Procedure*

1. Alerting

While enroute to the incident, the first responding ambulance shall alert the base hospital providing medical control (hereinafter referred to as Medical Control) and advise preliminarily:

a. The location and type of incident (fire, vehicle accident, aircraft crash, etc.)

b. Possible delaying factors, if known (pin-ins, decontamination, difficult terrain, etc.) and

c. Numbers of casualties, estimate.

This very preliminary information will be that which is obtained at the time of the original dispatch and/or from other radio frequencies monitored by the first ambulance enroute to the incident site.

2. Communications

a. Radio Frequency

Ambulances are directed to use the usual Medical Network Radio Channel.

b. Original Dispatch

The first ambulance to the scene will be dispatched by County Communications' EMERGENCY CENTER. The first ambulance dispatched will alert Medical Control.

c. Additional Ambulances/Mutual Aid

Requests for additional ambulances or mutual aid must be initiated by the Incident Commander and will be dispatched by County Communications' EMERGENCY CENTER.

3. Ambulance Procedure on Site

a. First Ambulance on Site

When this plan is activated, the Incident Commander will appoint the first ambulance crew on the scene to Medical Branch operations as Transportation Group Supervisor and/or medical scene evaluator(s)/extrication assistant(s). The ambulance unit will be used for an on-scene Medical Network Radio station.

The ambulance crew person assigned as Transportation Group Supervisor shall update the medical report from the scene, advising Medical Control *as soon as possible.*

b. Second Ambulance on Site

The second and subsequent ambulances dispatched to the incident site will report to the Staging Area. The Staging Manager will be signalled visually or by radio to direct incoming ambulances to the Ambulance Stand-by Area and will ask the ambulance crew(s) to off-load the following types of equipment for the Treatment Area as indicated and as appropriate:

1) Backboards
2) Scoop stretchers
3) Sandbags
4) Splints
5) Portable oxygen tanks, masks, and cannulas
6) Portable suction
7) Advanced Life Support supplies:
   a) MAST Suits
   b) I.V.'s

Under no circumstances should any ambulance surrender their gurney or folding stretchers (flats). This equipment must stay with the ambulance at all times, including after unloading patients at a designated hospital, since ambulance transportation is impossible without this equipment.

4. On-Site Ambulance Dispatch

   a. Ambulance Stand-by Area

   The Ambulance Stand-by Area Manager will visually direct ambulances from the ambulance stand-by area to the loading area.

   b. Ambulance Destination

   The Transportation Group Supervisor will, in collaboration with Medical Control, direct the loaded, departing ambulances to a receiving hospital. The ambulance(s) so directed will be committed to the facility designated by the Transportation Group Supervisor.

   c. Radio Logs/Records

   Medical Control is responsible for maintaining radio logs detailing the unit numbers; numbers of casualties loaded; and designated receiving hospital(s).

5. Reassignment of Ambulances/Completion of Incident

   Ambulances dispatched to an incident site are assigned to that incident. After unloading patients at a receiving hospital, ambulances will want to determine if they are to be reassigned to the incident site or get a new assignment.

   Ambulances will contact the Transportation Group Supervisor on the Medical Network Radio frequency being used by Medical Control/Transportation for the incident. The Transportation Group Supervisor will advise if the inquiring ambulance should return to the incident site, or be released from the incident.

When the ambulance is released from the incident, they are to contact the EMERGENCY CENTER (County Communications), and go in-service. (909-B)

F. *Medical Control*

1. Radio Identification

At the time this Multiple Casualty Incident Plan is initiated, the hospital providing Medical Control for the first responding ambulance shall assume "Medical Control" for all personnel or medical care rendered at the incident.

2. Frequency Designation

The usual Medical Network Channel will be assigned to the Medical Operations when this plan is initiated. Radio traffic should be kept to a minimum, so that air time for on-scene medical operations is not interferred with. Hospitals shall communicate with each other on another radio channel or on the telephone.

3. Plan Activation/Alerting

a. Initial Alert

The designated base hospital, Medical Control will first receive notice that this multiple casualty plan is initiated when the first ambulance dispatched to the incident radios the information on the usual Medical Network Radio Channel.

That initial alert from the first responding ambulance will provide preliminary information as follows:

1) Location and type of incident (fire, vehicle accident, hazardous materials, etc.)
2) Possible delaying factors if known (pin-ins, decontamination, difficult terrain, etc.)
3) Possible numbers of casualties, estimate.

b. On Scene Medical Report

Upon arrival and after the initial scene assessment, the first arriving ambulance person (Transportation Group Supervisor designee) will advise Medical Control, confirming that the incident is indeed/or is not as initially reported.

If it is a Multiple Casualty Incident, a cursory report of total injured and severity of injuries (e.g., major-minor) will be provided to Medical Control in this advisory.

c. Implementation Standing Orders

Medical Control shall authorize all in-field Emergency Medical Technicians/Paramedics to go on Standing Orders upon receipt of information that indicated patient delay through Medical/Field Control would adversely hinder patient care.

d. Alert Relay

After initial alerting by the first dispatched ambulance, Medical Control will contact all hospitals which may become receiving hospitals. The potential receiving hospitals will be contacted by means other than the usual or "working" Medical Network Radio Channel.

When the potential receiving hospitals are advised of the developing incident, Medical Control shall obtain the following data:

1) Intensive Care Unit bed count, Operating Room availability, how many critical/non-critical patients a particular receiving hospital can assume care of and any other pertinent information for the type of incident they are being alerted for (e.g., burns, hazardous materials contamination, etc.).

e. Information Updates

The Transportation Group Supervisor at the incident site will provide Medical Control with frequent updates on casualty count and other pertinent data.

f. Ambulance Destination

Medical Control, in collaboration with the Transportation Group Supervisor, will determine ambulance destination based upon the data obtained by Medical Control, after Medical Control has contacted and alerted the other hospital(s).

g. Radio Log Maintenance

Medical Control will be responsible for the maintenance of a radio log when this plan is activated. As each ambulance leaves the incident site for its ordered destination, Medical Control's radio log will reflect the ambulance unit number; number of casualties noted by most major injury; the receiving hospital and the ambulance's Estimated Time of Arrival at that receiving hospital. This Medical Control radio log will be available for later determination of casualty disbursement.

h. Advertising Receiving (Destination) Hospital

Each loaded transporting ambulance shall be responsible for advising any hospital (other than that which is designated as medical Control) on an alternate low level Medical Network Radio channel that an (your) ambulance has been dispatched from the incident site to the receiving hospital (e.g., 1, 4 or 8). At this time, the ambulance will provide that receiving hospital with the following information:

1) Number of patients being transported.
2) Basic or chief complaint(s) and vital signs as indicated.
3) Estimated time of arrival to the facility.

G. *Extrication Group Supervisor*

1. Identification

The Extrication Group Supervisor is a firefighter who is responsible for site safety, initial patient SIMPLE TRIAGE AND RAPID TREATMENT, and disentanglement of victims/patients and their removal to the treatment/tagging area.

2. Victim Management

The Extrication Group Supervisor is responsible for the management of victims "where they are found" at the incident site. "Where they are found" could be in the "hazard zone," that is within the vehicle(s), aircraft, or collapse area, as the case may be. Extrication manages the easily treated life-threatening problems, taking spinal or other precautions; and completes extrication efforts before the victims, now patients, are moved to the Treatment Area for more definitive care.

3. Patient Movement

Extrication includes the actual moving of victims or patients from "where they are found" to the Treatment Area.

4. Initial Triage/Priorities

In this plan, triage is an ongoing process that does not require a specialized person or persons. Triage begins on site with the SIMPLE TRIAGE AND RAPID TREATMENT (S.T.A.R.T.) system, refer to algorithm on page 68. This system will be used by the first rescuers. Those who are not ambulatory will be triaged by rescuers and noted by marking a number "1" or "2" denoting "delayed" or "immediate." Further triage is done in the treatment area where definitive triaging will be done by treatment staff.

The Extrication Group Supervisor is responsible for determining whether to provide initial triage "where they are found" or to move the victim(s) quickly to the Treatment Area.

If there is no safety hazard "where they are found," victims, now patients, can be moved in an orderly fashion, by priorities, to the Treatment Area.

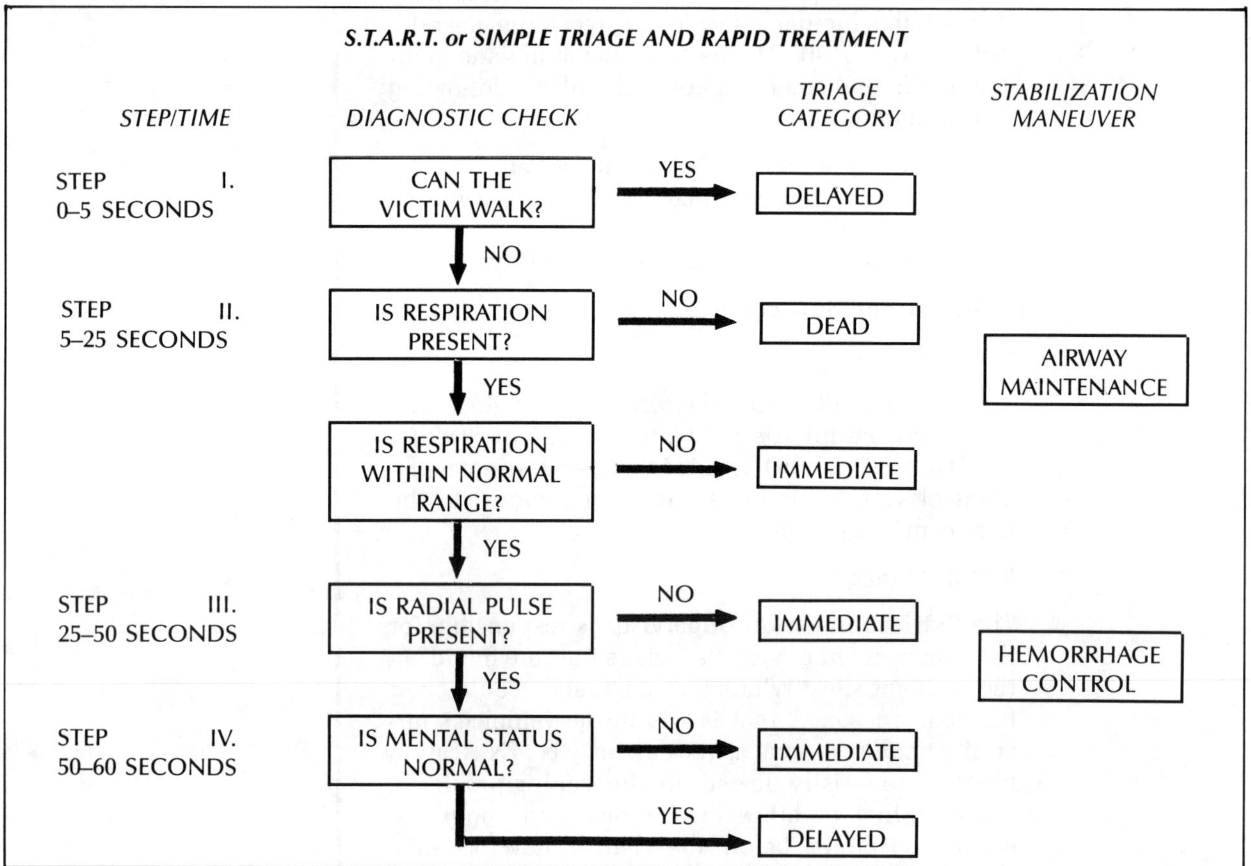

Priorities shall be indicated by marking a number "1" or "2" on each patient's forehead prior to moving for treatment. This "1" will designate the more serious patient(s) requiring immediate removal. Normally, at least one paramedic will assist with this initial survey and "1" and "2" classification, using the S.T.A.R.T. (SIMPLE TRIAGE AND RAPID TREATMENT) criteria (refer to diagram).

5. Advanced Life Support

Trapped victims requiring prolonged extrication should receive some advanced life support care by paramedics as required. On the other hand, should a hazard exist, such as leaking fuel, fire, or threatened collapse, victims will be removed rapidly to the Treatment Area on a first come basis. Definitive triage, treatment and tagging will take place in the Treatment Area.

I. *Treatment Group Supervisor*

1. Identification

The Treatment Group Supervisor, minimally an Emergency Medical Technician I, optimally a paramedic, has a general knowledge of medical care; but does not personally provide direct casualty/patient care.

2. Specific Assignments

   The Treatment Group Supervisor must have specific assignments for personnel assigned to his team. Personnel will need to be closely supervised to assure that patients are continuously monitored and proper treatment is provided. This close supervision is particularly critical if personnel unfamiliar with "field medical care" or disaster operations are working within this area.

J. *Treatment Area*

1. Function

   The function of the Treatment Area is to provide definitive basic and advanced life support for stabilization and continuing care of patients until they can be transported to a medical facility. Most paramedical personnel will be assigned to this area.

2. Location/Responsibility For

   The Treatment Group Supervisor is responsible for the establishment and operation of a Treatment Area, the proximity of which is governed by terrain, circumstances of the accident or incident, and existing safety hazards at the accident or incident site. The Treatment Area should be readily accessible, but isolated from any dangerous conditions associated with the incident.

3. Identification/Announcement

   The Treatment Area shall have a readily identifiable entrance. Signs, traffic cones, or other markers may be utilized to make the entrance obvious. The Treatment Area location and entrance must be announced or made known to all members of the medical group.

4. Treatment Area Tagging Zone

   Tagging the patients upon entering the Treatment Area is essential to insure proper patient flow from the Treatment Area through to hospital care. The triage tag shall be applied high up on an arm of the patient.

   During tagging, patient assignment to Immediate or Delayed Zones in the Treatment Area will be noted by tearing off the appropriate bottom section of the triage tag. The Treatment Group Supervisor will designate a medical care provider to make this DELAYED or IMMEDIATE designation.

5. Treatment Zones

   The Treatment Area shall be divided into two separate zones.

a. Immediate

The Immediate Zone is for the treatment of critical patients requiring advanced life support.

b. Delayed

A Delayed Zone shall be established for non-critical patients. Patients in this area will receive basic life support and continuous monitoring. If the condition of a patient in the Delayed Zone changes and requires advanced life support, the patient should be moved to the Immediate Zone for more definitive care.

K. *Transportation Group Supervisor*

1. Identification

The Transportation Group Supervisor is responsible for providing and coordinating patient transportation. This coordination requires the Transportation Group Supervisor to have in-depth knowledge of field medical care and to be familiar with local hospital capabilities.

2. Communications

a. Incident Command/Medical Control

The Transportation Group Supervisor must have communications capability on two separate radio systems, the fire tactical channel in use at the incident site for maintaining contact with the Medical Branch Director and the Incident Commander and the Medical Network Radio System's Operational Channel (first ambulance on site) for communicating with Medical Control.

b. Ambulance Stand-by Area

The Transportation Group Supervisor will also communicate with the Ambulance Stand-by Manager by radio or by hand signal, or summon ambulances for patient loading and transportation.

c. Ambulance Direction

The Transportation Group Supervisor will collaborate with Medical Control (the designated base hospital for the incident) and will determine ambulance destination when the ambulance is loaded and ready to leave the incident site.

d. Medical Control Communications

The Transportation Group Supervisor is the *only* person authorized to communicate directly with Medical Control from the incident site once the Multiple Casualty Plan has been activated and the Transportation Group Supervisor is appointed.

COMMUNICATION TO MEDICAL CONTROL MAY NOT BE DELEGATED.

e. Medical Control Alert

Medical Control will have been alerted by the first ambulance enroute to the incident site and Medical Control will have queried the local hospital(s) for data upon which to base a decision or decisions for patient loaded ambulance dispatch from the incident site.

f. Medical Control Advisory

When an ambulance is loaded and prepared to depart from the Transportation Area, the Transportation Group Supervisor will advise Medical Control of:

1) The number of patients loaded,
2) General patient complaint of each patient, and
3) Patient category as to critical or non-serious.

g. Alternate Medical Control

If for some reason, the base station hospital that would normally provide Medical Control is uncommunicative, the following steps should be taken:

1) Contact another base station hospital to act as Medical Control.
2) If no base station hospital is available, a non-base station hospital may act as Medical Control. In this event, the County's Chief Medical Officer or his designee shall be notified immediately, by the Incident Commander.

h. On-Site Ambulance Dispatch

When Medical Control has been advised of the nature of the patient load, Medical Control will determine the hospital destination and immediately advise the Transportation Group Supervisor. The Transportation Group Supervisor will advise the transporting ambulance.

3. Operational Area/Loading Priorities

The Transportation Group Supervisor shall set up operations close to the exit of the Treatment Area and very closely to where ambulance loading will take place.

The Transportation Group Supervisor will work closely with the Treatment Group Supervisor and the Extrication Group Supervisor in determining which patients are to be transported first. The Transportation Group Supervisor will provide the appropriate transportation, air or ground.

4. Ambulance Loading

   It is important that patient movement to hospital care not be delayed at the ambulance loading area; therefore, in large incidents, the Transportation Group Supervisor should direct the controlled loading of two or three ambulances simultaneously.

5. Helispot Transport

   If patients are to be taken from the incident site by aircraft as well as by ground transportation, the Transportation Group Supervisor must ask the Medical Branch Director to arrange for the appointment or designation of an Air Branch Director and Helispot Manager. Once appointed, the Air Branch Director will coordinate closely with the Medical Branch Director and/or Transportation Group Supervisor.

6. Transportation Aides

   The Transportation Group Supervisor will need several aides, depending on the size of the incident. All aides, or members of Transportation Team will wear identification vests, and shall have clipboards, paper and pencils available for maintaining information.

L. *Morgue Group Supervisor*

1. Identification

   A Deputy Sheriff-Coroner shall be assigned to establish and maintain a temporary morgue and to carry on the necessary coroner's investigation, recovery, reception and processing of human remains.

2. Investigation Task Force

   It is the policy of the Sheriff-Coroner's Office to investigate all death cases which are classified as coroner's cases, and in multiple death cases an Investigation Task Force may be initiated to:

   a. Identify the dead;
   b. Determine the cause of death; and
   c. Discover the cause of death to prevent a recurrence of the event(s) which resulted in the death(s).

3. Remains Left "Where Found"

   a. Authority

      It is absolutely essential that any and all human remains be left "where found" at any accident or incident site. The *ONLY* exception will be upon order of the Extrication Group Supervisor who will determine if existing hazards preclude leaving the remains "where found" for the coroner's investigation. If indeed the remains are likely to be destroyed, the Safety Officer will order their removal.

b. Removal Procedure/Tagging of Dead

If, in the Safety Officer's judgment a body or bodies should be moved from "where found," that body or bodies shall be triage tagged and all possible data concerning "where found" shall be noted on the triage tag. An example could be "row 18, aisle seat, not belted" or "face up under left wing". In any case where removal is ordered, the Extrication Group member making the removal and marking the triage tag shall put the Extrication Group member's name on the tag, to assist the coroner's investigation.

4. Coroner's Investigation

A Deputy Coroner will be in charge of the recovery of the dead at the scene of a multiple casualty incident and in the course of his investigation will:

a. Note the position of each body, or part thereof, at or near the incident site and photograph and mark that body or part.

b. Each body or part thereof will be numbered at the scene and a secure and waterproof label will be attached. The label will be attached directly to the body, or part. One Sheriff-Coroner's case number will be utilized and each body or part thereof will be numbered consecutively. (Example: 84-6400-306)

c. Each body, together with all its clothing and personal possessions must be placed in a body tag. This includes identifying items found in close proximity to that body.

d. Emergency ambulances shall *not* be used for the removal of bodies from the incident site.

e. Only Deputy Sheriff-Coroners will mark bodies and collect clothing and personal possessions.

5. Reception and Processing Remains

The investigation, recovery, reception and processing of human remains is fully addressed in the Sheriff's Investigation Bureau's policy and procedure manual. Any and all questions concerning this policy are to be directed to the Sheriff-Coroner's Investigation Bureau-Coroner's Section.

M. *Law Enforcement/Traffic Control Branch*

1. Law Enforcement/Traffic Control Branch Director

The Law Enforcement Traffic Control Branch Director's responsibility is managing both the internal and perimeter components of the law enforcement function at the scene of a multi-casualty incident. This person is ideally the ranking officer in the law enforcement agency with jurisdiction who can make the decisions necessary to manage all of the law enforcement tactical elements.

The Law Enforcement/Traffic Control Branch Director shall, upon request of the Information Officer, mobilize the law enforcement members of the Survivor Data Teams for receiving hospitals. Refer to Survivor Data, a responsibility of the Information Officer.

2. First Officer On Scene

   Absenting those occasions where law enforcement will assume Incident Command, the first law enforcement officer from the agency having jurisdiction at the incident site will report to the Incident Commander at the Command Post. This first officer will remain at the command post until relieved by a supervisor or tactical manager. (Rank will vary with the agency.)

   The first officer, supervisor, tactical manager or Watch Commander will act as the Law Enforcement Traffic Control Branch Director, subordinant to the Incident Commander, addressing the law enforcement traffic control function(s).

3. Perimeter Enforcement Group Supervisor

   A uniformed law enforcement supervisor or officer directly responsible to the Law Enforcement Traffic Control Branch Director, who will establish necessary perimeter security, supervise perimeter controls, and maintain communications with the Law Enforcement Traffic Control Branch Director. The Perimeter Enforcement Group Supervisor will call for barricades or other appropriate material to assist in perimeter control.

4. Controlled Zone Entry Point(s)

   The Perimeter Enforcement Group Supervisor shall, at his discretion, designate the incident site a "controlled zone" and set up controlled zone entry points from which only authorized personnel and vehicles may enter that incident site.

5. Controlled Zone Entry Pass

   A Controlled Zone Entry Pass will be issued to all authorized non-uniform non-emergency responders by an officer or officer(s) at the zone entry point(s). If the agency or the jurisdiction does not have a controlled zone entry pass system set up, the Sheriff-Coroner has forms and directions available through the Sheriff's Watch Commander.

   The Controlled Zone Entry Pass is an NCR form

with space for the entrant's name, vehicle identification, a zone number designation, the date, and the name and identification number of the issuing officer. The Internal Enforcement Group Supervisor and members will then know that the entrant is authorized. Complete details appear in Sheriff's Office Policy No. 13.

6. Internal Enforcement Group Supervisor

   The Internal Enforcement Group Supervisor, directly responsible to the Law Enforcement Traffic Control Branch Director, will manage the jurisdiction's investigative and internal security element within the inner perimeter of the incident site.

   Duties will range from supervising the investigation of the accident or incident to security of the site; emergency responder's equipment; victim's property not immediately retrieved; and crash debris which may be evidence.

7. Traffic Enforcement Group Supervisor

   The Traffic Enforcement Group Supervisor, responsible to the Law Enforcement/Traffic Control Branch Director, will maintain traffic flow around the incident site; and divert non-essential traffic away from the site. Although the extent of traffic restriction and/or diversion will depend on a number of factors, the primary consideration shall be protection of the public, and emergency responders and equipment responding to the incident.

N. *Information Officer*

1. Identification/Responsibilities

   The Information Officer is appointed by the Incident Commander, and the Information Officer's responsibility is to formulate and release information about the incident to the news media and other appropriate agencies.

2. Coordination of Activities

   The designated Information Officer will coordinate activities with the Incident Commander to check restrictions to be set upon release of data; and application of 409.5 P.C. at this incident site. The Information Officer will establish liaison with the incident jurisdiction's Public Information Officer and with other responding agencies' representatives (State and Federal) to provide uniformity of news releases.

3. Survivor Data

   If in the Incident Commander's opinion the incident warrants the Information Officer will be responsible for activating a survivor coordinating point and for the mobilization of pre-designated Survivor Data Teams. If the Survivor Data Teams are mobilized, it is the Information Officer's responsibility, through amateur radio, to maintain liaison with these teams.

The mechanics are for the Incident Commander to advise his radio dispatch point (EMERGENCY CENTER, CDF Command Center or Watsonville) to notify the Sheriff-Coroner's Office, or officer coverage at both Dominican and Watsonville Community Hospital; and the Santa Cruz Police Department for coverage at Community Hospital of Santa Cruz. Coverage of course may be requested for one, two or three acute care hospitals in the County. In any case, amateur radio personnel must be notified at the same time, to provide for coverage at the hospital(s) and to provide a liaison officer at the incident site.

4. Survivor Data Coordinating Point

In any multi-casualty incident where there are large numbers of injured survivors who will be dispersed to local area hospitals, a localized all survivor coordinating point must be established. Such a point will be established at the American National Red Cross Chapter in whose jurisdiction the incident occurs. If the incident occurs in the area from the North County line to Borregas Creek, the Santa Cruz Chapter headquarters shall be designated the Survivor Data Coordinating Point. If the incident occurs in South County from Borregas Creek to the Pajaro River, the Watsonville Chapter shall be the Survivor Data Coordinating Point.

The respective Red Cross Chapters will staff telephones at their respective headquarters if they are designated the Survivor Data Coordinating Point. The Red Cross Chapter headquarters will answer relatives inquiries and can, in the majority of cases, direct relatives to the hospital where a specific victim has been taken.

5. Survivor Data Teams

Three Survivor Data Teams consisting of a Law Enforcement Officer, a Red Cross volunteer and an Amateur Radio person will obtain the names of the victims received at the respective hospitals from hospital personnel, and will relay that data to the designated Red Cross Chapter headquarters for dissemination to relatives.

The information can either be relayed by Amateur Radio or by telephone. In any case, the teams can maintain liaison with the Information Officer at the incident site through Amateur Radio and the Amateur Radio person at local hospitals may also relay ancillary non-emergency communications from the hospital to the incident site, when necessary.

VII. WAR CAUSED DISASTER OPERATIONS

The Multiple Casualty Incident Plan does not apply to war-caused disaster operations.

VIII. NATURAL DISASTER OPERATIONS

A. General Response

While this Multiple Casualty Incident Plan is NOT designed to accommodate a County-wide disaster but to coordinate multi-agency response to the *single-site* disaster which could overcome the day-to-day emergency medical response system, it is not inconceivable that because of this magnitude and impact of the incident that an affected entity may proclaim a local emergency. In such a case, the entity could activate their emergency operating center to coordinate support for field operations.

B. Operational Concepts

1. Command Authority

Operational requirements of the Multiple Casualty Incident Plan dictate that Command Authority is vested in the Incident Commander, and all Emergency Operating Center personnel shall be considered staff support for the respective appointees at the incident site.

2. Carrier Support

A commercial passenger carrier will send investigators to the incident site and will provide both victim identification and public information support. Freight carriers will provide salvage information and provide security for freight, damaged and undamaged.

3. Federal/State Support

The Federal Aviation Administration and any other Federal service immediately affected by the incident will be called upon to respond; and if the entity proclaims a local emergency, State agencies can then assist, providing that local resources are called upon first.

4. Fiscal Operations Support

The entity's fiscal operations officer shall be available to advise the entity's Chief Executive Officer or Director of Emergency Services to support on-site operations.

5. Social Services Support

   The County's Human Resources Agency can be called upon to assist in coordinating the response of volunteer social services organizations who may be called upon to shelter/house/clothe or feed evacuees or persons whose homes are damaged or destroyed.

6. Other Emergency and Support Services

   If the County Emergency Operating Center is activated as the result of a major accident or incident this plan is activated for, other County emergency and non-emergency responders may be called upon to provide coordinated support.

   Medical and Health, Flood, Landslide Hazard Mitigation and Private Property Damage Assessment, Operational Information, Public Works/Engineering, Transportation, Supply, and Public Information and Education may be called upon to assist if necessary.

IX. INDIRECT DISASTER OPERATIONS

Should the EMERGENCY OPERATING CENTER be activated as the result of an accident or incident that would activate the Multiple Casualty Incident Plan, operating services called upon to support field operations will be obligated to maintain certain records.

1. Daily Staff Journal

   Each County Department or Service mobilized will be required to keep a Daily Staff Journal, recording all activities the Department or Service engages in during Emergency Operating Center Operations (Refer Information Management, Annex #1, Attachment #7 to Annex #1, Direction and Control, Santa Cruz County Emergency Plan).

2. Fiscal Records

   Depending upon a Department or Service's fiscal operations, a County Department or Service is obligated to maintain records of personnel time and material expenditures as the result of a particular incident.

3. Damage Reporting/Damage Assessment

   Participating County Departments and Services will be expected to maintain reports of damaged equipment. If the incident mobilized for results in damage to public or private property, appropriate damage reporting procedures shall be initiated from the Emergency Operating Center. These procedures will be exactly as those employed for any natural disaster operation.

X. RESOURCES AND SUPPORT SYSTEMS

An entity's Emergency Operating Center and operational support services may be activated should the situation warrant. All available support of both regular and volunteer agencies shall, upon order of the Chief Executive Officer, be made available to the Incident Commander.

XI. AUTHENTICATION

This Multiple Casualty Incident Plan is hereby adopted and will be supported by:

____

Director Emergency Service

____

Health Officer/Medical Director

APPENDIX K

FEMA NATIONAL, REGIONAL, AND STATE OFFICES

The United States is divided into 10 federal regions, each with a regional office of the Federal Emergency Management Agency. It is from these offices—and the governor's office or the state's emergency services office—that disaster information may be obtained.

Below is a list of FEMA's regional offices, their addresses, and the states for which they are responsible.

NATIONAL OFFICE

500 C Street, S.W.
Washington, DC 20472
(202) 646-4600 / operational 24 hours a day

REGIONAL OFFICES

REGION I—Maine, Massachusetts, Vermont, New Hampshire, Connecticut, Rhode Island
442 J. W. McCormack POCH
Boston, MA 02109
(617) 223-4741

REGION II—New York, New Jersey, Puerto Rico, Virgin Islands
26 Federal Plaza
New York, NY 10278
(212) 264-8980

REGION III—Pennsylvania, West Virginia, Virginia, Washington, DC, Maryland, Delaware
Liberty Sq. Bldg., 2nd Floor
105 S. Seventh St.
Philadelphia, PA 19106
(215) 597-9416

REGION IV—Kentucky, Tennessee, North Carolina, South Carolina, Georgia, Alabama, Mississippi, Florida
Suite 700, 1371 Peachtree St., N.E.
Atlanta, GA 30309
(404) 881-2400

REGION V—Minnesota, Wisconsin, Michigan, Ohio, Indiana, Illinois
300 S. Wacker Drive, 24th Floor
Chicago, IL 60606
(312) 353-8661

REGION VI—Texas, Oklahoma, Arkansas, Louisiana, New Mexico
Federal Regional Center, Rm. 206
800 North Loop 288
Denton, TX 76201
(817) 387-5811

REGION VII—Missouri, Iowa, Nebraska, Kansas
911 Walnut Street, Room 300
Kansas City, MO 64106
(816) 374-5912

REGION VIII—Colorado, Utah, Wyoming, Montana, North Dakota, South Dakota
Denver Federal Center, Bldg. 710
Box 25267
Denver, CO 80225-0267
(303) 235-4811

REGION IX—California, Arizona, Nevada, Hawaii, Pacific Trust Territories
Building 105 Presidio of San Francisco
San Francisco, CA 94129
(415) 556-8794

REGION X—Washington, Oregon, Idaho, Alaska
Federal Regional Center
130 228th St., S.W.
Bothell, WA 98021-9796
(206) 481-8800

STATE OFFICES

Region I

CONNECTICUT
Office of Civil Preparedness
360 Broad Street
Hartford, CT 06105
(203) 566-3180

MAINE
Bureau of Civil Emergency Preparedness
State Office Building, Station 72
Augusta, ME 04333
(207) 622-6201

MASSACHUSETTS
Massachusetts Civil Defense Agency
P.O. Box 1496
400 Worcester Road
Framingham, MA 01701
(617) 237-0200

NEW HAMPSHIRE
New Hampshire Civil Defense Agency
State Office Park South
107 Pleasant Street
Concord, NH 03301
(603) 271-2231

RHODE ISLAND
Rhode Island Emergency Management Agency
State House
Providence, RI 02903
(401) 421-7333

VERMONT
Office of Civil Defense
Department of Public Safety
Waterbury State Complex
103 S. Main St.
Waterbury, VT 05676
(802) 244-8721

Region II

NEW JERSEY
New Jersey State Office of Emergency Management
P.O. Box 7068
W. Trenton, NJ 08625
(609) 882-2000

NEW YORK
State Emergency Management Office
Public Security Bldg.
State Campus
Albany, NY 12226
(518) 454-2159

PUERTO RICO
Office of Civil Defense
P.O. Box 5127
San Juan, PR 00906
(809) 724-0124

VIRGIN ISLANDS
Office of Civil Defense & Emergency Services
P.O. Box 1208
Charlotte Amalie
St. Thomas, VI 00801
(809) 774-2244

Region III

DELAWARE
Division of Emergency Planning and Operations
P.O. Box C
Delaware City, DE 19706
(302) 834-4531

DISTRICT OF COLUMBIA
DC Office of Emergency Preparedness
Municipal Center, Rm. 5009
300 Indiana Ave., N.W.
Washington, DC 20001
(202) 727-6161

MARYLAND
Maryland Emergency Management and Civil Defense Agency
Sudbrook Lane & Reistertown Rd.
Pikesville, MD 21208
(301) 486-4422

PENNSYLVANIA
Pennsylvania Emergency Management Agency
Transportation and Safety Bldg., B-151
Harrisburg, PA 17120
(717) 783-8150

VIRGINIA
Office of Emergency & Energy Services
310 Turner Rd.
Richmond, VA 23225
(804) 323-2899

WEST VIRGINIA
Office of Emergency Services
Room EB-80
Capitol Building
Charleston, WV 25305
(304) 348-5380

Region IV

ALABAMA
Alabama Emergency Management Agency
220 N. Hull St.
Montgomery, AL 36130
(205) 261-3519

FLORIDA
Division of Emergency Management
1720 S. Gadsden Street
Tallahassee, FL 32301
(904) 488-1900

GEORGIA
Georgia Emergency Management Agency
P.O. Box 18055
Atlanta, GA 30316
(404) 656-5500

KENTUCKY
Kentucky Disaster & Emergency Services
EOC Building
Boone Center, Parkside Drive
Frankfort, KY 40601
(502) 564-8680

MISSISSIPPI
Mississippi Emergency Management Agency
P.O. Box 4501, Fondren Station
Jackson, MS 39216
(601) 353-9100

NORTH CAROLINA
Division of Emergency Management
116 W. Jones Street
Raleigh, NC 27611
(919) 733-3867

SOUTH CAROLINA
Emergency Preparedness Division
1429 Senate Street
Columbia, SC 29201
(803) 758-2826

TENNESSEE
Tennessee Emergency Management Agency
3041 Sidco Drive
Nashville, TN 37204
(615) 252-3300

Region V

ILLINOIS
Illinois Emergency Services and Disaster Agency
110 E. Adams Street
Springfield, IL 62706
(217) 782-2700

INDIANA
Indiana Dept. of Civil Defense and Emergency Management
100 N. Senate Avenue
Indianapolis, IN 46204
(317) 232-3830

MICHIGAN
Emergency Services Division
Department of State Police
11 S. Capitol Ave.
Lansing, MI 48913
(517) 373-0617

MINNESOTA
State of Minnesota Division of Emergency Services
State Capitol, Room B-5
St. Paul, MN 55155
(612) 296-2233

OHIO
Disaster Services Agency
2825 W. Granville Road
Worthington, OH 43085
(614) 889-7150

WISCONSIN
Wisconsin Division of Emergency Government
P.O. Box 7865
Madison, WI 53707
(608) 266-3232

Region VI

ARKANSAS
Arkansas Office of Emergency Services
P.O. Box 758
Conway, AR 72032
(501) 329-5601

LOUISIANA
Louisiana Office of Emergency Preparedness
Dept. of Public Safety
P.O. Box 66536, Audubon Station
Baton Rouge, LA 70896
(504) 342-5470

NEW MEXICO
Office of Civil Emergency Preparedness
P.O. Box 4277
Santa Fe, NM 87501
(505) 473-2476

OKLAHOMA
Oklahoma Civil Defense Agency
Sequoyah-Will Rogers Building Tunnel
P.O. Box 53365
Oklahoma City, OK 73152
(405) 521-2481

TEXAS
Division of Disaster Emergency Services
Texas Dept. of Public Safety
P.O. Box 4087, N. Austin Station
Austin, TX 78773
(512) 465-2000

Region VII

IOWA
Office of Disaster Services
Hoover State Office Building
Level A-29
Des Moines, IA 50319
(515) 281-3231

KANSAS
Division of Emergency Preparedness
P.O. Box C-300
Topeka, KS 66601
(913) 233-9253

MISSOURI
State Emergency Management Agency
P.O. Box 116
Jefferson City, MO 65102
(314) 751-2321

NEBRASKA
Nebraska Civil Defense Agency
1300 Military Rd.
Lincoln, NE 68508
(402) 473-1410

Region VIII

COLORADO
Colorado Division of Disaster Emergency Services
DOC, Camp George West
Golden, CO 80401
(303) 273-1624

MONTANA
Montana Disaster and Emergency Services
P.O. Box 4789
Helena, MT 59604
(406) 444-6911

NORTH DAKOTA
North Dakota Disaster Emergency Services
P.O. Box 1817
Bismarck, ND 58505
(701) 224-2111

SOUTH DAKOTA
South Dakota Emergency and Disaster Services
State EOC
State Capitol
Pierre, SD 57501
(605) 773-3231

UTAH
State Division of Comprehensive Emergency Management
1543 Sunnyside Ave.
Salt Lake City, UT 84108
(801) 533-5271

WYOMING
Wyoming Disaster & Civil Defense
P.O. Box 1709
Cheyenne, WY 82001
(307) 777-7566

Region IX

ARIZONA
Division of Emergency Services
5636 E. McDowell Rd.
Phoenix, AZ 85008
(602) 244-0504

CALIFORNIA
California Office of Emergency Services
P.O. Box 9577
Sacramento, CA 95823
(916) 427-4201

HAWAII
Office of the Director of Civil Defense
Department of Defense
3949 Diamond Head Rd.
Honolulu, HI 86816
(808) 734-2161

NEVADA
Division of Emergency Services
2525 S. Carson St.
Carson City, NV 89701
(702) 885-4240

AMERICAN SAMOA
Territorial Emergency Management Coordination
P.O. Box 3296
Pago Pago, American Samoa 96799
011-684-633-2331

GUAM
Office of Civil Defense
Territory of Guam
P.O. Box 2877
Agana, Guam 96910
011-671-477-9841

NORTHERN MARIANA ISLANDS
Office of Civil Defense
Commonwealth of Northern Mariana Islands
Office of the Governor
Saipan, Mariana Islands 96950
011-670-9274

TRUST TERRITORY OF THE PACIFIC
Office of Planning and Statistics
Trust Territory Headquarters
Saipan, Mariana Islands 96950
011-670-9333

Region X

ALASKA
Alaska Division of Emergency
P.O. Box 2267
Palmer, AK 99645
(907) 376-3061

IDAHO
Bureau of Disaster Services
650 West State Street
Boise, ID 83720
(208) 334-3460

OREGON
Oregon Division of Emergency Management
Capitol Building, Room 43
Salem, OR 97310
(503) 378-4124

WASHINGTON
Department of Emergency Management
4220 E. Martin Way
Olympia, WA 98504
(206) 459-9191

INDEX

INDEX

A

B

C

D

E

F

G

H

I

J

K

L

M

N

O

P

R

S

T

U

V

W